MW01069612

PRINCIPLES OF RISK ANALYSIS

DECISION MAKING UNDER UNCERTAINTY

PRINCIPLES OF RISK ANALYSIS

DECISION MAKING UNDER UNCERTAINTY

CHARLES YOE

CRC Press
Taylor & Francis Group
Boca Raton London New York

CRC Press is an imprint of the
Taylor & Francis Group, an **informa** business

CRC Press
Taylor & Francis Group
6000 Broken Sound Parkway NW, Suite 300
Boca Raton, FL 33487-2742

Printed in the United States of America on acid-free paper
Version Date: 20110804

International Standard Book Number: 978-1-4398-5749-6 (Hardback)

Visit the Taylor & Francis Web site at
http://www.taylorandfrancis.com

and the CRC Press Web site at
http://www.crcpress.com

Contents

Preface

I did not want to write this book. I wanted to buy it. Risk analysis is mature enough that it needs a principles text. There are many wonderful books available on the subject of risk. In fact, for years, in the training I have done, I used to schlep a dozen of them around for students to peruse. These I called the starter library. I urged people to buy them. Eventually I stopped carrying the books with me and started looking for that one book that would introduce students and professionals to the integrated topic of risk analysis. I never found it. So, I decided to go ahead and write it.

Risk analysis is a very parochial subject and practice. There are many tribes of risk practitioners, and they speak many dialects. To be honest, I am not entirely sure the field is ready for anything one might call a principles text. We may never have enough agreement on the principles to so ordain them as a community of practice.

Nonetheless, it has been my great fortune to have worked with a lot of people on many different applications of the risk analysis decision-making paradigm. This has included natural disasters, engineering, food safety, food defense, environmental issues, animals and plants, trade, quality management, business, finance, terrorism, defense applications, research, and other risk analysis applications. No matter how much the words and models changed from one application to the next, I found the basic principles were rather constant. Everyone was struggling to figure out how best to make good decisions when there are so many things we just don't know for sure.

My major accomplishment in all of this was simply to learn the jargon of each field; then I stole liberally from the other fields, adapting ideas, methodologies, and models from one field to another. And it worked. A good approach to a risk problem in food safety is often a good approach to an engineering risk problem. Sometimes you only need to change an egg to a Tainter gate and salmonella to metal fatigue and you are halfway home. The toolbox is the same. We're using the same math, the same probability. We all have to identify problems and evaluate solutions. We all have people waiting to hear what we have learned about the problems that affect them and the solutions they will rely upon.

This book was written to inform rather than to impress, although I hope the information you gain from it impresses you. Over many years of educating, training, and consulting in risk analysis, students and clients have taught me what people find most difficult about it. First, there are the "what is it" questions. Then there are the "how do you do this" questions. That is basically the outline for this book. Chapters 1–5 explain what risk analysis is, in a conceptual, big-picture way. These are the principles. Chapters 6–18 introduce the tools, techniques, and methodologies that will be most useful to new and intermediate risk analysts. These are the applications of the principles.

I have tried to keep the language of the book simple. The narrative is supported by many figures and a generous helping of text boxes. This is not the academic tome that provides an encyclopedic listing of the literature. I have consciously tried to limit the references to a few well-chosen resources an interested reader will find accessible and useful for learning and growing in risk analysis. I hope you will find it a helpful steppingstone to the truly excellent and more rigorous risk analysis literature that is out there.

The risk analysis paradigm has risen to prominence in recent years because of its intentional treatment of uncertainty in decision making. Chapter 1 introduces you to the basic language of risk analysis and its three tasks of risk management, risk assessment, and risk communication. Chapter 2 explains what uncertainty is and why it is important to decision making. The next three chapters present conceptual models that detail the basic tasks that comprise risk management (Chapter 3), risk assessment (Chapter 4), and risk communication (Chapter 5). You need to know something about each of the risk analysis tasks even if you think of yourself as primarily a manager, assessor, or communicator. I have tried hard to make these three chapters model-neutral, i.e., they do not address chemical, engineering, food safety, environmental, or any other specific approach to risk analysis. It is a general approach that presents the common principles and best practices of risk analysis that can be readily applied to any application of risk analysis.

Chapters 6–18 will help you apply the principles of risk analysis. Chapters 6–8 address concepts that are absolutely essential to the practice of good risk management. Problem identification is, perhaps, the most overlooked aspect of risk analysis. I am often amazed by how sloppy we can be in defining the problems we are trying to solve. If we do not identify the right problem properly, our prospects for desirable outcomes are greatly diminished. So, problem identification is the first application and the focus of Chapter 6.

Brainstorming is one of the most efficient and effective ways to get a lot of ideas out on the table in a short period of time. Good risk management frequently relies on divergent thinking in its early rounds, so brainstorming is the subject of Chapter 7. The "3× Yeah" technique is a great way to generate divergent ideas, and it works.

No matter what the problem is, someone is always going to care about costs. What will it cost to take action or to not take action? How much risk management can we afford? How do we decide how much residual risk we can tolerate? A few basic economics principles provide the answers to these and many other questions, and they are found in Chapter 8.

Chapters 9–16 feature risk assessment applications. Risk assessment does not have to be quantitative, nor does it have to take months and millions of dollars. Risk analysis is a process that can be scaled to the time and resources available. Qualitative risk assessment is especially useful for two reasons. Sometimes it is sufficient to meet the information needs of risk managers, and sometimes it is your only option. Qualitative risk assessment techniques are found in Chapter 9, and you should be able to begin using one or more of these techniques when you finish the chapter.

Model-building skills are a scarce resource for many organizations. Risk assessment and new software tools have made model building a far more common occurrence, however, so Chapter 10 addresses the art and practice of model building. You will find a process here, 13 steps to follow that have served me well over the years, as well as a review of some valuable skills for modelers.

Chapters 11–14 focus on probability-related topics. If you're going to do risk analysis, someone must know a little something about probability. Chapter 11 begins with a review of essential probability concepts. It is followed by one of the more valuable chapters in this book. Chapter 12 describes a process for choosing a probability distribution to represent the uncertainty that is present in your decision problem.

This assumes you have some data available to you. When you do not have data, you may want to use probability elicitation techniques, which are the subject of Chapter 13. Probabilistic risk assessment models often make use of the Monte Carlo process. Chapter 14 provides a peek at the two-step Monte Carlo process used in many risk assessment models.

Probabilistic scenario analysis combines the powerful techniques of scenarios and probabilistic methods to produce a bundle of tools and techniques that comprise the most common forms of probabilistic risk assessment models in Chapter 15. Separating what we know from what we don't know—and then keeping track of what we don't know—is one of the hallmarks of risk analysis. Sensitivity analysis, the topic of Chapter 16, should be an essential part of every risk assessment, whether qualitative or quantitative. A wide variety of useful techniques are surveyed in this chapter.

Chapters 17 and 18 are risk communication applications. Chapter 17 addresses the critical need to present and understand the quantitative results of risk assessment. This chapter is probably equal parts assessment, management, and communication, truth be told. The final chapter addresses the need to develop effective risk communication messages.

There are also two appendices. Appendix A provides an introduction to the risk language of ISO 31000. The language of risk is still somewhat unsettled, and the International Organization for Standardization (ISO) uses the language differently than much of the rest of the risk analysis community practice. The differences are largely semantic, but semantics are important. Appendix B provides an introduction to three of the tools found in Palisade Corporation's DecisionTools® Suite. This is the software used to prepare this book and the software of choice for many risk assessors. Now, it is time for a confession.

This book is not a unified theory of risk analysis; it is a modest first attempt to articulate some principles. Read this book and you will know what risk analysis is, and you will be ready to begin to practice it. It is my hope that many better principles texts will follow as we figure out what those principles are. This leads to another interesting conundrum: "What comes first, the course or the text?" It is not easy to shop a textbook when there are few if any courses on the subject being offered in our colleges and universities. Yet, it is hard to find a discipline that does not include risk as a topic in its courses and texts outside the fine arts. I like to think the discipline of risk analysis is poised for significant, if not explosive, growth in higher education, and we simply need some good basic texts so our students can get a toehold on what this risk analysis is all about. Then they can move more confidently into their tribes and dialects. I hope this book might contribute to the education of risk analysts in many disciplines.

In truth, however, this book is really for practitioners and professionals who want to know more about risk analysis and how it can help them make decisions under uncertainty. Although my work as a risk analyst began in the 1970s, my career in risk analysis training began in the 1980s with the U.S. Army Corps of Engineers. Since that time I have trained thousands of people from more than 100 countries in dozens of nations around the world in a wide variety of risk analysis applications. This book is based on my work and those training experiences. A truly global community taught me what people want to know about risk analysis, and that is what I endeavor to pass along to you, the reader.

I must be honest here. This book, though written by me, is the work of many, many people. Because I never intended to write this book, I did not keep careful notes these last few decades. If we have ever met and you read this book and find yourself thinking, "I agree," there could be a good reason for that. You may be reading what you gave me years ago as a seed of an idea that has grown since I first heard you. I have listened and observed carefully over the years.

In that time I have stuffed so many handouts into files; saved papers, reports, and official guidance; kept piles of marble-covered notebooks; and underlined and highlighted the books of so many smart people that I am no longer sure I have ever had an idea I can truly call my own. To all of you, I say thank you. I only apologize for my inability to recognize your contribution to my own learning more formally.

Still, there are people I want to recognize and thank. I begin with my wife, Lynne, and my sons, Jonathan, Adam, and Jason, who have always supported and encouraged me. Notre Dame of Maryland University honors the teaching profession every day and has done so in Baltimore, Maryland, for well over a century. Nearly 25 years ago they took a chance that I might be able to teach one day, and I am eternally grateful to them for doing so and for helping me learn how to teach. David Moser is my friend and longest continuous colleague in this risk arena. I did not know I was even doing risk analysis until I began to attend the risk workshops he and Yacov Haimes offered in water resources management. It was there that I began to meet people who identified themselves as "risk people." You have all influenced me. Nell Ahl introduced me to food safety, and I will always be grateful. She is a special lady. Richard Williams has always been a good friend, well worth listening to. I am especially grateful for his comments and suggestions for Chapter 8. Sam Ring has inspired me more than once. Ken Orth taught me the value of lists and simple things. Mark Walderhaug is a kindred spirit I would have enjoyed working alongside.

Wes Long and Brad Paleg did more to help me learn how to help people learn than they will ever realize. The online risk classes we developed through a USDA grant have brought the world together to learn risk analysis more times than I can count. I learned so much from them along the way. The people at JIFSAN have been instrumental in helping me take risk analysis training to the world and in bringing the world to College Park, Maryland, each summer. It would be criminal not to single out my friend, Judy Quigley, for her constant, steady presence. Being able to sit occasionally in the shadow of Bob Buchanan has been a tremendous advantage to me.

When I was debating whether to undertake this book or not, the first thing I did was send out an e-mail to some friends to ask what needs to be in a risk analysis 101 textbook. In return I got many excellent suggestions and a great deal of encouragement. So I want to thank Boris Antunović, Mary Bartholomew, John Boland, Kim Callan, Sherri Dennis, Doug Eddy, Sue Ferenc, Steven Gendel, Leon Gorris, Robert Griffin, Benjamin Hobbs, Sandra Honour, Audrey Ichida, Hong Jin, Janell Kause, Lip Tet Ng, James Schaub, Dominic Travis, and Cody Wilson for your ideas and your support.

Finding a publisher for a textbook without a proven academic market is a lonely pursuit made easier by my friend Mickey Parish, who introduced me to Stephen Zollo, a senior editor in food science and technology. Steve was the perfect person for me and many other authors, I suspect. David Fausel, my project coordinator, answered a zillion questions in a timely and always patient manner. Prudence Board was a dream to work with. A special thank-you to Melissa Parker for her hard work in getting my references to Chicago in time!

Perhaps my most profound thanks go to the thousands of people I have worked with on projects and in the classes, workshops, seminars, agency courses, online classes, and on-site training. Your curiosity opened my eyes a thousand times. Your questions helped me know what people want to know and need to know. Through working with you, I received the gift of the opportunity to learn, and that is some of the best fun there is. Thank you, one and all.

About the Author

Charles Yoe is a professor of economics at Notre Dame of Maryland University and an independent risk analysis consultant and trainer. Working extensively for U.S. and other government agencies as a consultant and risk analyst, his wide range of risk experience includes international trade, food safety, natural disasters, public works, homeland security, ecosystem restoration, resource development, navigation, planning, and water resources. As a consultant to private industry his work includes a discrete but wide variety of concerns. He has trained professionals from more than one hundred countries in risk analysis and has conducted customized risk training programs for government agencies and private industry in more than two dozen nations. He can be reached at cyoe1@verizon.net.

1

The Basics

1.1 What Is Risk?

Risk is a measure of the probability and consequence of uncertain future events. It is the chance of an undesirable outcome. That outcome could be a loss (fire, flood, illness, death, financial setback, or any sort of hazard) or a potential gain that is not realized (new product did not catch on as hoped, your investment did not produce expected benefits, the ecosystem was not restored, or any sort of opportunity missed). What usually creates the "chance" is a lack of information about events that have not yet occurred. We lack information because there are facts we do not know, the future is fundamentally uncertain, and because the universe is inherently variable. Let's call all of this "uncertainty" for the moment.

Given the presence of a hazard or an opportunity, there are two important components to a risk: chance or probability and an undesirable outcome or consequence. Risk is often described by the simple equation:

$$\text{Risk} = \text{Probability} \times \text{Consequence} \tag{1.1}$$

Consider this expression a mental model that helps us think about risk rather than an equation that defines it. What this expression is conveying is not so much that this is the manner in which all risks are calculated (they are not) as much as that both of these elements must be present for there to be a real risk. If an event of any consequence has no probability of occurrence, there is no risk. Likewise, if there is no consequence or undesirable outcome, then there is no risk.

A hazard is the thing that causes the potential for an adverse consequence. An opportunity causes the potential for a positive consequence. If a population, an individual, or some asset of interest to us is not exposed to the hazard or opportunity, then there will be no consequence and no risk. The range of possible consequences (loss of life, property damage, financial loss or gain, improved environmental conditions, product success, and the like) is vast, but even similar types of consequences can vary in frequency, magnitude, severity, and duration.

It is not likely that many professional risk analysts would agree with such simple definitions. There are any number of alternative definitions in use or found in the literature. Some purists prefer to define risk entirely in terms of adverse consequences, ignoring the chance of gains that may not be realized. These risks of loss are sometimes called pure risks. Some definitions specify the nature of the consequences. The U.S. Environmental Protection Agency (EPA), for example, "considers risk to be the chance of harmful effects to human health or to ecological systems resulting from exposure to an environmental stressor" (EPA 2010).

HAZARD

In a general sense, a "hazard" is anything that is a potential source of harm to a valued asset (human, animal, natural, economic, social). It includes all biological, chemical, physical, and radiological agents or natural/anthropogenic events capable of causing adverse effects on people, property, economy, culture, social structure, or environment.

THE LANGUAGE IS MESSY

The language of risk is relatively young and still evolving. The seeds of risk analysis are sown across many disciplines, and each has found it useful to define the terms of risk analysis in a way that best serves the needs of the parent discipline. The EPA, for example, identifies 18 variations on the meaning of risk in their "Thesaurus of Terms Used in Microbial Risk Assessment," which eponymously takes a narrow focus on the concept of risk (http://www.epa.gov/waterscience/criteria/humanhealth/microbial/thesaurus/).

Frank Knight (1921) is credited with the first modern definition of risk. Kaplan and Garrick (1981) attempted to unify the language with their famous triplet. There is not yet any one universally satisfactory definition of risk nor of many of the other terms used in this book. ISO 31000 (2009), for example, offers quite a different lexicon than the one used in this book. An appendix is devoted to the language differences in ISO 31000. There is more agreement on the practice of risk analysis than there is on its language.

Storms, hurricanes, floods, forest fires, and earthquakes are examples of natural hazards. When humans and human activity are exposed to these hazards there are risks with consequences that include loss of life, property damage, economic loss, and so on. There are human-made hazards by the scores: tools, weapons, vehicles, chemicals, technology, and activities that can pose similar risks to life, property, environment, economies, and so on. Health hazards comprise their own category and include pathogens, disease, and all manner of personal health difficulties and accidents that can arise. These risks of adverse consequences are traditional examples of risk.

Less widely accepted as risks, among the risk analysis community's members, are potential gains or rewards. Would anyone say they risk a promotion or an inheritance? Probably not, as this is not the traditional use of the word. Nonetheless, when there is some uncertainty that the gain will be realized, it qualifies as a risk under the definition used here. The International Organization for Standardization (ISO 2009) defines risk as the effect of uncertainty on an organization's objectives. This is clearly broad enough to include uncertain opportunities for gain. Risks of uncertain gain are often called speculative risks.

For those who prefer to think of risks only as adverse consequences, it takes only a small convolution of thought to say that not realizing the gain/promotion/inheritance

A FEW PROPOSITIONS ABOUT RISK

- Risk is everywhere
- Some risks are more serious than others
- Zero risk is not an option
- Risk is unavoidable

Therefore, we need risk analysis to:

- Describe these risks (risk assessment)
- Talk about them (risk communication)
- Do something about the unacceptable ones (risk management)

is the adverse consequence. In any event, loss and uncertain potential gains are considered risks throughout this book. Know that some would prefer to distinguish and separate risks and rewards more carefully.

Thus, we have pure risks, which are losses with no potential gains and no beneficial result, and speculative or opportunity risks, which are generally defined as risks that result in an uncertain degree of gain. They are further distinguished by the fact that pure risk events are beyond the decision maker's control and are the result of uncontrollable circumstances, while speculative risks are the result of conscious choices made in decision making. These two types of risks lead to two distinct risk management strategies: risk avoiding and risk taking. Risk managers select options that will enable them to reduce unacceptable levels of pure risk to acceptable or tolerable levels. Risk managers also choose to take risks when they select one alternative course of action or option from among a set of alternatives. So risk managers function as risk avoiders when they decide how best to reduce the adverse consequences of risk and as risk takers when they decide how best to realize potential gains in the future. Uncertainty makes all of this necessary; there is no risk without uncertainty.

There is very little we do that is risk free, although risks certainly vary in the magnitudes of their consequences and the frequencies of their occurrences. A leaky ballpoint pen is not in the same class of risks as an asteroid five miles in diameter colliding with Earth.

Risk is sometimes confused with safety. In the past we have tried to provide safety, and getting to safety has been the goal of many public policies. The problem with a notion like safety is that someone must decide what level of chance or what magnitude of consequence is going to be considered safe. That is a fundamentally subjective decision, and subjective decisions rarely satisfy everyone. Risk, by contrast, is measurable, objective, and based on fixed criteria.

Safety has been defined in a number of legislative and administrative frameworks* as a "reasonable certainty of no harm," a phrase extended in some contexts to include "when used as intended." The very language chosen suggests the existence of a residual risk, and if there is a residual risk, then safety in any absolute sense is a psychological fiction.

* The Food Quality Protection Act of 1996 is one such example.

An alternative to looking for safety and providing margins of safety is to look objectively for risk and to manage it when it is not acceptable. That means we have to be able to objectively describe these risks for ourselves and others. Then we need to be able to communicate that information to one another. Finally, we need a means of determining when a risk is not acceptable and needs to be avoided or managed to some level we can tolerate. This is basically the risk analysis process.

Because uncertainty gives rise to risk, the essential purpose of risk analysis is to help us make better decisions under conditions of uncertainty. This is done by separating what we know about a decision problem from what we do not know about the problem. We use what we know and intentionally address those things we do not know in a systematic and transparent decision-making process that includes effective assessment, communication, and management of risks. Risks may be described using good narrative stories or through a complex combination of sophisticated quantitative and probabilistic methods.

Many people in many different disciplines have long ago figured this all out. They also articulated these ideas in the language of their own disciplines, and that has given birth to a wonderfully chaotic language of risk. Many of these discipline-based uses of risk analysis have deep enough roots that practitioners are sometimes reluctant to consider other views of this new composite discipline. There may be emerging consensus about some ideas, but there is little or no universal agreement about the language of risk analysis. That makes it difficult for anyone trying to understand the essence of risk analysis to get a clear view of just what this is all about.

Risk analysis is a framework for decision making under uncertainty. It was born spontaneously, if not always simultaneously, in many disciplines. It has evolved by fits and starts rather than by master design. Its practice is a wonderful mess of competing and even, at times, contradictory models. Its language borders on a babel of biblical proportions. And still it has begun to become something we can all recognize.

This book makes no pretense toward unifying, standardizing, or exemplifying the language, definitions, or models of risk analysis. What it does modestly attempt is to distill the common elements and principles of the many risk tribes and dialects into serviceable definitions and narratives. Once grounded in the basic principles of risk analysis, the reader should feel free to venture forth into the applications and concepts of the many communities of practice to use their models and speak their language. Now, with this simple understanding of risk and this caveat in mind, let's consider a few critical questions.

1.2 What Is Risk Analysis?

Risk analysis is a process for decision making under uncertainty that consists of three tasks: risk management, risk assessment, and risk communication, as shown in Figure 1.1. We can think of it as the process of examining the whole of a risk by assessing the risk and its related relevant uncertainties for the purpose of efficacious management of the risk, facilitated by effective communication about the risk. It is a systematic way of gathering, recording, and evaluating information that can lead to recommendations for a decision or action in response to an identified hazard or

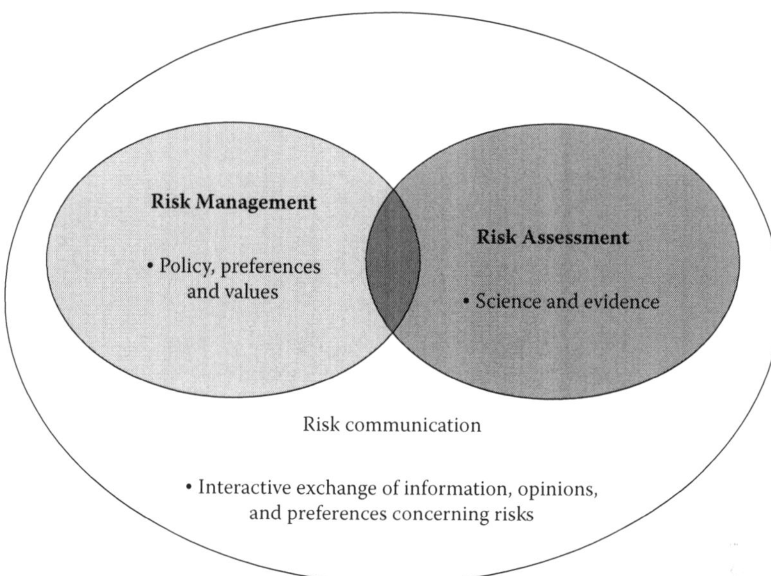

FIGURE 1.1 Three tasks of risk analysis.

opportunity for gain. Risk analysis is not a science; it is not certain; it is not a solution; it is not static.

Risk analysis is evolving into a paradigm for decision making under conditions of uncertainty. We may be uncertain about one or more aspects of the likelihood or the consequence of the risk(s) of concern to us. More troubling, we may be unsure how effective our risk management efforts will be.

People tend to not be terribly analytical when making decisions. Human reasoning is fallible. Risk analysis influences our thinking by making it more analytical. This simultaneously limits the "damage" our fallible human reasoning can inadvertently do when making decisions. Risk analysis is a useful and an evolving way to think about and solve risky and uncertain problems. It is science-based decision making, but it is not science. This is true in part because the uncertainty is sometimes substantial, but also because risk analysis honors social values. In fact, it is not a stretch to think of risk analysis as the decision-making interface between science and values.

What makes risk analysis a paradigm? That question will be answered in some detail throughout this book. But let's consider a few features that distinguish this paradigm.

First, it is based on good science. Scientific facts, evidence, and good analytical techniques are hallmarks of risk analysis. In best practice, risk analysis relies on the best available science. Risk analysis separates what we know (the science) from what we don't know (the uncertainty), and it focuses appropriate attention on what we don't know and how that might affect decision outcomes and, therefore, the decision itself. It aspires to get the right science into the decision process and then to get that science right. The risk assessment task is always based as much as possible on sound evidence, whether that evidence is qualitative or quantitative, known with certainty or shadowed by uncertainty. Done well, it uses the best available analytical techniques and methods.

Second, risk analysis considers social values. As important as science is, it is not the sole basis for decision making. Social values enter the risk analysis process through the risk management task. Risk analysis incorporates both good science and social values when making decisions under uncertainty.

Third, risk analysis addresses uncertainty explicitly. Few, if any, decisions are ever made with complete information and certainty. Lacking complete information and facing sometimes considerable uncertainty rarely absolves us of the need to make a decision. Risk analysis has evolved explicitly for these kinds of decision problems. It is a paradigm that copes well with soft data and that tolerates ambiguity both in analysis and decision making. Risk assessors address uncertainty in the assessment of risks; risk managers address it in their decision making; and risk communicators convey its significance to interested parties as appropriate.

Fourth, the purpose of this paradigm is to begin to make good decisions by finding and defining the right problem. If the problem is not properly identified, little that follows will aid a successful solution. Risk analysis seeks the needed information from a variety of sources. In the process of doing so, it involves many people in its efforts to identify and resolve that which we do not know about the problem.

Fifth, because of its focus on uncertainty, risk analysis is well suited to continuously improving decisions. As uncertainty is reduced over time and problems are better understood, new and better solutions may come into view. Risk analysis is flexible and can be updated. Every risk management decision is conditional on what is known and what is not known at the time the decision is made. Risk analysis has an eye on uncertainty, and this enables it to deal with a future-focused vision of the next solution as well as the current one. Reducing that which is not known about a situation and ever-changing social values ensures that many risk management decisions are part of an evolutionary decision-making process.

Risk analysis provides information to decision makers; it does not make decisions. It is neither a magic bullet nor a black box. The risk analysis paradigm helps establish the balance between the expediency of decision making and having all the information. It does not remove subjectivity and judgment from decision making. If anything, it shines a light on these things and forces them to consider what is not known with certainty.

Risk analysis has been defined here as a process with three tasks. These tasks will be discussed in considerable detail in subsequent chapters. For now let's content ourselves with some informal characterizations of these tasks.

Risk management is a process of problem identification, requesting information, evaluating risks, and initiating action to identify, evaluate, select, implement, monitor, and modify actions taken to alter levels of unacceptable risk to acceptable or tolerable levels. The goals of risk management are often said to include scientifically sound, cost-effective, integrated actions that reduce risks while taking into account economic, environmental, social, cultural, ethical, political, and legal considerations. More informally, risk management is the work one has to do to pose and then answer the following kinds of questions:

- What's the problem?
- What information do we need to solve it, i.e., what questions do we want risk assessment to answer?
- What can be done to reduce the impact of the risk described?
- What can be done to reduce the likelihood of the risk described?
- What are the trade-offs of the available options?
- What is the best way to address the described risk?
- (Once implemented) Is it working?

Risk assessment is a systematic process for describing the nature, likelihood, and magnitude of risk associated with some substance, situation, action, or event that includes consideration of relevant uncertainties. Risk assessment can be qualitative, quantitative, or a blend (semiquantitative) of both. It can be informally described by posing and answering the following questions that build on the Kaplan and Garrick triplet (1981):

- What can go wrong?
- How can it happen?
- What are the consequences?
- How likely is it to happen?

Risk communication is the open, two-way exchange of information and opinion about risks intended to lead to a better understanding of the risks and better risk management decisions. It provides a forum for the interchange of information with

REWARDS

When the definition of risk is expanded to include uncertain potential gains, it challenges the conventional language. In particular, risk management must include modifying risks as well as mitigating them. Some of the management questions here must be altered somewhat.

- What's the opportunity?
- What information do we need to attain it, i.e., what questions do we want risk assessment to answer?
- What can be done to increase the positive impact of the opportunity risk described?
- What can be done to increase the likelihood of the desired outcomes?

The other questions remain the same for either risk or reward. In a similar vein, the risk assessment questions can be modified to ask what can go right, if that helps you understand the concepts.

The alternating emphasis on potential gains could become tedious and cumbersome if constantly continued. Alternative discussions of potential gains will be restricted to only the most critical topics.

all concerned about the nature of the risks, the risk assessment, and how risks should be managed. Risk communication may be informally characterized by its own set of questions (Chess and Hance 1994):

- Why are we communicating?
- Who are our audiences?
- What do our audiences want to know?
- How will we communicate?
- How will we listen?
- How will we respond?
- Who will carry out the plans? When?
- What problems or barriers have we planned for?
- Have we succeeded?

Even a brief review of the literature will reveal a staggering range of definitions for these tasks. As with the definitions of risk, the discipline of origin and the nature of the risk have a powerful influence over the words used to define these tasks. However, once the words are distilled to their essence, the ideas represented by the questions above capture the spirit of most definitions in use today. The differences are, in my opinion, more semantic than substantive, but semantics are very important to some people. Any good risk analysis approach will identify hazards and opportunities, characterize the risks, recognize and address uncertainty, summarize conclusions, recommend options, and document the basis for all recommendations and decisions.

1.3 Why Do Risk Analysis?

In a word, "uncertainty" is the reason we do risk analysis. There is very little in life that is risk free, and risk is everywhere. In a certain world, decision making may not be a seamless process: We would still argue values, priorities, and trade-offs, for example; but it would be much easier than it is now. In the uncertain world in which we live, the circumstances of our lives, the problems we face, and the evidence available to us as well as the outcomes of our decisions are often unknown. We have come to realize the value found in managing, assessing, and communicating about risks to make better decisions and to better inform affected publics and stakeholders about the nature of the risks they face and the steps we take to manage them.

There are other compelling reasons as well. Our decision-making processes and approaches to problems used in the past have paid amazing dividends. We have done much to make the world less risky through modern medicine, engineering, finance, environmental management, and the like. Even so, substantial and persistent problems remain. New risks appear all the time. Clearly, as well as our decision processes have served us in the past, they have not been sufficient to rid the world of risks. So we do risk analysis to intentionally make our lives less risky, to wisely take risks when warranted and, hopefully, to reduce unacceptable risks to levels that we can at least tolerate.

Traditional approaches to decision making have relied on such things as precedent, trial and error, expert opinion, professional judgment, compromise, safety assessment, standards, precaution, inspection, zero tolerance, ignorance, and a host of other more-or-less structured decision-making strategies. Recurring problems and unrealized opportunities persist, and these traditional approaches have proven insufficient. They have been unable to detect and resolve many current problems. They have been slow to effectively deal with the growing complexity and rapid pace of change in society. Few of these traditional approaches have effectively integrated science and social values in decision making. They do not deal especially well with uncertainty.

Science-based risk analysis activities have been shown to be effective in reducing risks, and they are becoming the standard operating procedure for many public and private-sector organizations. Risk analysis adds value to decisions by improving the quality of our thinking before a decision is made. Uncertainty is ubiquitous, and every organization develops its own culture of uncertainty. At one extreme, this culture is dominated by risk analysis; at the opposite extreme, we are oblivious to what we do not know. Intentionally considering the relevant uncertainty in a decision problem improves decision making.

One of the principal reasons we do risk analysis is to help provide and ensure a safer living and working environment for people. Risk analysis has also been used extensively to help protect animals, plant life, ecosystems, infrastructure, property, financial assets, and other aspects of modern society.

Risk analysis has become essential to economic development. The Technical Barriers to Trade (TBT) and Sanitary Phytosanitary (SPS) Agreements of the World Trade Organization, for example, establish risk assessment as a legitimate means for establishing protective trade practices when the life, health, and safety of a sovereign nation's people are at risk. Risk analysis and risk assessment are being used more and more frequently by international organizations.

1.4 Who Does Risk Analysis?

Many risk analysis practices have been around for centuries. It is only in the last half century or so, however, that the practice of risk analysis has begun to become more formalized and structured. Government agencies use risk analysis as the basis for regulation, resource allocation, and other risk management decisions. Private industry, sometimes following government's lead and sometimes leading government, is making more frequent and widespread use of risk analysis as well. To understand who is using risk analysis it is helpful to begin with a brief history of its evolution.

1.4.1 A Brief Historical Perspective on Risk Analysis

Risk analysis was not possible until we became able to think intentionally about probabilities and the facts of what can go wrong, how it can happen, and its consequences. Anyone who wants to know the history of risk analysis needs to read "Against the Gods: The Remarkable Story of Risk" (Bernstein 1996). If there is not time to read the book, then read Covello and Mumpower's (1985) "Risk Analysis and Risk Management: An Historical Perspective." These are simply the best single works of their genres done on the subject. This section owes a great debt to each.

We have always faced problems and have solved them, more or less successfully, since we have walked the planet. The authors cited in the previous paragraph detail this history delightfully. It was the possibility of risk assessment, however, that opened the door for risk analysis, and risk assessment has been made possible by the confluence of many events throughout history. These include the development of probabilistic thinking, which enables us to thoughtfully consider the "chance" dimension of risk and the evolution of science, which enables us to analyze and understand the "undesirable outcomes" that can occur. Our ability to think about and to understand probability and consequences in risk scenarios made risk assessment possible.

The rise of decision sciences, especially in the last century, has enhanced the role of the manager in the analysis of risk. Our growing interest in finding effective ways to deal with uncertainty in the universe has magnified the importance of both the assessment and management tasks. The fact that we still face many old as well as a growing number of new and emerging problems not solved by our old decision-making paradigms has opened the door for risk analysis at this point in time. Growing emphasis on the involvement of the public and stakeholders in public policy decisions has created a role for risk communication. So let us begin a brief look at the historical development of our ability to think about probability and the development of scientific methods to establish and demonstrate causal links and connections between adverse effects and different types of hazards and activities.

Undoubtedly, risk assessment began when some unknown *Homo sapiens* picked up something, ate it, fell sick, and died. "Don't eat that!" must have been the mental note all around him made. Risk analysis had begun.

History is filled with scientific footnotes that suggest aspects of risk assessment. The Asipu in 3200 BCE (Covello and Mumpower 1985) plied the Tigris-Euphrates Valley offering guidance for risky ventures. Hippocrates (460–377 BCE), centuries later, studied the toxicity of lead. Socrates (469–399 BCE) experienced the risks of hemlock, and Aristotle (384–322 BCE) knew that fumes from charcoal could be dangerous. Pliny (23–79 CE) and Galen (131–201 CE) explored the toxicity of mercury in their medical studies. The point being, we humans have long been engaged with aspects of risk, especially identifying those things that can do us harm.

Fast forward to the later Renaissance period in Europe where gambling, always a popular pastime, piqued an interest in the more formal study of probability. If anyone succeeded in figuring out the odds of various games of chance, it was not documented until Girolamo Cardano (1500–71) wrote *Liber de Ludo Aleae* (Book on games of chance). This is one of the earliest works to explore statistical principles of probability. His book, which focused on chance, is the first to express chance as a fraction. Odds began to appear soon after.

Blaise Pascal's (1623–62) wager is broadly considered to be one of the first examples of decision science. In it he undertook the age-old question of God's existence. God is, or God is not? Which way should we incline, and how shall we live our lives? Considering two states of the world (God is, God is not) and two alternative behaviors (live as a Christian, live as a pagan), Pascal used probability and concluded that the expected value of being a Christian outweighed the expected value of paganism.

The use of basic statistics is also relatively new to humans. John Graunt (1620–74) undertook a study of births and deaths in 17th-century London to learn how many people may be available for military service. He used raw data in new ways (sampling and statistical inference) that formed the basis for modern statistics. He published his famous life expectancy tables in 1662 and changed data analysis forever.

Risk management made a formal appearance in Edward Lloyd's 1687 London coffeehouse. By 1696, Lloyd's List of ship arrivals and departures along with conditions abroad and at sea was a risk management standard for everyone in the British maritime industry. Ships' captains compared notes on hazards in one corner of the coffeehouse, and it grew into the headquarters for marine underwriters, a precursor of the modern insurance industry. Flower and Jones (1974) report that London's insurance industry would help protect you from house-breaking, highway robbery, death by gin-drinking, death of horses, and would provide assurance of female chastity—no doubt the great risks of the day!

Jacob Bernoulli (1654–1705) began to integrate ideas about information and evidence into the growing body of thought on probabilities. He noted we rarely know a probability before an event (a priori) but can often estimate a probability after an event (a posteriori). This, he noted, implies changing degrees of belief as more information is compiled; so the past is only part of reality. Thomas Bayes (1701–61) extended this work and wrote of using new information to revise probabilities based on old information. The world was beginning to discover tools and to think that perhaps uncertainty can be measured and variability described. Many others (Laplace, Chebyshev, Markov, von Mises, and Kolmogorov, to name a few) followed, and the quantitative universe was gradually being revealed.

Meanwhile, our knowledge of disease and our powers of scientific observation were also making great leaps. Edward Jenner (1749–1823) observed that milkmaids got cowpox but not smallpox. John Snow (1813–58) figured out that cholera was transmitted by contaminated water, by studying what today we might call a GIS (geographic information system) map of a cholera outbreak. The microscopic world was beginning to come into focus.

The Industrial Revolution marked a change in the public sector's role in the management of risks. Concerns about occupational disease and the need to protect workers and the public from toxic chemicals gave rise to the field of public health. Toxicology was one of many emerging sciences, and the idea of a no observed adverse effects level (NOAEL) was born in the last century. This is the dose of a chemical at which there are no statistically or biologically significant increases in the frequency or severity of adverse effects between the exposed population and its appropriate control. This was clearly a firm step in the direction of risk analysis, combining science with a value judgment.

Early efforts to determine a safe level of exposure to chemicals were based on laboratory animal tests to establish a NOAEL. To leap the uncertain hurdles of extrapolating from animals to humans and from the high doses of a chemical given to animals to the low doses to which humans were exposed, the scientific community approximated a safe level by dividing the NOAEL by an uncertainty or safety factor to establish the acceptable daily intake (ADI):

$$\text{ADI} = \text{NOAEL}/\text{uncertainty factor} \tag{1.2}$$

In the 1950s, the U.S. Food and Drug Administration (FDA) used a factor = 100 to account for the uncertainty.

Risk assessment per se began with radiation biology in the middle of the 20th century. The Japanese survivors of World War II atomic bomb blasts made the dangers of radiation eminently clear. This new technology raised concerns about how the incidence of human cancer is influenced by exposure to small doses of radiation.

The National Academies of Science (NAS) in the United States struggled with this radiation question, and the first formal risk assessment, "Reactor Safety Study: An Assessment of Accident Risks in U.S. Commercial Nuclear Power Plants," NUREG 75/014, better known as the Rasmussen Report, was prepared for the Nuclear Regulatory Commission (NRC 1975). This was, among other things, a study of core meltdowns at nuclear power plants that used a no-threshold model to estimate cancer deaths following a nuclear reactor accident.

More formal notions of risk were finding their way into the public-sector mentality. The Delaney Clause was a 1958 amendment to the Food, Drugs and Cosmetics Act of 1938 that was an effort to protect the public from carcinogens in food. It is often cited as an effort to establish a zero tolerance for policy purposes. When scientific methods were a bit cruder than they are now, it was easier to equate an inability to detect a hazard with a notion of zero risk.

As science improved, it became clear that zero risk was not a policy option, and the notion of de minimis risk took root. A de minimis risk is a risk so low as to be effectively treated as negligible. Mantel and Bryan (1961) suggested that anything that increases the lifetime risk of cancer by less than 1 in 100 million was negligible. The FDA later relaxed this to 1 in 1 million. The EPA proposed to adopt a uniform "negligible risk" policy for all carcinogenic residues in food in 1988. The Occupational Safety and Health Administration (OSHA) regulated all carcinogens in the workplace to the lowest level feasible. The point to be taken for our purposes is that society was beginning to get used to the idea that we would have to live with some nonzero level of risk.

Government agencies were routinely doing risk assessment by this time, and the early pioneers of risk assessment describe a rather ad hoc process. In the 1980s the National Research Council was asked to determine whether organizational and procedural reforms could improve the performance and use of risk assessment in the federal government. In 1983 they published their response, *Risk Assessment in the Federal Government: Managing the Process*, better known as "the Red Book" because of its cover (NRC 1983). This is one of the seminal publications in risk assessment and it identified the four steps of risk assessment as:

- Hazard identification
- Dose-response assessment
- Exposure assessment
- Risk characterization

This has been the foundation model for risk assessment that has been modified and evolved many times since.

Risk assessment came before the U.S. Supreme Court in two cases during the Carter administration. The Industrial Union Department, AFL-CIO v. American Petroleum Institute, 448 U.S. 607 (1980) case considered whether quantitative cancer risk assessments could be used in policy making. One federal agency, OSHA, said no, while

NATIONAL FLOOD PROGRAM

Risk analysis was creeping into the public consciousness in a number of ways, although no one called it by that name at the time. In 1936 the U.S. government passed the Flood Control Act of 1936, which established a national flood control program. This program assessed, communicated, and managed risk. Following Hurricanes Katrina and Rita, the U.S. Army Corps of Engineers renamed this program Flood Risk Management.

the EPA and FDA said yes. The majority opinion established that risk assessment is feasible and that OSHA must do one before taking rule-making action to reduce or eliminate the benzene risk. Later, in the American Textile Manufacturers Institute v. Donovan, 452 U.S. 490 (1981) case, the Supreme Court reaffirmed the Benzene case finding and added that safe does not mean zero risk. With this last hurdle cleared, risk assessment moved more confidently into the government's policy arena.

Internationally, risk assessment was also growing in credibility. The General Agreement on Tariffs and Trade's (GATT) Uruguay Round on multilateral trade negotiations (1986–94) was instrumental in the global spread of risk analysis. Specifically, two agreements—on Sanitary and Phytosanitary Measures (SPS) and on Technical Barriers to Trade (TBT)—paved the way for risk analysis in the World Trade Organization (WTO).

The SPS agreement recognizes the right of governments to protect the health of their people from hazards that may be introduced with imported food by imposing sanitary measures, even if this meant trade restrictions. The agreement obliges governments to base such sanitary measures on risk assessment to prevent disguised trade protection measures.

Following the lead of the WTO, many regional trade agreements, including the North American Free Trade Agreement (NAFTA), incorporate risk analysis principles into their agreements. Both the Food Agricultural Organization (FAO) and the World Health Organization (WHO), two United Nations agencies, lend extensive support to the use of risk analysis principles globally. The International Organization for Standardization (ISO 2009) uses risk assessment and risk management in much of its guidance and standards (ISO 31000).

In recent years many nations have begun to make extensive use of risk analysis in their regulatory and other government functions. Risk analysis is now well established in both the private and public sectors around the world.

THREE SISTERS

The SPS agreement has influenced the international standards of the Codex Alimentarius (for food), the World Organization for Animal Health (OIE), and the International Plant Protection Convention (IPPC), all of which have adopted risk analysis principles for their procedures.

1.4.2 Government Agencies

Government agencies are widely adopting risk analysis principles to varying extents. Some agencies have begun to redefine their missions and modes of operation in terms of risk analysis principles. Risk analysis has become their modus operandi. Other agencies have added risk analysis principles to their existing methodologies and tools for accomplishing their mission (see accompanying text box). *Risk-informed decision making* is a term of art often used to describe the use of risk analysis in some government agencies. States and local governments are adopting risk analysis approaches at varying rates. Natural and environmental resource agencies as well as public health and public safety agencies tend to be the first to adapt risk analysis principles at the nonfederal levels of government.

Internationally, risk analysis has proliferated in some communities of practice. Food safety, animal health, plant protection, engineering, and the environment are some of the areas in which other national governments are likely to have established the practice of risk analysis. The global economic recession that began in 2008 has propelled economic and financial regulatory agencies to move more aggressively toward risk analysis.

1.4.3 Private Sector

The insurance industry may represent the oldest and most explicit application of risk management in the private sector. Other safety-oriented professions and businesses like engineering, construction, and manufacturing have long been devoted to safety assessments. Many have now begun a more explicit consideration of risk. The private financial sector has also been an innovator in risk-related areas. Security has taken on a growing number of risk applications as technology has expanded the notion of and need for risk analysis.

All links in the food chain have been devoting increased attention to food-safety risk analysis. The medical community is increasingly involved with risk reduction as well. Although formal risk analysis has not yet penetrated the private sector to the extent it seems to have done in the public sector, the number of companies and facilities that show interest in risk analysis continues to grow. ISO 31000 marks a landmark effort to standardize many risk management notions for industry. As public policies increasingly reflect the influence of risk analysis, it is inevitable that the private-sector interest will continue to grow.

1.5 When Should We Do Risk Analysis?

Risk analysis is for organizations that make decisions under conditions of uncertainty. Figure 1.2 provides a schematic illustration of the kinds of decision contexts where risk analysis adds the most value to decision making. This value depends on how much uncertainty the organization faces and the consequences of making a wrong decision.

In the lower right quadrant there is little uncertainty and the consequences of being wrong are minor. This kind of decision making does not require risk analysis. Any convenient means of decision making will do here.

Selected U.S. Agencies Using Risk Analysis Principles

Animal and Plant Health Inspection Service
http://www.aphis.usda.gov/

Bureau of Economic Analysis (BEA)
http://www.bea.gov/

Bureau of Reclamation
http://www.usbr.gov/

Centers for Disease Control and Prevention (CDC)
http://www.cdc.gov/

Coast Guard
http://www.uscg.mil/

Congressional Budget Office (CBO)
http://www.cbo.gov/

Consumer Product Safety Commission (CPSC)
http://www.cpsc.gov/

Corps of Engineers
http://www.usace.army.mil/

Customs and Border Protection
http://www.cbp.gov/

Defense Advanced Research Projects Agency (DARPA)
http://www.nsf.gov/

Department of Defense (DOD)
http://www.defenselink.mil/

Department of Energy (DOE)
http://www.energy.gov/

Department of Homeland Security (DHS)
http://www.dhs.gov

Director of National Intelligence
http://www.dni.gov

Economic Research Service
http://www.ers.usda.gov/

Endangered Species Committee
http://endangered.fws.gov/

Environmental Protection Agency (EPA)
http://www.epa.gov/

Federal Aviation Administration (FAA)
http://www.faa.gov/

Forest Service
http://www.fs.fed.us/

Federal Bureau of Investigation (FBI)
http://www.fbi.gov/

Fish and Wildlife Service
http://www.fws.gov/

Food and Drug Administration (FDA)
http://www.fda.gov/

Food Safety and Inspection Service
http://www.fsis.usda.gov/

Foreign Agricultural Service
http://www.fas.usda.gov/

Geological Survey (USGS)
http://www.usgs.gov/

Government Accountability Office (GAO)
http://www.gao.gov/

National Aeronautics and Space Administration (NASA)
http://www.nasa.gov/

National Marine Fisheries
http://www.nmfs.noaa.gov/

National Oceanic and Atmospheric Administration (NOAA)
http://www.noaa.gov/

National Park Service
http://www.nps.gov/

National Science Foundation
http://www.nsf.gov/

National Security Agency (NSA)
http://www.nsa.gov/

National Transportation Safety Board
http://www.ntsb.gov/

National War College
http://www.ndu.edu/nwc/index.htm

National Weather Service
http://www.nws.noaa.gov/

Natural Resources Conservation Service
http://www.nrcs.usda.gov/

Selected U.S. Agencies Using Risk Analysis Principles	
Nuclear Regulatory Commission http://www.nrc.gov/	Risk Management Agency (Department of Agriculture) http://www.rma.usda.gov/
Oak Ridge National Laboratory http://www.oro.doe.gov/	Securities and Exchange Commission (SEC) http://www.sec.gov/
Occupational Safety and Health Administration (OSHA) http://www.osha.gov/	Superfund Basic Research Program http://www.niehs.nih.gov/research/supported/sbrp/
Office of Management and Budget (OMB) http://www.whitehouse.gov/omb/	Tennessee Valley Authority http://www.tva.gov/
Office of Science and Technology Policy http://www.ostp.gov/	

When there is a lot of uncertainty but the consequences of an incorrect decision are minor, it would be sufficient to do a modest level of risk analysis. This may entail little more than sifting through the uncertainty to assure decision makers that the decision and its outcome are not especially sensitive to the uncertainties. In some instances it may be sufficient to establish that one or the other factors of the "probability × consequence" product is sufficiently small as to render the relevant risks acceptable.

When the consequences of making a wrong decision rise, so does the value of risk analysis. In an environment with relatively less uncertainty but with serious consequences for wrong decisions, risk analysis is valuable as a routine method for decision making. As the uncertainty grows in extent, risk analysis becomes the most valuable, and more-extensive efforts may be warranted.

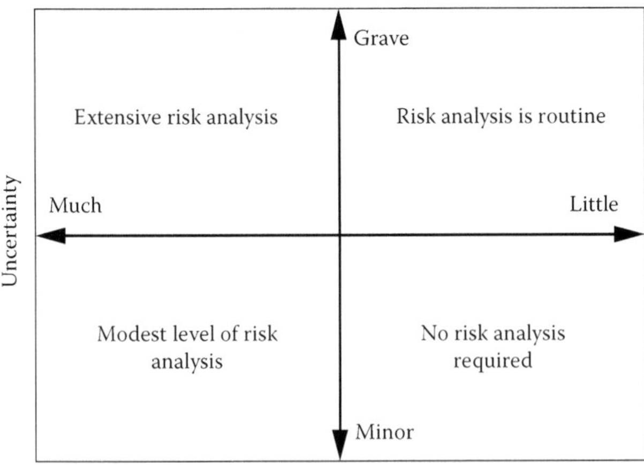

FIGURE 1.2 When to use risk analysis.

Some organizations would be wise to always be doing risk analysis. In fact, a slowly growing number of organizations define themselves as risk analysis organizations, meaning that what they do as an organization is manage risks, assess risks, and communicate about risks. Others use risk analysis as a framework, tool, or methodology for specific situations. Stewards of the public trust would be well advised to use risk analysis for decision making.

Risk analysis, as a way of doing business, is especially useful for organizations that have some or all of the following elements of decision making in common (adapted from NRC 2009):

- A desire to use the best scientific methods and evidence in informing decisions
- Uncertainty that limits the ability to characterize both the magnitude of the problem and the corresponding benefits of proposed risk management options
- A need for timeliness in decision making that precludes resolving important uncertainties before decisions are required
- The presence of some sort of trade-off among disparate values in decision making
- The reality that, because of the inherent complexity of the systems being managed and the sometimes long-term implications of many risk management decisions, there may be little or no short-term feedback as to whether the desired outcome has been achieved by the decisions.

Every organization has its own unique reality. They have a history, a mission, personnel, resources, policies, procedures, and their own way of doing things. If you drop risk analysis down into any organization, the context of that organization is going to affect the way the risk analysis paradigm is going to look and work. For example, there is no one risk analysis model followed by the U.S. government agencies mentioned previously. In fact, it is probably fair to say that risk analysis looks different in every organization that uses it.

Take the FDA as an example. Risk analysis is vigorously pursued in several of its Centers including the Centers for Food Safety and Applied Nutrition, Veterinary Medicine, Devices and Radiological Health, and Drug Evaluation and Research. Each of these defines the terms of risk analysis differently and applies the concepts in different ways and to varying extents. They have developed their own risk-related tools and techniques. This is a strength of the paradigm. It is a remarkably flexible and robust way to think about and to solve problems, so be assured there is no one best way to practice risk analysis.

1.6 Organization of Book

This book is organized into two broad sections. The first 5 chapters provide a generic introduction to the principles of risk analysis. The next 13 chapters provide details on how to apply these principles.

There are hundreds of very good books already in print devoted to risk analysis. Most tend to focus, relatively quickly, on a rather narrow aspect or practice of the discipline. Many are written from a particular disciplinary or topical perspective, like engineering, finance, environment, public health, food safety, water, and so on. This book avoids a narrow focus on any one field in favor of distilling principles, integrating topics, and stressing the application of the principles in a generic fashion.

Risk analysis can become complex. Some books can become overwhelming for those new to risk analysis because they introduce so many ideas so fast. This text focuses narrowly on the most basic principles of the risk analysis paradigm, i.e., risk, uncertainty, risk management, risk assessment, and risk communication.

Like good risk analysis, this book proceeds in an iterative fashion. Each of the next four chapters unpacks and explains important concepts introduced in this chapter. Chapter 2 takes up the notion of uncertainty in more detail. Uncertainty is the primary reason for risk analysis, and its pervasiveness is what has caused the use of risk analysis to spread so quickly in recent years. It's important for risk managers, risk assessors, and risk communicators to have a sound and common understanding of uncertainty.

Chapter 3 develops the risk management component and the job of the risk manager. In best practice, every risk analysis task begins and ends with risk management. The tasks of risk management are presented in a generic fashion, free from any particular preexisting risk management model. Chapter 4 unpacks the risk assessment component. This is where the analytical work gets done for any risk management activity. As with the risk management chapter, the risk assessment tasks are presented model free. Chapter 5 explores the risk communication component in greater detail. The generic tasks of both the internal and external risk communication responsibilities are presented. These three chapters together describe the risk analysis paradigm.

Chapters 6, 7, and 8 expand on the risk management task. Chapter 6 emphasizes the importance of problem-identification in risk management by expanding on the nature of problems and opportunities and by offering several techniques that have been useful in identifying them. This critical step is one of the most overlooked and underemphasized in my risk analysis experience.

Chapter 7 is about brainstorming. Good risk management requires divergent thinking at many points along the process. Many, if not most, well-educated professionals are justifiably leery of processes, especially those that seem trendy and fashionable. When the goal is to generate ideas and different perspectives, brainstorming just works. No risk manager should be without a technique or two to draw on.

The final risk management chapter summarizes economics for risk mangers. Someone is always going to care about costs. With complex decision problems, there will always be conflicting values. This chapter focuses primarily on the economic aspects of opportunity cost and trade-offs. The ability to correctly identify all the relevant costs and to understand how trade-offs are made at the margin is an indispensable skill for risk managers in any field.

The risk assessment task is emphasized in Chapters 9 through 16. Many essential details will be found in them. These chapters help fill the risk assessor's toolbox. Qualitative risk assessment tools are provided in Chapter 9. A generic qualitative assessment process, adaptable to virtually any kind of application, is followed by a discussion of several other techniques and methodologies in use today. These include risk narratives, evidence maps, ranking techniques, risk matrices, and others.

The subject of Chapter 10 is modeling. It begins by considering different types of models and then focuses on a 12-step model-building process that can be followed for any kind of quantitative model. This section builds on the work of others and my own experience in building risk assessment models.

The next four chapters address varying aspects of probability, all essential to decision making under uncertainty. A review of basic probability concepts used by risk assessors is found in Chapter 11. It presents a pragmatic distinction between the frequentist and subjectivist views of probability without taking an advocacy position. This is followed by a discussion of probability essentials, like where they come from and what the most important axioms and propositions are. All of this is done with an eye on why you need to know this to do risk assessment.

Chapter 12 is, I hope, one of the more useful chapters in this applications part of the text. It begins by describing the different functions that can be used to present probability distribution information. The real heart of the chapter is devoted to helping risk assessors—whether experienced or inexperienced—choose the right probability distribution to represent their knowledge uncertainty or the natural variability in the world they are modeling. The method presented has been pieced together over a number of years of experience, where at times I was the only person who cared about the distribution and at other times the acceptance of a model stood on the credibility of the distribution(s) chosen.

Chapter 13 introduces the topic of probability elicitation, a specific form of expert elicitation that is growing rapidly. One of the most common uncertainties encountered stems from a lack of data about quantities of critical interest to our risk assessments. Experts are increasingly being used to fill in gaps in dose-response curves, to estimate probabilities of failure and unsatisfactory performance, to forecast the likelihoods of future events from sea level rise to terrorist attacks, and virtually anything you can imagine. The problem is that even experts are not so reliable at estimating subjective probabilities. Thus, it is important to have a good grasp of the issues that can arise in subjective probability elicitations. That grasp can be found in Chapter 13.

The Monte Carlo process may be the most commonly used approach to probabilistic risk assessment. Every good risk assessor needs to know a little about this process and how it works. Chapter 14 provides the reader with a peek behind the Monte Carlo process curtain. It will also help you think about how many iterations you need and whether you should generate them using Monte Carlo or Latin Hypercube sampling techniques.

Chapter 15 builds on the work in earlier chapters to present an especially useful bundle of risk assessment tools called probabilistic scenario analysis. This chapter begins by defining scenarios and describing some of the most common types used in risk assessment. It then focuses on tree structures as one of the more useful tools for structuring scenarios. Once the probability tools and techniques of the earlier chapters are layered on top of the scenario tools, the risk assessor has a very powerful suite of tools to use to assess risk.

Techniques for addressing uncertainty are woven throughout the chapters of this text. Chapter 16, however, discusses sensitivity analysis, which can be described as the uncertainty table stakes for every risk assessment. Exploring the significance of the uncertainty encountered in a decision problem for the decision-making process is the absolute bare minimum standard for any risk assessment. A variety of qualitative and quantitative sensitivity analysis techniques are presented or summarized.

The last two chapters expand on the risk communication task. Chapter 17 focuses on presenting and using the results of a risk assessment. This content could just as easily be considered an expansion of either the risk management or risk assessment tasks. It is considered part of the risk communication task here because communication of complex quantitative and probabilistic information to decision makers and the general public remains a substantial hurdle in the risk analysis process.

The final chapter is about developing risk communication messages for the public. Beginning with a basic communication model, it quickly differentiates the challenges of communicating about risk, especially during the high-anxiety circumstances of a crisis. Message development and message mapping are introduced as useful entry points into this rich field of risk communication.

Appendix A is provided to aid any reader who comes to this book steeped in the risk dialect of ISO 31000 (2009). The language of risk analysis is, indeed, messy, and it is still evolving. Rather than attempting to resolve the long-standing matter of defining terms, this book opts for distilling the principles from the many dialects of risk that are spoken in all its fields of practice. To aid that process, the ISO language is worth a special discussion, and it is found in Appendix A. Appendix B provides an introduction to Palisade Corporation's DecisionTools® Suite software that was used in the preparation of this book. Files used in the creation of this book and additional exercises as well as a free student version of the software are available at http://www.palisade.com/bookdownloads/yoe/principles/.

1.7 Summary and Look Forward

Risk is the chance of an undesirable outcome. That outcome may be a loss or the failure to attain a favorable situation. In a certain world there is no risk because every outcome is known in advance. It is uncertainty that gives rise to risk.

Safety is a subjective judgment, while risk analysis is, in principle, an objective search for the risks in any given situation. Risk analysis is the framework or, if you prefer, the discipline used to manage, measure, and talk about risk. It has three components: risk management, risk assessment, and risk communication. Risk analysis is now possible because of the confluence of many scientific developments. Its use in the United States and internationally is growing steadily, and applications are found in a wide variety of fields.

The language of risk analysis is evolving. It would be comforting to think that it is evolving toward some consensus definitions and a common terminology. That is not yet the case, and this book makes no attempt to resolve the language differences. What it does do is attempt to distill the principles common to the many different dialects of risk that are spoken in the fields of applied risk analysis.

The next chapter gets to the heart of risk analysis, i.e., decision making under uncertainty. A primary role of the risk analyst is to separate what we know from what we do not know and then to deal honestly, intentionally, and effectively with those things we do not know. The job of the risk analyst is to be an honest broker of information in decision making. Knowledge uncertainty and natural variability, two fundamental concepts essential to understanding risk analysis, are the focus of Chapter 2.

REFERENCES

Bernstein, Peter L. 1996. *Against the gods: The remarkable story of risk.* New York: John Wiley & Sons.

Chess, C., and B. J. Hance. 1994. *Communicating with the public: Ten questions environmental managers should ask.* New Brunswick, NJ: Center for Environmental Communication.

Covello, Vincent T., and Jeryl Mumpower. 1985. Risk analysis and risk management: An historical perspective. *Risk Analysis* 5 (2): 103–120.

Environmental Protection Agency. 2010. Thesaurus of terms used in microbial risk assessment. http://www.epa.gov/waterscience/criteria/humanhealth/microbial/thesaurus/T51.html.

Flower, Raymond, and Michael Wynn Jones. 1974. *Lloyd's of London: An illustrated history.* Newton Abbot, England: David and Charles.

International Organization of Standardization. 2009. *Risk management—principles and guidelines.* Geneva, Switzerland: International Organization of Standardization.

Kaplan, Stanley, and B. John Garrick. 1981. On the quantitative definition of risk. *Risk Analysis* 1 (1).

Knight, Frank. H. 1921. *Risk, uncertainty and profit.* Chicago: University of Chicago Press.

Mantel, N., and W. R. Bryan. 1961. Safety testing of carcinogenic agents. *Journal of the National Cancer Institute* 27:455–470.

National Research Council. 1983. Committee on the Institutional Means for Assessment of Risks to Public Health. *Risk assessment in the federal government: Managing the process.* Washington, DC: National Academies Press.

National Research Council. 2009. Committee on Improving Risk Analysis Approaches Used by the U.S. EPA. *Advancing risk assessment 2009.* Washington, DC: National Academies Press.

Nuclear Regulatory Commission. 1975. *Reactor safety study: An assessment of accident risks in U.S. commercial nuclear power plants.* Washington, DC: Nuclear Regulatory Commission.

2

Uncertainty

2.1 Introduction

Because risk analysis focuses on decision making under uncertainty, we must clearly understand what uncertainty is. At the most basic level, when we are not sure, then we are uncertain. Uncertainty arises at two fundamentally different levels. First, there is the macrolevel of uncertainty. We all make decisions in a changing and uncertain decision environment. This means the systems, processes, social values, and outcomes of concern to us may be uncertain. Second, there is the microlevel of uncertainty. This is the uncertainty that pertains to specific decision contexts and their relevant knowledge, data, and models. These latter uncertainties receive most of the attention in risk assessment.

If there was no uncertainty there would be no question about whether or when a loss would occur and how big it would be. Likewise, we would always know how an opportunity would turn out. Uncertainty is the reason for risk analysis. Risk assessors have to understand uncertainty because they are, in a sense, the first responders to uncertainty. It is the assessors who identify data gaps, holes in our theories, shortcomings of our models, incompleteness in our scenarios, and ignorance about some quantities and variability in others. It is an important part of the risk assessor's* job to address the uncertainty in individual assessment inputs.

Think of the assessor's job as separating what we know from what we do not know about a decision problem context and then being intentional about addressing the things we do not know in decision making. There are always things we know with certainty. We can measure distances; we know atomic structures of chemicals; our physical world is loaded with facts. But every decision problem comes with a "pile" of things we do not know. The risk assessor, along with the risk manager, has to identify that pile and what is in it.

This chapter focuses in a conceptual way on the pile of things we don't know. In order to know how best to address the "things" we don't know, the assessor must first understand the nature of those things in the pile of unknowns (see Figure 2.1). We will sort the original pile of unknowns, or uncertainty, into two distinct sources of not knowing: natural variability and knowledge uncertainty. Knowledge uncertainty is, in turn, divided into three main piles: scenarios and theory, models, and quantities. The quantities, in their turn, are separated into types of quantities first proposed by

* The possessive case used for risk assessors and risk managers will always be in the singular form for the sake of simplicity. It should be understood, however, that both can be multiple in numbers at times.

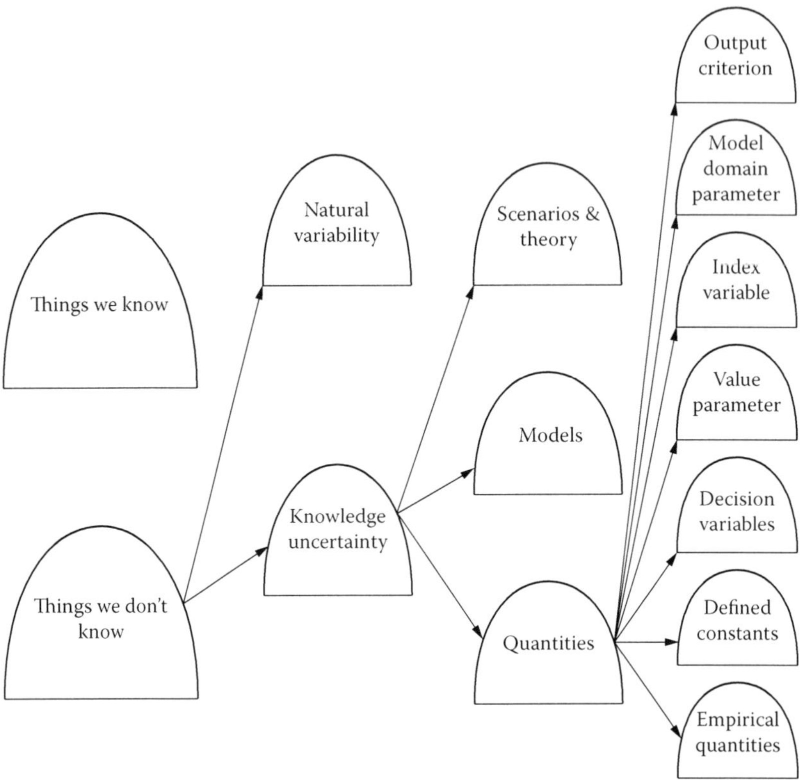

FIGURE 2.1 Separating what we know from what we don't know and sorting out the unknowns.

Morgan and Henrion (1990). This conceptual sorting activity enables us to choose most appropriately from the various tools, techniques, and methodologies available for addressing uncertainty in the assessment and management tasks. Each small pile of unknowns is best addressed by a specific set of tools.

Risk managers need to understand uncertainty because they are the final arbiters of it in the decision-making process. It is the risk manager's responsibility to consider and weigh the cumulative effects of uncertainty in assessment outputs on decision outcomes.

Risk communicators must understand uncertainty if they are to explain it and its significance to others. That task requires risk communicators to understand input, output, and outcome uncertainty well enough to make it understandable by diverse audiences.

This chapter begins by considering the macrolevel of uncertainty that is sometimes overlooked in risk analysis. Then it settles into a consideration of the microlevel uncertainty issues that have occupied so much of the risk assessor's concerns. At that point in the chapter we will reengage the structure shown in Figure 2.1.

An important distinction will be drawn between knowledge uncertainty and natual variability, and the many piles of Figure 2.1 will be explored and discussed. The causes of uncertainty in empirical quantities are considered to round out the discussion of uncertainty as depicted in the figure.

Before beginning we offer one point here, trivial to some, more important to others, but true nonetheless. People often speak of uncertain values, and this is, strictly speaking, not correct. The values themselves are not uncertain; it is the assessor and the manager who are uncertain about what a factual value is or what a value that reflects a preference should be. Bear that in mind as you read and when you hear people speak of uncertain values. It will help you keep your head on straight as you learn these concepts.

2.2 Uncertainty from 75,000 Feet

One of the emerging constants in the modern world is uncertainty. Growing social complexity and an increasingly rapid pace of change are normal parts of the decision-making landscape, and they contribute a great deal to the uncertain environment in which we all operate. Risk analysis offers a viable alternative to clinging to a deterministic style of decision making in this uncertain environment.

The world grows more complex. Think of complexity, as used here, in a social sense. It refers to such things as the size of a society, the number of its parts, the distinctiveness of those parts, the variety of specialized social roles that it incorporates, the number of distinct social personalities present, and the variety of mechanisms for organizing these into a coherent, functioning whole. Augmenting any of these dimensions increases the complexity of a society (Tainter 1996).

For over 99% of human history we lived as low-density foragers or farmers in egalitarian communities of no more than a few dozen persons and even fewer distinct social roles. In the 21st century we live in societies with millions of different roles and personalities. Our social systems grow so complex as to defy understanding. Consequently, our systems of problem solving have grown more complex.

We face an increasingly rapid pace of change in almost every arena. Scientific breakthroughs make commonplace things that once were impossible to conceive. Much of this change is driven by rapid advances in technology. Technology changes social values and beliefs as well as the way we live and work. The level of complexity in our social, economic, and technological systems is increasing to a point that is too turbulent and rapidly changing to be wholly understood or predicted by human beings. Japan's earthquake/tsumami/nuclear disaster of 2011 is the perfect example of this complexity.

We see rapid increases in social, economic, and technological connectivity taking place around the world. Social movements, e.g., environmentalist, women's rights, World Trade Organization (WTO) opposition, and the like, are global in their pervasiveness. We are increasingly a global economy. Fashions are designed in New York and approved in London; patterns are cut in Hong Kong; clothes are made in Taiwan; and the final product is sold across Europe and North America. A computer virus spreads around the world in hours. A human virus spreads in weeks or months.

Relentless pressure on costs is now a fixture in all public decision making. Patterns of competition are becoming unpredictable. It is getting harder and harder to understand and anticipate who the competition is for a job, for U.S. grain, for land use, and so on. For businesses and government agencies alike, customer/client profiles are changing rapidly and unpredictably. We see quickly increasing and diversified customer demands. There is a growing role for one-of-a-kind production, and rapid sequences of new tasks in business and government are becoming more routine. A media explosion is just one of the consequences of an increase in the number and speed of communication channels.

As a result of these and other changes, we have entered a world where irreversible consequences, unlimited in time and space, are now possible. This is or should be extremely important to risk managers in both the public and private sectors. Decades after the accident at Chernobyl, some of the victims haven't even been born yet. Some of the wicked problems* risk managers face can have a long latency period. Many of our country's landscape-scale ecosystem restoration problems—like those in the Florida Everglades, Coastal Louisiana, and the Columbia River basin—as well as global concerns like greenhouse gases, climate change, and sea level rise provide clear examples of problems that took decades to emerge and be recognized. The implications of the solutions being formulated may similarly take decades to be understood.

A new phenomenon of "known unawareness" has entered our lexicon. Donald Rumsfeld in November 2006 summarized this truth to scattered laughter when he said: "There are known knowns. These are things we know that we know. There are known unknowns. That is to say, there are things that we now know we don't know. But there are also unknown unknowns. These are things we do not know we don't know." No one is laughing anymore. As a society we are beginning to realize that despite all we know, the unknown far outweighs what is known. Knowledge is as much to create more questions as it is to provide definitive answers.

Clearly, scientists now know much more about BSE (mad cow disease) than when it was first found in cattle in 1986. Even now, decades after the disease's discovery, its origins, its host range, its means of transmission, the nature of the infectious agent, and its relation to its human counterpart new variant Creutzfeldt-Jakob disease remain mostly unknown. We have begun to suspect that there are some risks for which there may be no narrative closure, no ending by which the truth is recovered and the boundaries of the risk established.

Although most of us live and work in nations, our interactions and our risks are increasingly global in nature. An oil spill in the Gulf of Mexico reverberates around the world. It becomes increasingly difficult to affix responsibility for problems and their solutions. Who is destroying the ozone, causing global warming, spreading BSE and AIDS? Where did the H5N1 or H1N1 viruses originate and how? Whose responsibility is it to fix these things?

Despite the world's rapid advances in all kinds of sciences, we are increasingly dominated by public perception. Public perception is a palpable force, and in some situations it is an irresistible one. When it comes to uncertainties and risks, acceptability often depends on whether those who bear the losses also receive the benefits. When this is not the case, the situation is often considered unacceptable. Risks and uncertain situations have a social context. It is folly to regard social and cultural judgments as things that can only distort the perception of risk. Without social and

* Wicked problems are complex problems that lack right and wrong solutions. Instead there are many candidate solutions and some are better and some are worse than others, but none is clearly best.

cultural judgments, there are no risks. Nonetheless, these social and cultural judgments are not always grounded in fact.

As a result, possibility is often accorded the same significance as existence in the public's view. This view can find and has found its way into public policy. This is in part because many things that were once considered certain and safe, and often vouched for by authorities, turned out to be deadly. The BSE experience in Europe, the SARS (severe acute respiratory syndrome) experience in Asia and elsewhere, and the melamine contamination from Chinese products provide vivid examples of this phenomenon. Applying knowledge of these experiences to the present and the future devalues the certainties of today. This is what makes conceivable threats seem so possible and what fuels our fears of that which is uncertain. It is also what makes criticism of a decision that masquerades as certain embarrassingly easy.

Responsibility in this more connected world has become less clear. Who has to prove what? What constitutes proof under conditions of uncertainty? What norms of accountability are being used? Who is responsible morally? Who is responsible for paying the costs? These questions plague decision makers nationally and transnationally.

We all live and operate in this uncertain reality. Yet many organizations and individuals cling stubbornly to a deterministic approach to decision making that belies the experience of public and private sectors the world over. Decision making needs a "culture of uncertainty." Risk analysis provides just such a culture.

The future is fundamentally unknowable. There must be recognition of the central importance of demonstrating the collective will to act responsibly and accountably with regard to our efforts to grapple with this fundamental uncertainty and the inevitable shortfalls that will occur despite every best effort to account for this uncertainty. In an uncertain world we cannot know everything, and we will make mistakes despite our best efforts to the contrary. This is the challenge that invites risk analysis to the fore. Social values are formed, change, and are re-formed against this backdrop of macrolevel uncertainty.

2.3 The Uncertainty on Your Desk

The uncertainty that has received the most attention in risk analysis is not the macrolevel uncertainty we see from 75,000 feet, nor is it the resulting uncertain environment in which we make decisions. It is the uncertainty that plagues our specific decision contexts. Anyone involved in real problem solving and decision making knows we rarely have all the information we need to make a decision that will yield a certain outcome. For any decision context, we can always make a pile of the things we know and a pile of the things we do not know. For risk analysis, we need to be able to take that pile of things we do not know and sort through it to better understand the nature and causes of the uncertainties we face. It is the nature and cause of the uncertainty that dictates the most appropriate tool to use to deal with it. The first and most important distinction to make in our pile of unknowns is that between knowledge uncertainty and natural variability.

2.3.1 Knowledge Uncertainty or Natural Variability?

You're headed for Melbourne, Australia, in November and are unsure how to pack because you do not know what the weather is like there at that time of year. For simplicity, let's focus on the daily high temperature. You do not know the mean high temperature for Melbourne in November. This is a parameter, a constant, with a true and factual value. That you do not know this fact makes the situation one of knowledge uncertainty. A true value exists and you do not know it. You are uncertain about a fact.

Suppose you learn from the Bureau of Meteorology, Australia, that this value is 21.9°C (71°F). The uncertainty has been removed. Now a new problem emerges. Even though you know the average temperature is 21.9°C, you have no way of knowing what the high temperature will be on any given day. In fact, you wisely expect the high temperature to vary from day to day.

Using our very loose definition at the start of this chapter, you say you are not sure what the temperature will be on any given day, so that must be uncertainty as well. And in a very general sense it is. However, and this is an important however, this value is uncertain for a very specific, common, and recurring reason; there is variability in the universe.

This variability is usually separated out from other causes of uncertainty in order to preserve the distinction in its cause for reasons that will soon be apparent. Hence, we'd say you are no longer uncertain about the mean high temperature, but you still do not know the high temperature on any given day because of natural variability. The temperature varies from day to day due to variation in the complex system that produces a high temperature each day. For a more formal distinction of these two concepts, we introduce the terms epistemic and aleatory uncertainty.

Epistemic uncertainty is the uncertainty attributed to a lack of knowledge on the part of the observer. It is reducible in principle, although it may be difficult or expensive to do so. Epistemic uncertainty, what was described in the previous example as knowledge uncertainty, arises from incomplete theory and incomplete understanding of a system, modeling limitations, or limited data. Epistemic uncertainty has also been called internal, functional, subjective, reducible, or model form uncertainty. *Knowledge uncertainty* is another easier to remember and perhaps more descriptive

NATIONAL RESEARCH COUNCIL (2009)

Uncertainty: Lack or incompleteness of information. Quantitative uncertainty analysis attempts to analyze and describe the degree to which a calculated value may differ from the true value; it sometimes uses probability distributions. Uncertainty depends on the quality, quantity, and relevance of data and on the reliability and relevance of models and assumptions.

Variability: Refers to true differences in attributes due to heterogeneity or diversity. Variability is usually not reducible by further measurement or study, although it can be better characterized.

term used to describe this kind of uncertainty that is used throughout this book when we refer specifically to epistemic uncertainty.

Some generic examples of knowledge uncertainty include lack of experimental data to characterize new materials and processes, poor understanding of the linkages between inputs and outputs in a system, and thinking one value is greater than another but being unsure of that. More obvious examples include dated, missing, vague, or conflicting information; incorrect methods; faulty models; measurement errors; incorrect assumptions; and the like. Knowledge uncertainty is, quite simply, not knowing. The most common example may be not knowing a parameter or value we are interested in for model building or decision-making purposes.

Aleatory uncertainty is uncertainty that deals with the inherent variability in the physical world. Variability is often attributed to a random process that produces natural variability of a quantity over time and space or among members of a population. It can arise because of natural, unpredictable variation in the performance of the system under study. It is, in principle, irreducible. In other words, the variability cannot be altered by obtaining more information, although one's characterization of that variability might change given additional information. For example, a larger database will provide a more precise estimate of the standard deviation, but it does not reduce variability in the population. Aleatory uncertainty is sometimes called variability, irreducible uncertainty, stochastic uncertainty, and random uncertain. The term adopted for usage in this book when we refer specifically to aleatory uncertainty is *natural variability*.

Some generic examples of natural variability include variation in the weight of potato chips in an eight-ounce bag, variation in the response of an ecosystem to a change in the physical environment, and variation in mean hourly traffic counts from day to day. There is also variability in any attribute of a population.

Knowledge uncertainty and *natural variability* are terms used by the National Research Council (2009). It will be convenient to use the term *uncertainty* to encompass both of these ideas, so that is the convention adopted in this book. However, this is by no means the usual convention, and the reader is advised to always clarify, when possible, and to try to carefully discern, when it is not, what the user of these terms means from the context of their usage.

To complicate matters, reality is often messy. Returning to our Melbourne example, we can see that at the outset we are dealing with both knowledge uncertainty and natural variability. It takes experience for a risk assessor to be able to comfortably label the reasons that a value may be unknown. It is not always possible and not always important to be able to separate knowledge uncertainty and natural variability. In general, the most important reasons for separating the effects of the two in a risk assessment are to select an appropriate tool for addressing them and to

AN IMPORTANT DISTINCTION

Natural variability cannot be reduced with more or better information.

Knowledge uncertainty can be reduced with more and better information through such means as research, data collection, better modeling and measurement, filling gaps in information and updating out-of-date information, and correcting faulty assumptions.

understand that simply devoting more resources to the risk assessment effort may reduce knowledge uncertainty, but it will not reduce variability. The only way to change the variability produced by a system is to change the system itself. This will not eliminate variability; it will produce a new form of, presumably, more favorable variability in the altered system. Risk assessment can reduce uncertainty. Risk management measures can alter variability.

2.3.2 Types of Uncertainty

To sort through and understand the nature of the things we do not know, we begin by differentiating knowledge uncertainty from natural variability. Natural variability is most often addressed through narrative descriptions of the variability, statistics, and probabilistic methods. Natural variability tends to apply to quantities only. Knowledge uncertainty is a bit more complex and needs additional sorting. The next tier of sorting separates our knowledge uncertainty into scenarios, models, and quantities, as seen in Figure 2.2.

Figure 2.2 provides an example of the types of knowledge uncertainty that might be encountered. It presents an ecological risk example. For now, think of scenarios as the stories we tell about risks. These are the narratives that describe what we believe to be true about the phenomena we study. This is where theory and knowledge of processes are most important. Models are used to give structure to scenarios and to perform calculations based on the inputs provided. Thus, we identify three broad basic types of knowledge uncertainty you can expect to encounter in risk assessment.

Scenario uncertainty results when the elements of a scenario or their relationships are unknown or incomplete. Gaps in theory and understanding are most likely to occur in the stories we tell about what can go wrong, how it happens, the consequences of it happening, and how likely it is to happen.

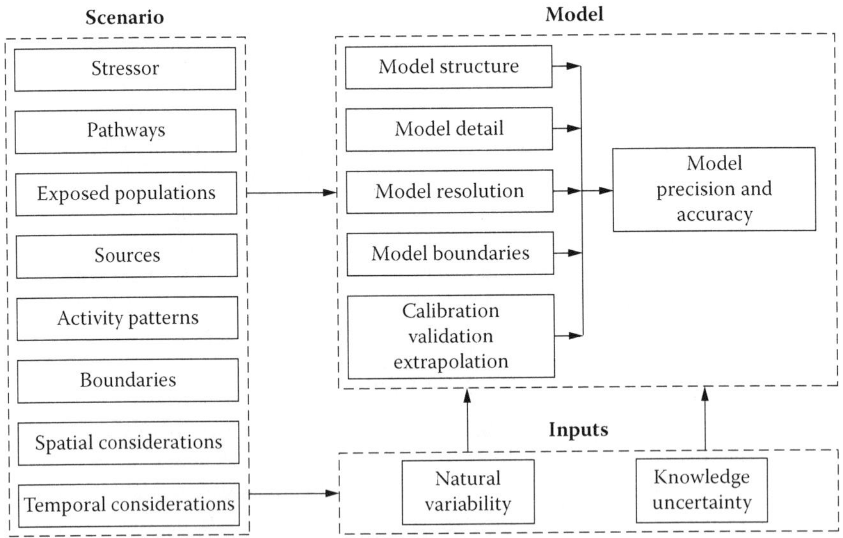

FIGURE 2.2 Sorting our knowledge uncertainty into scenarios, models, and input quantities.

KNOWLEDGE UNCERTAINTY DEFINITIONS

Scenario uncertainty: Uncertainty in specifying the risk scenario that is consistent with the scope and purpose of the assessment

Model uncertainty: Uncertainty due to gaps in scientific knowledge that hamper an adequate capture of the correct causal relations between risk factors

Parameter/input uncertainty: Uncertainty involved in the specification of numerical values (be it point values or distributions of values) for the factors that determine the risk

***Source*: WHO (2006).**

In the case of an ecosystem scenario, we might misunderstand the stressors that affect a habitat. Not knowing the relevant activity patterns of a locally threatened species could be another source of scenario uncertainty. We may also fail to understand all the relevant pathways in an ecosystem.

Model uncertainty reflects the bias or imprecision associated with compromises made or lack of adequate knowledge in specifying the structure and calibration (parameter estimation) of a model. Model structure typically refers to the set of equations or other functional relationships that comprise the specified scenario for the model. Model detail refers to the inclusion or omission of specific phenomena as well as the simplicity or complexity with which they are represented. Model resolution refers to the temporal or spatial scale at which information can be distinguished, e.g., minutes vs. hours vs. years. Model boundaries describe the fidelity with which the desired scenario is captured by the model. Ideally, the precision and accuracy of the model predictions will be assessed as part of the validation exercise. In other words, how well does our model capture reality?

Quantity or input uncertainty is encountered when the appropriate or true values of quantities are not known (knowledge uncertainty). These quantities are of enough importance to warrant additional discussion. That discussion will cause us to briefly revisit natural variability.

2.3.3 Quantity Uncertainty

The most commonly encountered uncertainty is quantity uncertainty. Quantities can be unknown because of knowledge uncertainty or because of natural variability. Morgan and Henrion (1990) offer a very useful taxonomy for those seeking to understand the basic types of quantity uncertainty shown in Figure 2.1. Before considering their taxonomy, we need to make an important distinction. Some quantities have a true or factual value, while others do not. Instead of a true value, they have a best or most appropriate value that reflects some subjective judgment. There may be significant consternation about the best or preferred value for these kinds of quantities, but they have no true value we can discover. The search for a true value is an objective one, while the search for a best value is subjective.

TRUE VALUES

The population of a city, number of colony-forming units per gram of material, percentage of channel bottom that is rock, mean strength of materials in a structure, mean daily stream flow, average weight of an adult striped bass, median serving size, specificity of a diagnostic test, closing price of a stock, contaminant concentration in a specific exposure: These are all quantities that have a true value.

In general, true values are looked up, measured, or estimated by some means. The means by which quantities with true values are estimated vary, and the best choice will depend on the cause of the value's uncertainty. Best or appropriate values are varied systematically (sometimes called parametric variation or sensitivity analysis) to examine the sensitivity of the model and its outputs to different chosen values.

Risk analysis can require a lot of information. Risk assessment, in particular, can involve a great deal of quantitative information that includes many parameters (numerical constants) and variables. The quantities used in risk assessment are frequently a major source of uncertainty. Having a way to think about these quantities and to talk about their uncertainty is critical to the success of any risk analysis.

Morgan and Henrion's (1990) classification of uncertain quantities includes:

- Empirical quantities
- Defined constants
- Decision variables
- Value parameters
- Index variables
- Model domain parameters
- Outcome criteria

The significance of the objective or subjective nature of the quantity uncertainty, as well as the type of quantity, will become most evident when one chooses a tool, technique, or methodology to treat the uncertainty appropriately. Look at the examples in Table 2.1, then read the explanations that follow to better understand Morgan and Henrion's taxonomy of quantities. The approach used to resolve uncertainty depends very directly on what is uncertain and why it is uncertain.

2.3.3.1 Empirical Quantities

Empirical quantities are the most common quantities encountered in a quantitative risk assessment; they have a true value. Empirical quantities are things that can be measured or counted. This includes distances, times, sizes, temperatures, statistics, and any sort of imaginable count. They have exact values that are unknown but measurable in principle, although it may be difficult to do so in practice. A full range of

TABLE 2.1

Uncertain Quantity Types and Examples

Types of Quantities	Selected Examples
Empirical quantities	Stream flow, eggs produced daily, vehicles crossing a bridge, temperaure, time to complete a task, prevalence
Defined constants	Pi, square feet in an acre, gallons in an acre foot, speed of light, size of a city
Decision variables	Acceptable daily intake, tolerable level of risk, appropriate level of protection, reasonable cost, mitigation goal
Value parameters	Value of a statistical life, discount rate, weights assigned in a multicriteria decision analysis, user-day values
Index variable	A particular year in a multiyear model, the location of an egg on a pallet, a geographic grid in a spatial model
Model domain parameters	Study area, planning horizon, industry segment, climate range
Outcome criteria	Mortalities, illness rates, infrastructure failures, fragility curves, costs, probabilities, benefit-cost ratios, risk-risk tradeoffs

methods from narrative descriptions through probabilistic methods are suitable for addressing uncertainty in these quantities.

2.3.3.2 Defined Constant

Defined constants have a true value that is fixed by definition. When these values are not known by the analyst, these quantities can end up in the pile of things we do not know. For example, there are 43,560 square feet in one acre and 325,851 gallons of water in one acre-foot of water. Defined constants provide the perfect opportunity to

DECISION RULE UNCERTAINTY

What is the best endpoint for your purposes? Imprecise or inappropriate operational definitions for desired outcome criteria, e.g., "risk," can be a subtle problem.

Concerned about a public health risk? Should you use the number of exposures, infections, illnesses, hospitalizations, or deaths? Which is a better criterion on which to base a decision: lifetime mortality risk, annual risk of mortality, risk to children or other subpopulations, or something entirely different?

Concerned about an economic issue? Should you maximize net benefits or minimize costs? Do you want to maximize market share or profits?

There is no right answer to these questions, only better or worse ones. Someone must decide what the decision criterion or rule will be to resolve this uncertainty.

point out the importance of understanding the nature of your unknowns. When you do not know one of these quantities you do not use sensitivity analysis or probabilistic methods; you look them up.

2.3.3.3 Decision Variables

This is a quantity which someone must choose or decide. Decision makers exercise direct control over these values; they have no true value. The person deciding this value may or may not be a member of the risk analysis team, depending on the nature of the variable. Policy makers may determine the values of some decision variables to ensure uniformity in decision making. An agency may decide it is unacceptable to increase the lifetime risk of cancer by more than 10^{-6}, for example. Thus, decision variable values are sometimes set by decision makers external to the risk analysis process.

In other instances, risk analysis team members may make these decisions. Examples could include determining a tolerable level of risk or design characteristics of risk management options that differentiate one option from another. Decision variables are subjectively determined. Uncertainty about them is most appropriately addressed through parametric variation and sensitivity analysis.

2.3.3.4 Value Parameters

These values represent aspects of decision makers' preferences and judgments; they have no true value. They are subjective assessments of social values that can describe the values or preferences of stakeholders, the risk manager, or other decision makers. Like decision variables, some of them may be decided by those external to a specific risk management activity, while others are decided by risk analysis team members.

Social values, like the monetary value of a statistical life or society's time preferences for consumption, are likely to be established corporately to ensure uniformity in decision making. Establishing decision-specific values, like assigning relative weights to different decision criteria, may be set by the team. Uncertainty about value parameters is most appropriately addressed through parametric variation and sensitivity analysis.

2.3.3.5 Index Variables

Index variables identify elements of a model or locations within spatial and temporal domains; they may or may not have a true value. A point in time can be referenced as a time step in a model, and a grid cell can be referenced using coordinates. If a very specific point in time or place in space are desired, there is a true value. Random or representative choices of index variables do not have true values and are subjectively determined. Uncertainty in index variables is most appropriately addressed through parametric variation and sensitivity analysis.

2.3.3.6 Model Domain Parameters

These values specify and define the scope of the system modeled in a risk assessment. These parameters describe the geographic, temporal, and conceptual boundaries (domain) of a model. They define the resolution of its inputs and outputs; they may or may not have true values. Scale characteristics are chosen by the modeler and most often have no true value in nature. They reflect judgments regarding the model domain and the resolution needed to assess risks adequately. Some risk assessments, however, may be restricted to specific facilities, towns, time frames, and so forth. These may have true values. Uncertainty about domain parameters may also be considered a form of model uncertainty. If the domain is the XYZ processing plant, it is trivially specific and objective. The hinterland affected by economic activity at the Port of Los Angeles is a much more subjective determination. These kinds of quantities are most appropriately addressed through parametric variation and sensitivity analysis.

2.3.3.7 Outcome Criteria

Outcome criteria are output variables used to rank or measure the desirability or undesirability of possible model outcomes. Their values are determined by the input quantities and the models that use them. Uncertainty in these values is evaluated by propagating uncertainty from the input variables to the output variables using one of several different methods. Generating the uncertainty about output criteria is the responsibility of the risk assessor; addressing it in decision making is the responsibility of the risk manager.

2.3.4 Sources of Uncertainty in Empirical Quantities

Empirical quantities are the most commonly encountered uncertain values with true values that must be measured or estimated. When good measurement data are available, there may be little or no knowledge uncertainty about the true value of a parameter or variable. Even when there is no knowledge uncertainty, we may have natural variability to address in the risk assessment. It is useful to continue the excellent conceptual framework of Morgan and Henrion (1990) to consider the different sources of uncertainty in empirical quantities. Understanding the reasons that you are uncertain about empirical quantities is essential to your ability to choose an effective treatment of that uncertainty in a quantitative risk assessment.

2.3.4.1 Random Error and Statistical Variation

No measurement can be perfectly exact. Even tiny flaws in observation or reading measuring instruments can cause variations in measurement from one observation to the next. Then there is the statistical variation that results from sample bias. If we take measurements on a sample, we only have an estimate of the true value of a population parameter. Classical statistical techniques provide a wide array of methods and tools for quantifying this kind of uncertainty, including estimators, standard deviations, confidence intervals, hypothesis testing, sampling theory, and probabilistic methods.

IT FEELS LIKE AN 8 TO ME

Much data are collected outside a laboratory and under less than ideal conditions. Which box of produce do we open and inspect? Where in the stream does the investigator insert the meter to read dissolved oxygen? How do you estimate how far away a workboat is on the open water? How quickly can you count the deer in a running herd?

Subjective judgments like these are notoriously suspect under uncontrolled conditions. Just as with faulty instruments, the solution is better calibration. Ideally, it should take place before measuring, but calibration is better late than never.

2.3.4.2 Systematic Error and Subjective Judgment

Systematic errors arise when the measurement instrument, the experiment, or the observer is biased. Imprecise calibration of instruments is one cause of this bias. If the scale is not zeroed or the datum point is off, the solution is better calibration of the instrument or data. If the observer tends to over- or underestimate values, a more objective means of measurement is needed or the observer needs to be recalibrated. The challenge to the risk assessor is to reduce systematic error to a minimum. The best solution is to avoid or correct the bias. When bias can be identified, e.g., the scale added 0.1 g to each measurement, it can sometimes be corrected for, i.e., by remeasuring or subtracting 0.1 g from each measurement.

The more difficult task concerns the biases that are unknown or merely suspected. Estimating the magnitude of these biases is very difficult and often requires a lot of subjective judgment, which, as the text box notes, can present its own problems. Bias in subjective human estimates of unknown quantities is a topic covered extensively in the literature; see, for example, Chapter 13 or O'Hagan et al. (2006).

2.3.4.3 Linguistic Imprecision

After all these years on the planet, communication is still humankind's number one challenge. We routinely use the same words to mean different things and different words to mean the same things. This makes communication about complex matters of risk especially challenging. If we say a hazard occurs frequently or a risk is unlikely, what do these words really mean? But the problems are more pervasive than that. Tasked with measuring the percentage of midday shade on a stream, a group of environmentalists engaged in a lengthy discussion of when midday occurs and how dark must a surface be to be considered shade.

The best and most obvious solution to this kind of ambiguity is to carefully specify all terms and relationships and to clarify all language as it is used. Using quantitative rather than qualitative terms can also help. Fuzzy set theory may be an alternative approach to resolving some of the more unavoidable imprecision of language in a more quantitative fashion.

2.3.4.4 Natural Variability

Many quantities vary over time, space, or from one individual or object in a population to another. This variability is inherent in the system that produces the population of things we measure. Frequency distributions based on samples or probability distributions for populations, if available, can be used to estimate the values of interest. Other probabilistic methods may be used as well.

2.3.4.5 Randomness and Unpredictability

Inherent randomness is sometimes singled out as a form of uncertainty different from all others, in part because it is irreducible in principle. The indeterminacy of Heisenberg's uncertainty principle is one example of inherent randomness. However, a valid argument could be made that this is just another instance of knowledge uncertainty because we simply have been unable to resolve this puzzle at the present time.

This cause of uncertainty identifies those uncertainties that are not predictable in practice at the current time. Examples include such things as when the next flood will occur on a stream or where the next food-borne outbreak will occur in the United States. Such events can be treated as a legitimately random process. The danger here is the personalist view of randomness that could emerge, where randomness is a function of the risk assessor's knowledge. Phenomena that appear random to one assessor may be the result of a process well known by a subject-matter expert. Strong interdisciplinary risk assessment teams combined with peer involvement and review processes provide a reasonable hedge against this sort of problem arising. Uncertainty about such quantities can be addressed by a full range of methods, from narrative descriptions through probabilistic methods.

2.3.4.6 Disagreement

Organizations and experts do not always see eye to eye on matters of uncertainty. Different technical interpretations of the same data can give rise to disagreements, as can widely disparate views of the problem. This is not to mention the real possibility of conscious or unconscious motivational bias.

Disagreements can sometimes be resolved through negotiation and other issue resolution techniques. Allowing the disagreements to coexist is also an option. Sensitivity analysis would consider the results of the analysis using each different perspective. A common approach for some disagreements is to combine the judgments using subjective weights.

2.3.4.7 Approximation

Uncertainty due to approximation is similar to what we earlier called model uncertainty in this chapter. The fact that the model is a simplified version of reality ensures that uncertainty will remain about the outcome criteria. We are only able to approximate the function of complex systems because of scenario, model, and quantity uncertainty. Methods for dealing with this source of uncertainty will depend on the specific limitations of the approximation.

2.4 Being Intentional about Uncertainty

To recap, the risk assessor's first responsibility in approaching a well-defined decision problem is, figuratively, to make a pile of the things that are known and a pile of the things that are not known about the problem. In that pile of things that are not known are things that vary and things about which we are uncertain.

Empirical quantities are the only things that vary, but not all empirical quantities are variable. Take that knowledge uncertainty pile of unknown things that are not variable and separate it into additional piles of scenarios/theory, models, and quantities. You might also want to consider what could possibly show up in the unknown unknowns lurking in that pile.

Identify the various types of uncertain quantities encountered. Go back to that pile of empirical quantities and identify the individual causes of uncertainty among those that are not variable. Making a pile of things you do not know and separating that pile into smaller like piles of unknown values is always a good starting point for addressing relevant uncertainties. Identify what you do not know; group like things; and figure out why they are unknown. Do this and you are well on your way to being intentional and rational about uncertainty.

The risk assessor's responsibility is to intentionally address the pile of things we do not know. Once you have separated the piles, this job begins by identifying the uncertainties that are likely to have a significant effect on model outputs, decision criteria, and, therefore, decisions and decision outcomes. Not every uncertain quantity is going to make a difference, and you need only address those that do. Identifying the significant uncertainties at the very outset can often simplify the work. Methods for doing this are sometimes obvious. Experts often know which uncertainties drive the decision-making process. When they do not, the methods for discerning these quantities, such as those found in Chapter 16, should be employed.

Once the significant uncertainties have been identified and classified, there are a number of methods available for addressing them that range from simply identifying the uncertainty and describing it, to making assumptions, to the use of more complex probabilistic techniques. Finally, the risk assessor must communicate the nature and relevance of the uncertainty to the risk manager and any other decision makers. Common methods for characterizing uncertainty include codes, categories, mathematics, and statistics, including probabilistic methods, narratives, and summaries. A good characterization of uncertainty should address both the quality and quantity of the data and the uncertainty. The response to the uncertainty and the rationale for that response should also be explained. This should include documenting all significant sources of uncertainty, identifying all processes and methods used to address the uncertainty, and linking the available evidence to all conclusions and decisions. Efforts to describe the uncertainty should always identify data gaps and areas for additional research.

Making this effort to be honest brokers of information, saying what we do and do not know, and stating the significance of the latter for decision making distinguishes risk analysis from other decision-support frameworks and tools. As Figure 2.3 shows, the uncertainty encountered in an assessment of a risk can be found in either the probability or the consequence. It can be due to knowledge uncertainty or natural variability.

We have now conceptually described the uncertainty under which risk managers must make decisions. There is a wide variety of tools, techniques, and methodologies

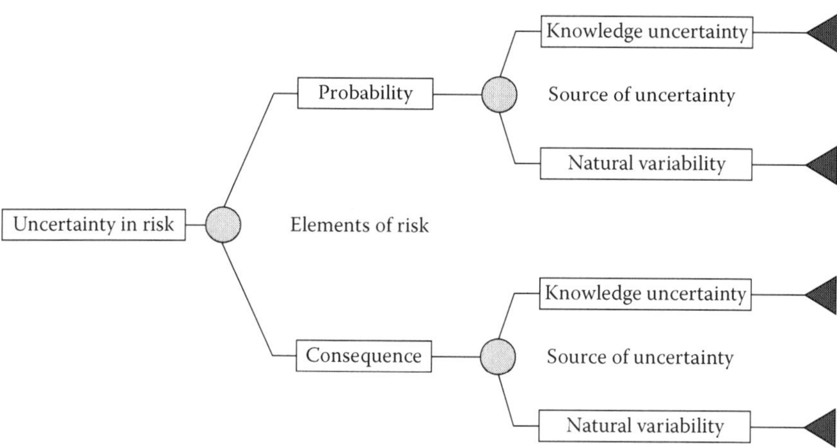

FIGURE 2.3 Source elements of uncertainty in risk analysis.

for addressing uncertainty in risk assessment and risk management. These are addressed in considerable detail in later chapters.

It is the risk assessors' responsibility to address significant uncertainties in their assessments. Some of the simpler tools they can use include narrative descriptions of the uncertainty, clarification of ambiguous language, negotiation for differences of opinion, and confidence ratings for their analyses. When the relevant uncertainty can lead to dramatically different futures and a few key drivers of this uncertainty can be identified, scenario planning is a useful technique. In more quantitative analyses, assessors can use parametric variation, bound uncertain values, use sensitivity analysis, or use probabilistic risk assessment, which can include both deterministic scenario analysis and probabilistic scenario analysis. There are numerous other more advanced techniques in use.

Risk managers are expected to address uncertainty in their decision making. When the uncertainty is great and the consequence of making a wrong decision is a concern, adaptive management strategies may be implemented. Adaptive management strategies are designed to reduce key uncertainties (through research, experiments, test plots, trial and error, and so on) to provide information to better inform managers about the risks and the efficacy of the risk management options before they are irreversibly implemented.

The precautionary principle is sometimes favored as an approach to decision making under uncertainty. Premise sets lay out sets of beliefs or assumptions (premises) about how the key uncertainties will resolve themselves so risk managers can choose the set of conditions they believe will prevail. The chosen premise set will then point in the direction of specific risk management options.

There are also a number of criteria that have been developed for choosing from among alternative risk management measures under uncertainty. They include the:

- Maximax criterion—choosing the option with the best upside payoff
- Maximin criterion—choosing the option with the best downside payoff
- Laplace criterion—choosing the option based on expected value payoff

- Hurwicz criterion—choosing an option based on a composite score derived from preference weights assigned to selected values (e.g., the maximum and minimum)
- Regret (minimax) criterion—choosing the option that minimizes the maximum regret associated with each option

Good risk analysis requires assessors, managers, and communicators to be intentional about dealing explicitly with uncertainty when carrying out their responsibilities.

2.5 Summary and Look Forward

Uncertainty is the reason for risk analysis. Risk analysis is, in a sense, the confluence of social values and science. Uncertainty at the macro level affects values through a constantly and rapidly changing social environment. Uncertainty at the micro level occurs in the specific details of the problems decision makers face, at the level of our scientific knowledge. The two levels of uncertainty can pose markedly different challenges to risk analysts.

Separating what we know from what we do not know is a primary responsibility of the risk assessor. In the "pile of things" we do not know about a given decision problem are things that reflect our knowledge uncertainty and things that are naturally variable. It is important to know the difference between the two. Uncertainty is, in principle, reducible, while variability is not. This can be important to how risks are assessed, managed, and communicated.

A major purpose of risk analysis is to push risk assessors and risk managers to be intentional in how they address uncertainty in analysis and decision making. There are helpful taxonomies to aid our thinking about how to identify uncertainties and their causes. These are important to know because different kinds and causes of uncertainty have different sets of appropriate treatments. It is the risk assessor's job to address uncertainty and variability in risk assessment inputs. It is the risk manager's job to address them in risk assessment outputs.

The next three chapters will carefully unpack and explain the basic activities that comprise the three components of the risk analysis model presented in Chapter 1. We begin with the risk management process, which is the cornerstone of the risk analysis process. Although there are many well-developed risk management models already in use, the approach taken here is not to put any one of these before the others so much as to find the common ground in all of them to aid your understanding and practice of the risk management process.

REFERENCES

Morgan, M. Granger, and Max Henrion. 1990. *Uncertainty: A guide to dealing with uncertainty in quantitative risk and policy analysis.* Cambridge, U.K.: Cambridge University Press.

National Research Council. 2009. Committee on Improving Risk Analysis Approaches Used by the U.S. EPA. *Advancing risk assessment 2009*. Washington, DC: National Academies Press.

O'Hagan, Anthony, Caitlin E. Buck, Alireza Daneshkhah, J. Richard Eiser, Paul H. Garthwaite, David J. Jenkinson, Jeremy E. Oakley, and Tim Rakow. 2006. *Uncertain judgments: Eliciting experts' probabilities*. West Sussex, U.K.: John Wiley & Sons.

Tainter, Joseph A. 1996. *Getting down to earth: Practical applications of ecological economics*. Washington, DC: Island Press.

World Health Organization. International Programme on Chemical Safety. 2006. Draft guidance document on characterizing and communicating uncertainty of exposure assessment, draft for public review. Geneva: World Health Organization. Accessed May 15, 2010. http://www.who.int/ipcs/methods/harmonization/areas/draftuncertainty.pdf.

3

Risk Management

3.1 Introduction

Risk management is decision making that evolves as uncertainty is reduced. In the past, many organizations have managed risks by prescribing policy, procedures, regulations, and other guidance, the rationale being that if you follow the "rules," then whatever results must be okay. That is not risk management. Risk management, done well, is intentional about its process, addresses uncertainty in decision making, and focuses on outcomes. Risk management is maturing. There are now thousands of people who identify themselves as risk managers when only a decade or two ago no one outside of the insurance industry used this title.

There is no shortage of risk management models. As with every other aspect of risk analysis, many disciplines and organizations have spawned their own particular view of how to do risk management. Describing the risk management process in a generic fashion is, therefore, a daunting challenge. It is impossible to define risk management in a way that will satisfy many, much less all people. The Society for Risk Analysis own risk glossary shies away from this task. It is not for a lack of definitions so much as it is the proliferation of definitions that are in use by organizations and in play in the professional and other literature. The U.S. EPA (2010) "Thesaurus of Terms Used in Microbial Risk Assessment," for example, identifies 12 different definitions for risk management.

It goes without saying that most organizations are quite fond of the nuances or parsimony of their own definitions and are not inclined to surrender it for another. No one seems to be clamoring for a universal definition, so do not look for one here. In place of a formal definition, the risk management component is described in some detail. That description will not be any more universally applicable than a definition would be, but we must begin somewhere, so we begin by identifying those risk management activities that are common to many definitions, models, and practice.

I am going to call a new initiative undertaken by an organization that practices risk analysis, a risk management activity. There are five basic parts to a risk management activity:

1. Identifying problems and opportunities
2. Estimating risk
3. Evaluating risk
4. Controlling risk
5. Monitoring risk

A SAMPLING OF RISK MANAGEMENT DEFINITIONS

The culture, processes, and structures that are directed toward the effective management of potential opportunities and adverse effects (Australia/New Zealand Risk Standard)

The sum of measures instituted by people or organizations to reduce, control, and regulate risks (German Advisory Council on Global Change)

Decision-making process involving considerations of political, social, economic, and technical factors with relevant risk assessment information relating to a hazard so as to develop, analyze, and compare regulatory and nonregulatory options and to select and implement appropriate regulatory response to that hazard; risk management involves three elements: risk evaluation, emission and exposure control, and risk monitoring (IPCS)

Coordinated activities to direct and control an organization with regard to risk (ISO/IEC Risk Management Vocabulary)

All the processes involved in identifying, assessing, and judging risks; assigning ownership; taking actions to mitigate or anticipate them; and monitoring and reviewing progress. Good risk management helps reduce hazard and builds confidence to innovate. (U.K. Government Handling Risk Report)

The process of analyzing, selecting, implementing, and evaluating actions to reduce risk (U.S. Presidential/Congressional Commission)

The process of evaluating alternative regulatory actions and selecting among them (NRC 1983)

The generic model is shown in Figure 3.1. It shows the five tasks in a continuous loop to capture the iterative nature of risk management. Risk management is making effective and practical decisions under conditions of uncertainty. As long as there is any uncertainty, a risk management decision is conditional, i.e., based on what was known and not known at the time of the decision. As the uncertainty is reduced in the future or as the outcomes of the management decision become known, it may be prudent to revise the decision, hence, the ongoing nature of risk management. Every decision is based on what we know now and is subject to further revision in the future; in that sense no decision is necessarily final as long as significant uncertainty remains. Expanding on and explaining the elements of the risk management model of Figure 3.1 is the primary work of this chapter.

You will find this to be a wide-ranging chapter, as befits the risk manager's job. I have distilled the most consistent elements of a great many risk management models (see for example FDA 2003, FAO 2003, PCCRARM 1997, ISO 2009), as well as my own experience, to the five broad categories of risk management activities, which are described here in some detail. To round out the discussion, a few specific risk management models are offered at the end of the chapter to illustrate how different organizations approach the risk management task, which is basically to make effective decisions about whether and how to manage risks with less than all the information desired.

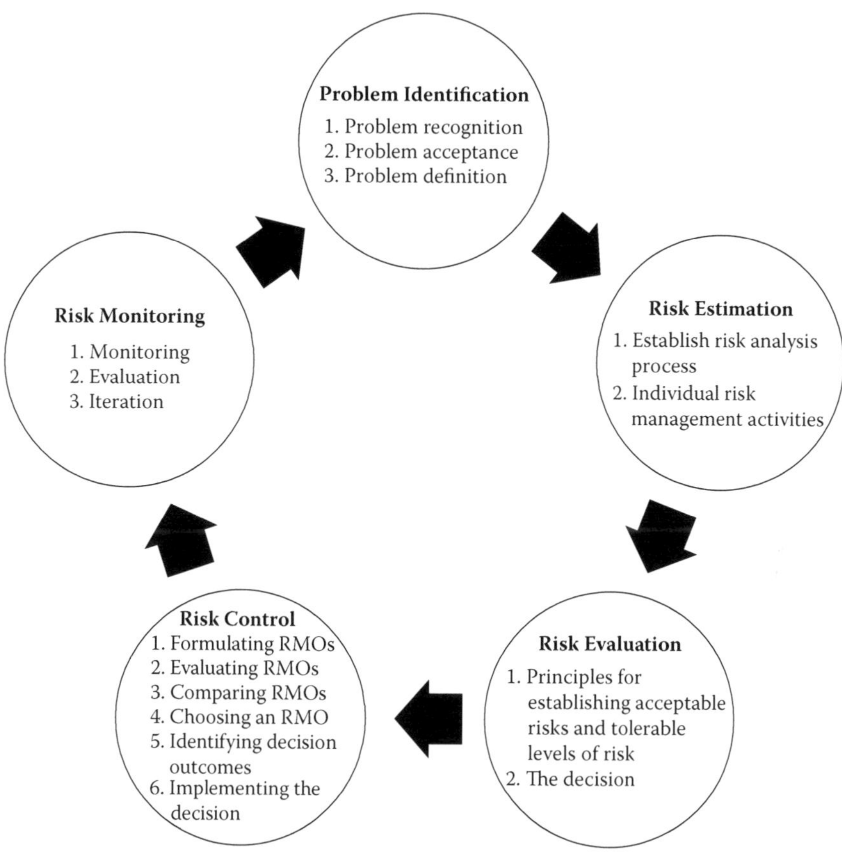

FIGURE 3.1 A generic risk management process comprising five tasks.

3.2 Identifying Problems and Opportunities

Something happens to start a risk management activity. That something is usually a problem that needs attention or an opportunity* that can be pursued. A risk management activity that may require decision making is usually triggered by some sort of event or it is initiated in response to accumulated information inputs.

Einstein is quoted as having said, "If I had one hour to save the world, I would spend 55 minutes defining the problem." This is the stake that good risk analysis drives into the ground at its outset that helps distinguish it from other decision-making paradigms. The purpose of risk analysis is to find the right problem and to solve it. Defining the problem (see Figure 3.2) provides a focal point for all of the risk manager's subsequent problem-solving efforts.

What often happens in organizations is that as soon as a problem arises we're so eager to solve it that we spend very little time understanding, refining, and communicating our understanding of it. As a consequence, organizations often treat the symptoms of problems rather than their causes. Worse, we often don't even know

* To avoid the awkward redundancy of saying problem/opportunity throughout this section, let it be understood that problem will stand for both kinds of risky situations.

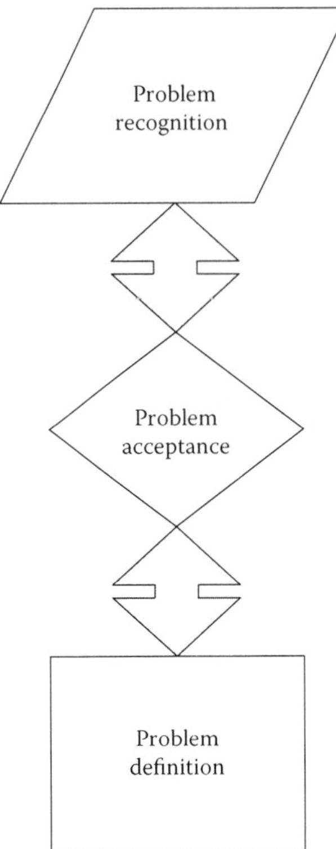

FIGURE 3.2 Problem identification steps.

TYPES OF RISK MANAGEMENT ACTIVITY TRIGGERS

Crisis: Real or perceived criticism from media, public outcry, adverse comments, changing public values or awareness, decreased consumer confidence, other

Science and technology: New knowledge or technology; emerging health problem; improved detection, surveillance, or method

Emerging or "on the horizon" concepts: Planned search, forecasting, scan of the risk landscape for natural and anthropogenic disasters and events, new imported products

Strategic plan: Strategic planning, social needs, opportunities, "beyond the horizon" concepts, historical precedents

These examples of the kinds of events and inputs that can trigger a risk management activity are from the U.S. FDA's Center for Food Safety and Applied Nutrition, *CFSAN's Risk Management Framework*, 2003.

when we are unclear about a problem, and frequently the result is that we solve the "wrong" problem correctly.

Problem identification, the risk manager's first major responsibility, is defined here as a three-part process (see Figure 3.2):

- Problem recognition
- Problem acceptance
- Problem definition

3.2.1 Problem Recognition

Problem recognition is the simple act of recognizing that a problem exists and gaining an initial understanding of the problem. This happens in one of two broad ways. Reactive or passive problem recognition is when a problem finds you. These are problems triggered by outside influences. Stakeholders bring you a problem or an event occurs that results in a problem you cannot ignore. Alternatively, there is proactive or intentional problem finding, in which management looks actively and often strategically for the most important problem(s) to solve.

Despite the seemingly obvious nature of this task, it is surprising how frequently organizations fail to recognize a problem. This is all the more true in a risk analysis context because risky problems often lurk unseen over the horizon or around the corner. They are frequently hidden from view, obscured by uncertainty and higher priorities, and occluded by smoke from the organizational brushfires that need constant tamping out. Anyone can recognize the problem that forces its way through your door and onto your desk at 4 p.m. on a Friday afternoon. It takes a risk manager to see the problems just over the horizon or just around the corner.

Recognizing the existence of opportunities for potential gain or betterment parallels the process of problem recognition. Fewer opportunities seem to break down the risk manager's door than do problems, however, and the search for opportunities is usually more active.

3.2.2 Problem Acceptance

Once a problem makes your radar screen, the question becomes, "Will you own it and do something about it?" The second step in problem identification is problem acceptance. This requires risk managers to articulate the problem they have found in enough detail to determine if it is a problem they're willing and able to address. Addressing a problem means allocating resources to its solution.

Risk managers must identify the resources required to address the problem in a timely manner. Then they must evaluate the adequacy of their available resources in the context of their program authorities, organizational mission, and vision. This obviously implies consideration of competing uses for the organization's resources. We cannot solve every problem.

Problem acceptance is a priority-setting step. It is deciding to act. Accepting a problem as one to be solved or an opportunity as one to be pursued is a significant organizational commitment. Our understanding of the problem is revised and refined beyond the initial recognition in this step. Risk managers must identify and commit to the time frame and resources required to address each problem they accept.

Choosing from a number of potential opportunities and deciding which are worth pursuing is a common problem in business decision making. Articulating and accepting the opportunities to be pursued parallels the problem acceptance step.

3.2.3 Problem Definition

The third step is problem definition. This is when the problem is fully articulated for the first time and linked to possible solutions. Opportunities are likewise articulated and linked to potential strategies that could realize the gains. Information needs begin to become clear and a risk management activity is initiated. This step encompasses a focused and intentional effort to provide a commonly understood description of the problem. It includes stakeholder input when appropriate.

If you can't clearly and concisely finish the sentence, "The problem is ...," then nothing that follows will be clear either. A written "problems and opportunities statement" is the desired output of this problem identification process. Your problems and opportunities statement provides the rationale or reason for your risk management activity. It should be considered a conditional statement that will change as you begin to gather information, reduce the initial uncertainty, and better understand the problem(s) and stakeholders' concerns. So date that piece of paper. Risk analysis is an iterative process and you can expect to revise and refine your problems and opportunities statement several times before you are done.

The stakeholders in any problem context will vary. For some problems the stakeholders may comprise the general public and many special interests. In other problem settings the stakeholders may be wholly contained within the organization. Stakeholders, however defined, should be involved in the problem identification process. The appropriate level of involvement will vary with the decision problem. Some problems will be identified for you by stakeholders; at other times they will have to be made aware of the existence of a problem.

Vet your problems and opportunities statement with your stakeholders. Publish it appropriately. Make it public if your stakeholders include the public. Show them your best thinking and ask, "Did we get the problem(s) right? What's missing? What is here that should not be? Do you have information about these problems and opportunities

SAMPLE PROBLEMS AND OPPORTUNITIES STATEMENT

Increasing resistance of *Campylobacter* in chicken to fluoroquinolones due to subtherapeutic use of antibiotic drugs in food-producing animals

Declining efficacy in the use of fluoroquinolones for the treatment of campylobacteriosis in humans

Reducing incidence of all campylobacteriosis in humans due to consumption of chicken

that would be helpful to share?" Stakeholders can be an effective ally in reducing uncertainty.

The output of this activity is a written problems and opportunities statement. Keep that statement up to date. Let people know how it changes and why it changes as it changes.

3.3 Risk Estimation

Estimating risks is the assessor's job. It can't be done without direction and guidance from the risk manager. Risk managers have an important, but limited, role in the science-based risk assessment process. The risk manager's positive decision-making role is found in the risk estimation activities (see Figure 3.3) that help describe the world as it actually is. That role includes establishing the organization's risk analysis process and managing individual risk management activities.

There are two groups of activities in the risk estimation part of risk management, as seen in Figure 3.3. The first, developing a risk analysis process, consists of one-time or periodic activities required to establish and maintain the risk analysis process. The other, individual risk management activities, consists of duties that recur in every risk management activity. These activities are addressed below at the level of detail shown in the figure.

3.3.1 Establish Risk Analysis Process

If the plethora of definitions for the basic terminology of risk analysis teaches us nothing else, it teaches us this: There is no one best way to do risk analysis. The most commonsense rule seems to be to use what works best for you. Think of risk analysis as a process that is firm in its principles but flexible in the details of how they are pursued.

The risk manager's job with respect to establishing a risk analysis process is basically to say, "This is how we do risk analysis here." This process establishes the risk management model the organization will use so that there is an agreed-upon framework for addressing risk problems and opportunities. It establishes the roles and responsibilities of everyone involved in the risk analysis process.

A significant piece of any risk analysis process is the risk assessment policy, which addresses the manner in which the many subjective judgments and choices that arise in the course of a risk assessment will be resolved to protect the integrity of the science and the decision-making process. Some predictable issues that will arise include how to deal with uncertainty and what assumptions to use when the available data are inconsistent. These are sometimes called "science policy" issues. It is wise to devise a means of resolving these kinds of problems before they are encountered in practice.

Establishing a risk assessment policy is the risk manager's responsibility. It needs to be a collaborative process for any organization that is engaged in making public policy or in the stewardship of public resources like public health, public safety, natural resources, or the environment. This collaboration should include risk managers, risk assessors, and risk communicators. It should provide appropriate opportunities

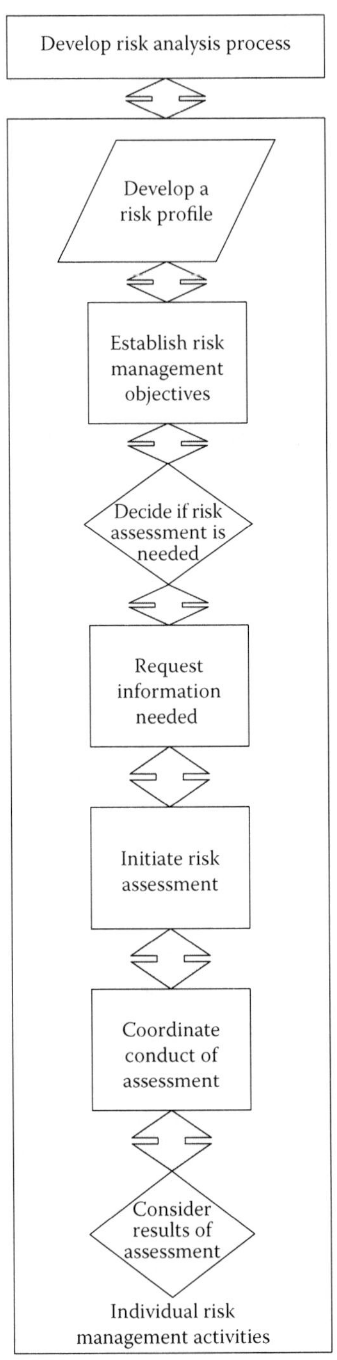

FIGURE 3.3 Risk estimation steps.

SCIENCE POLICY

"Science-policy choices are distinct from the policy choices associated with ultimate decision making.... The science-policy choices that regulatory agencies make in carrying out risk assessments have considerable influence on the results" (NRC 1983).

When the science is unclear, what assumptions are to be made and by whom? A good risk assessment policy addresses these questions and all questions of so-called default assumptions.

for input and feedback from relevant stakeholders. The risk assessment policy should be documented and made publicly available to ensure consistency, clarity, and transparency. For most organizations outside the public sector, establishing a risk assessment policy is an internal affair.

One of the hallmarks of best-practice risk analysis is insulating the science from the policy. In the early days of risk analysis, many thought the risk management and risk assessment tasks must be totally separated from one another. This is not true. It is best when these functions are separate and handled by different people with the appropriate skill sets required by their jobs. However, it is absolutely essential that managers and assessors communicate, cooperate, and even, at times, collaborate throughout the risk analysis process.

Risk managers begin and end the risk assessment process. They may collaborate in identifying a problem, in formulating risk management options, or on other tasks throughout the risk management activity. They will cooperate in the conduct of the risk assessment and must communicate continuously throughout the iterative risk analysis process.

The output of this task is a well-defined risk analysis process that will guide the organization. That process should include a specific risk management model and a risk assessment policy. Both of these should be carefully documented and publicized to all those with a legitimate interest in the organization's risk analysis process.

A SCALABLE PROCESS

A good risk analysis process is perfectly scalable. You can use it when you have 30 minutes and no budget as easily as you can over years with millions of dollars.

One of its greatest values is that it provides a systematic, science-based approach to solving problems. The risk management elements identified in a conceptual model can be completed in any time frame with any budget.

3.3.2 Individual Risk Management Activities

Risk managers have several specific preliminary risk management responsibilities to complete before, during, and after the risk assessment. Identifying the right problem to solve is only the starting point. Additional responsibilities include:

1. Develop a risk profile
2. Establish risk management objectives
3. Decide if a risk assessment is needed
4. Request needed information
5. Initiate the risk assessment
6. Coordinate the conduct of the assessment
7. Consider the results of the assessment

Each of these activities is considered in turn in the following sections.

3.3.2.1 Develop a Risk Profile

Once the problems have been articulated in a problems and opportunities statement, it is time to find out quickly what is and is not known about the decision problem. A risk profile frames problems and opportunities in their risk context* and provides as much information as possible to guide subsequent risk assessment, risk management, and risk communication activities. It also provides the first formal identification of the uncertainty in your decision problem.

The risk profile is the risk manager's responsibility. Managers need not do it alone or at all, for that matter, but they need to see that it is done. Profiling a risk will almost surely mean consulting and collaborating with risk assessors, and it often will involve stakeholders. Think of this as the point at which the risk management activity team provides a situation report that fleshes out what can go wrong, how it can happen, the consequences of it happening, and how likely it is to happen. The profile presents the current state of knowledge related to the issue(s) identified by the risk managers in a concise form at the outset of the risk management activity. The profile will also include consideration of potential risk management options identified to date.

The profile step is important for several reasons beyond the fact that it identifies data gaps by separating what we initially know about a problem or opportunity from what we do not know. It develops the risk analysis team's knowledge and understanding of the problem and may evolve the problem definition further. It also provides the basis for some very important preliminary risk management tasks including:

- Identifying risk management objectives
- Deciding whether or not to initiate a risk assessment
- Identifying the questions to be answered by risk assessment

* Profiles vary by the hazard or opportunity, and these provide the risk context. Engineering risks are framed in terms of the relevant engineering considerations; food safety problems are framed in a food safety context; financial risks are placed in their proper context, and so on.

INFORMATION YOU MIGHT FIND IN A RISK PROFILE

- Latest statement of the problem
- Description of the hazard or opportunity involved
- How assets are exposed to the hazard
- Frequency, distribution, and levels of occurrence of the hazard
- Identification of possible risks from the available scientific literature
- Nature of values at risk (human health, economic, cultural, etc.)
- Distribution of the risk and benefits from the risky activity
- Characteristics of available risk management options
- Current risk management practices relevant to the issue
- Public perceptions of the potential risks
- Information about possible risk management (control) measures
- Preliminary identification of important scientific data gaps that may prevent or limit a risk assessment
- International implications of risk management
- Risk management objectives
- Decision to pursue a risk assessment
- Questions to be answered by risk assessment

One of the most important functions of the risk profile is to reduce and better define the uncertainty relevant to the decision problem. When a problem is initially identified, it is likely that the uncertainty is going to be great. As the first formal information-gathering step in the process, the risk profile is often effective in reducing uncertainty and identifying the greatest remaining data gaps. A risk profile sometimes provides enough information to make a risk management decision.

The term *risk profile* is used extensively by the food safety risk analysis community, for one example. It may be an unfamiliar term to other communities of practice. However, the initial data-gathering step is or should be universal in any risk management activity. Finding out what is already known or readily knowable about the identified risks precedes the risk assessment. In fact, it is an essential step in deciding whether a risk assessment is even needed or in some instances whether it is doable.

The output of this step is a documented risk profile that includes the initial sorting and assessing of the things that are not known about the identified problems and opportunities. Documentation may be in a brief report, an organized sheaf of papers, or an electronic folder of information sources and memoranda. A formal document is not always required.

3.3.2.2 Establish Risk Management Objectives

Objectives say what we desire to see happen and when. It is easy to confuse objectives with strategies, which describe how we intend to achieve the objectives. It is the risk manager's job to write the risk management objectives. They should be specific and conceptually measurable.

Once the risk has been profiled and the decision context is better understood, risk managers need to determine their broad risk management objectives. The problems

A GOOD OBJECTIVE IS:

Specific: It is clear and free from ambiguity.
Flexible: It can be adapted to new or changing requirements.
Measurable: Its achievement can be documented by some objective means.
Attainable: It can be reached at the end of a course of action.
Congruent: It is in harmony with other objectives.
Acceptable: It is welcome or pleasing to key stakeholders.

and opportunities statement describes why a risk management activity has been initiated. The objectives state in broad and general terms what the risk managers intend to do about the problems and opportunities they face. These objectives should reflect the most important social (or organizational) values in the decision process.

Objectives do not identify specific risk management options, they are not solutions to the problem(s) identified. They identify the intended purposes of the risk management activity. An objective is a clear statement of a desired end that risk management options are intended to accomplish.

Where do these objectives and constraints come from? Values! They reflect what is important to people. You can find values in what concerns the public, the experts, and our institutions (law, regulations, guidance, policy, organizational missions).

Consider a risk management objective related to a health risk. The objective may be to reduce or eliminate the health risk. An objective does not say how that can or should be done, only that it is an objective to do so. Objectives related to economic values might include increasing jobs, income, and profits or minimizing costs. Objectives related to other public values might include things like protecting children or the environment.

Objectives reflect the most important social (or organizational) values in the decision-making process. They identify the things risk managers are trying to do. Sometimes there are important things we're trying not to do. These things we'll call constraints. Examples of constraints include not creating new risks, avoiding the loss of jobs or income, and avoiding negative impacts on endangered and threatened species.

A formal and written objectives and constraints statement is the desired output of this task. Consider it conditional and subject to change as uncertainty is reduced and you iterate your way through the risk management activity.

SAMPLE OBJECTIVE WORDS

Eliminate, reduce, minimize/maximize, enhance, harmonize, identify, define, describe, increase/decrease, raise/lower, strengthen/weaken, avoid, adapt, blend, reconcile, coordinate, affirm, diminish, weaken, promote, raise, complement, strengthen

OBJECTIVES ARE NOT:

Absolute targets: It does not specify a particular level of achievement.
Management options: It does not prescribe a specific course of action.
Government goals: It is not a political or governmental objective.
Risk assessment tasks: For example, developing a dose-response curve is not an objective.
Resource constraints: It does not address time, money, or expertise.

Discerning the relevant values and public opinion requires risk managers to develop a strategy for the external communication process. This process will ordinarily include providing information to and gathering data from outside experts and the public. This information may include opinions as well as facts. A transparent and open risk management process requires communication with all relevant stakeholders, and it is integral to all risk management decisions. This communication should include input and feedback opportunities for stakeholders in the early stages of the risk management activity.

It is important to make the communication process known to those with an interest in communicating. If an official process has been defined for you, all external communication interactions must comply with standard administrative procedures. Even these standard operating procedures should be made as transparent as possible. At a minimum, this means sharing what is known when it is known. Communication must be timely. Risk managers, with the assistance of risk communicators and public involvement experts, must decide when and how to communicate. An open communication process shares what is being done to find answers to the important questions of risk managers and stakeholders. Providing public access to your data and models enhances both transparency and openness. A good external communication process is as inclusive of stakeholders as possible.

This is one of the critical ways in which social values are appropriately reflected in the risk analysis process. Stakeholder input is essential for identifying good objectives and constraints. Like the problems and opportunities statement, this statement should be published and vetted as appropriate to the decision problem's context. Seek input to the formation of these objectives and ask for feedback on your statement. Of course, not every risk management activity will require the same kind of external communication process. Private organizations making internal decisions may require

SAMPLE OBJECTIVES AND CONSTRAINTS STATEMENT

Reduce adverse human health effects related to antimicrobial resistance to antibiotic drugs.
Reduce the number of cases of human illness due to fluoroquinolone-resistant *Campylobacter* in chicken.
Improve animal welfare in chicken production.
Do not weaken the economic viability of chicken production.

no public involvement, while some government organizations may require extensive public involvement. These issues are taken up again at greater length in Chapter 5, "Risk Communication."

The success of a risk management activity is defined by the extent to which objectives are met and constraints are avoided. That makes preparing this statement one of the most critical steps in the risk management process. These are the things we must do and must avoid doing to succeed in solving the problems and attaining the opportunities we've identified. If we do not meet our objectives to at least some extent, our risk management activity has failed. If we violate our constraints, our risk management activity has failed.

The chain of logic is simple in best-practice risk management. If you meet your objectives and avoid your constraints, you will have solved your problems and attained your opportunities. Opportunities and constraints provide a sound foundation for formulating and, later, evaluating risk management options.

So far, we are describing a rather broad and open risk management process. Not every risk management activity will require such breadth and openness. Some risk management activities are laser-focused on recurring issues of interest to only a few people. The process we are describing works as well for these activities as it does for public policy making. Risk management is a perfectly scalable process. Objectives can be identified in five minutes or five months.

3.3.2.3 Decide the Need for a Risk Assessment

Do you need a risk assessment? Not every risk management activity requires a risk assessment. Every risk management activity requires science-based evidence, but there will be times when there is enough knowledge in a room full of experts to

WHAT'S IN A RISK ASSESSMENT?

Want to start an argument? Go to a conference or listserv of risk people and ask the above question.

Risk assessment, like everything else about risk analysis, has many different definitions. A significant point of division for many seems to be whether risk assessment includes analysis that enables risk managers to evaluate the risks. This could include, for example, benefit-cost analysis. Some insist that such information is not and should not be part of risk assessment. They consider this to be risk management information that is used to evaluate the acceptability of a risk or a risk management option.

This narrow view may work for certain kinds of risk, like public health risks. But it falls apart for other kinds of risk, like risks of financial or economic losses and gains.

For our purposes, it is not so important *where* the necessary decision-making information is included as that it *be* included. So be aware that, in some interpretations, risk assessment includes information from the natural sciences only, while in others it may include much more extensive information.

know how to solve a well-defined problem. Other times the risk profile will produce sufficient information to enable risk managers to know how to solve their problems.

When an issue requires immediate action or when a risk is well described by definitive data, a risk assessment will not be needed. If the risk managers already know what decision they are going to make, a sham assessment is not needed to justify a foregone decision. A relatively simple problem with little uncertainty, and where the consequences of a wrong decision are minor, does not require a risk assessment. When the cars are speeding by, stay on the sidewalk. If the milk has turned sour, throw it way. Do not build in the floodplain. There are many instances where there is no need for a risk assessment.

A risk assessment can be useful when there are little data and much uncertainty or when there are multiple values in potential conflict; they clarify the facts. Risk assessments are useful for issues of great concern to regulators or stakeholders or when continuous decision making is in order. Risk assessment can be used to guide research by filling significant data gaps and reducing significant uncertainties. They are useful for establishing a baseline estimate of a risk or for examining the potential efficacy of new risk management options. They can also be helpful in international disputes. Practical issues that can affect the decision to do a risk assessment include:

- The time and resources available
- The urgency of a risk management response
- Consistency with responses to other similar issues
- The availability of scientific information

Deciding to do a risk assessment is a distinct result of the risk profiling task. A risk assessment should be requested when two conditions are met:

1. The risk profile fails to provide sufficient information for decision making.
2. The risk profile suggests there is sufficient information to complete a risk assessment.

Sometimes there is so much uncertainty and such sparse data that it is not even feasible to attempt a risk assessment. In these situations, risk managers may make a preemptive decision based on caution or some other set of values. Alternatively, the risk profile results can be used to direct research toward filling the most critical data gaps so that risk assessment can then proceed. The decision of whether or not to do a risk assessment is often based on the results of the risk profile. That decision is the desired output of this activity. The remainder of this risk estimation discussion assumes that a risk assessment will be completed.

3.3.2.4 Request Information Needed

If the risk profile does not provide enough information to decide how to solve the problems or pursue the opportunities, risk managers must ask for the information they need to do so. They are going to need specific kinds of information in order to be able to meet their objectives and avoid their constraints, thereby solving the problems and

MY EXPERIENCE

I have worked on many risk management activities and risk assessments and am often called in as a consultant, usually not because things have been going well. When it is my turn to speak, I hand out 3 × 5 index cards and ask everyone present to right down the question(s) they believe they are trying to answer through their risk assessment.

I then collect the cards and read them aloud. Amazingly, I have yet to have two cards identify the same question(s). How do we know what data to collect, what models to build, what analysis to do when we do not even agree on what question(s) we are trying to answer?

Getting the question(s) right is the next most critical step after problem identification.

attaining the opportunities. No one is better positioned to know what this information is than the person who will make those decisions, the risk manager.

Some of the information risk managers will need is likely to be scientific, evidence-based, factual information. This will be provided through risk assessment and possibly other evaluations. Some of the information they need will be more subjective in nature, e.g., who is concerned with this issue and how do they feel about it? This will be obtained through other means, including a good risk communication program.

It is absolutely essential, however, that risk managers explicitly ask for the information they know they are going to need to make a decision. It is not sufficient to request a risk assessment based on a specific problem. If the managers do not ask the right question(s), they may not get the right information back from the risk assessment. Risk assessments that are not guided by questions to answer may produce information managers do not want to have, or they may fail to produce the information needed to make a good decision.

The importance of the risk manager's questions can hardly be overstated. They guide the risk assessment and other evaluations required to provide the necessary information. Once answered, they provide the information needed to make decisions. These questions need to come from risk managers, often with input from assessors and stakeholders.

Risk analysis supports decision making by using science and evidence to identify what we know and what we do not know. It integrates this knowledge and uncertainty with social values to meet objectives and avoid constraints and thereby to solve problems. When the initial risk profile is completed, it is time to ask the most basic of all questions: "What do we know and what do we need to know?"

Risk managers ask questions. Risk assessors answer risk questions. Other evaluations answer other questions. If we do not ask the right questions, the analysis that follows may well not meet decision makers' needs. These questions must be available at the start of a risk assessment. They must be specific and they should be specified by risk managers.

It is essential that they be written down. They are not real and concrete until one can articulate them in precise words on paper. The questions will almost surely be refined by negotiation among managers, assessors, and possibly stakeholders. The questions will evolve and change as our understanding of the problem and the decisions to be made evolve. Consequently, they must always be kept up to date and they must be known to everyone who is working on the risk assessment.

The desired output of this task is a written set of questions to be answered by the assessors and other analysts of the risk analysis team. Many process problems begin with missing, incomplete, inappropriate, or just plain bad questions. To make sure they get the information they need to make a decision, risk managers need to ask for it directly.

An organization with a well-defined mission and recurring issues is likely to develop standard information needs for these kinds of problems. That makes this part of the risk manager's job less burdensome when those recurring information needs become general knowledge or a standard operating procedure (SOP). Everyone knows how to approach these recurring problems, and the information needs are often institutionalized. However, every organization faces enough unique situations that this task of getting the questions right should never be overlooked.

It is impossible to anticipate all the kinds of information risk managers may require for decision making early in the process. In general, four broad categories of questions can be anticipated. Risk managers will usually want to ask questions about:

1. Objectives and constraints
2. Risk characterization
3. Risk mitigation
4. Other values

Risk managers may require additional information in order to know how best to meet their objectives and avoid violating their constraints. How will you achieve/avoid them? How will you measure success toward them? What kinds of information do you need to have in order to formulate options that achieve your objectives and avoid your constraints? Some questions can be expected to focus on these kinds of concerns. They can also overlap the other question categories.

SAMPLE QUESTIONS

How many annual cases of fluoroquinolone-resistant campylobacteriosis currently occur in the United States due to eating chicken?

How many annual cases of fluoroquinolone-resistant campylobacteriosis will occur in the future due to eating chicken if there are no changes in the current usage of fluoroquinolone (FQ) drugs?

How many annual cases of fluoroquinolone-resistant campylobacteriosis will occur in the future due to eating chicken if the use of FQ drugs in all food-producing animals is prohibited?

How many annual cases of fluoroquinolone-resistant campylobacteriosis will occur in the future due to eating chicken if FQ-treated chicken is sent for use in processed chicken products?

Risk characterization questions are trickier to discuss at this point because the risk assessment steps have not yet been introduced, and this is one of them. For now, think of this as the step in the risk assessment where all the various bits of information are pulled together to characterize the likelihood and consequences of the various risks you are assessing. Risk managers must direct assessors to characterize risks in ways that are going to be of most use to them.

Suppose a risk to public health is caused by disease and the objective is to reduce the adverse human health effects of this disease. How should assessors characterize the risks? Do managers want to know the probability of contracting this disease for a given exposure? Is the exposure of interest an annual one or a lifetime one? Might it be more useful to have the numbers of people affected by the disease? If so, are managers interested in the numbers of infections, illnesses, hospitalizations, or deaths? Are there any special subpopulations of interest to managers? Risk managers need to take great pains to ask questions at the characterization level that, when answered, will give them the information they need to make a decision. An experienced risk assessment staff will know all the details they must address leading up to answering those questions. A less experienced staff may require additional questions about other assessment steps to guide them through the risk characterization.

It is wise to think of risk holistically when posing risk characterization questions. There may be separate questions about existing and future risks, residual risk, transferred risk, and transformed risk. A residual risk is the risk that remains after a risk management option is implemented. When a risk management option reduces risk at one point in time or space for one kind of event or activity while increasing risk at another time or space for the same event or activity, this is called a "transferred risk." When a risk management option alters the nature of a hazard or a population's exposure to that hazard, this is called a "transformed risk." These concepts are not as readily applied to risks of uncertain potential gain.

Risk mitigation questions are another category of questions that will usually be appropriate to ask. What does a risk manager need to know to formulate and choose the best risk management option? What are others doing to manage this risk? What else can be done to manage this risk? How well is it likely to work? For example, how many illnesses will we have if there is a vaccination program? Specific questions about the efficacy of risk management options are important to ask.

Finally, there are, for lack of a better term, values questions. These focus on obvious values of importance that are not included in the objectives and constraints. Someone will almost always care about costs, benefits, environmental impacts, authority, legal considerations, and the like. Values questions may also include stakeholders' concerns and their perceptions. An important subcategory of questions should address the uncertainty encountered in the risk management activity and its implications for the findings of the risk assessment and the answers to the risk manager's questions.

Once the questions are prepared, assessors and managers need to discuss them and what they mean. When a risk manager asks, "What is the risk of …?" it may seem a perfectly clear question, but the assessor is left to decide what is meant by risk, for example. Is it a probability or a consequence the risk manager wants to know? Communication between managers and assessors is necessary to gain a clear common understanding of the questions. It may be necessary to negotiate the list of questions at times. Some questions may be incomplete, unreasonable, or impossible to answer. When that is so, risk assessors have to tell the manager these things.

Some important questions may be missing and, if so, they should be added. The questions will be clarified, modified, deleted, and added to throughout the risk management activity.

As mentioned previously, not all questions are science questions. It may take more than a risk assessment to answer all of the risk manager's questions. Some activities will require legal analysis, benefit-cost analysis, consumer surveys, market assessment, and the like.

At this point we are up to three important pieces of paper that are essential to the successful completion of the risk management activity. They are:

- A problems and opportunities statement
- An objectives and constraints statement
- A list of questions the risk manager would like answered

If you vet the contents of these three pieces of paper with your stakeholders, you have the beginnings of an excellent risk communication process. These three pieces of paper and the process you went through to prepare them also provide an excellent basis for the eventual documentation of your risk management activity.

3.3.2.5 Initiate the Risk Assessment

With a decision to do a risk assessment in one hand and the questions to be answered by the assessment in the other, it's time to initiate the risk assessment. It is the risk manager's responsibility to provide the resources necessary to get the risk assessment done. In general, that means assembling an appropriate team of experts to carry out the task, providing them with sufficient time, budget, and other necessary resources, and interacting with them extensively enough to instruct them clearly on the information needed for decision making. All of this is to take place while maintaining a functional separation between risk assessment and risk management activities.

An independent interdisciplinary team of scientists and analysts is preferred for conducting risk assessment. In routine situations, in-house experts and personnel are sufficient for a risk assessment team. In more structured or international environments,

FUNCTIONAL SEPARATION

Functional separation means separating the tasks carried out as part of risk assessment from those carried out by risk management at the time they are performed. Some organizations may have separate offices to conduct the two tasks. In some situations, the same individual(s) may be responsible for management and assessment. This occurs most often in resource-poor situations, but it may also occur with routine and simple issues.

It is important that safeguards be in place to ensure that management and assessment tasks be carried out separately from each other, even if they are performed by the same individuals. Management and assessment are fundamentally different. The objective assessment needs to remain objective, and the subjective judgment needs to remain apart from it.

risk assessments may be carried out by an independent scientific institution, an expert group attached to an institution, or an expert group assembled for the express purposes of the risk assessment.

Risk managers are responsible for supporting the work of the risk assessment team and other evaluations by ensuring that they have the necessary resources including personnel, budget, and a reasonable schedule. In general, a good risk assessment policy will have established guidelines for much of this administrative work on a once-and-for-all basis prior to the actual risk assessment. The roles and responsibilities of key personnel, the manner in which different organizational units interact, milestones, methods for communicating and coordinating—all of these administrative matters are the responsibility of the risk managers.

3.3.2.6 Consider the Results of the Risk Assessment

After initiating the risk assessment, assessors go off and complete their work in a risk assessment, the subject of the next chapter. When the risk assessment is completed and submitted to the risk manager, the major question at this step in the risk management process is: "Did risk managers get answers to their questions that they can use for decision making?"

The risk assessment should clearly and completely answer the questions asked by the risk managers to the greatest extent possible. Those answers should identify and quantify sources of uncertainties in risk estimates and in the answers provided to risk managers. Whenever the uncertainty might affect the answer to a critical question and, consequently, the risk manager's decision, this information must be effectively communicated. Hence, in addition to getting answers to their questions, risk managers must also know the strengths and weaknesses of the risk assessment and its outputs.

It is not necessary for the risk managers to understand all the details of the risk assessment, but they must be sufficiently familiar with the risk assessment techniques and models used to be able to explain them and the assessment results to external stakeholders. To understand the weaknesses and limitations of the risk assessment it is important to:

- Understand the nature, sources, and extent of knowledge uncertainty and natural variability in risk estimates.
- Understand how the answers to critical questions might be changed as a result of this uncertainty.
- Be aware of all important assumptions made during the risk assessment as well as their impact on the results of the assessment and the answers to the questions.
- Peer review may be a useful tool for discovering implicit assumptions of the risk assessment that may have escaped the assessors' awareness.
- Identify research needs to fill the key data gaps in scientific knowledge to improve the results of the risk assessment in future iterations.

If the assessment has adequately met the information needs of the risk manager, it is complete. If the assessment has failed to provide the necessary information for any reason, another iteration of the assessment may be in order.

3.4 Risk Evaluation

The risk assessment is now complete and it is time to evaluate the risk following the steps shown in Figure 3.4. Is the risk acceptable? This is the first significant decision for the risk manager to make. It requires the risk manager to be able to distinguish two important ideas: acceptable risk and tolerable risk. An acceptable risk is a risk whose probability of occurrence is so small or whose consequences are so slight or whose benefits (perceived or real) are so great that individuals or groups in society are willing to take or be subjected to the risk that the event might occur. An acceptable risk requires no risk management; it is, by definition, acceptable. A risk that is not acceptable is, therefore, unacceptable and by definition must be managed.

It is conceptually possible to take steps to reduce an unacceptable level of risk to an acceptable level. More often than not, however, unacceptable risks are managed to tolerable levels. A tolerable risk is a nonnegligible risk that has not yet been reduced to an acceptable level. The risk is tolerated for one of three reasons. We may be unable to reduce the risk further; the costs of doing so are considered excessive; or the magnitude of the benefits associated with the risky activity are too great to reduce it further.

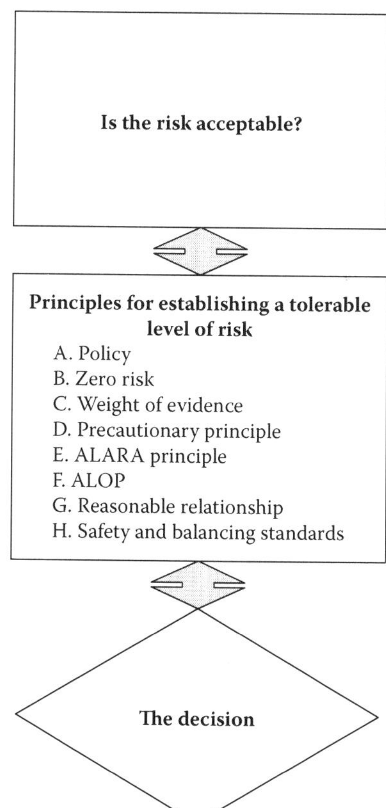

FIGURE 3.4 Risk evaluation steps.

OPPORTUNITIES FOR GAIN

What is an acceptable uncertain potential gain? Does it make sense to talk about a tolerable level of such risk? It's impossible to anticipate every potential gain situation, but for a vast majority of them, the concepts of acceptable and tolerable risk hold up pretty well.

Let's consider the potential for economic gain as the endpoint. Acceptability is going to be defined by both the consequence, for example a net negative or net positive outcome, and its probability. A low probability of a large positive outcome may be acceptable, while a high probability of a small positive outcome may not be, or vice versa. In situations where the combination of a consequence and its probability are not acceptable, in a desirable sense, the risk can be managed to the point that it becomes tolerable. This is done by taking steps to increase the likelihood of a desirable outcome or by increasing the magnitude of the potential beneficial consequences.

Risk taking is essentially different from risk avoiding. Risk-taking decisions are conscious decisions to expose one's self to a risk that could have otherwise been avoided. Consequently, managing uncertainty prior to decision making or during evolutionary decision making is a significant risk management strategy for opportunity risks.

A tolerable risk is not an acceptable risk; it is still unacceptable, but its severity has been reduced to a point where it is tolerated.

If a risk is initially judged to be unacceptable, risk managers will seek to determine a level of risk that can be tolerated. Several principles (see Figure 3.4) have been used to determine a tolerable level of risk (TLR). Once a methodology for establishing the TLR is chosen, it is the risk manager's responsibility to determine the TLR as part of the risk evaluation activities. This determination overlaps considerably with subsequent risk control activities.

Bear in mind that risk managers are not, at this point, being asked to evaluate the effectiveness of any specific risk management options that may have been assessed in the risk assessment. That particular evaluation task is considered later under the risk control activities. It is also helpful to bear in mind that we are describing an iterative process in what amounts to a linear narrative. It will often be necessary to double

INPUT AND FEEDBACK

Determining what risk is acceptable and establishing a tolerable level of risk (TLR) for risks that are unacceptable are decisions that cannot be made without input from stakeholders and the public. In some decision contexts, this may require a rather extensive public involvement program. In others it may be a simple risk communication task. Offering opportunities for input and providing feedback on views about what is acceptable, unacceptable, and tolerable is a critical part of an effective risk communication program.

back on the process and repeat a few steps. So, although the description here might suggest that the risk and all risk management options are assessed in a risk assessment that is then handed, complete, to risk managers, the real process is not nearly so simple. Very often, for example, the TLR is less an explicit determination than it is a default result of what is possible to do.

Determining whether a risk is acceptable or not is a matter of subjective judgment. It is not a scientific determination. There is potential for the uncertainty about the best thing to do to increase at this point if the risk manager's information requests have not included the views of key stakeholders or if the risk communication program has not yet provided these stakeholders with opportunities for input and feedback. There will, of course, be uncertainty about the results of the risk assessment. There can also be uncertainty about the social values that must be weighed in the decision-making process. The principles described in the following section can be used to help determine whether the assessed risk is acceptable or not. They can also be used to find a TLR when the risk is unacceptable.

3.4.1 Principles for Establishing Acceptable and Tolerable Levels of Risk

There is no magic bullet to be found in this section. Deciding whether an assessed risk is acceptable or not and determining a tolerable level of risk for risks that cannot be rendered acceptable are fundamentally searches for subjective targets. Does the risk manager seek the highest possible level of protection, a desirable level of protection, an achievable level of protection, or something that is practical (implementable) or affordable? Does equity matter? Must there be a consistent level of protection, or is the economic efficiency of a level of protection more important? There is no one answer that will satisfy everyone. Therefore the process by which this decision is reached may be as important as the decision rule that is used to reach it. To determine an acceptable or tolerable level of risk, managers must take into account the scientific evidence, the uncertainty, and the values evident in their objectives and constraints. Several principles have been used by risk managers, and they are reviewed here briefly.

3.4.1.1 Policy

Some decisions have already been made for the risk manager by persons higher in the decision-making hierarchy. These may be the owners of a company, upper management, Congress, the president, or other elected officials. In the United States, for example, Congress and the president may pass authorizing legislation that prescribes what an agency can and must do. In that case, the risk manager's job is to figure out the best way to do it.

Some risk issues may be resolved by a court decision. Decision contexts initiated as a result of administrative or other legal proceedings are often circumscribed by the entity that orders the decision action. Courts at all levels of jurisdiction are increasingly being drawn into policy decisions that could affect the principles for determining an acceptable or tolerable risk.

THE DELANEY CLAUSE

The Delaney Clause is a part of the 1958 Food Additives Amendment (section 409) to the 1954 Federal Food, Drug and Cosmetic Act (FFDCA). This clause governs regulation of pesticide residues in processed foods. It establishes that no residues from pesticides found to cause cancer in animals will be allowed as a food additive. This means that tolerance levels must be based only on the risk of carcinogenicity and that the benefits of the pesticide may not be considered. This clause was considered to have set a zero-risk standard.

Decisions made in the public sector, especially by agencies and organizations acting as stewards of a public asset or trust, will often be constrained by policy. Working with a government agency often means dealing with their policy restrictions and requirements. International treaties and agreements may identify solutions or limit options as well.

3.4.1.2 Zero Risk

Banning risky activities has been a popular approach in years gone by. Making actions that involve any risk at all taboo and declaring them forbidden has been tried in the past when it was once possible to imagine zero risk. Years ago, the limits of our knowledge and of scientific detection made it possible to find comfort in laws that appeared to legislate safety as a matter of zero risk. See the Delaney Clause text box for an example.

By the middle of the 1980s, decision makers began to abandon the notion of zero risk in favor of more realistic versions of negligible risk. The 1-in-a-million standard seems to have captured our imagination early. This evolved from and morphed into a notion of de minimis risk, a numerical value of risk too small to be bothered about. You can think of negligible and de minimis as "practically zero" without doing any real damage to the concepts.

DE MINIMIS RISK

A careful reading of official documents about the de minimis principle, as well as of relevant journal articles, shows that it is usually explained along one of the following three formulations. The *specific-number* view says a risk is de minimis provided its probability falls below a certain number, e.g., 10^{-6}. The *nondetectability* view says a risk is de minimis provided that it cannot be scientifically established whether or not the risk has in fact materialized. The *natural-occurrence* view for an anthropogenic risk says a risk is de minimis provided that its anthropogenic risk does not exceed the natural occurrence of this type of risk.

***Source*: Peterson (2002).**

Society and policy makers have, by and large, abandoned the idea that zero risk is a realistic measure of acceptable risk. Establishing a level of de minimis risk remains a viable concept for determining acceptable and tolerable levels of risk in certain settings.

3.4.1.3 Weight of Evidence

In an uncertain world, the truth is not always easy to see. Data gaps and conflicting evidence often obfuscate risk management decisions. In a weight-of-evidence approach to evaluating risk, risk managers assess the credibility of conflicting evidence about hazards and risks in a systematic and objective manner. A formal weight-of-evidence process may rely on a diverse group of scientists to examine the evidence to reach consensus views. The evidence must be of sufficient strength, coherence, and consistency to support an inference that a hazard and a risk exist.

We tend to like once-and-for-all resolution of problems on the basis of compelling scientific evidence. Evaluating the weight of evidence is an ongoing activity that attempts to balance positive and negative evidence of harmful effects based on relevant data. Thus, the evaluation of risk is conditional on the available evidence and subject to change as new evidence becomes available. When there is uncertainty about the nature of a risk, a weight-of-evidence approach may be useful in establishing whether it is acceptable or tolerable.

3.4.1.4 Precautionary Principle

Precaution may be described in this context as refraining from action if the consequences of the action are not well understood. It is prudent avoidance. The precautionary principle is broadly based on the notion that human and ecological health are irreplaceable human goods. Their protection should be treated as the paramount concern for regulatory organizations and government. All other concerns are secondary.

WINGSPREAD STATEMENT

We believe there is compelling evidence that damage to humans and the worldwide environment is of such magnitude and seriousness that new principles for conducting human activities are necessary.

While we realize that human activities may involve hazards, people must proceed more carefully than has been the case in recent history. Corporations, government entities, organizations, communities, scientists, and other individuals must adopt a precautionary approach to all human endeavors.

Therefore, it is necessary to implement the Precautionary Principle: When an activity raises threats of harm to human health or the environment, precautionary measures should be taken even if some cause and effect relationships are not fully established scientifically.

In this context the proponent of an activity, rather than the public, should bear the burden of proof.

"The Wingspread Statement on the Precautionary Principle," Science and Environmental Health Network, accessed April 23, 2011, http://www.sehn.org/state.html

The precautionary principle is controversial and heavily influenced by culture and uncertainty. In a very loose and informal sense, the precautionary principle suggests that when there is significant uncertainty about a significant risk, we should err on the side of precaution, if we are to err at all. That means that activities that could give rise to catastrophic outcomes should be prohibited. It also means that if inaction could give rise to catastrophic outcomes, we should act, not wait. The precautionary principle is generally considered to be most appropriate in the early stages of an unfolding risk problem, when the potential for serious or irreversible health consequences is great, or when the likelihood of occurrence or magnitude of consequence is highly uncertain. The desire for precaution is usually positively related to the amount of uncertainty in a decision problem. The precautionary principle can be invoked for decision making when uncertainties are large or intractable.

3.4.1.5 ALARA Principle

ALARA is an acronym for As Low As Reasonably Achievable. Technology and cost present two realistic constraints on what it is possible to achieve in terms of risk reduction. If a risk is not yet as low as is reasonably achievable, it is not acceptable according to this principle. One popular criterion for establishing a tolerable level of risk is to get risk as low as we are capable of making it. Then what choice do we have but to tolerate what risk remains?

Sometimes the ALARA principle is used to take risks even lower than an acceptable level of risk. Minimizing risks even below levels that would be acceptable is sometimes justified based on the presumption that what constitutes "acceptable risk" can vary widely among individuals.

Best available technology (BAT) is a related concept. It differs in a potentially significant way, however, as BAT says to use the best available with no further qualification. ALARA introduces the idea of reasonableness, and this opens the management door to the consideration of other factors like cost and social acceptability. BAT does not consider these other factors.

ALOP EXAMPLE

The Healthy People 2010 goals for national health promotion and disease prevention called on federal food safety agencies to reduce foodborne listeriosis by 50% by the end of the year 2005.... It became evident that, in order to reduce further the incidence to a level of 0.25 cases per 100,000 people by the end of 2005, additional targeted measures were needed. The *L. monocytogenes* Risk Assessment was initiated as an evaluation tool in support of this goal.

***Source*: U.S. FDA. http://www.fda.gov/Food/FoodSafety/FoodSafetyPrograms/ActionPlans/ListeriamoncytogenesActionPlan/default.htm.**

3.4.1.6 Appropriate Level of Protection

An appropriate level of protection (ALOP) defines or is defined by the risk society is willing to tolerate. Despite the promising sound of this principle, it is little more than circular reasoning because it presumes one has found a way to identify the holy grail of what is "appropriate" for society. In fact it is often little more than a statement of the degree of protection that is to be achieved by the risk management option implemented. Policy (see text box) or a rigorous public involvement program provide alternative ways to define the ALOP.

The significant contribution of this concept is that it flips the focus from risk to protection, where we might think of protection as akin to different degrees of safety. The factors used to determine an ALOP typically include:

- Technical feasibility of prevention and control options
- Risks that may arise from risk management interventions
- Magnitude of benefits of a risky activity and the availability of substitute activities
- Cost of prevention and control versus effectiveness of risk reduction
- Public risk reduction preferences, i.e., public values
- Distribution of risks and benefits

3.4.1.7 Reasonable Relationship

This principle suggests that costs of risk management should bear "a reasonable relationship" to the corresponding reductions in risks. It is not a benefit-cost analysis but it is an attempt to balance nonmonetary benefits (i.e., risk management outputs and outcomes) and the monetary costs of achieving them. Cost effectiveness and incremental cost analysis are often used as the basis for determining the reasonableness of this relationship.

3.4.1.8 Safety and Balancing Standards

Safety maintains deep roots within the risk analysis paradigm. A great many safety standards have been used to establish the tolerable level of risk. Safety standards encompass a bundle of standard-setting methods that rely ultimately on some degree of subjective judgment. For example, the zero-risk standard mentioned previously is one possible safety standard. Zero just happens to be one of many potential thresholds that can be established to define safety. Any nonzero level of risk can be stipulated as safe, acceptable, or tolerable. In fact, the term *tolerable level of risk* (TLR) has been dangled as one such tantalizing threshold standard in some of the literature. If we could develop a TLR for dam safety or for food safety or for transportation modalities, policy making would be much easier.

Many determinations of a TLR require a subjective balancing decision. Risks of uncertain potential gain or benefits may be best served by using some type of balancing standard. For example, risk-benefit trade-off analysis generally implies that greater benefits mean we are willing to accept a greater level of risk in exchange for those benefits. The risk-benefit trade-off explains why we are all willing to assume the risk of driving in a modern society.

Comparative risk analysis (CRA) ranks risks for the seriousness of the threat they pose. It began as an environmental decision-making tool (USAID 1990, 1993a, 1993b, 1994; EPA 1985, 1992a, 1992b, 1994; World Bank 1994) used to systematically measure, compare, and rank environmental problems or issues. It typically results in a list of issues or activities ranked in terms of relative risks. The most common purpose of comparative risk analysis is to establish priorities for a government agency. The concept is perfectly adaptable to any organization.

Benefit-cost analysis (BCA) is another kind of balancing standard used to determine what is acceptable or tolerable. BCA attempts to identify and express the advantages and disadvantages of a risk or risk management option in dollar terms. It is considered a useful measure of economic efficiency.

In addition to threshold and balancing standards, procedural standards are sometimes used to define what is acceptable or tolerable. Procedural standards typically identify an agreed upon process, which is often the result of negotiation or a referendum of some sort. If the agreed-upon process is followed, then the results of that process are considered acceptable or at least tolerable.

3.4.2 The Decision

If the assessed risk is judged by any one of these or any other method to be acceptable, there is little more for the risk manager to do. However, an unacceptable risk must be managed. The ideal would be to manage it to an acceptable level, and when that cannot be done it should be managed to a tolerable level. There are six broad strategies for managing risk. These are:

1. Risk taking
2. Risk avoidance
3. Reduce the probability of the risk event (prevent)
   Increase the probability of a potential gain (enhance)
4. Reduce the consequence of the risk event (mitigate)
   Increase the consequence of a potential gain (intensify)
5. Insurance (pooling and sharing)
6. Retain the risk

NO ONE SPEAKS THIS CAREFULLY

Beware. I've gone to some effort to try to carefully differentiate risk management strategies in the text. In my experience, no one speaks quite this carefully. In fact, mitigation, management, control, treatment, avoidance, prevention, and probably several other terms are all used interchangeably. So if you take pains to speak carefully and precisely, do not assume others hear you with the same precision. Take the time to clarify your meanings.

Risk managers may choose to take a risk when it presents an opportunity for gain that is acceptable or at least tolerable. When it comes to losses with no chance of gain, it is usually preferable to avoid such a risk whenever possible. If avoidance is not practical, we can try to manage either or both of the two dimensions of risk. Risk prevention reduces the likelihood of exposure to a hazard or otherwise reduces the probability of an undesirable outcome. Conversely, efforts can be made to increase the likelihood of gain from an opportunity risk. This is an enhancement strategy.

Risk mitigation allows that risky events will occur, so it seeks to reduce the impact of the risk by reducing the consequences of the event. Increasing the magnitude of a potential positive consequence, intensification, is another opportunity risk management strategy. A fifth option is to pool the risks into a larger group and share these risks over a greater spatial or temporal extent. Arguably, we could introduce a sixth strategy, retain the risk. When no viable option for managing the risk can be found, we have no option but to put up with the risk as is. This is retaining the risk. As this does nothing to lessen the risk or its impacts, many would elect not to call it a strategy for managing risk.

If the risk manager's role in the risk assessment can be described as a positive one, then the manager's role shifts to a normative one in these risk evaluation tasks. Here the risk manager describes the world as it "ought" to be. This is a subjective deliberative decision. This normative role continues into the manager's risk control responsibilities.

3.5 Risk Control

Presuming the risk has been judged to be unacceptable during risk evaluation, the risk manager's job now becomes reducing the risk to an acceptable level or at least to a tolerable level. Risk control is a term of art used to avoid greater confusion with the risk management strategies described previously. It may be misleading to suggest that we can control some risks. It some cases, it is more likely that we struggle to manage them. However, calling this risk management activity "risk management" might cause even greater confusion. So be forewarned not to interpret control too literally in the current context. The basic tasks during this risk control phase of the manager's job are shown in Figure 3.5.

The extent to which the public and stakeholders are engaged in this phase may show the greatest range of variation of any risk management activity. Private risk management decisions may not involve anyone outside the organization. Collective decision-making processes can involve extensive public involvement programs for the risk control activities.

3.5.1 Formulating RMOs

What does success look like? Risk management options (RMOs) are strategies that describe specific ways your risk management objectives can be achieved. These strategies are subordinate to your objectives. An RMO is relevant only to the extent that it helps you meet your objectives. Best-practice risk management recognizes that objectives can be achieved in a variety of ways and formulates alternative strategies that reflect these different ways.

Laws, authorities, policies, budget priorities, and politics may limit what you can actually do. None of these should limit the things you think, however. Formulate

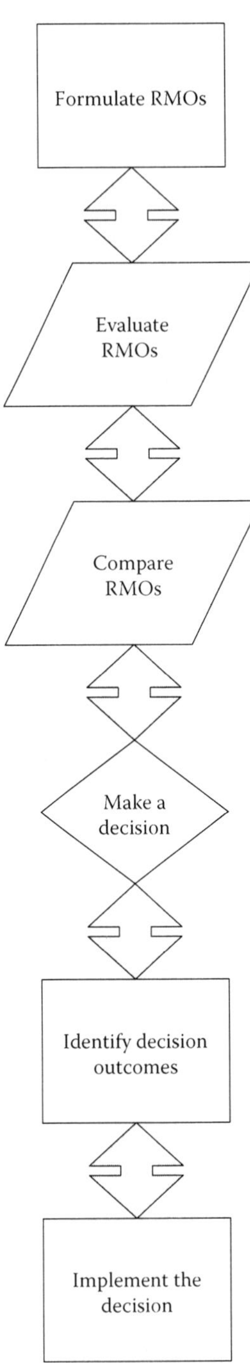

FIGURE 3.5 Risk control steps.

THE PROBLEM IS, WE NEED A LEVEE

Many problems are initially identified in terms of a solution. Risk analysis focuses on getting the problem right. The problem may be flooding, unrestricted land development, or any number of other things. A levee is one possible solution. It is common for many problem-solving processes to begin with someone identifying the solution before the problem is clearly identified. Keep track of and consider that solution, but do not let it prevent you from identifying alternative means of achieving your objectives or from properly identifying the problem.

RMOs comprehensively and creatively without respect to any limitations. Thinking creatively and comprehensively about solutions to risk problems is one area in which there is room for substantial improvement for many organizations. Getting risk managers to consider a broad array of risk management options has not been the easiest thing to do. One major reason for this is that we tend to favor solutions we are familiar with or that we have the authority and ability to act upon. There is a certain obvious appeal to this sort of thinking.

If there is an effective way to mange a risk that your organization can't implement, others may implement it voluntarily if the idea is good enough. Or perhaps there are ways to motivate those who can implement a good idea to do so. Bear in mind that good ideas for achieving worthy objectives are valid reasons for organizations to be granted new authorities.

RMOs may be formulated by the risk analysis team with input from stakeholders and decision makers. They may be imposed from above by higher authorities. They may be suggestions from the public or new scientific or technological developments. The ideas can come from Congress, agency staff, industry, government officials at all levels, academia, "the public," television, science fiction, your left frontal lobe, or a bottle of beer. They are the children of perspiration, inspiration, and imitation. An RMO may be proposed at any point in the risk analysis process. Some processes may begin because someone has framed a problem (incorrectly, I hasten to add) in terms of a solution.

The process of identifying RMOs is simplified by considering a few option formulation steps. If risk managers have done a good job identifying objectives and constraints, the simplest way to begin is to identify measures that achieve each objective. A measure is a feature that constitutes part of a strategy or risk management option. Think of a feature as some physical change or an activity, where an activity is a change in the way we do something. So for objective 1, we identify as many measures that could contribute to this objective as possible, then repeat this process for each objective and constraint.

Second, you formulate or construct RMOs from these measures. Think of the measures as building blocks and RMOs as the "structures" you build to solve problems and attain opportunities.

The third step is to reformulate RMOs. Like the rest of the risk analysis process, RMO formulation is iterative. Once an option is formulated, see if you can refine it. Once evaluation of the options begins, it can be very effective to reformulate or tweak the options to improve their performance.

KEY POINT

When developing RMOs, quantity counts more than quality in the initial stages. You cannot be sure you have the best option unless you have considered many options. Avoid the temptation to fall in love with your first idea. Formulating alternative RMOs is an essential step in good risk management.

There are a few generic criteria one might consider in a general sense when formulating RMOs. To be a viable solution, an RMO should, at a minimum, meet the following criteria:

- *Completeness*: Are all the necessary pieces accounted for and included in the option?
- *Effectiveness*: Can the objectives be better met by a different mix of measures? Can constraints be better avoided?
- *Efficiency*: Is there a less costly option? Can the beneficial effects of the RMO be increased at no additional cost? Can adverse effects be decreased?
- *Acceptability*: Is everybody (reasonably) happy?

3.5.2 Evaluating RMOs

Earlier we spoke of evaluating the risks that are assessed in the risk assessment. Don't confuse that with evaluating the risk management options that are being considered for use in managing unacceptable risks to a tolerable level. In some cases the performance of these RMOs may have been assessed simultaneously with the risks themselves in the risk assessment. In other situations, RMOs will not even be identified until after the risks have been assessed and found to be unacceptable. The actual sequence of events will depend on the information needs of the risk manager and the nature of the risk in any given decision context.

No matter which sequence your own risk management activity might follow, there comes a time when the formulation of RMOs is complete enough that you need to begin to evaluate these ideas. This is part of the nonscientific part of the risk analysis process. Values, beliefs, and biases all enter the process here, and appropriately so. This is where risk managers begin to weigh their policy options and where they really begin to earn their pay!

After RMOs have been comprehensively formulated, to get from a number of options to the best option you must:

- Evaluate options
- Compare options
- Make a decision (select the best option)

These can be discrete steps or all mixed together; they are usually iterated. Up until now, the emphasis has most appropriately been on generating as many serviceable ideas for managing an unacceptable risk as are possible. Only now do we begin to go through those ideas and evaluate them to judge which are viable solutions and which are not. Risk managers may not be directly involved in the analytical steps described here. Ordinarily they will at least identify the decision criteria and they will also make the subjective judgments leading up to and including the choice of the best RMO.

Evaluation of RMOs is a deliberative analytical process. Evaluation looks at each RMO individually and considers it on its own merits. Think of this evaluation step as a pass/fail decision that qualifies some options for serious consideration for implementation as a solution and rejects others. One of the simplest ways to evaluate an RMO is to examine the effects it would have on the risk management objectives and constraints. The underlying presumption, once again, is that if we achieve our objectives and avoid violating the constraints, we will solve our problems and realize our opportunities. That is our definition of a successful risk management process.

It is, of course, common practice to focus carefully on the management of risks during the evaluation process. In a good risk management process, risk reduction can be expected to be prominently displayed among the risk management objectives.

The effects of an RMO can be identified by comparing two scenarios as shown in Table 3.1. Identifying the existing levels of risk defines one scenario. Reestimating those risk levels with an RMO in place and functioning is the second scenario. The differences between these scenarios can be attributed to the effectiveness of the RMO, all other things being equal. The table here shows a hypothetical evaluation using a scenario without any additional risk management activity (without) and a scenario with a new RMO in place (with) as the basis for the evaluation.

There are currently 50,000 illnesses, and implementing RMO 1 will reduce that total to 20,000. The change is a reduction of 30,000 illnesses. The changes are what the risk managers will evaluate. If the objectives and constraints were to reduce adverse health effects, minimize costs, and avoid reductions in benefits and job losses, then a subjective judgment needs to be made about whether this particular option does this in a manner satisfactory enough to warrant serious consideration for implementation as a solution. The process is repeated for each individual RMO. All scenario comparisons would use the same "without" condition scenario as the starting point.

Uncertainties affecting estimates of the evaluation criteria must be considered at this step. For simplicity, the values presented in Table 3.1 are shown as point estimates. In actual fact they may be probabilistic estimates reflecting varying degrees of natural variability and knowledge uncertainty.

Note that a plan is not being compared to other plans at this point. We are simply separating our RMOs into two piles. One pile "qualifies" for serious consideration for

TABLE 3.1

Evaluating a Plan via Comparison Scenarios without Additional Risk Management and with Additional Risk Management

Effect	Future Without	Future With	Change
Annual illnesses	50,000	20,000	–30,000
Cost	$0	$150 million	+$150 million
Benefits	Unchanged	Reduced	Decrease
Jobs	Unchanged	Lose 2,000	–2,000

implementation and the other pile does not. The reject pile can either be reformulated to improve their performance or dropped from further consideration. The qualified RMOs will later be compared to one another.

Evaluation requires evaluation criteria. The risk management objectives and constraints are a logical source of such criteria, but risk managers are free to evaluate on any basis that serves their decision-making needs. At times the evaluation criteria may be a subset of the objectives and constraints or a set of criteria quite different from them. The details of the evaluation process are less important to us now than the presence and use of a consistent rational process. The risk manager's role may include identifying evaluation criteria and selecting qualified plans. The latter tasks are sometimes delegated to the assessors, as this is not the final risk management decision. Assessors analyze the RMO's contributions to the evaluation criteria for the managers.

3.5.2.1 Comparison Methods

There are at least three different ways to compare scenarios: gap analysis, before-and-after comparison, and with-and-without comparison. The latter is generally preferred as the most objective comparison, and it is recommended for use in the RMO evaluation task, described above. The three methods are shown in Figure 3.6 and described below.

A risk estimate is used to stand for any evaluation criterion relevant to risk managers. A common first step is to describe the baseline risk condition. This is often assumed to be the existing risk extended over time with no change. Although a serviceable assumption for some risks, it is not likely to serve all, if many, risks. Evaluations of changes in risks over time are desirable but are relatively seldom done.

The "without" condition describes the most likely future condition of the evaluation criterion in the absence of any intentional change in risk management. This scenario shows the future without additional risk management. Every RMO is to be evaluated against this same "without" condition.

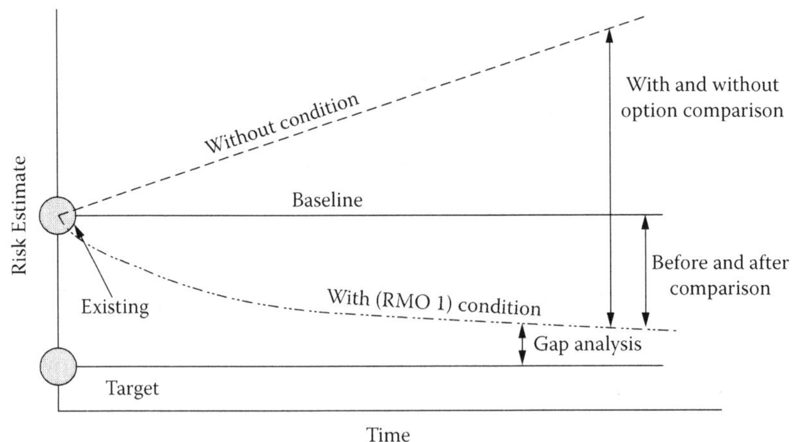

FIGURE 3.6 Three methods for comparing scenarios.

The "with" condition describes the most likely future condition of the evaluation criteria with a specific RMO in place. Each intervention (e.g., RMO 1) has its own unique "with" condition. Therefore RMO1 will have a different "with" condition than RMO 2, etc.

The best evaluation method compares the "with" and "without" condition levels of the evaluation criteria for each RMO. The resulting analysis provides values like those shown in Table 3.1. These values serve as the basis for qualifying an RMO for further consideration or not.

Note that before-and-after comparison, popularized by the National Environmental Policy Act (NEPA) process and also quite common in the food safety and some public health fields, could yield significantly different insights about the efficacy of an RMO. There are several definitions of gap analysis. Here, gap analysis refers to the difference between a prescribed target level of risk (or other effect) and what you are able to attain in reality. This graphic represents a hypothetical example; other trends in all the scenarios are possible.

3.5.3 Comparing RMOs

A good RMO formulation process will produce numerous alternative solutions to a problem. A successful evaluation process will identify several of these as viable solutions. At this point it is necessary to compare the qualified solutions to identify the best one from among them. Comparison is an analytical step. It means contrasting the merits among the various RMOs. Selection is based on weighing the differences among the compared RMOs such as those shown in Table 3.2.

A good comparison process identifies differences among the RMOs that matter to people. It also makes the trade-offs among the options clear. A simplified example of a comparison summary is shown in Table 3.2. For simplicity, it ignores the complication of expressing uncertainty about these estimates. Understand that uncertainty may be an important part of an actual comparison. Each column represents a risk management option that has been qualified by the evaluation process. The rows represent decision criteria that have been identified as important to decision makers.

Table 3.2 shows how different RMOs make different contributions to the risk management objectives (assuming, for convenience, that they are reflected in the criteria chosen). RMO 2 reduces the number of illnesses more than any other option does. It also costs more. A summary table like this enables decision makers to see the differences among the options, and it makes the trade-offs more evident. Again, you are cautioned that these determinations are more difficult to make when the

TABLE 3.2

Comparing Plans by Contrasting the Differences in Their Effects on Decision Criteria

Effect	RMO1	RMO2	RMO3
Annual illnesses	–30,000	–40,000	–10,000
Illnesses remaining	20,000	10,000	40,000
Cost	+$150 million	+$500 million	+$100 million
Benefits	Decrease	Decrease	No change
Jobs	–2,000	0	–500

uncertainties in these estimates are reflected. Methods for doing this are discussed in later chapters.

Risk managers will direct this process, though they will rarely do the analysis. A critical management step is identifying the criteria to be used in the comparison. The comparison provides the analytical summary of the information that will form the foundation for a decision. Thus, the risk manager's main role in comparison is often to request and understand the information that will be used to make a decision.

Comparisons are easiest when all effects can be reduced to a single, common metric. This, conceptually, could be lives saved, illnesses prevented, jobs created, or any metric at all. In benefit-cost analysis, that common metric is monetary. Many, if not most, comparisons involve noncommensurable metrics. These situations will involve trade-off techniques. Those techniques can range from simple ad hoc decisions to sophisticated multicriteria decision analysis techniques. An example of the latter follows.

3.5.3.1 Multicriteria Decision Analysis

A decision is always easier to make when you consider only one dimension of the problem and when you are the only decision maker. Risk management decisions, however, are often complex and multifaceted. They can involve many risk managers, each with a different responsibility for the RMO, as well as stakeholders with different values, priorities, and objectives. They often involve complex trade-offs of risks, benefits, costs, social values, and other impacts because of the values in conflict as a result of the many perspectives represented by the stakeholders to a decision. One of the most predictable sources of uncertainty in any public decision-making process and in many private ones is what weights should be assigned to the decision criteria. Multicriteria decision analysis (MCDA) is a bundle of techniques and methodologies that enables analysts to reduce the varied effects of different RMOs to a single metric that enables more direct comparison. Figure 3.7 shows the steps in a typical MCDA process (Yoe 2002). Through the evaluation step in the figure, this process tracks closely with the risk management process described here. What MCDA adds is a useful methodology for evaluating options.

Multicriteria decision problems generally involve choosing one of a number of alternative solutions to a problem based on how well those alternatives rate against a set of decision criteria. The criteria themselves are weighted in terms of their relative importance to the decision makers. The overall "score" of an alternative is the weighted sum of its ratings against each criterion. The ultimate value of MCDA, both as a tool and a process, is that it helps us to identify and understand conflicts and the trade-offs they involve.

A risk management activity defines the problem and, done well, fits the MCDA process neatly. It provides alternative means to solve the problem as well. Decision criteria would be identified, quite possibly, from the objectives and constraints during the evaluation and comparison processes. The last four steps of a generic MCDA process are often executed by a variety of user-friendly software tools.

A simple example based on the comparison in Table 3.2 is illustrated in the following discussion using Criterium Decision Plus.*

* Trademark product of InfoHarvest Inc.

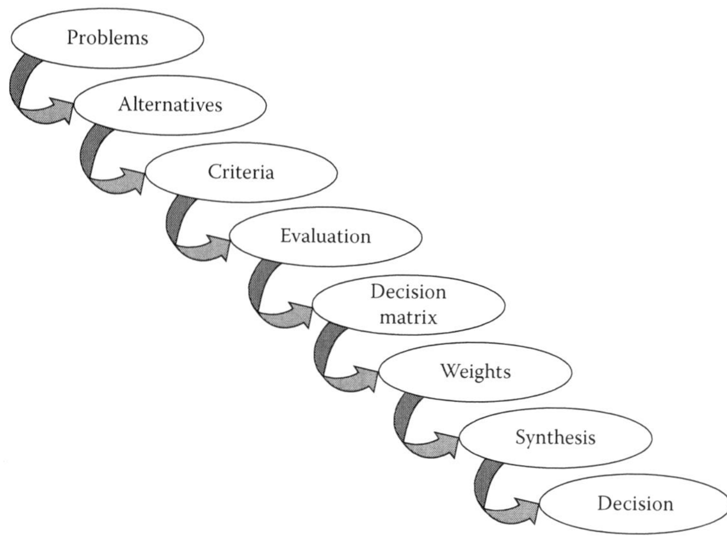

FIGURE 3.7 Multicriteria decision analysis process.

The process begins with a simple model as shown in Figure 3.8. On the far left is the decision objective. In the middle are the criteria that will be used to make the decision, and alternative solutions are shown on the far right.

Someone must specify the relative importance of the four criteria in the decision-making process. In the hypothetical example shown in Figure 3.9, assume that the weights shown reflect the decision makers' preferences. Looking at the verbal descriptions, we see that "illnesses reduced" is a critical criterion. In contrast, benefits associated with the risky activity and jobs lost as a result of an RMO are considered important.

Measurements for each alternative's contribution to each criterion are entered as well. These are simply the data from the comparison in Table 3.2. The MCDA process can accommodate estimates of the uncertainty in these values although they are not used in this example. The weights assigned by the risk manager and criteria values developed by the assessors are combined using the Simple Multi-Attribute Rating Technique (SMART) to produce scores for the three RMOs, as shown in Table 3.3.

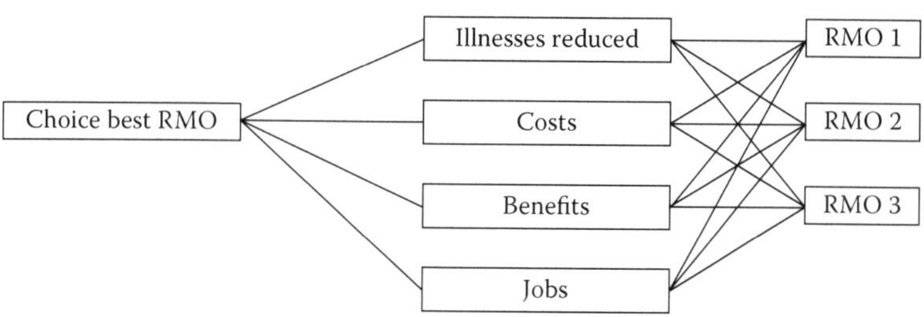

FIGURE 3.8 A simple MCDA model.

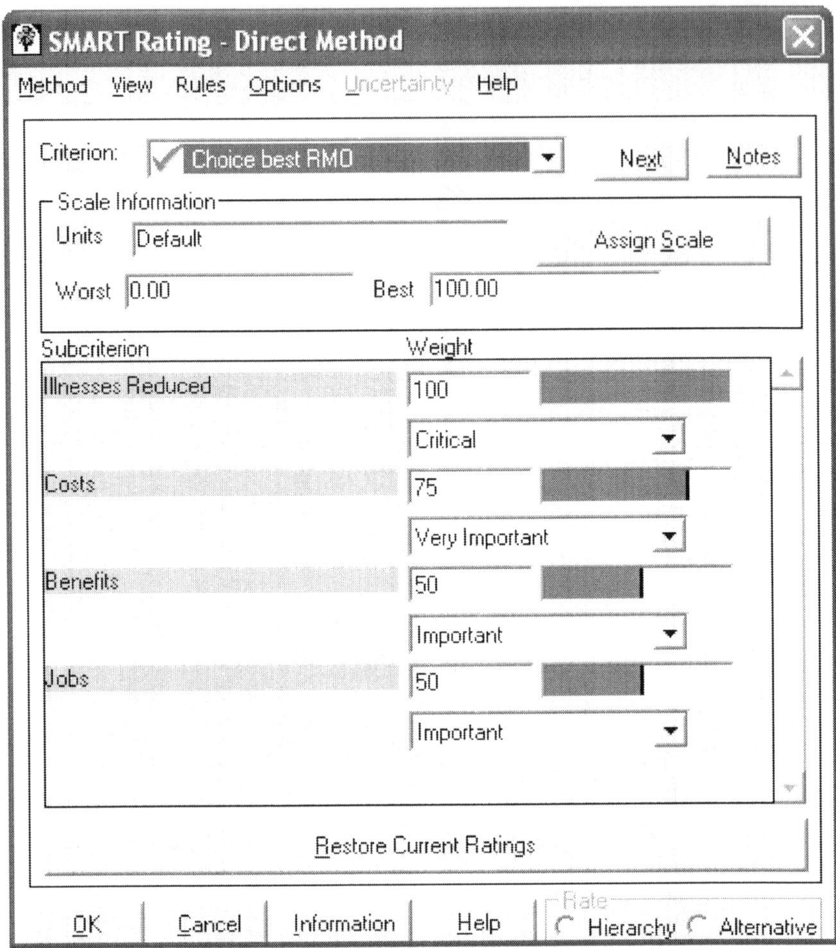

FIGURE 3.9 Assigning subjective weights to MCDA decision criteria.

RMO 3 is the "best" plan based on the available data and their relative weights. Figure 3.10 illustrates the trade-offs visually. RMO 3 makes some positive contributions to all four criteria. Its contributions to illnesses reduced, however, is the smallest of the three alternative solutions. MCDA does not produce answers or decisions. It produces information that can be helpful in identifying strengths and weaknesses of alternatives in light of the social values expressed in the analysis.

TABLE 3.3

Consolidated SMART Rating for Each of Three Risk Management Options

Risk Management Option	Option Score
RMO 1	.409
RMO 2	.473
RMO 3	.518

ADAPTIVE MANAGEMENT

Adaptive management is a risk management strategy that is useful when significant uncertainties can be expressed as testable risk hypotheses. Although there are many definitions, it usually consists of a series of steps that include the following:

- Identify known uncertainties at the time a decision is made
- Include experiments that can be used to test hypotheses about the known uncertainties among the design features in the RMO
- Measure and monitor the results of the experiments to test the identified hypotheses
- Modify predictive models based on what is learned
- Use the revised models to identify adjustments to the RMO actions over time to increase the likelihood that management objectives will be attained

Adaptive management refers to actions that are taken to learn about and manage the risks of interest. The U.S. Department of the Interior's "Technical Guide to Adaptive Management" is an excellent online resource (U.S. Department of Interior 2009).

3.5.4 Making a Decision

Choosing the best risk management option is the risk manager's next decision. This should be done only after taking the relevance of the remaining key uncertainties to the risk manager's options into account. Did uncertainty prevent assessors and others from providing risk managers with the information they needed to make a decision? Did uncertainty prevent assessors from estimating the risk or did it severely limit

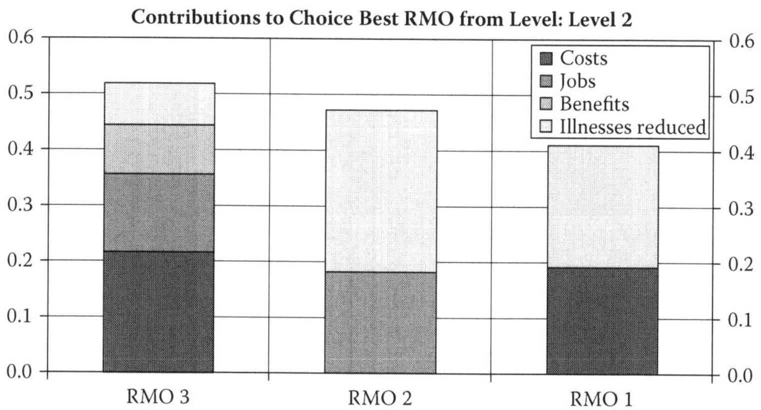

FIGURE 3.10 Contribution of options to decision criteria.

their confidence in the risk estimates? Does the uncertainty mean the efficacy of the different risk management options is in doubt?

If the uncertainty, for any reason, is significant enough to affect the nature of the answers to the risk manager's questions or to affect the choice of a course of action, risk managers must intentionally address that circumstance. That might be done through additional research, an additional iteration of the risk assessment, decisions phased to take advantage of the gradual resolution of key uncertainties, or an adaptive management approach to risk management.

The risk manager's role in the evaluation and comparison tasks is likely to be somewhat limited. Making a decision based on the work done in these steps will usually be the risk manager's responsibility. In some decision contexts, the ultimate decision makers may be elected leaders or other personnel removed from or above the risk analysis process. Even in these instances, however, it is usual for risk managers to make a recommendation based on their experience and intimate knowledge of the problem.

Risk management as described in this chapter is, in one sense, an iterative screening process based on scientific and other criteria. Making a decision, specifically selecting a recommended RMO, is the final screening activity for a given risk management activity. It is in the risk control activities that the risk manager's job shifts from the normative role of describing the world as it ought to be to taking action, which is the policy dimension of the risk manager's job.

It is not unusual for some organizations to rely on default decision rules. For example, some businesses will choose the option with the minimum payback period. Doing nothing is sometimes the default action for an organization, especially one affected by the National Environmental Policy Act (NEPA)—a safeguard that attempts to ensure that any action taken is preferable to taking no action at all.

The manner in which decisions are made cannot be fairly generalized; they will vary from organization to organization, and even within an organization they may vary from situation to situation. Good decisions are strategic; they meet objectives,

DECISION MAKING FOR OPPORTUNITY RISKS

The concepts of acceptable and tolerable risk differ between pure and opportunity risks. When we consider these terms from the perspective of an opportunity risk, an acceptable risk is one with a negligible probability of a negative outcome or with positive consequences so large that it offsets the chance of a negative outcome. Alternatively, the negative consequences may be so slight that individuals or groups in society are willing to take or be subjected to the risk. Investing in a project that has zero chance of negative net environmental benefits would be an acceptable risk.

A tolerable opportunity risk is one that is not acceptable. Risk taking is essentially different from risk avoiding. Risk-taking decisions are conscious decisions to expose one's self to a risk that could have otherwise been avoided. Consequently, managing uncertainty prior to decision making or during evolutionary decision making is a significant risk management strategy for opportunity risks.

avoid constraints, solve problems, and attain opportunities. Selecting an RMO is, to the extent that the RMO establishes a residual risk level, equivalent to choosing a TLR, although there may be instances where a TLR is determined first and then RMOs are formulated to attain that specific level of risk. The same decision rules reviewed previously may be used for this task. No matter which way it is handled, the process and the decision itself should be documented.

3.5.5 Identifying Decision Outcomes

One of the things that distinguishes risk management from other management approaches and decision-making methodologies is its focus on uncertainty. When decisions are made with less than perfect information, it is important to ask, "Is the decision working?" The answer to this question may not be evident in the near term when uncertainty is great, probabilities of occurrence are small, or time frames are long. On the other hand, we may learn quickly if our solution is working or not.

To deal effectively with uncertainty at this level of the process, the risk manager needs to identify one or more desired outcomes of the risk management option so we can verify that the solution is working. These outcomes should relate back to the risk management objectives. We want to be able to measure the impact of our risk decision(s) on public health, the company's bottom line, ecosystems, economic activity, or other appropriate outcomes. To do this we need outcomes that are measurable in principle. In some cases the outcome may never, in fact, be measured, but if there is any question about the effectiveness of the RMO, it could be measured. There is no effective way to discern methods that work from those that do not without a performance measure.

3.5.6 Implementing the Decision

Implementing an RMO means acting on the decision that was made. It requires risk managers to identify and mobilize resources necessary to actualize the RMO. Implementing a decision will very often expand the definition of who is a risk manager.

Implementation may require the cooperation of many people outside the relatively small circle of people who have worked on a risk management issue. The details of the RMO's must often be implemented by a great many people. A plan to reduce traffic accidents may involve highway engineers, automobile manufacturers, drivers, and others. Reducing the number of illnesses from *Salmonella* enteritidis in shell eggs will involve farmers, food processors, transportation companies, retailers, and consumers. Many parties can own a piece of the responsibility for implementing risk management options. The specific manner of implementation will, of course, vary markedly with the nature of the risk problem and its solution.

What can we do to ensure that the various risk managers will cooperate and implement the chosen risk management strategy? This commitment is best achieved throughout the risk management process. Best practice calls for an explicit public involvement plan as part of the risk communication process for gaining commitment to the RMO. Stakeholders and the public can be expected to have an interest in the risk management decision. At a minimum, they need to know what the decision

WHO OWNS THE RISK?

Although we have spoken of risk managers as if they are all members of the same organization that is rarely the case for decision making in the public sector. The success of an RMO may depend on many different people managing their piece of the risk.

A flood risk management (FRM) decision, for example, may require approval by and funding from Congress and the president. The U.S. Army Corps of Engineers must diligently construct all FRM structures approved by Congress. The county government may be expected to manage land use in flood hazard zones as part of the plan. State government may be responsible for operating and maintaining the FRM structures, and individual residents of the floodplain may be expected to obtain flood insurance and obey evacuation orders.

A food safety risk analysis decision may involve producers, processors, wholesalers, retailers, and even consumers in the management of a risk.

Once a plan has been selected, the number of risk managers may increase markedly. They, of course, will not all have the full range of responsibilities described in this chapter.

is and how it will affect them. Most stakeholders will want to know how the decision was made and especially how trade-offs of interest to them were resolved. Risk managers must see that this communication takes place in a timely and effective manner.

3.6 Risk Monitoring

Good RMOs can fail through faulty implementation or unravel because of false assumptions. The most brilliant strategy can be undermined if communication breaks down. Risk analysis is an evidence-based process. What is the evidence our RMO is working? If we were charged in a court room with successfully managing the risk, would there be enough evidence to convict us?

How do we know our solution works? Hubbard (2009) suggests that if we cannot answer this question, the most important thing a risk manager can do is find a way to answer it and then adopt an RMO that does work if the one currently in place does not. Figure 3.11 shows the steps comprising this last set of risk management activities: monitoring, evaluation, and iteration.

3.6.1 Monitoring

It is important to provide feedback to the organization and its stakeholders on how well they are achieving their objectives. Risk managers are responsible for monitoring the outcomes of their decisions to see if they are working. There are actually three distinct things that may be monitored in any given situation. These are decision information, decision implementation, and decision outcomes.

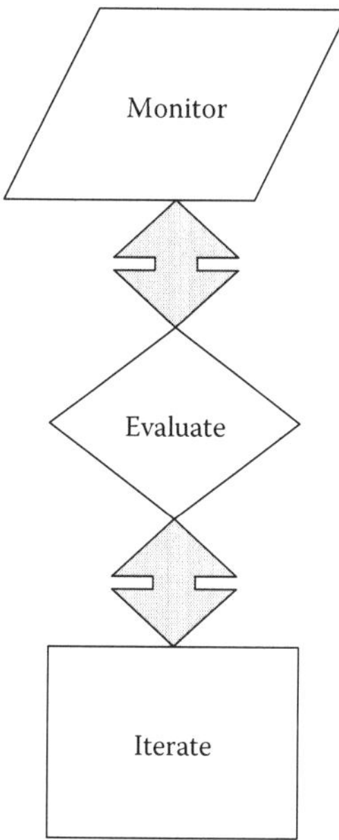

FIGURE 3.11 Risk monitoring steps.

Some risk management decisions may not yield immediately observable outcomes. Actions taken now to change conditions in the distant future are not observable, e.g., measures taken to ameliorate effects of future sea level rise. Some risk problems are so uncertain that the risks themselves may be considered speculative. It is difficult to observe the reduction of risks of rare events. In these kinds of situations, it may be important to monitor information to see if data gaps are being filled. Were the underlying assumptions of the risk assessment valid? Is the risk assessment consistent with the external data? If not, are the inconsistencies known and justified? As uncertainty is reduced, a new iteration of the risk management activity may be warranted. A new risk assessment, for example, may lead to better solutions in those instances where it is not yet possible to observe the effects of the RMO.

Monitoring actual implementation of the RMO is likely to be important in the near term after an organization decides to implement a specific measure. Are people doing what they are supposed to be doing? Audits can answer these questions for processes under the direct control of an organization, but when implementation requires large groups of stakeholders or the public to take or avoid certain activities, other methods of monitoring will be required. If everyone is doing what they need to do to get the RMO to work, then it is time to start monitoring outcomes.

A good risk management process will identify the outcomes to monitor in order to judge the success of an RMO. Monitoring means to watch, keep track of, or check for a special purpose. In this instance, that purpose is to determine if the desired risk reductions and other outcomes of an RMO are being achieved. In other words, "Are we meeting our risk management objectives?"

Decision makers do not always ask, "What measurable effect has our risk management option had?" The monitoring part of the risk management activity requires the risk manager to consider:

- What will we measure?
- How often will we measure?
- For how long will measurements be taken?
- Who will measure?
- What will they do with the measurements?
- Who will decide if the results are good, bad, or indifferent?
- How much will measurement cost?
- Who will pay?

It may not be necessary to begin to make these measurements immediately, but the desired outcomes (see section 3.5.5) need to be identified before the RMO is implemented. Risk managers need to articulate for themselves and others what success looks like and how it will be measured. All of this should be tied directly to the risk management objectives so all can see how well they are being achieved and to allow for corrective action if necessary.

3.6.2 Evaluation and Iteration

Once the monitoring information is gathered, it must be evaluated. This process should compare results and objectives to decide whether the RMO is successful. This means looking at the results and judging them as satisfactory or not. One way to do this is to compare them with risk management expectations based on the risk assessment and other data. Are the desired risk reductions being achieved? Have you attained the potential benefits from your opportunities? An alternative evaluation can mean contrasting your results with what you believe is possible from other actions. This evaluation is part of the risk manager's postimplementation responsibility.

If the evaluation step produces unsatisfactory results, the risk management decision should be modified. That modification most often will take the form of a new iteration of some or all of the risk management process. It could mean beginning again from the problem identification task, revising and updating the risk assessment, formulating new RMOs, and modifying the decision or its implementation strategy. The public and stakeholders should be kept informed of any and all postimplementation findings and changes in the risk management option.

Hubbard (2009) discusses four potential objective evaluations of risk management. The first is statistical inference based on a large sample. This is often the hard way to establish the effectiveness of an option. For example, if the RMO is intended to reduce the risk of rare events, it could take a very long time indeed to compile a sample sufficient for drawing conclusions. The ability to perform risk management experiments is even more rare. Comparing results of experiments to establish the best measures is virtually unheard of in risk management.

Second, one can seek direct evidence of cause-and-effect relationships between our RMOs and lower risk. This approach is reasonably common in certain applications. We have repeated evidence of public works projects producing the desired effects as well as of safety devices functioning as designed. Each time airport security catches a hazard at check-in we have evidence. When a seat belt restrains a passenger, there is a clear cause-and-effect relationship.

A third method is component testing of risk management options. This method looks at the gears of risk management rather than at the entire machine. Sometimes it is possible to examine how components of the RMO have fared under controlled experiments or prior experience even if we cannot evaluate the RMO as a whole. Thus, if a pasteurization step in a food process achieves the desired log reduction in pathogens, we can have some confidence in the RMO that includes such a step.

A check of completeness is Hubbard's fourth suggestion. This technique does not measure the validity of a particular risk management method. Instead, it tries to address the question of whether the RMO is addressing a reasonably complete list of risks or risk components. It is not possible to manage a risk that no one has identified. Hubbard counsels risk managers to consider any list of considered risks to be incomplete.

To better ensure completeness, four perspectives should be considered: internal completeness, external completeness, historical completeness, and combinatorial completeness. Internal completeness requires the entire organization to be involved in risk identification. External completeness involves all stakeholders in identifying risks. Historical completeness considers more than recent history. It goes back as far as possible to consider potential situations of risk. Finally, the risk manager should consider combinations of events to help explore the unknown unknowns of risk.

3.7 Risk Management Models

Very few people have actually been educated or formally trained to be risk managers. There are an infinite variety of ways to approach all or some subset of the tasks described in this chapter. It's helpful to have mental models that inform people about how an organization handles the risk management task. In the world of risk management, there are a few relatively generic models and many more organization- or application-specific models.

One of the earliest models of risk management, shown in Figure 3.12, comes from the so-called Red Book (NRC 1983). In the early days of risk analysis, risk assessment was the centerpiece of the risk analysis process. Figure 3.12 shows that risk assessment is supported by research. Risk management in a government regulatory context is rather crudely depicted as a matter of formulating and choosing the regulatory option to use to respond to the assessed risks.

It is not much of a stretch to suggest that in the early days of risk analysis the general recognition of the existence of a risk initiated the conduct of a risk assessment. Risk management was more of a reaction to the risk assessment than the proactive, directive and foundational step it is becoming today.

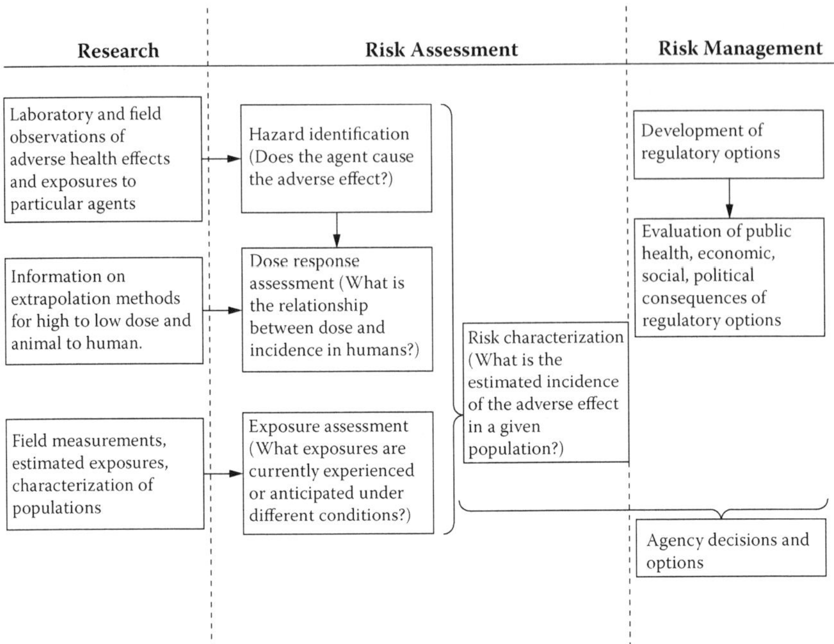

FIGURE 3.12 Elements of risk assessment and risk management (*Source*: National Research Council 1983).

One of the first more evolved generic risk management models offered in the United States was developed by the Presidential/Congressional Commission on Risk Assessment and Risk Management in 1997. It is shown in Figure 3.13. It begins with defining the problem and decision context and proceeds through a series of seven mostly distinct steps in an iterative fashion.

Once the problem is defined, risks are identified and assessed, RMOs are formulated, and the best option is chosen in the decisions step. Implementation occurs in a series of actions, and the success of the RMO is subsequently evaluated. This can lead to another iteration of the risk management process. At the center of the process is stakeholder involvement.

One of the more widely applied risk management models was developed by the International Organization for Standardization (ISO), ISO 31000, "Risk Management—Principles and Guidelines." The basic model is shown in Figure 3.14. It is not specific to any industry or sector, and it can be applied to any type of risk, whatever its nature, whether it has positive or negative consequences. The draft model shown lends itself more to the simpler summary purpose of this chapter. The final ISO risk management model differs slightly in appearance but not at all in substance. Be forewarned that ISO defines terms rather differently but at its heart is consistent with the process described in this chapter.

The ISO model has five steps and two ongoing processes, as seen in Figure 3.14. It begins by establishing the decision context. This is followed by three steps that comprise the risk assessment process. Risks are identified and then qualitatively or quantitatively described in an analytical step (not to be confused with the overall risk

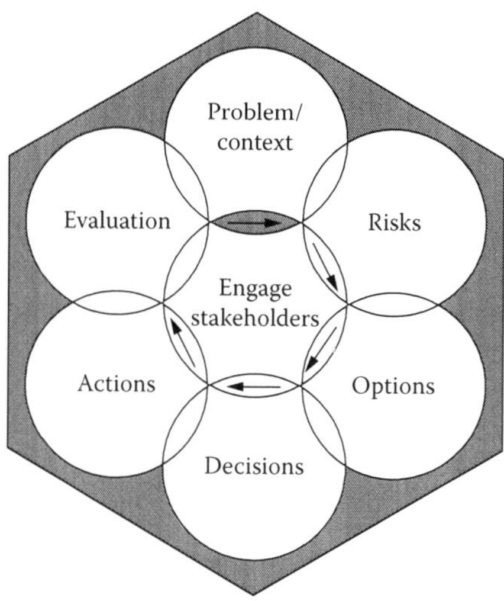

FIGURE 3.13 Risk management framework of the Presidential/Congressional Commission on Risk Assessment and Risk Management.

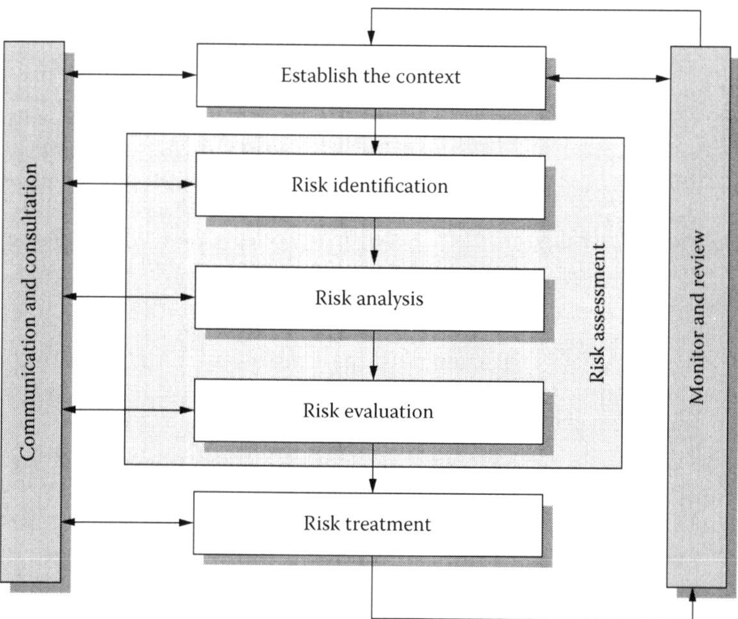

FIGURE 3.14 International Organization for Standardization risk management framework.

analysis process) that produces information that enables risk evaluation. Risk treatment involves selecting one or more options for modifying risks and then implementing those options. Communication and consultation among managers and assessors as well as with external stakeholders takes place throughout the process. Monitoring and reviewing embodies the iterative nature of the risk analysis process as well as the kinds of tasks discussed in the body of this chapter.

In addition to these generic models, there are infinite varieties of application/organization-specific risk management models. One such model is presented in Figure 3.15 as an example. This model is a microbiological risk management model (FAO 2003) to be applied to food safety problems. The details of this model are less important than the greater point that there is no one right way to do risk management. There are generic models that can be adapted for specific use; the ISO 31000 is a good example. In addition, there are any number of organization/industry/application-specific models. The best of all of these embody most if not all of the tasks described previously in this chapter.

What is most important for any organization that seeks to do risk analysis is to develop its own risk management model or adapt and adopt one of the existing models. Risk management is a process. People must know and use the process; that means the organization must have one!

3.8 Summary and Look Ahead

To do successful risk analysis you must have a risk management process. Then you must spend time working your process. Risk management begins and ends the risk analysis process. The risk manager's job may be described by the responsibilities they have in identifying problems and opportunities, estimating risk, evaluating risk, controlling risk, and monitoring risk. You have to spend time on each of these activities to do good risk analysis.

Ultimately, the risk manager's job is to make effective and practical decisions under conditions of uncertainty. Establishing a process that ensures that the best available evidence under the circumstances of the decision context are gathered, analyzed, and considered is the risk manager's responsibility. Carefully considering the significant uncertainties encountered in a risk management activity and seeing that these are carefully communicated to all interested parties is a primary responsibility of the risk manager.

The next chapter continues our unpacking of the components of the risk analysis model. The practice of risk assessment is endlessly varied because of the broad and growing number of applications of the risk analysis paradigm. Risk assessment is where the initial focus of the evolving risk analysis paradigm was concentrated. In fact, there are many practitioners who would argue vociferously that risk assessment is still the heart of risk analysis. Consistent with the aim of this primer, the common elements of many of these risk assessment models are identified and presented.

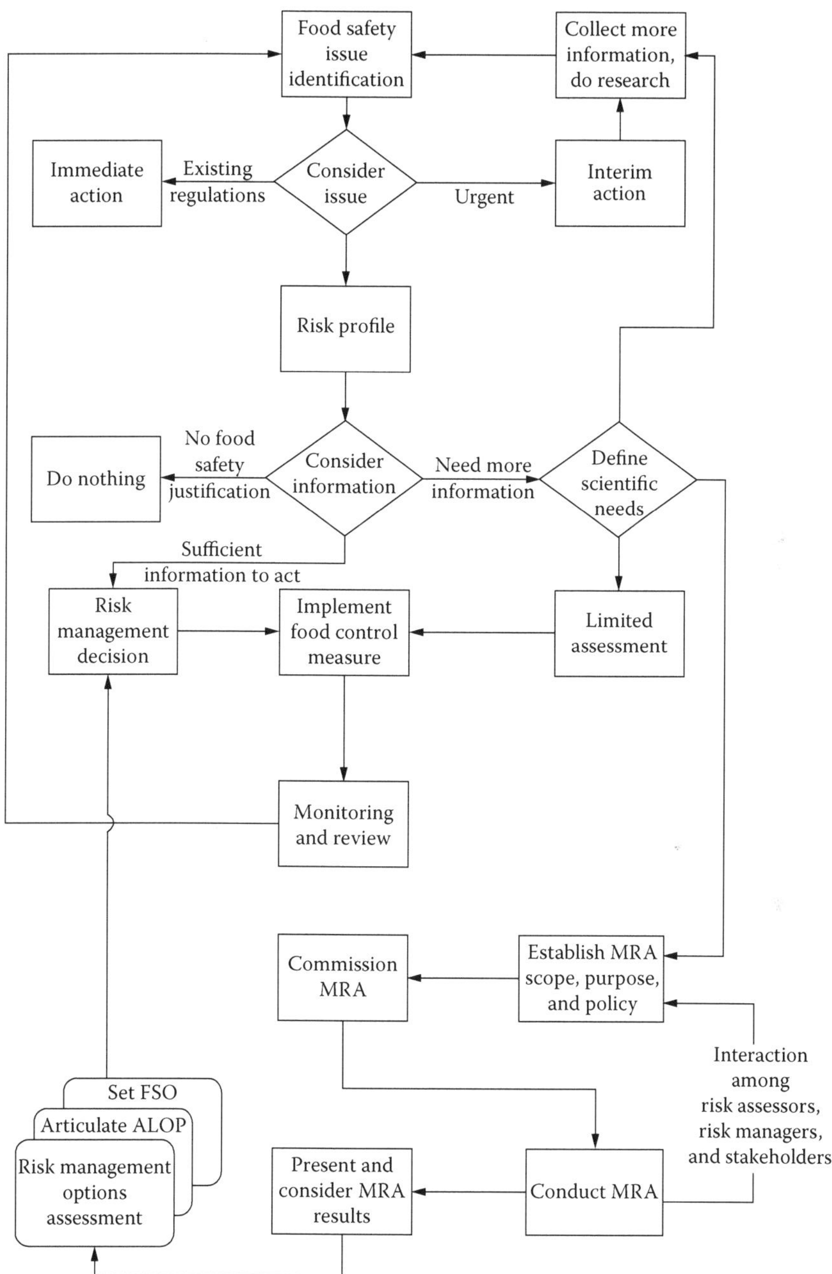

FIGURE 3.15 Microbiological risk management model (*Source*: FAO 2003).

REFERENCES

Environmental Protection Agency. 1985. Office of Air Quality Planning and Standards. *Compilation of air pollution emissions factors*. Washington, DC: EPA.

———. 1992a. Risk Assessment Forum. *Framework for ecological risk assessment*. Washington, DC: EPA.

———. 1992b. Prepared for the Technical Workgroup, Ostrava (former Czechoslovakia), and EPA by IEC, Inc., and Sullivan Environmental Consulting. *Project Silesia: Comparative risk screening analysis*. Washington, DC: EPA.

———. 1994. Prepared for the Technical Workgroup, Katowicc, Poland, and USEPA by IEC, Inc., and Sullivan Environmental Consulting. *Project Silesia: Comparative risk screening analysis*. Washington, DC: EPA.

———. 2010. *Thesaurus of terms used in microbial risk assessment*. http://www.epa.gov/waterscience/criteria/humanhealth/microbial/thesaurus/T51.html.

Food and Agricultural Organization. 2003. United Nations. *Guidelines for microbiological risk management*. Orlando, FL: Codex Committee on Food Hygiene.

Food and Drug Administration. 2003. Center for Food Safety and Applied Nutrition. *CFSAN's risk management framework*. College Park, MD: CFSAN.

Hubbard, Douglas W. 2009. *The failure of risk management: Why it's broken and how to fix it*. Hoboken, NJ: John Wiley and Sons.

International Organization of Standardization. 2009. *Risk management—principles and guidelines*. Geneva, Switzerland: International Organization of Standardization.

National Research Council. 1983. Committee on the Institutional Means for Assessment of Risks to Public Health. *Risk assessment in the federal government: Managing the process*. Washington, DC: National Academies Press.

Peterson, Martin. 2002. What is a de minimis risk? *Risk Management* 4 (2): 47–55.

Presidential/Congressional Commission on Risk Assessment and Risk Management. 1997. *Framework for environmental health risk management final report*. Vol. 1 and 2. Washington, DC: www.riskworld.com. http://riskworld.com/nreports/1997/risk-rpt/pdf/EPAJAN.PDF.

U.S. Agency for International Development. 1990. *Ranking environmental health risks in Bangkok, Thailand*. Washington, DC: AID.

———. 1993a. Environmental Health Division, Office of Nutrition and Health. *Environmental health assessment: An integrated methodology for rating environmental health problems*. Washington, DC: AID.

———. 1993b. Office of Health, Bureau for Research and Development. *Environmental health assessment: A case study conducted in the city of Quito and the county of Pedro Moncayo, Pichincha Province, Ecuador*. WASH Field Report No. 401. Washington, DC: AID.

———. 1994. *Comparing environmental health risks in Cairo, Egypt*. Washington, DC: AID.

U.S. Department of the Interior. 2009. *Adaptive management, the U.S. Department of the Interior technical guide*. Washington, DC: DOI.

World Bank. 1994. *Thailand: Mitigating pollution and congestion impacts in a high-growth economy*. Washington, DC: World Bank.

Yoe, Charles. 2002. *Trade-off analysis planning and procedures guidebook*. IWR report 02-R-2. Alexandria, VA: Institute for Water Resources.

4

Risk Assessment

4.1 Introduction

Risk assessment is the science-based component of risk analysis that answers the risk manager's questions about the risks. It provides the objective information needed for decision making, including a characterization of the relevant uncertainty that could influence the decision. An assessment is done to gain an understanding of the risk(s) and to measure and describe them to the fullest extent possible. A good risk assessment meets the needs of risk managers for decision making. It meets the manager's specific needs for timing, quality, and comprehensiveness. It provides an objective, unbiased treatment of the available evidence in well-organized and easy to understand documentation that clearly links the evidence to its conclusions. It also describes and addresses uncertainty in intentional ways.

Risk assessment is based on orderly reasoning. It is a set of logical, systematic, evidence-based analytical activities designed to provide risk managers with the best possible identification and description of the risk(s) associated with the decision problem. Evidence can be considered to include anything that helps assessors discern the truth about a matter of concern to them. It is a methodical process with specific steps that provide for a thorough and consistent approach to the assessment of risks. It also provides a thorough appreciation for the uncertainties that attend those risks. Because it includes the best available scientific knowledge, it is science based.

This chapter begins by considering what makes a good risk assessment. This is followed by a few definitions. Risk assessment models and techniques vary widely from one application to the next. The common elements of these have been distilled into a generic set of risk assessment activities, the description of which comprises the bulk of the chapter. Several specific risk assessment models are presented to illustrate the manner in which some communities of practice have formalized the assessment process. The chapter concludes with a brief description of the differences between qualitative and quantitative risk assessment.

4.2 What Makes a Good Risk Assessment?

Get the questions right, then answer them. Good risk assessment begins with the questions to be answered by the risk assessment. In best practice, risk assessors will understand the entire decision context and help ensure that risk managers get the questions right. Good risk assessment answers the questions clearly and concisely.

WHAT MAKES A GOOD RISK ASSESSMENT?

Here are the 14 aspects of a good risk assessment as described in this section:

Question	Sensitivity
Team	Relevant risks
Magnitude of effort	Qualities
Point of view	Results
Science/assumption	Evaluation
Data	Education
Uncertainty	Documentation

Risk assessment needs to be functionally separated from the risk management task. This is usually done by having different people perform these two tasks. An assessment should never be designed to provide analysis to support a predetermined answer. If risk managers know what they want to do, a risk assessment is not necessary. Not every decision requires a risk assessment. Those that do, begin with the questions.

Risk assessment is usually a team sport. Evidence-based analysis requires subject-matter experts. It is unusual, but certainly not unheard of, for a single person to be able to complete a risk assessment. As risk analysis grows in acceptance, the number of routine risk assessment applications likewise increases. Some of them can be handled by one person. The more complex and unique risk assessments are never completed by a single person. Good teams are at least multidisciplinary. Better teams are interdisciplinary. The best teams are transdisciplinary.

Multidisciplinary teams ensure that the needed expertise is available. Experts on these teams tend to function as experts in isolation of one another's disciplines. The knowledge tends to be integrated by one or a few individuals. This stands in contrast to the interdisciplinary team, where all the experts integrate their knowledge with that of others. Engineers understand something about economics, and economists understand a little engineering. Biologists know a little about what the statistician is doing, and the statistician knows some biology. The team itself is integrating the knowledge of its member experts. An interdisciplinary team works more efficiently and effectively than a multidisciplinary team. A transdisciplinary team dissolves the boundaries among disciplines and moves beyond integration to assimilation of perspectives. In the process they are often able to construct knowledge and understanding that transcends the individual disciplines. Transdisciplinary teams are preferred, but they are still rare.

The best teams spend time together working on substantive issues of common interest. Good assessment teams are collaborative and effective. Roles and responsibilities are well defined and conscientiously executed. The team answers the risk manager's questions.

The magnitude of a good assessment effort is commensurate with the resources available and in proportion to the seriousness of the problem. The effort should reflect the level of risk. Risk analysis in general and risk assessment in particular are perfectly scalable processes. A good risk assessment process can be completed in an hour if that is all the time you have or in a couple of years if that is what is warranted. The assessment process does not have to take months or years and millions of dollars, but there may be a lot of uncertainty in a quick one.

The process itself is often as important as the result. The process provides a basis for trust as well as for information. The process aids the understanding of the problem and its solutions. The process has to be sufficient to allow for answers to the questions posed by risk managers.

A good risk assessment has no point of view. It yields the same answers to the same questions regardless of who finances or sponsors the assessment. Although a question from the risk managers may, appropriately, reflect a point of view, the answer never should. It is not the assessor's job to protect the children, to make a product look profitable, to punish or reward anyone. It is to provide objective evidence-based answers to the questions they have been asked.

On a related note, assessors should not pursue their own curiosity in a risk assessment. Nor should they ever pursue a desired answer to any question. Good risk assessment avoids value judgments, and when it cannot, it identifies those value judgments explicitly.

Good risk assessment separates what we know from what we do not know, and it focuses special attention on what we do not know. Risk assessment is not pure science. The existence of uncertainty often prevents it from being so, but good risk assessment gets the right science into the assessment and then it gets that science right. Science provides the basis for answering the risk manager's questions. Honesty about uncertainty provides the confidence bounds on those answers.

Good science, good data, good models, and the best available evidence are integral to good risk assessment. Assessors need to tie their analysis to the evidence and to take care to ensure the validity of the data they use. It's both useful and important to know that not all data are quantitative. Likewise, data are not information. Skilled assessors are needed to extract the information value from data in ways that are useful and meaningful to risk managers. The answers to the risk manager's questions stand or fall on the quality of the information used to answer the questions.

It is the way that risk analysis handles the things we do not know that makes it such a useful and distinctive decision-making paradigm. In a good risk assessment, all assumptions are clearly identified for the benefit of other members of the assessment team, risk managers, and anyone else who will read or rely upon the results of the risk assessment. Risk assessors should not rely on their own default assumptions. If any default assumptions are to be used, they should be identified in the organization's risk assessment policy, prepared by risk managers.

There is uncertainty in every decision context. Risk assessors must recognize the uncertainty that exists. Moreover, they need to identify specific uncertainties that influence the answers to questions, describe their significance in meaningful ways, and then address the significant uncertainty appropriately.

There has always been uncertainty in decision making. In the past, including the recent past, it has been commonplace to overlook or ignore the existence of uncertainty, often to the regret of those affected by decisions made this way. Admitting the things that one does not know when making a decision has often been perceived as a weakness. We like confident and bold decision makers. But we also like decisions that produce good outcomes, and the two are not always compatible. Uncertainty analysis is a strength of good risk assessment, not a weakness. Good risk assessment addresses knowledge uncertainty and natural variability in the risk assessment inputs.

Good risk management addresses the variation, i.e., the uncertainty, in risk assessment outputs.

Sensitivity analysis should be a part of every risk assessment, qualitative or quantitative. Testing the sensitivity of assessment results, including the answers to the risk manager's questions, to changes in the assumptions assessors made to deal with the uncertainties they encountered is a minimum requirement for every assessment. The scenarios used to describe the risks we assess must reflect reality. That means they should be based on good science and field experience. Risk assessors need to understand how answers to the risk manager's questions might change if the risk assessment inputs change.

The risk assessment should address all the relevant risks. Risk is everywhere. Zero risk is not an option for any of us. Risk assessment is different from safety analysis, although we will find safety analysis to be a handy tool for the risk assessor's toolbox. To distinguish risk assessment from safety analysis, we need to consider risk broadly and focus on the risks of interest. These may include:

- Existing risk
- Future risk
- Historical risk
- Risk reduction
- New risk
- Residual risk
- Transferred risk
- Transformed risk

It will not always be necessary to consider each of these kinds of risk but it is rarely adequate to consider only one of these dimensions of a risk. Good risk assessment considers both the explicit and implicit risks relevant to the questions posed by risk managers.

Good risk assessments share some qualities in common. First, they are unbiased and objective. They tell the truth about what is known and not known about the risks. They are as transparent and as simple as possible but no simpler. Practicality, logic, comprehensiveness, conciseness, clarity, and consistency are additional qualities desired in a risk assessment. Of course, a risk assessment must be relevant. To be relevant it must answer the questions risk managers have asked.

Risk assessments often produce more estimates and insights than scientific facts. The assessment results provide information to risk managers; they do not produce decisions. Risk managers make decisions. The best assessments evaluate their own assumptions and judgments and convey that information to risk managers and other interested parties. A good process makes the assessment open to evaluation. It is often wise to submit a controversial or important risk assessment to an independent evaluation or peer review.

Good risk assessments can have educational value. They often identify the limits of our knowledge and in so doing guide future research. They can help direct resources to narrowing information gaps. They help us learn about the problems, our objectives, and the right questions to ask. Completed risk assessments may be conducive to learning about similar or related risks.

RISK ASSESSMENT LANGUAGE IS ALSO MESSY

The World Organization for Animal Health (OIE) defines risk assessment as follows: "The evaluation of the likelihood and the biological and economic consequences of entry, establishment, or spread of a pathogenic agent within the territory of an importing country." It has four steps:

1. Release assessment
2. Exposure assessment
3. Consequence assessment
4. Risk estimation

The International Plant Protection Convention of the UN defines risk assessment as follows: "Determination of whether a pest is a quarantine pest and evaluation of its introduction potential."

ISO Guide 73:2009, definition 3.4.1, defines risk assessment as the overall process of risk identification, risk analysis, and risk evaluation.

There are no generally agreed upon definitions for risk assessment. Fortunately, the ability to develop useful and serviceable definitions for specific organizations and applications has rendered the need for a single generic definition moot. Feel free to adopt, adapt, or invent your own definition if it helps you describe the risk(s) of interest to you.

There may be more than one audience for the risk assessment. Each audience is likely to have different information needs. This makes documentation an important part of the risk assessment process. Effective documentation tells a good story well. It is explained in simple terms and is readable by the intended audience. A good document is clear and spells important details out in terms the audience can understand. Scientific details are often most appropriately presented in technical appendices. Most important, a good risk assessment lays out the answers to the risk manager's questions clearly, well, and simply.

4.3 Definitions

At its simplest, risk assessment is estimating the risks associated with different hazards, opportunities for gain, or risk management options. Many definitions of risk assessment simply identify the steps that comprise the assessment process for that application. No one definition is going to meet the needs of the many and disparate uses of risk assessment. Nonetheless, it can be informative to consider a few formal definitions.

The seminal definition may be found in *Risk Assessment in the Federal Government: Managing the Process* (NRC 1983, 19). This book, known widely as the Red Book, for its cover, represents the first formal attempt to provide a description of the risk assessment process. The risks of primary interest at the time were chemical risks found in the human environment. Risk assessment was initially defined as follows: "Risk

RISK ASSESSMENT

When asked to name the riskiest things they do, most people will quickly identify driving. A simple risk assessment process can be demonstrated by asking the four questions used to define risk assessment in Chapter 1:

What can go wrong? One could have an accident.
How can it happen? The driver could be impaired, road or weather conditions could be hazardous, or the car could be poorly maintained.
How likely is it? Knowing yourself, you might say, "Not very likely."
What are the consequences? Property damage, injury, fatalities, or perhaps only delay and annoyance could result from an accident.

That is a risk assessment process. Is it good enough to run an insurance company or to design highways? Of course not! However, it is perfectly adequate to demonstrate the idea of a scalable and systematic process. Given more time, resources, and importance, all of these answers could be expanded and quantified.

assessment can be divided into four major steps: hazard identification, dose-response assessment, exposure assessment, and risk characterization."

The steps are described by the NRC as follows:

Hazard identification: the determination of whether a particular chemical is or is not causally linked to particular health effects
Dose-response assessment: the determination of the relation between the magnitude of exposure and the probability of occurrence of the health effects in question
Exposure assessment: the determination of the extent of human exposure before or after application of regulatory controls
Risk characterization: the description of the nature and often the magnitude of human risk, including attendant uncertainty

STEM

Cox (2002) says a risk can be defined by answering the following four questions:

- What is the *source* of the risk?
- What or whom is the ***target*** that is at risk?
- What is the adverse ***effect*** of concern that the source may cause in exposed targets?
- By what causal ***mechanism*** does the source increase the probability of the effect in exposed targets?

This definition has been and continues to be the focal point for many definitions of risk assessment. The Codex Alimentarius, for example, defined risk assessment for food safety purposes in 2004 as follows: "Risk Assessment: A scientifically based process consisting of the following steps: (i) hazard identification, (ii) hazard characterization, (iii) exposure assessment, and (iv) risk characterization" (FAO/WHO 2004, 45).

The roots of the original definition are clearly evident in this one, which has simply broadened the notion of using a dose-response assessment to characterize the consequences of exposure to a hazard.

The NRC definition as broadened by Codex has a lot of appeal for risk assessors. It begins by identifying the hazard, which is the thing or activity that can cause harm. The hazard characterization step describes the nature of that harm and the conditions required to cause it. The exposure assessment describes the manner in which people or other assets of value can become exposed to the hazard under conditions that will cause harm. The last step, risk characterization, pulls together the information in the

DEFINITION OF RISK ASSESSMENT STEPS

These definitions from a food-safety perspective can be generalized to many other applications.

Hazard identification: The identification of biological, chemical, and physical agents capable of causing adverse health effects and that may be present in a particular food or group of foods.

Hazard characterization: The qualitative or quantitative evaluation of the nature of adverse health effects associated with biological, chemical, and physical agents that may be present in food. For chemical agents, a dose-response assessment should be performed. For biological or physical agents, a dose-response assessment should be performed if the data are obtainable.

Dose-response assessment: The determination of the relationship between the magnitude of exposure (dose) to a chemical, biological, or physical agent and the severity or frequency of associated adverse health effects (response).

Exposure assessment: The qualitative or quantitative evaluation of the likely intake of biological, chemical, and physical agents via food as well as exposures from other relevant sources.

Risk characterization: The qualitative or quantitative estimation, including attendant uncertainties, of the probability of occurrence and severity of known or potential adverse health effects in a given population based on hazard identification, hazard characterization, and exposure assessment.

***Source: Codex Alimentarius Commission, Procedural Manual* (FAO/WHO 2004).**

three preceding steps to describe the probability that the risk will occur as well as the severity of the consequences.

The Presidential/Congressional Commission on Risk Assessment and Risk Management (1997) defined risk assessment more generally. It said that risk assessment is the systematic, scientific characterization of potential adverse effects of human or ecological exposures to hazardous agents or activities. It is performed by considering the types of hazards, the extent of exposure to the hazards, and information about the relationship between exposures and responses, including variation in susceptibility. Adverse effects or responses could result from exposures to chemicals, microorganisms, radiation, or natural events.

This definition did not catch on with federal agencies in the United States because it did not meet the widely varying needs of the agencies using risk assessment. You may have also noticed that these definitions do not really address the assessment of opportunity risks. For the general purposes of this book, risk assessment is defined as a systematic evidence-based process for describing (qualitatively or quantitatively) the nature, likelihood, and magnitude of risk associated with some substance, situation, action, or event that includes consideration of relevant uncertainties. The generic assessment steps adopted for this book are shown in Figure 4.1.

Risk assessment is a continuously evolving process with a stable core that takes many forms, as evidenced by the previous definitions and the models that follow. The core of risk assessment may best be described by the four informal questions introduced in Chapter 1:

- What can go wrong?
- How can it happen?
- What are the consequences?
- How likely is it to happen?

4.4 Risk Assessment Activities

The great variety of risk assessment models, methods, and applications makes it difficult to speak about risk assessment in a way with which all will agree. I describe the risk assessment component by breaking it down into eight generic risk assessment activities that appear to varying extents in one form or another in all best-practice risk assessment. They are shown in Figure 4.2 and are addressed in the sections that follow.

The steps, though presented in a linear fashion, are not always followed so. Some tasks will have been begun in the risk profile. When a risk assessment is done, the profile can be considered the first iteration of the assessment. Other tasks may be initiated simultaneously. For example, the consequence and likelihood assessments may be done concurrently. The efficacy of risk management options (RMOs) may be considered piecemeal in the steps that precede it. It is less important that the steps be accomplished in a rigid fashion than that all of the steps get done at least once.

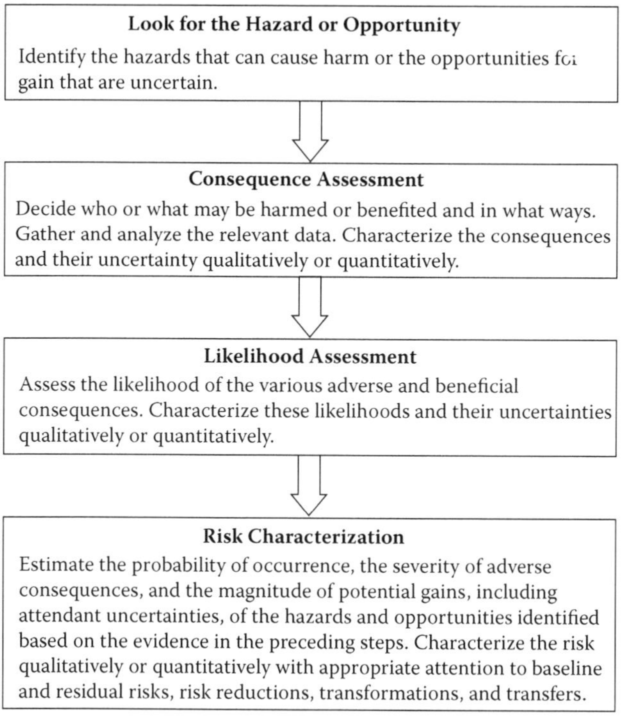

Look for the Hazard or Opportunity
Identify the hazards that can cause harm or the opportunities for gain that are uncertain.

⇓

Consequence Assessment
Decide who or what may be harmed or benefited and in what ways. Gather and analyze the relevant data. Characterize the consequences and their uncertainty qualitatively or quantitatively.

⇓

Likelihood Assessment
Assess the likelihood of the various adverse and beneficial consequences. Characterize these likelihoods and their uncertainties qualitatively or quantitatively.

⇓

Risk Characterization
Estimate the probability of occurrence, the severity of adverse consequences, and the magnitude of potential gains, including attendant uncertainties, of the hazards and opportunities identified based on the evidence in the preceding steps. Characterize the risk qualitatively or quantitatively with appropriate attention to baseline and residual risks, risk reductions, transformations, and transfers.

FIGURE 4.1 Four generic risk assessment components.

4.4.1 Understand the Questions

In case you had not picked up on the importance of questions from earlier in the chapter, risk assessors need to understand what information they are being asked to provide back to risk managers. Assessors are often involved in helping risk managers to identify the questions* that need to be answered for risk managers to achieve their risk management objectives and solve their problems. Whether assessors are involved or not, obtaining a preliminary set of the questions to be answered by the risk assessment is the essential starting point for any risk assessment.

Assessors should review the questions with the managers to make sure they have a common understanding of the meaning of the questions and the information required to answer them. If the assessors know it is going to be impossible to answer some of the questions, they need to tell the risk managers this. A revised set of questions needs to be negotiated and approved by the managers. If important questions are missing, assessors should argue for their inclusion.

Some organizations face recurring situations or they handle a specific kind of risky situation as a matter of routine. In these instances the specific questions to be answered may be well known and long established. For example, the Animal Plant Health

* We speak of assessors answering questions. Questions are just one common way of eliciting information that is needed for decision making. Questions need not literally be questions. Information needs can be expressed in any number of ways.

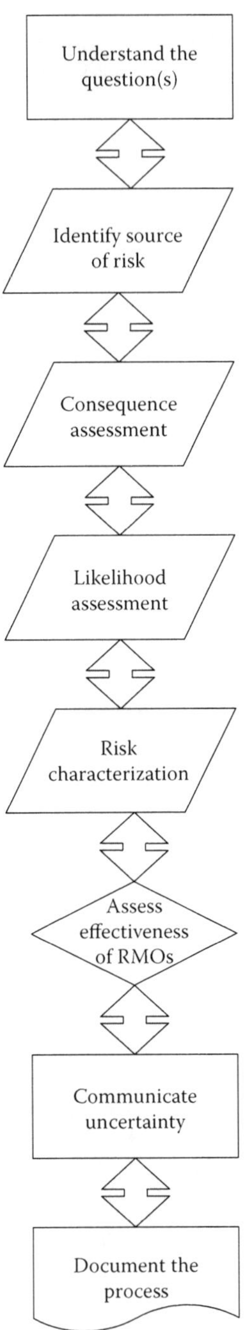

FIGURE 4.2 Eight risk assessment tasks.

SOME REAL QUESTIONS

What is known about the dose-response relationship between consumption of *Vibrio parahaemolyticus* and illnesses?

What is the frequency and extent of pathogenic strains of *Vibrio parahaemolyticus* in shellfish waters and in oysters?

What environmental parameters (e.g., water temperature, salinity) can be used to predict the presence of *Vibrio parahaemolyticus* in oysters?

How do levels of *Vibrio parahaemolyticus* in oysters at harvest compare to levels at consumption?

What is the role of post-harvest handling on the level of *Vibrio parahaemolyticus* in oysters?

What reductions in risk can be anticipated with different potential intervention strategies?

***Source*: Food and Drug Administration (2005).**

Inspection Service (APHIS) of the U.S. Department of Agriculture routinely processes requests from countries that would like to export their plants and plant products to the United States. APHIS has developed a standardized risk assessment process that relies on a well-defined set of questions that is used for all such routine requests.

The U.S. Army Corps of Engineers (USACE) routinely addresses flood problems across the country. No two of them are alike, but flood risk management investigations are similar enough that many of the information needs have become standardized. Expected annual flood damage estimation is an example of one such measure that has been standardized in guidance documents and practice.

The food-additive safety assessment process has a well-established procedure for assessing health risks of chemicals added to food. It begins by undertaking toxicity studies of the substance, if needed. These studies are used to determine the "No Observed Adverse Effect Level" (NOAEL). A safety or uncertainty factor is used to extrapolate the NOAEL results from animals to humans. Dividing the NOAEL by the safety factor yields an Acceptable Daily Intake (ADI). This is the maximum amount the average human can consume daily for a lifetime with no adverse health effect. Simultaneous efforts to calculate the Estimated Daily Intake (EDI) of the substance are undertaken, and the two are compared via the ratio, EDI/ADI. Values greater than 1 require risk management. The data input requirements are well established.

Thus, the risk manager's data needs may already be well established for some risk assessments. These will usually be available in official policy, guidance, or practice. In other unique instances, risk managers will have to carefully articulate each of their data needs. The majority of risk assessments may well fall between these two extremes. Regardless of the individual circumstances of an assessment, the risk manager's questions should be written down and understood by all. These questions guide the risk assessment.

SELECTED SOURCES OF SCIENTIFIC INFORMATION FOR RISK ASSESSMENTS

- Published scientific studies
- Professional literature
- Specific research studies designed to fill data gaps
- Administrative data and internal documents
- Gray literature—unpublished studies and surveys carried out by academia, government, industry, and NGOs
- National and other monitoring data
- National human health surveillance and laboratory diagnostic data
- Disease outbreak investigations
- National, published, and proprietary surveys, inventories, and the like
- Expert panels to elicit expert opinion where specific data sets are not available
- Risk assessments carried out by others
- International databases
- International risk assessments

4.4.2 Identify the Source of the Risk

Data collection and analysis begin in earnest when risk assessors identify, understand, and describe the source of the harm that could occur or the gain that may be realized. The source of the risk—the hazards that can cause harm or the opportunities for gain that are uncertain—may already have been identified, as the decision context was established by the managers, with or without the assistance of the assessors, in the risk profile step. Note that it is usual for assessors to participate in the preparation of a risk profile.

In many risk assessment models, this step is called "hazard identification." For EPA Superfund risks, this is the process of determining whether exposure to a chemical agent can cause an increase in the incidence of a particular adverse health effect (e.g., cancer, birth defects) and whether the adverse health effect is likely to occur in humans. For food safety concerns, it may mean identifying a pathogen-commodity pair that is of concern. For engineering projects, it may mean describing an earthquake or coastal storm risk, or it could be determining the demand on a structural component relative to its capacity. For APHIS, it may include identifying a pest of potential quarantine concern. For a pharmaceutical firm, it may be impurities or irregularities in the production of a drug. It could also mean identifying the potential gains from an investment in tourism or the gain associated with opening a new store or launching a new product line.

Identifying the source of the risk is more than simply naming a hazard or opportunity. It also includes understanding the background, context, and aspects of the hazard or opportunity relevant to the problem being addressed and communicating that to others. The extent of this process will vary from situation to situation. For example, identifying a food-borne pathogen may be very straightforward for well-known microbiological hazards yet far from fully developed for emerging or new

TWO CONSEQUENCE CAVEATS

Managers and assessors need to remain vigilant against imprecise language. It is easy to ask, "What is the risk associated with eating oysters contaminated with pathogenic *Vibrio parahaemolyticus*?" But what are the consequences of concern to risk managers? Is it annual deaths, hospitalizations, illnesses, or exposures? Should people be segregated by age, gender, ethnicity, or other factors? Do we want estimates of the probability of death and illness? If so, should these be per exposure or annually? Or are other measures like loss of life expectancy, working days lost, or quality adjusted life years (QALY) appropriate?

Managers and assessors also need to think broadly about consequences. A narrow focus on consequences can cause managers to overlook important impacts of both the risks and the RMOs. For example, if the assessment is motivated by public health consequences, it is easy to overlook important impacts on trade, industry, and consumers.

The cures for these mistakes are found in a good risk management process and those "three pieces of paper" it produces.*

* The "three pieces of paper" are the problems and opportunities statement, the objectives and constraints statement, and the list of questions to be answered.

microbiological hazards. Economic opportunities may need to be supported with market studies and cost details. Technological risks need to be explained in a narrative fashion that facilitates understanding by all interested parties.

The risk assessor should think comprehensively about risks. This means identifying all of the decision-relevant risks. It is all too easy to focus too narrowly and too quickly on a single risk when there may in fact be more than one. It also means considering all the relevant dimensions of a risk, including not only the existing risk, but residual, new, transformed, and transferred risks. This is primarily a qualitative analysis. Importantly, this is where risk assessors begin to carefully identify and separate what we know about a risk from what we do not know.

4.4.3 Consequence Assessment

In this activity, risk assessors characterize the nature of the harm caused by a hazard and the gain possible with an opportunity. Assessors identify who or what may be harmed or benefited and in what ways by the sources of risk identified in the previous step. This activity might be described as the cause-effect link in the risk assessment. What undesirable effects do the hazards have? What desirable effects might the opportunities offer? Risks affect human, animal, and plant life, public health, public safety, ecosystems, property, natural and cultural resources, human systems (political, legal, education, transportation, communication, and the like), infrastructure, economies, international trade and treaties, financial resources, and so on. Carefully identifying the consequences and linking them to the hazards and opportunities is an essential early step in any risk assessment. This activity may be likened to the hazard characterization step of the Codex model.

The consequences of most importance should already be reflected in the "three pieces of paper" developed by risk managers before the risk assessment was initiated. In an iterative process, such as risk assessment, it is to be expected that, as uncertainty is reduced, some aspects of the assessment will become better understood. This may necessitate revising the assessment of relevant consequences.

Effectively managing a risk requires a broad understanding of the relevant losses, harm, consequences, and potential gains to all interested and affected parties (NRC 1996). Consequences may be characterized qualitatively or quantitatively. No matter which type of characterization is used, it is essential to catalogue the significant uncertainties encountered in describing and linking the consequence to the source hazards and opportunities.

4.4.3.1 Dose-Response Relationships

The earliest risk assessments used dose-response relationships to characterize the consequences of human health risks. Dose-response relationships, often represented as curves, remain the primary health-consequence model used to characterize the adverse human health effects of chemicals, toxins, and microbes in the environment. There is extensive literature on these dose-response relationships that includes its own journal, *Dose-Response*. A discussion of consequence assessment is not complete without a brief mention of this often-used relationship.

Consider the conceptual representation of a dose-response curve shown in Figure 4.3, which was adapted from Covello and Merkhofer (1993), where they described the effect of a risk agent on a large population. First, notice that the dose of the risk agent increases from left to right on the horizontal axis. The response, on the vertical axis, is not specified. In practice, this axis may be the probability of illness

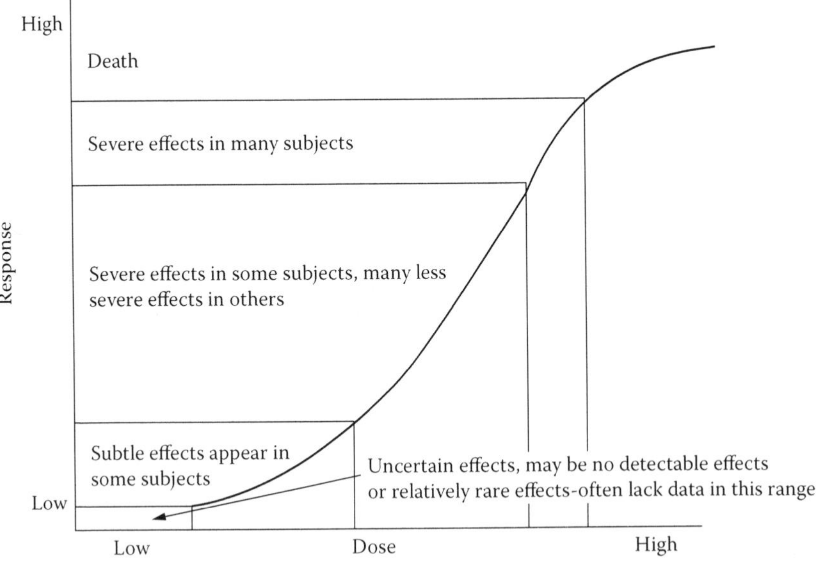

FIGURE 4.3 A stylized dose-response relationship. (Adapted from Covello and Merkhofer 1993.)

or of some other adverse health effect, the number of excess tumors (above those observed in a control group, for example), or virtually any adverse health effect. The metrics on the vertical axis are most often developed for a representative individual of the general population, the general population itself, or any subpopulation of interest.

Five different conceptual distinctions are made in the figure to aid the general understanding of the relationship. First, at sufficiently low doses there may be significant uncertainty about the response. "No threshold" models assume there is no absolutely safe level of exposure. However, for many risk agents there may be no effects at very low doses regardless of the duration of the exposure. At this level of exposure when the response is, say, the probability of cancer for a representative member of the population, the shape of the curve in this range of doses may be highly uncertain.

At somewhat higher levels of exposure, subtle adverse health responses may be detected in some subjects. Using the probability of cancer as an alternative interpretation of this range of doses, we would see an increasing probability of an adverse health effect.

As the dose increases more, we might see the beginning of some severe effects in some members of the population with a growing number of less severe effects. The alternative interpretation of the cancer risk model simply has an increasing likelihood of cancer. In that fourth dose region, we would expect to see increasing numbers of severe health effects among members of the population. Both incidence and severity increase with these relatively high doses of the risk agent. If the dose increases high enough and lasts long enough, all members of the population are at risk of an adverse response. The final response range corresponding to the highest dose levels results in death.

This conceptual model is not to be taken literally. Response will not always be a continuum from no effects to death, as the probability of cancer alternative explanation illustrates. The doses themselves will vary as well. Chemical exposures, for example, may be measured in mg/kg/body weight daily for a lifetime. Microbial doses may be measured in the number of cells or colony-forming units. Acute exposures may be measured in any number of physical metrics, depending on the nature of the risk agent.

4.4.4 Likelihood Assessment

Risk assessors analyze the manner in which the undesirable consequences of hazards or the desirable consequences of opportunities occur so they can characterize the likelihoods of the sequence of events that produce these outcomes. Risks can't be directly observed or measured because they are potential outcomes that may or may not occur. Uncertain occurrence is a necessary, but not sufficient, condition for risk.* Probability is the most common language of uncertainty. Thus, qualitatively or quantitatively assessing the probability/likelihood of the various adverse and beneficial consequences associated with the identified risks is necessary for risk assessment. This is most analogous to the exposure assessment task of the NRC Red Book model.

Assessing the likelihoods of the consequences associated with identified risks can often be aided by developing a risk hypothesis. A risk hypothesis is a model

* If risk is the chance of an undesirable outcome, an undesirable outcome is the other necessary condition.

or scenario that credibly explains how the source of the risk can lead to the consequences of concern by identifying the appropriate sequence of uncertain events that must occur for this to happen. The likelihood assessment characterizes the chance of that sequence of events occurring. Think of this step as estimating the probability that a risk target (person or thing) will be exposed to the hazard that can cause harm. Alternatively, it is estimating the probability that an opportunity does (or does not) yield a favorable outcome.

Three simple risk hypotheses are illustrated in Figures 4.4–4.6. The first example in Figure 4.4 is a risk model for estimating expected annual flood damages. The upper right quadrant assumes that property damage (consequence), measured in dollars, increases with water depth. The upper left quadrant shows the volume of water flow (hazard) required to reach the corresponding depths of water. On the lower left, the likelihoods of the various flows being equaled or exceeded in a year are shown. These three relationships* together yield the fourth one (damage-frequency) in the lower right quadrant, which when integrated provides a measure of property damages called expected annual damages. The likelihood characterization is derived from the middle two quadrants.

A second risk hypothesis is seen in Figure 4.5. This shows the presumed relationship between a human activity (logging), the stressors it creates, and the adverse effects that can result in a forest environment (EPA 1998). Likelihood assessment would require estimating the likelihood of each of these model elements.

The risk hypothesis embodied in the FDA risk assessment on *Vibrio parahaemolyticus* in raw oysters (FDA 2005) is shown in Figure 4.6. The likelihood characterization is more complexly woven through elements of this hypothesis.

It is essential to identify the significant uncertainties and to analyze their potential impact on the likelihoods of the risks. This may be done qualitatively or quantitatively, in concert with the level of detail in the overall risk assessment.

4.4.4.1 Exposure Assessment

The likelihood characterization includes the special subset of exposure assessments that virtually all health-related risk assessments require. An exposure assessment estimates the intensity, frequency, and duration of exposure to a risk agent. Exposure assessments identify the relevant pathways by which a human or other population is exposed to a hazard. The exposure assessment is often the most difficult part of a risk assessment, because the pathways can be both numerous and complex. They are also plagued by knowledge uncertainty and natural variability.

Exposures can be monitored directly, when direct measurement of the individual's exposure by instruments is possible. Otherwise we are restricted to measuring factors that affect exposure rather than the exposure itself. These are indirect methods of monitoring. Spatial and temporal variations are often important considerations in exposure assessments. Models used to capture the relevant aspects of an exposure pathway can be quite complex.

A general exposure equation adapted from EPA (Covello and Mumpower 1985) is:

* Adapting the terminology of the NRC definition of risk assessment, the depth-damage curve is a dose-response relationship. The depth-flow and flow-frequency curves comprise an exposure assessment. The damage-frequency curve is a risk characterization.

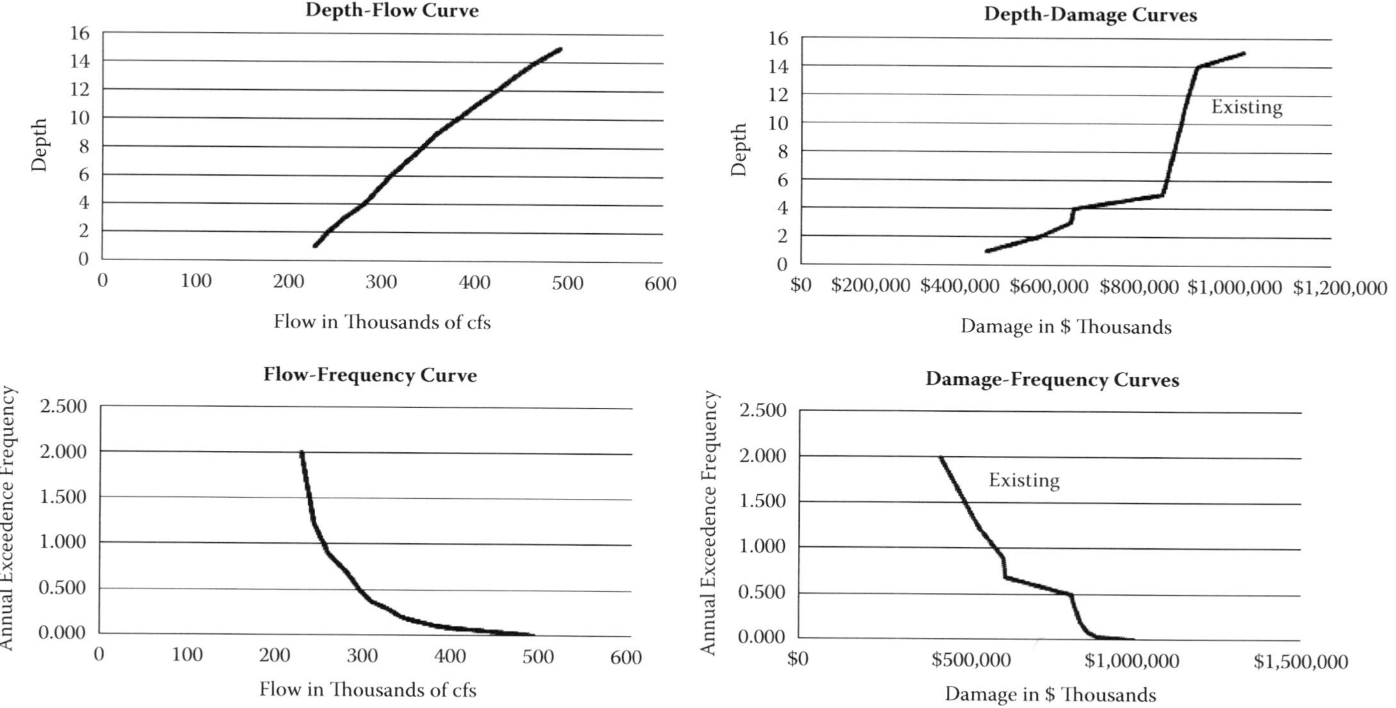

FIGURE 4.4 Hydroeconomic model for flood risk.

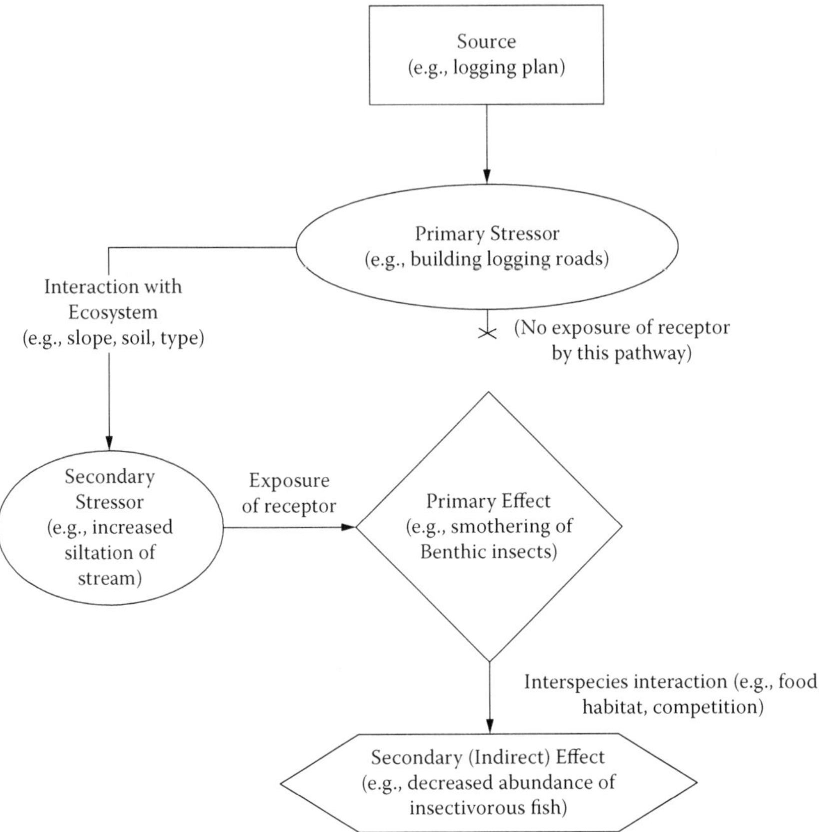

FIGURE 4.5 Conceptual model for logging forest products. (*Source*: EPA 1998.)

$$\text{Intake} = \frac{\text{Concentration} \times \text{Contact rate} \times \text{Exposure frequency} \times \text{Exposure period}}{\text{Body weight}}$$

Here intake is defined as mg/kg of body weight. The concentration of the risk agent (e.g., a chemical) in a medium during the exposure period is multiplied by the amount of contaminated medium contacted per unit of time or per event to get the amount of risk agent per exposure. The exposure frequency (days per year, for example) is multiplied by the exposure period (number of years) to get the duration of the exposure. The product of these two values (amount of risk agent and duration of exposure) is divided by body weight to get a dose to which one is exposed. This dose may then feed into a dose-response relationship, as discussed previously.

In the event of a microbial exposure, a general exposure equation might be:

$$\text{Intake} = \text{Concentration of pathogen per medium weight} \times \text{Total weight of medium}$$

The resulting number of cells or colony-forming units (CFUs) provides an estimate of the dose, which might have an associated probability of an adverse health effect if a dose-response relationship is available.

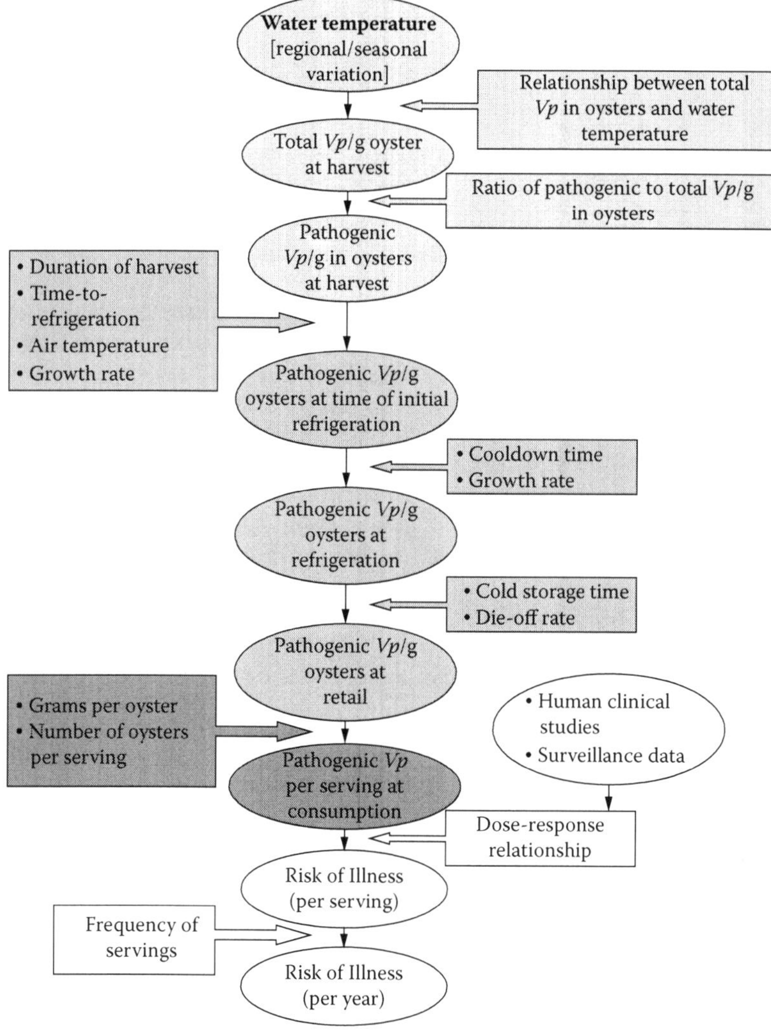

FIGURE 4.6 Schematic representation of the *Vibrio parahaemolyticus* risk assessment model. (*Source*: FDA 2005.)

4.4.5 Risk Characterization

Risk characterization is where the risk manager's questions get answered. Risk characterization typically includes descriptions of the probability of occurrence and the severity of adverse consequences associated with hazards, as well as the magnitude of potential gains from opportunities identified based on the evidence and analysis in all the preceding steps. Risks can be characterized qualitatively or quantitatively.

Characterizations include one or more estimates of risk and their accompanying risk descriptions. A risk estimate is an estimate of the likelihood and severity of the adverse effects or opportunities, which addresses key attending uncertainties. Quantitative estimates are numerical in nature and are preferred over narrative qualitative estimates. Risk estimates should include all the relevant aspects of the risk, which may encompass existing, future, historical, reduced, residual, new, transformed, and transferred risks. A risk description is a narrative explanation and depiction of a risk that bounds and defines a risk for decision-making purposes. The story that accompanies the risk estimate is what places it in a proper context for risk managers and others to understand.

EXAMPLE: RISK DESCRIPTION AND RISK ESTIMATE

Risk per Annum. The "risk per annum" is the predicted number of illnesses (gastroenteritis alone or gastroenteritis followed by septicemia) in the United States each year. As shown in Summary Table 7, for each region, the highest number of predicted cases of illnesses is associated with oysters harvested in the summer and spring and the lowest in the winter and fall. Of the total annual predicted *Vibrio parahaemolyticus* illnesses, approximately 92% are attributed to oysters harvested from the Gulf Coast (Louisiana and non-Louisiana states) region in the spring, summer, and fall and from the Pacific Northwest (intertidal) region in the summer. The lower numbers of illnesses predicted for the Northeast Atlantic and Mid-Atlantic oyster harvests are attributable both to the colder water temperatures and the smaller harvest from these regions. The harvesting practice also has an impact on the illness rate. Intertidal harvesting in the Pacific Northwest poses a much greater risk than dredging in this region (192 vs. 4 illnesses per year). This is likely attributable to elevation of oyster temperatures during intertidal exposure leading to *Vibrio parahaemolyticus* growth.

SUMMARY TABLE 7

Predicted Mean Annual Number of Illnesses Associated with the Consumption of *Vibrio parahaemolyticus* in Raw Oysters

	Mean Annual Illnesses [a]				
Region	**Summer**	**Fall**	**Winter**	**Spring**	**Total**
Gulf Coast (Louisiana)	1,406	132	7	505	2,050
Gulf Coast (non-Louisiana)[b]	299	51	3	193	546
Mid-Atlantic	7	4	<1	4	15
Northeast Atlantic	14	2	<1	3	19
Pacific Northwest (Dredged)	4	<1	<1	<1	4
Pacific Northwest (Intertidal)[c]	173	1	<1	18	192
TOTAL	1,903	190	10	723	2,826

[a] Mean annual illnesses refers to the predicted number of illnesses (gastroenteritis alone or gastroenteritis followed by septicemia) in the United States each year.

[b] Includes oysters harvested from Florida, Mississippi, Texas, and Alabama. The time from harvest to refrigeration in these states is typically shorter than for Louisiana.

[c] Oysters harvested using intertidal methods are typically exposed to higher temperature for longer times before refrigeration compared with dredged methods.

Source: Food and Drug Administration (2005).

EXAMPLE: EFFECTIVENESS OF RMOS

SUMMARY TABLE 9

Predicted Mean Number of Illnesses per Annum from Reduction of Levels of Pathogenic *Vibrio parahaemolyticus* in Oysters

Region	Predicted Mean Number of Illnesses per Annum			
	Baseline	Immediate Refrigeration[a]	2-log Reduction[b]	4.5-log Reduction[c]
Gulf Coast (Louisiana)	2,050	202	22	<1
Gulf Coast (Non-Louisiana)	546	80	6	<1
Mid-Atlantic	15	2	<1	<1
Northeast Atlantic	19	3	<1	<1
Pacific Northwest (Dredged)	4	<1	<1	<1
Pacific Northwest (Intertidal)	192	106	2	<1
TOTAL	2,826	391	30	<1

[a] Represents refrigeration immediately after harvest; the effectiveness of which varies both regionally and seasonally and is typically approximately 1-log reduction.
[b] Represents any process that reduces levels of *Vibrio parahaemolyticus* in oysters 2-log, e.g., freezing.
[c] Represents any process that reduces levels of *Vibrio parahaemolyticus* in oysters 4.5-log, e.g., mild heat treatment, irradiation, or ultrahigh hydrostatic pressure.
Source: Food and Drug Administration (2005).

There are different approaches to risk characterization. The choice of an approach depends on the needs, objectives, and questions of the risk managers. Any good risk characterization will convert the scientific evidence base into a statement of risk that answers the manager's questions. It is convenient to think of the different risk characterization approaches as running along a continuum from qualitative to quantitative. Between the two lie semiquantitative risk characterizations, which are sometimes useful.

It is during the risk characterization that the overall importance of the various uncertainties encountered throughout the risk assessment begins to come into focus. Risk characterization should include sensitivity analysis or formal uncertainty analysis commensurate with the nature of the risk assessment.

4.4.6 Assess Effectiveness of RMOs

In most cases, risk assessors will be asked to estimate risk reductions attributable to the risk mitigation options under consideration. In some situations, assessors or others may be asked for additional evaluations of the RMOs along with the evaluation of RMO efficacy. These evaluations might include economic costs and benefits, environmental impacts, social impacts, legal ramifications, and the like.

Typically, the existing level of risk is assessed, and if it is judged to be unacceptable, it will be reduced to a tolerable level if it cannot be eliminated. Often the unspoken "default" level of tolerable risk is as low as reasonably achievable. In other instances risks are reduced to the point where the costs of further reductions clearly outweigh the benefits of additional risk reductions.

In some decision settings the trade-offs among risk reduction, cost, and other criteria are more complex, and an array of RMOs may be under consideration. In these situations it is usually desirable to use a "with" and "without" risk management option comparison, as described in the previous chapter, along with a more formal trade-off analysis.

Considering residual, new, transferred, and transformed risk at the time that risk reductions are estimated is likely to be efficient. In some instances RMOs will be reasonably well formulated at the time that the risk assessment is initiated. In other situations, due to the iterative nature of risk assessment, RMOs may not even be identified until well after the risk assessment has begun. In some cases it may not even be appropriate to begin to formulate RMOs until after the risk has been initially assessed and judged to be unacceptable. For these reasons, this RMO assessment step, often considered an integral part of risk characterization in some descriptions of risk assessment, is separated out here. Uncertainties concerning the performance or efficacy of an RMO should be investigated and documented so that risk managers and other interested parties may be made aware of them.

4.4.7 Communicate Uncertainty

It is not enough for risk assessors to identify and investigate the significance of the uncertainties identified throughout the risk assessment. They must communicate its significance for decision making to risk managers. Methodologies for effectively conveying information about what is known with certainty and which remaining uncertainties could affect the risk characterization or the answers to the risk manager's questions need to be developed and carried out. It is better to have a general and incomplete map, subject to revision and correction, than to have no map at all (Toffler 1990). But those using the risk assessment map to make decisions must know its limitations.

Characterizing the significance of the key uncertainties in a risk assessment is critical to informed decision making. The NRC (1994) said, "Uncertainty forces decision makers to judge how probable it is that risks will be overestimated or underestimated." This is important for risk managers to understand, as they determine the need for and appropriate choice of an RMO.

Characterizing uncertainty can also support the informed consent of those affected by risk management decisions. When people are asked to live behind a levee or near a nuclear power plant, to get a vaccination for a seasonal flu, or to board an airplane, they have a right to know the limitations of the risk management measures taken on their behalf as well as the limitations of the information on which those measures were based. Characterizing uncertainty is essential to the transparency of a risk assessment. Transparency enhances the credibility of the process, improves the defensibility of actions taken or not taken, and empowers affected individuals to make better choices for themselves in response to the risks that remain.

Uncertainty analysis also identifies important data gaps, which can be filled to improve the accuracy of the risk assessment and, hence, support improved decision making. Risk assessors should communicate their degree of confidence in the risk assessment they have done so that risk managers can take this into consideration for decision making. To do this, risk assessors should explicitly address natural variability

ASSUMPTIONS

No risk assessment can be completed unless the evidence is supplemented with assumptions. Explicit assumptions are those that assessors consciously make. In principle, they can be readily documented. Implicit assumptions are those that escape the conscious awareness of the assessors. They may be based on the culture of the organization; the beliefs of the assessors; the basic assumptions, principles, and theories of the different disciplines employed; and so on. They are rarely documented. An independent review of a risk assessment by a multidisciplinary review panel can often be effective in picking up implicit assumptions, because the implicit assumptions of one discipline or person often conflict with those of another discipline or person. All significant assumptions, whether explicit or implicit, need to be conveyed to the risk managers and other users of the assessment.

and knowledge uncertainty and their potential impacts on the risk estimate in every risk characterization, qualitative or quantitative. All assumptions should be acknowledged and made explicit. The impacts of these assumptions on the risk characterization and subsequently the manager's use of the risk assessment in decision making are to be thoroughly discussed. Assessors should describe the strengths and limitations of the assessment along with their impacts on the overall assessment findings. Assessors should also say whether they believe the risk assessment adequately addresses the risk manager's questions.

The International Programme on Chemical Safety (IPCS 2008) has proposed four tiers or levels of uncertainty analysis, which provide a useful way to think about this activity. These are:

- Tier 0: Default assumptions
- Tier 1: Qualitative but systematic identification and characterization of uncertainty
- Tier 2: Quantitative evaluation of uncertainty making use of bounding values, interval analysis, and sensitivity analysis
- Tier 3: Probabilistic assessment with single or multiple outcome distributions reflecting uncertainty and variability

Each of these tiers entails different responsibilities for the risk assessors. In best practice, default assumptions will rarely be used. Uncertainty can be discussed in the absence of quantitative data. It is always possible to tell decision makers what is known with certainty, what we suspect based on incomplete data, and what we assume based on inadequate or missing data. The key is to adopt a systematic approach to communicate the uncertainty. This will help to ensure that the job is done adequately.

Quantitative risk assessments lend themselves to numerical characterizations of the uncertainty attending risks. Deterministic risk characterizations can be supplemented by offering high/low, optimistic/pessimistic, or more formal statistical confidence interval estimates of decision criteria and risks estimated in the assessment. Using interval estimates for uncertain inputs and tracking their impact on critical outputs is a feasible way to identify and communicate what is most important. In probabilistic risk assessment, the challenges of communication are greater, because although there is usually more useful information, it is quantitatively complex and problematic for many risk managers and stakeholders who lack training in interpreting and understanding probabilistic data. Examples of both qualitative and quantitative risk assessment techniques are found in later chapters.

The basic problem in risk assessment is that our data are incomplete and we are uncertain about many things. None of this absolves us of the need to make decisions, however. Risk assessment is a process through which complex, incomplete, uncertain, and often contradictory evidence and scientific information are made useful for decision making (NRC 1983). As a decision-support framework, risk assessment fills the gap between the available evidence and the RMOs being considered to respond to the risks identified. Risk managers must understand those gaps, how they were bridged, and their significance for decision making. It is the assessor's job to explain all of this to them.

4.4.8 Document the Process

Risk assessment is initiated to support decision making that solves problems and realizes opportunities. Substantial resources are dedicated to risk assessment, and it is essential that we carefully and effectively document its findings. It is equally important to document the basis for the risk management actions taken or not taken. Think of documentation as a set of different communications that authenticate and support the results of the risk management activity. The hope is that risk management decisions will be directly linked to the evidence found in the risk assessment documentation.

Assessors are well advised to document the assessment process as it progresses rather than to wait until it has been finished to write up a report. Risk assessment generally progresses in an iterative fashion. Our understanding of problems evolves as the assessment progresses. Analysis is refined as data gaps are filled and models

TELLING YOUR STORY

Storytelling is underestimated as an effective communication skill. Stop listing facts and dumping data into reports and tell a simple story well. We all remember engaging stories from our childhood, and they had three key elements in common:

An engaging beginning ... Once upon a time
An interesting middle ... Consider a talking mirror
A satisfying ending ... They all lived happily ever after

Tell the story by structuring the facts so they have a narrative quality. Let your documentation be a journey with a narrative theme. Good stories are simple; let the facts speak for themselves.

are built. Numerous people will be involved at many points along the way. It is often easier to have assessors document their findings as they go, revising them as new data and analysis warrant.

Documentation need not be restricted to a written report. Nontraditional risk assessment documentation methods might include:

- Interactive Web sites
- Interactive CDs
- Video reports
- Workshops
- Chat rooms
- Wikis
- Discussion groups
- Electronic files
- Training in the use of the risk assessment model
- Live online briefings
- Page limits on written documents

Identify your audience and choose a suitable documentation format.

4.5 Risk Assessment Models

The generic risk assessment activities you've been reading about have been standardized for a variety of applications and communities of practice (COP). The food-safety community, for example, has been aggressive in trying to harmonize risk assessment methods in part to facilitate international trade. They, like other COPs, have promulgated models for use by their constituents. A few of these models are presented in the following discussion to illustrate the diverse range of ways in which these rather generic risk assessment activities are being formalized for specific applications. The details of the model are less important than the overarching point that risk assessors exercise a great deal of latitude in the specific ways they do risk assessments.

The Codex Alimentarius represents the international food-safety community. They employ a familiar risk assessment model, shown in Figure 4.7. Within or alongside of this framework, COPs have developed distinctive models and methodologies for different hazards like food-additive chemicals, pesticides, microbiological hazards, food nutrients, antimicrobial resistance, and genetically modified organisms.

Chemical food additives are evaluated using a safety assessment, which on the surface looks quite different from the generic model of this chapter. To review, its six steps comprise the following: test toxicity, identify a NOAEL, choose a safety factor, calculate the ADI, estimate the EDI, and characterize the risk with the ratio EDI/ADI.

A conceptual application of the model is presented in Figure 4.8. Toxicity studies, most often based on animal data, are used to identify a level of exposure to a chemical, usually measured in a lifetime dose that causes no adverse effects. This is equivalent to a hazard characterization. In our example, this level is 5 mg per kg of body weight daily for a lifetime.

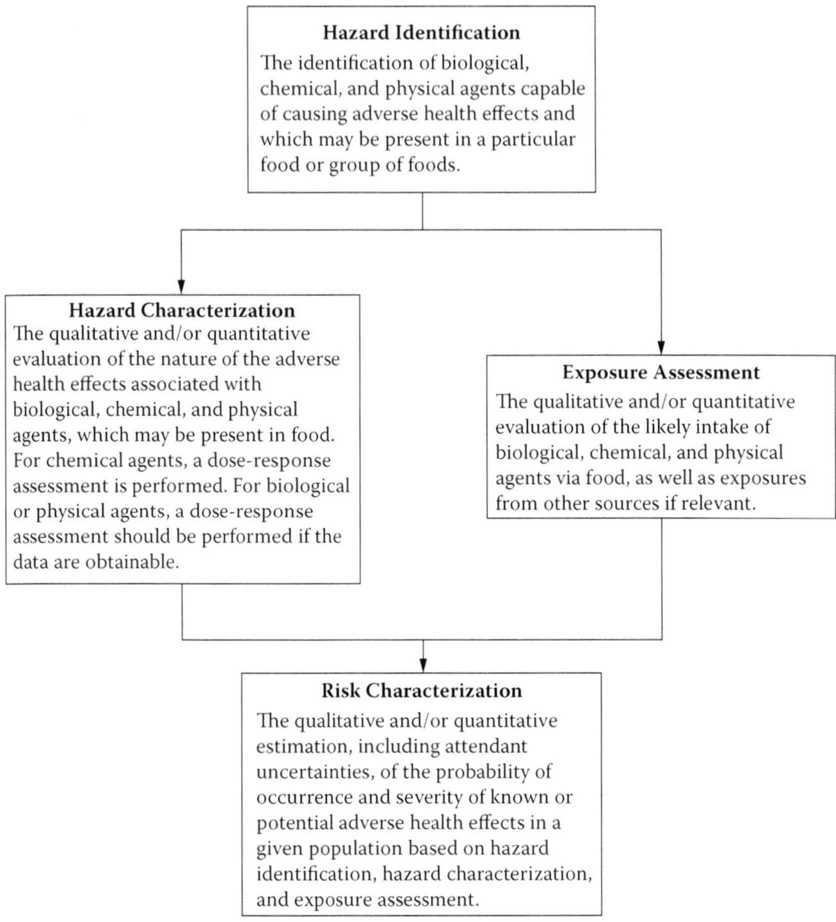

FIGURE 4.7 Generic Codex description of the risk assessment components. (*Source*: FAO 2006.)

To extrapolate from animal studies and their typically high doses to humans and their typically low doses, an uncertainty factor (in our example, 100) is used to identify an ADI. Thus, a NOAEL of 5 mg/kg/day/lifetime divided by 100 yields an ADI of 0.05 mg/kg/day/lifetime.

A survey of consumption behavior yields the daily consumption of the additive for a high-end consumer, say the 90th percentile consumer of this additive, and this is used as the EDI. This constitutes the exposure assessment. The risk characterization is completed by simply comparing the EDI to the ADI. There is no effort to explicitly identify the likelihood of an adverse outcome in this assessment model.

Pesticide chemical risks are somewhat similar, although the language changes a little:

- Identify pesticide residue of interest
- Undertake toxicity studies of substance if needed
- Determine the "no observed adverse effect level" (NOAEL)

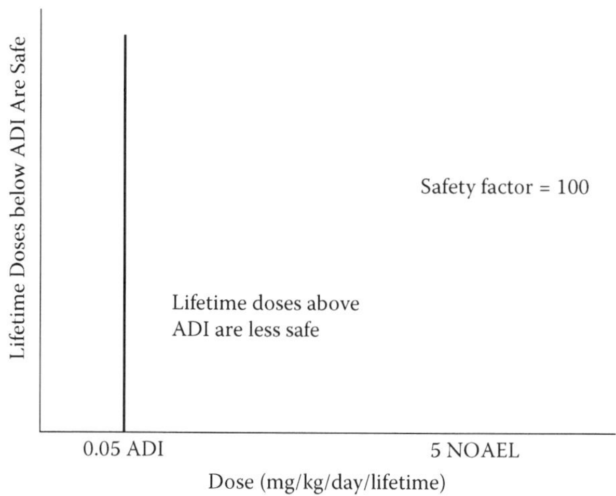

FIGURE 4.8 Representation of the food-additive safety assessment model.

- Select a safety factor or uncertainty factor to extrapolate results from animals to humans
- Calculate the acceptable daily intake (ADI)
- Identify a suitable index of residue levels to predict residue intake—usually the maximum residue limit (MRL)
- Estimate the dietary intake of the residue (exposure assessment)
- Compare exposure to ADI (when exposure exceeds ADI, some sort of risk mitigation is required)

Note that although the language differs in each, they all exhibit elements of the previously described generic process. The hazard is identified; the consequences and likelihoods are assessed; and it is all pulled together in some sort of characterization of the risk.

Antimicrobial-resistant risk assessment is used to evaluate the safety of new animal drugs with respect to concerns for human health. Exposing bacteria in animals to antimicrobial drugs could increase the number of resistant bacteria to the point where it reduces the efficacy of antimicrobial drugs prescribed for human health. This model, shown in Figure 4.9 and taken from FDA Guidance Document 152 (FDA 2003), suggests a qualitative approach for identifying new drugs as potentially high, medium, or low risks for human health.

Food safety is not the only COP to have developed standardized mental models to guide thinking about risk assessment. The U.S. EPA (1998) developed the model shown in Figure 4.10 for ecological risk assessment. Note that it differs from the four-step definition of the NRC Red Book but has clear roots in that model as well. The Red Book model was developed principally for assessing human health risks due to chemicals in the environment. The ecological model expands the notion of hazards to include a broad class of stressors and it includes adverse effects on ecosystems. It has three main steps: problem formulation, analysis, and risk characterization. The

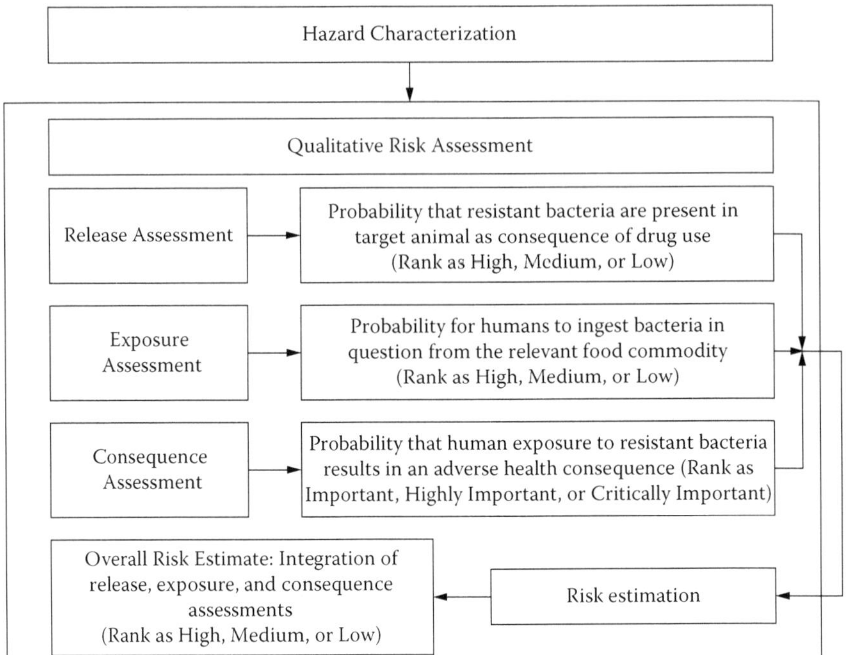

FIGURE 4.9 Components of a qualitative antimicrobial-resistance risk assessment. (*Source*: FDA 2003.)

analysis step is divided into characterizations of exposure and ecological effects, the latter of which is evocative of the hazard characterization step. The point here is that although the models vary in their language and details, they remain firmly committed to the principles articulated in the activities identified earlier in the chapter.

If you search Internet images using the phrase "risk assessment model," you'll see thousands of different models in millions of hits. Many risk assessment problems are so unique that they cannot be usefully fit to any of the existing mental models for risk assessment. It is always wise to familiarize yourself with any standardized assessment models used by your COP. More important, you should always feel free to adapt these models or to develop your own mental model when it suits your decision-making needs to do so. If you flounder at times, keep coming back to the four informal questions: What can go wrong? How can it happen? What are the consequences? How likely is it? Find a way to ask and answer these question and you will be doing risk assessment, formal model or not.

Very often an organization has a model with a well-established structure. Consider the model in Table 4.1, which is used to estimate the costs of a dredging project that includes marsh creation for disposal of the dredged material. It is not difficult to imagine that risk managers may be concerned with the risk of a cost overrun. There is no need here to develop a conceptual model. In this instance, the structure of the model is well established and we need only reach into the risk assessors' toolbox for the appropriate techniques for assessing this risk using our four informal questions. For now it is sufficient to understand that there is a large class of problems that require

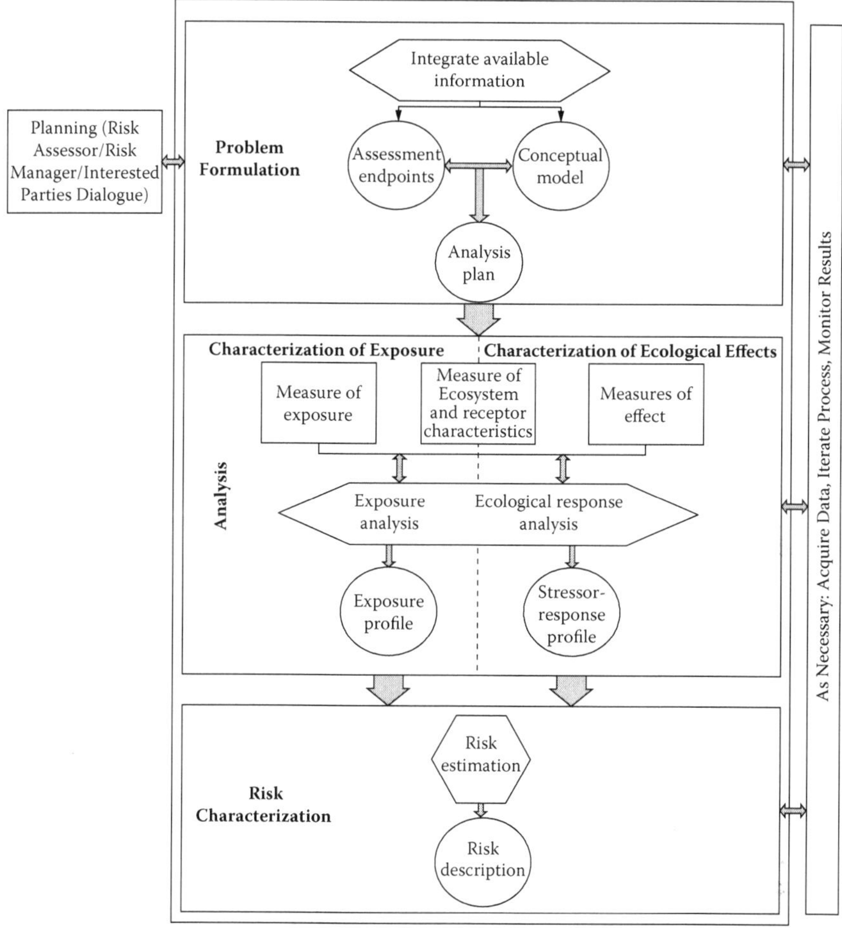

FIGURE 4.10 Ecological risk assessment framework, with an expanded view of each phase. (*Source*: EPA 1998.)

no specific risk assessment model. Often it is sufficient to pay appropriate attention to the uncertainty that has always been present in our work. In many instances, risk assessment can mean doing what you have always done, with the exception of paying close attention to the things you do not know in your work. Using the generic risk assessment activities described here should provide you with a serviceable model when a formal one is not available.

4.6 Risk Assessment Methods

Any self-contained systematic procedure conducted as part of a risk assessment is a risk assessment method (Covello and Merkhofer 1993). These methods are conveniently divided into qualitative and quantitative methods. There has been a

TABLE 4.1

Channel Modification Dredging Cost Estimate

Account Code	Description	Quantity	Unit[a]	Unit Price	Amount
01	Lands and damages	0	LS	...	...
02	Relocations				
	Lower 20 pipeline, 653 + 00	427	LF	$843.66	$359,979
	Remove 8" pipeline, 678 + 00	986	LF	$47.85	$47,197
02	*Total—relocations*				*$407,176*
06	Fish and wildlife facilities (mitigation)				
	Oyster reef creation	0	ACR	...	...
06	*Total—fish and wildlife facilities (mitigation)*				...
12	Navigation, ports, and harbors				
	Mobe and demobe	1	LS	$500,000	$500,000
	Pipeline dredging, Reach 1	576,107.00	CY	$2.43	$1,398,788
	Pipeline dredging, Reach 2	1,161,626.68	CY	$2.76	$3,209,691
	Pipeline dredging, Reach 3A	1,532,227.12	CY	$3.72	$5,693,450
	Pipeline dredging, Reach 3B	708,252.02	CY	$2.89	$2,049,398
	Scour pad, Reach 1	16,484	SY	$16.62	$273,906
	Geotubes, 30', Reach 1	1,345	LF	$221.03	$297,192
	Geotubes, 45', Reach 1	4,601	LF	$291.00	$1,338,995
	Scour pad, Reach 3	39,059	SY	$16.62	$649,029
	Geotubes, 45', Reach 3	13,848	LF	$291.00	$4,029,879
12	*Total—navigation, ports, and harbors Subtotal*				*$19,440,328*
30	Engineering and design	8%			$19,847,504
31	Construction management	6%			$1,587,800
	Total project cost				$1,190,850
					$22,626,155

[a] LF = linear feet; LS = lump sum; CY = cubic yards; SY = square yards, ACR = acres.

misperception on the parts of some and a bias on the parts of others that have suggested that qualitative risk assessment is not a valid form of risk assessment. I think it fair to say that quantitative risk assessment is preferred whenever there are data adequate to support it. It is equally fair to say that qualitative risk assessment is a valid and valuable form of risk assessment. Quantitative assessments use numerical expressions to characterize the risks; qualitative assessments do not. Examples of each are provided in later chapters.

4.6.1 Qualitative Risk Assessment

The fundamental need is to manage risk intentionally and to do that better than has been done in the past. Quantitative risk assessment is not always possible or necessary, so qualitative risk assessment is often a viable and valuable option. It is especially useful:

- for routine noncontroversial tasks
- when consistency and transparency in handling risk are desired
- when theory, data, time, or expertise are limited
- when dealing with broadly defined problems where quantitative risk assessment is impractical

A qualitative risk assessment process compiles, combines, and presents evidence to support a nonnumerical estimate and description of a risk. Numerical data and analysis may be part of the input to a qualitative risk assessment, but they are not part of the risk characterization output. Qualitative assessment produces a descriptive or categorical treatment of risk information. It is a formal, organized, reproducible, and flexible method based on science and sound evidence that produces consistent descriptions of risk that are easy to explain to others. Its value stems from its ability to support risk management decision making. If you can answer the risk manager's questions and describe the risk in a narrative or categorically, then a qualitative assessment is sufficient. Uncertainty in qualitative assessments is generally addressed through descriptive narratives. Specific qualitative risk assessment methodologies can be found in Chapter 9.

4.6.2 Quantitative Risk Assessment

Quantitative risk assessment relies on numerical expressions of risk in the risk characterization. Numerical measures of risk are generally more informative than qualitative estimates. When the data and resources are sufficient, a quantitative assessment is preferred, except where the risk manager's questions can be adequately answered in a narrative or categorical fashion.

Quantitative assessments can be deterministic or probabilistic. The choice depends on the risk manager's questions, available data, the nature of the uncertainties, the skills of the assessors, the effectiveness of outputs in informing and supporting decision makers, and the number and robustness of the assumptions made in the assessment.

Generally, quantitative risk characterizations address risk management questions at a finer level of detail and resolution than a qualitative risk assessment. This greater detail introduces the need for a more sophisticated treatment of the uncertainty in the risk characterization than is found with qualitative assessment.

Being honest information brokers and saying what we know and do not know means that the analysis will be more complex and consequently so will the decision making. More complex risk problems require more complex models. The complexity of the results depends on the information that risk managers need to make a decision and the methods used throughout the assessment.

4.7 Summary and Look Ahead

Risk assessment is where the evidence is gathered. It is the primary place where we separate what we know from what we do not know and then deal intentionally with those things we do not know. It is where we get the right science focused into the analysis and take pains to get that science right.

In general terms, risk assessment is the work you must do to answer four informal questions. What can go wrong? How can it happen? What are the consequences? How likely is it? The four primary steps that comprise a risk assessment and answer these questions are: identify the hazards and opportunities, assess the consequences, assess the likelihood, and characterize the risk. There are any number of application-specific refinements of these notions in widespread use. Recurring kinds of problems lend themselves well to the development of standardized approaches to assessing these risks.

In the best practice of risk assessment, risk managers will identify specific questions they want the risk assessment to answer. Risk assessors answer these questions and characterize the uncertainty in their assessment in ways that support informed decision making. The answers to these questions can be provided in a qualitative or a quantitative manner, depending on the needs of the risk management activity.

Estimating and describing risks in the risk assessment is the critical analytical step in risk analysis. But if we are not able to communicate this often complex information to risk managers, stakeholders, and the public, all will have been for naught. The next chapter addresses the risk communication component of risk analysis. Until relatively recently, risk communication has been treated like the stepchild of risk analysis, too often an afterthought or an add-on. More recently it has begun to receive more of the attention it rightfully deserves.

REFERENCES

Covello, Vincent T., and Miley W. Merkhofer. 1993. *Risk assessment methods: Approaches for assessing health and environmental risks*. New York: Plenum Press.

Covello, Vincent T., and Jeryl Mumpower. 1985. Risk analysis and risk management: An historical perspective. *Risk Analysis* 5 (2): 103–119.

Cox, Louis Anthony, Jr. 2002. *Risk analysis foundations, models, and methods*. Boston: Kluwer Academic.

Environmental Protection Agency. 1998. Guidelines for ecological risk assessment. EPA/630/R-95/002F. *Federal Register* 63 (93): 26846–26924.

———. 2010. Risk assessment portal, basic information. http://www.epa.gov/riskassessment/basicinformation.htm#risk.

Food and Agricultural Organization and World Health Organization. 2004. United Nations. *Codex Alimentarius Commission, procedural manual*. 14th ed. Rome, Italy: FAO.

Food and Agricultural Organization. 2006. United Nations. FAO Food and Nutrition Paper 87. *Food safety risk analysis: A guide for national food safety authorities*. Rome, Italy: FAO.

Food and Drug Administration. 2003. Guidance for industry evaluating the safety of antimicrobial new animal drugs with regard to their microbiological effects on bacteria of human health concern. No. 152. Rockville, MD: Center for Veterinary Medicine. http://www.fda.gov/downloads/AnimalVeterinary/GuidanceComplianceEnforcement/GuidanceforIndustry/UCM052519.pdf.

———. 2005. Quantitative risk assessment on the public health impact of pathogenic *Vibrio parahaemolyticus* in raw oysters. College Park, MD: Center for Food Safety and Applied Nutrition. http://www.fda.gov/Food/ScienceResearch/ResearchAreas/RiskAssessmentSafetyAssessment/ucm050421.htm.

International Programme on Chemical Safety. 2008. *Uncertainty and data quality in exposure assessment, part 1 and part 2*. Geneva, Switzerland: World Health Organization.

National Research Council. 1983. Committee on the Institutional Means for Assessment of Risks to Public Health. *Risk assessment in the federal government: Managing the process*. Washington, DC: National Academies Press.

———. 1994. Committee on Risk Assessment of Hazardous Air Pollutants. *Science and judgment in risk assessment.* Washington DC: National Academies Press.

———. 1996. Committee on Risk Characterization. *Understanding risk: Informing decisions in a democratic society*. Washington, DC: National Academies Press.

Presidential/Congressional Commission on Risk Assessment and Risk Management. 1997. "Framework for environmental health risk management." Final report, vol. 1. Washington, DC: www.riskworld.com. http://riskworld.com/nreports/1997/risk-rpt/pdf/EPAJAN.PDF.

———. 1997. "Framework for environmental health risk management." Final report, vol. 2. Washington, DC: www.riskworld.com. http://riskworld.com/nreports/1997/risk-rpt/volume2/pdf/v2epa.pdf.

5

Risk Communication

5.1 Introduction

Risk communication is one of the three components of risk analysis. It is prominently featured in many risk analysis models. The model presented in this book in Chapter 1 showed it as a great sea of communication in which risk management and risk assessment float. My own experience, however, suggests that while this is the goal, it is not always the reality. In fact, if I may summarize the history of risk analysis in three nonscientific figures, I would choose the "story" told in Figure 5.1.

The ideal is shown in a popular alternative version of the risk analysis model on the bottom left. Here the three tasks are coequal, with modest overlap but functional integrity. Not so terribly long ago, risk analysis might have been described by the figure on the top left. Risk assessment was the tail wagging the dog. Risk management was an afterthought, and risk communication was scarcely on the horizon. In fact, there were many models of risk analysis that did not even mention it! Risk communication was for years the bastard child of risk analysis, seldom talked about and often ignored or treated poorly. Other than a few devoted adherents, it struggled for any recognition at all.

In the recent past, risk management has grown in importance and stature, and the descriptive model might be that shown on the top right of Figure 5.1. While risk management has come of age and now guides risk assessment, risk communication too often remains the weak sister in actual practice. That always comes at a cost. Because of those costs, risk communication is finally coming into its own, and it is now increasingly recognized in models and, more important, by organizations as being at least as important as the assessment and management tasks. Like the other components, its definition is difficult to pin down in words that will satisfy everyone. The term means different things to different people, and a wide variety of definitions can be found in the literature and organizational guidance of different institutions.

Despite the variations in definitions, there is a growing consensus on a set of core principles for risk communication that include the following:

- It is an interactive exchange of information and opinion.
- It takes place throughout the risk analysis process.
- It concerns risk, risk-related factors, and risk perceptions.
- It involves risk assessors and risk managers as well as affected groups and individuals and interested parties.
- It includes an explanation of the risk, possibly an explanation of the risk assessment, and the basis for the risk management decision.

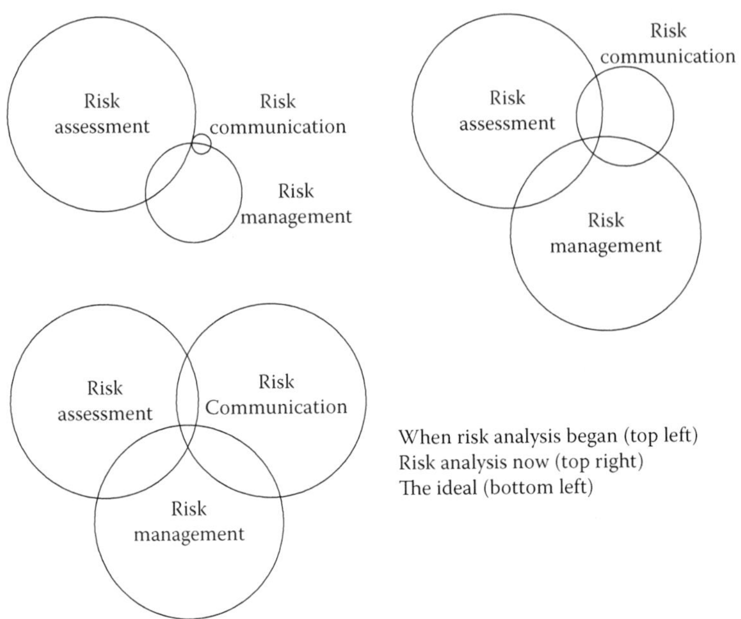

FIGURE 5.1 The role of risk communication in risk analysis.

There are many reasons to communicate about risks, including the goals of achieving a consensus understanding of the magnitude of the risk and developing credible and acceptable risk management responses. Risk communication improves understanding of the risk and risk management options (RMOs). It enhances trust and confidence in the decision-making process and promotes the participation and involvement of interested parties. Done well, it can strengthen working relationships among stakeholders.

Risk communication is needed to explain actions to avoid or take risks, and it is needed to explain the rationale for the chosen RMO. The effectiveness of a specific option needs to be communicated to people so they understand their own risk management responsibilities and know what actions they must take to reduce the risk. The benefits of an RMO as well as the costs of managing the risk and who will bear them include the additional information to be conveyed to interested parties. Risk communication needs to pay special attention to describing the risks that remain after the RMO is implemented. The uncertainty that could affect the magnitude of the risk or the efficacy of the RMO must be carefully communicated to stakeholders and the public. This should include the weaknesses, limitations of, or inaccuracies in the available evidence. It should also include the important assumptions on which risk estimates are based so that stakeholders can understand the sensitivity of both risk estimates and the efficacy of an RMO to changes in those assumptions and how those changes can affect risk management decisions.

Risk communication is not about everyone coming to consensus or an agreement. Neither is it intended to get everybody on the "same page." It is, however, about providing people with meaningful opportunities for input before decisions are made and for feedback as evidence is accumulated. It is about listening to and understanding people's concerns so they can be considered in decision making and so the public will respect the process even if they disagree with some of its decisions and outcomes.

Risk communication theory and practice are well documented in a very rich risk communications literature. (For example, see the works by Chess, Covello, Fischhoff, Hance, Johnson, Krimsky, Sandman, Slovic, and others presented in the references at the end of this chapter.) How one frames the risk communication component for the purposes of risk analysis is of some importance because the scope and role of risk communication is rapidly advancing. To some, the risk communication component is relatively narrow and focuses on risk and crisis communication. I think this is far too narrow a definition, even while recognizing its adequacy for many situations. A more proactive expanded view of risk communication is provided in Figure 5.2.

Risk communication can first be split into internal and external tasks. Internal risk communication takes place within the risk analysis team. It begins with the coordination between the assessors and managers that has been described in the previous two chapters. This aspect of risk communication is arguably little different from the kind of good organizational communication that is part of any effective organizational management philosophy. I make a distinction because it involves communicating about uncertainty, and talking about what we don't know is not something many organizations do well. The coordination is ongoing throughout a risk management activity.

Developing and conducting an effective risk communication process is not something that happens by accident. Good risk communication cannot be an afterthought or an add-on. Neither is it a hypodermic needle injection into the activity at prescribed

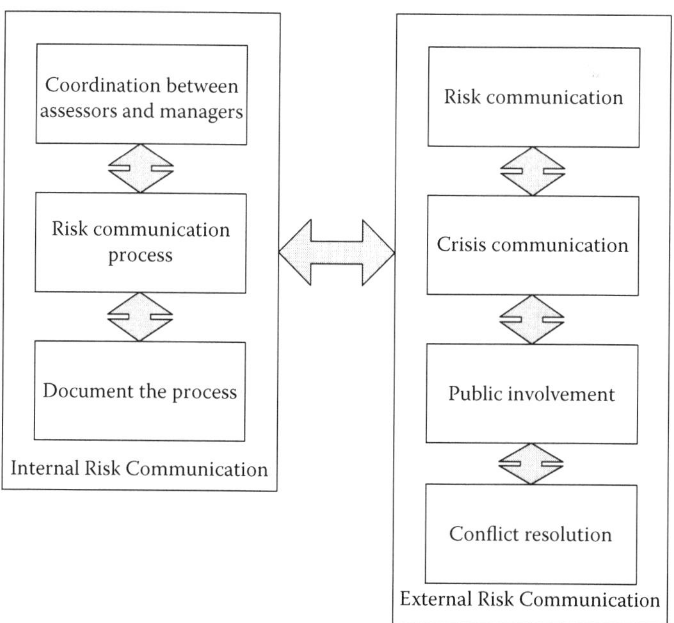

FIGURE 5.2 Components of the internal and external risk communication tasks.

BEST PRACTICES FOR RISK COMMUNICATION

1. Infuse risk communication into policy decisions.
2. Treat risk communication as a process.
3. Account for the uncertainty inherent in risk.
4. Design risk messages to be culturally sensitive.
5. Acknowledge diverse levels of risk tolerance.
6. Involve the public in dialogue about risk.
7. Present risk messages with honesty.
8. Meet risk perception needs by remaining open and accessible to the public.
9. Collaborate and coordinate about risk with credible information sources.

***Source: Effective Risk Communication: A Message-Centered Approach,* by Timothy L. Sellnow et al. (2009).**

or periodic points in time. It needs to be a dynamic ongoing process. To be so, it must be designed.

Documenting the process is usually described (including in this book) as a management or assessment process, and it is. There is little that is more fundamental to the internal risk communication process, however, than documenting the results of the activity and the decisions made from them. These internal risk communication tasks are often given little attention in the literature, which tends to favor the external risk communication processes shown on the right side of Figure 5.2.

Most texts on risk communication focus on risk and crisis communication, as will this chapter. Nonetheless, my work with risk analysis organizations around the world suggests that there is a growing recognition of the need to expand the public's role in the risk analysis process. It makes sense to me to expand the definition of risk communication activities to include public-involvement and, in some cases, conflict-resolution activities. I am not sure there is anything close to a consensus on that idea just yet, however! Conflict resolution methods are sufficiently specialized and distinct from the other risk communication topics as to be largely ignored in this chapter. Their relevance is, however, recognized. Public involvement, risk, and crisis communication are sufficiently interconnected as to be considered in this chapter, although the discussion of public involvement is rather brief. I think this treatment may be a fair and honest compromise between where I think risk communication is or ought to be headed and how the real risk communication experts might describe the task.

The chapter begins with definitions. It then briefly considers the three internal risk communication tasks of Figure 5.2. We move into the area of external risk communications and begin a discussion of risk and crisis communication by considering the important distinction between the hazard and outrage dimensions of risk, which interact to define very distinct risk communication strategies. Risk perceptions, the next topic, are important for understanding the disconnect between producers and consumers of risk information.

Just as strategies for risk communication vary, so do the audiences for these strategies. The next discussion covers the importance of knowing the audiences for your communications. Psychographic information about these audiences is then added to the discussion. A basic communication model is presented to lay the framework for stressing the role of stress in risk communication before the three M's of risk communication are introduced. Some of the critical differences between crisis and risk communication are considered before the chapter turns to the challenge of explaining risk to nonexperts. This leads into a discussion about explaining uncertainty, a task as critical to risk assessors and risk managers as it is to risk communicators. The chapter ends with a short consideration of public involvement.

To prepare you for what follows I offer two key words. The keyword for the internal risk communication task is *uncertainty*. The keyword for the external risk communication task is *emotion*. Get a handle on how to communicate uncertainty to people who are feeling strong emotions and the world of risk communication will be your oyster.

5.2 Definitions

No one formal definition of risk communication will meet the needs of all practitioners of risk communication. The Codex Alimentarius offers a definition quite close to the consensual core offered above. It says risk communication is: "The interactive exchange of information and opinions throughout the risk analysis process concerning risk, risk-related factors and risk perceptions, among risk assessors, risk managers, consumers, industry, the academic community and other interested parties, including the explanation of risk assessment findings and the basis of risk management decisions."

A shorter and simpler definition from the USDA is also useful. Risk communication is "an open, two-way exchange of information and opinion about risk leading to better understanding and better risk management decisions" (University of Minnesota 2006).

An informal definition is implicitly offered by the ten questions shown in the text box. Risk communication is the work you have to do to answer those ten questions.

COMMUNICATING WITH THE PUBLIC: TEN QUESTIONS TO ASK

1. Why are we communicating?
2. Who is our audience?
3. What do our audiences want to know?
4. What do we want to get across?
5. How will we communicate?
6. How will we listen?
7. How will we respond?
8. Who will carry out the plans? When?
9. What problems or barriers have we planned for?
10. Have we succeeded?

***Source*: Chess and Hance (1994).**

Risk communication can be divided into two distinct tasks. First, there is the internal risk communication between managers and assessors that is absolutely critical to a successful risk analysis. The importance of this task is often overlooked in the literature. Second is the external risk communication task by which the risk analysis team communicates with the world around them.

5.3 Internal Risk Communication

5.3.1 Coordination between Assessors and Managers

The internal risk communication task is essentially to ensure effective interaction between managers and assessors. Three rules of thumb are suggested for this task for the managers and assessors:

- Collaborate early
- Coordinate often
- Cooperate always

Early risk analysis experience and, subsequently, models showed the need to separate the roles of managers and assessors. In their zeal to ensure the integrity of the science-based foundation of risk analysis, some early practitioners were somewhat manic about this separation and stretched it almost to the point of no contact. That is most emphatically not best practice. Although the integrity of the science needs to be relentlessly protected, managers and assessors need to interact constantly throughout the risk analysis process despite the fact that they have very clear and different individual responsibilities.

Referring to the risk management activities described in Chapter 3, the extent of the interaction between risk assessors and risk managers is suggested below using a scale of no, minimal, moderate, and maximum interaction based on the author's experience.

1. Problem identification
    a. Problem recognition: *minimal interaction*
    b. Problem acceptance: *no interaction*
    c. Problem definition: *maximum interaction*
2. Risk estimation
    a. Establish risk analysis process: *minimal interaction*
    b. Develop a risk profile: *maximum interaction*
    c. Establish risk management objectives: *maximum interaction*
    d. Decide on the need for risk assessment: *no interaction*
    e. Request information: *moderate interaction*
    f. Initiate risk assessment: *moderate interaction*
    g. Coordinate the conduct of the assessment: *moderate interaction*
    h. Consider the results of the assessment: *maximum interaction*
3. Risk evaluation
    a. Is the risk acceptable?: *no interaction*

   b. Establish tolerable level of risk: *moderate interaction*
   c. Risk management strategies: *maximum interaction*
4. Risk control
   a. Formulating risk management options (RMOs): *moderate interaction*
   b. Evaluating RMOs: *moderate interaction*
   c. Comparing RMOs: *no interaction*
   d. Making a decision: *no interaction*
   e. Identifying decision outcomes: *moderate interaction*
   f. Implementing the decision: *no interaction*
5. Monitoring
   a. Monitor: *moderate interaction*
   b. Evaluate: *moderate interaction*
   c. Modify: *moderate interaction*

Turning to the risk assessment activities of Chapter 4 and using the same subjective scale, the author's judgments are:

1. Understand the question(s): *maximum interaction*
2. Identify the source of the risk: *no interaction*
3. Consequence assessment: *no interaction*
4. Likelihood assessment: *no interaction*
5. Risk characterization: *no interaction*
6. Assess effectiveness of RMOs: *moderate interaction*
7. Communicate uncertainty: *maximum interaction*
8. Document the process: *moderate interaction*

Some of the most critical points of interaction occur at the beginning and end of the risk management activity. Identifying problems, objectives, and the initial list of questions together are essential early interactions, as is preparing a risk profile. These tasks will also provide stakeholders with opportunities for input and feedback for these three important tasks in best-practice risk communication.

Revising the risk assessment questions together is an especially important interaction. This clarifies the information needs of the risk managers and is an essential step in managing the expectations of both managers and assessors. Interaction is clearly needed to set reasonable assessment schedules, budgets, and milestones together. The two parties are to do their own jobs, but they should brief each other often. After the risk assessment is completed, understanding the results of the assessment and significant uncertainties together are critical interactions. Interaction continues, but the manager's role increases relative to the assessor's after the assessment is completed. The risk manager's role is preeminent when asking, "Is the current level of risk acceptable" or "What level of risk is tolerable?"

Managers and assessors should formulate risk management options together. They also need to coordinate the evaluation of options together to ensure that managers have the information they need to decide which measures are best. This may happen before, during, or after the risk assessment. Assessors are responsible for the analytical work in the evaluation of RMOs, while risk managers do the deliberating in these tasks.

To a great extent, this internal communication task is just good organizational management. It is not unique to risk analysis. What is somewhat unique in risk analysis is the role of communicating effectively about those things that are uncertain and potentially important for decision making.

5.3.2 Risk Communication Process

Designing the external risk communication process, outlined in the remainder of this chapter, is a critical part of the risk analysis team's internal communications task. Creighton's (2005) book on public participation provides an excellent blueprint for those looking to provide a larger and more active role for the public. A narrower and more traditional risk communication program can be designed following the templates in risk communication handbooks like those of Lundgren and McMakin (2009) and Heath and O'Hair (2009).

5.3.3 Documenting the Process

Telling the story of the risk management process and the risk assessment is also an important part of the internal risk communication task. The decision process must be carefully documented to provide a defensible rationale for actions taken or not taken as a result of the risk analysis process. Risk managers, ideally with the assistance of risk communication experts, should carefully plan the documentation of the process. This topic, introduced in Chapter 3, is taken up later in this chapter. The interested reader can also see Yoe and Orth (1996) for more ideas about how to tell your story effectively.

5.4 External Risk Communication

The external communication tasks generally describe how the risk analysis team (managers, assessors, and communicators) interact with their various publics and external stakeholders. These interactions can overlap with the internal tasks, as may be the case for identifying problems and objectives as well as preparing the risk profile and the initial list of questions, when external input is likely to be important. The extent to which this may happen will depend on how involved the public is in the risk management activity. Four broad tasks have been identified as part of the external risk communication process. These are:

- Risk communication
- Crisis communication
- Public involvement
- Conflict resolution

GOALS OF RISK COMMUNICATION

1. Promote awareness and understanding of the specific issues under consideration during the risk analysis process, by all participants.
2. Promote consistency and transparency in arriving at and implementing risk management decisions.
3. Provide a sound basis for understanding the risk management decisions proposed or implemented.
4. Improve the overall effectiveness and efficiency of the risk analysis process.
5. Contribute to the development and delivery of effective information and education programs, when they are selected as risk management options.
6. Foster public trust and confidence in the safety of the food supply.
7. Strengthen the working relationships and mutual respect among all participants.
8. Promote the appropriate involvement of all interested parties in the risk communication process.
9. Exchange information on the knowledge, attitudes, values, practices, and perceptions of interested parties concerning risks associated with food and related topics.

***Source*: United Nations (1998).**

An external communication program will not always require all four elements. For the purposes of the current discussion we'll focus more narrowly on a more traditional risk communication process and will return to considerations of public involvement at the end of the chapter.

The external risk communication process can have many different goals. Three reasonably common, if not universal, generic goals (Food Insight, 2010)* are:

1. Tailor communication so it takes into account the emotional response to an event.
2. Empower the audience to make informed decisions.
3. Prevent negative behavior and encourage constructive responses to crisis or danger.

Unlike the basic unidirectional, "We tell them what we did," communication model, risk communication is two-way (listening and speaking) and multidirectional. It uses multiple sources of communication, and it actively involves the audience as an information source. Risk analysts can learn from individuals, communities, and organizations.

The desired outcomes of effective risk communication will vary from problem to problem, but there are some generic outcomes that recur with regularity. First, it can

* I would like to acknowledge the Food Insight materials, sponsored by the International Food Information Council Foundation, as a major source of information for much of the discussion in Section 5.4 of this chapter.

decrease deaths, illness, injury, and other adverse consequences of risks by informing people and changing behaviors. Alternatively, it can increase the positive outcomes of opportunities. It fosters informed decision making concerning risk and empowers people through useful and timely information to make their own informed decisions. It prevents the misallocation and wasting of resources and keeps decision makers well informed. Good risk communication builds support for risk management options and can aid the successful implementation of an RMO. It also can counter or correct rumors.

Risk communication is not spinning a situation to control the public's reaction, nor is it public relations or damage control. It is more than how to write a press release or how to give a media interview. It is not always intended to make people "feel better" or to reduce their fear. It is multidirectional communication among communicators, publics, and stakeholders that considers human perceptions of risk as well as the science-based assessment of risk. It includes activities before, during, and after an event. It is during these activities that risk communication can broaden to include public involvement. Risk communication is an integral part of an emergency response plan. Aware of the many dimensions of risk communication, let's drill down a little deeper to understand it better.

5.4.1 Risk and Crisis Communication

This section covers the first two elements of the external risk communication task. The discussion begins with some important background information that addresses the dimensions and perceptions of risk before it turns to the importance of knowing and engaging one's audiences. The value of psychographic information is considered as a lead-in to the consideration of risk, stress, and the communication model. The three M's of risk communication are then discussed. Risk and crisis communication are juxtaposed for distinction, and then the discussion turns to the challenges of explaining risk and uncertainty to nonexperts. Risk comparisons are the final topic of discussion.

5.4.1.1 Risk Dimensions

Virtually everything we do involves risk, and zero risk is unachievable. Risk communication is complicated by the fact that people interpret risk in very different ways,

HAZARD AND OUTRAGE

Let's divide the "risk" people are worried about into two components. The technical side of the risk focuses on the magnitude and probability of undesirable outcomes: an increase in the cancer rate, a catastrophic accident, dead fish in the river, or a decline in property values. Call all this "hazard."

The nontechnical side of the risk focuses on everything negative about the situation itself (as opposed to those outcomes). Is it voluntary or coerced, familiar or exotic, dreaded or not dreaded? Are you trustworthy or untrustworthy, responsive or unresponsive? Call all this "outrage."

***Source*: Sandman (1999).**

especially experts and the public. Risk involves both facts and feelings, and these competing dimensions of risk—the objective vs. the subjective—give rise to some unique communication challenges.

Peter Sandman describes these two elements of risk as hazard and outrage (see text box). These two elements shape the perceptions of risk. In general, hazard (the something that can go wrong, the likelihood of it happening, and its factual consequences) is what the assessors and scientists are primarily concerned with. Experts think about these hazards, and they know things that others do not. Predictive microbiologists know the conditions under which a pathogen may grow or die off. Engineers understand the hydrographs of rivers. Financial advisers understand the subtle details of their derivatives. Toxicologists know how much of a chemical is toxic. Most of the rest of us do not.

The public is concerned less with the science, numbers, and facts of the risk and more with the personal and social context of the risk. The public feels things about the risks, and they believe things to be true or not, often without respect to the facts of a situation. The public is less concerned with the details of the probabilities than they are with a subjective evaluation of the relative importance of what might be lost. They do not care about pathogen growth; they care that their daughter got sick. They could care less about the hydrograph; they do care that their first floor was damaged by the flood. The details of the derivatives are of less interest than the college fund that was lost.

These two distinct dimensions of a risk can lead to a disconnect between the scientist/risk professional and the public. Scientists tend to focus on what they know and think, while the public focuses on what they feel and believe. Both dimensions of a risk are important, but for different reasons. They are very different aspects of a risk. Sometimes the public worries when perhaps the scientists would say they shouldn't, e.g., about irradiated foods. Other times they may not worry about things scientists think they should, e.g., the oncoming hurricane.

Risk communication that is based wholly on explaining the facts of the risk may well miss the greater concerns of the public, which tend to be the social and personal meaning of the risk. There is a whole lot more to risk communication than explaining the results of your risk assessment. This disconnect between producers (scientists) and consumers of risk information gives rise to four kinds of risk communication strategies (adapted from Sandman and Lanard 2003), as shown in Figure 5.3.

When people are outraged but the actual hazard is low, the appropriate communication strategy is outrage management. Its goal is to reduce outrage so people don't take unnecessary precautions. The opposite situation—when the danger is high but the public is not very concerned—is precaution advocacy. Its goal is to increase concern for a real hazard in order to motivate people to take preventive action.

When both the outrage and danger are high, crisis communication is in order. It acknowledges the hazard, validates the concerns of the public, and gives people effective ways to act to manage their risk. Situations of low hazard and low outrage are well served by ordinary public relations communications. These communications are brief messages that reinforce whatever appeals are most likely to predispose the audience toward your goals.

These different situations and their associated strategies give rise to an obvious need to understand the perceptions of the public. Risk professionals who do not realize that the public perceives risks differently than they do are in danger of choosing an ineffective risk communication strategy.

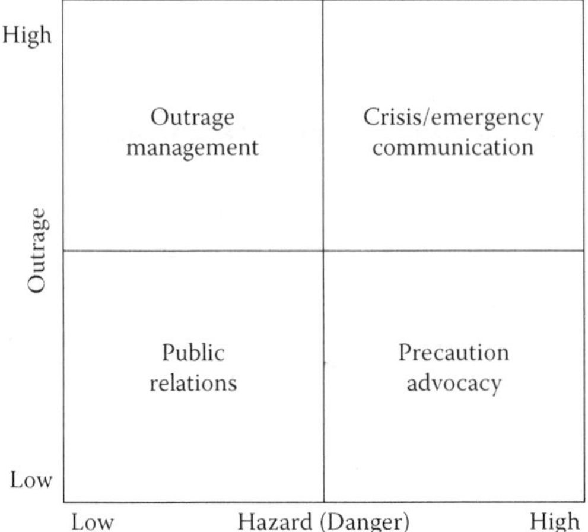

FIGURE 5.3 Four different types of risk communication strategy. (*Source*: Sandman and Lanard 2003.)

5.4.1.2 Risk Perceptions

What scares people the most? What kinds of conditions increase stress and anxiety? The public takes a rather complex array of factors into account when they form their perception of a risk. The correlation between hazard or actual danger and the public's outrage is not always as high as we might like. Psychometric research has done a great deal to help explain this disconnect.

Slovic in the 1980s published seminal research in the perception of risk that helped explain people's extreme aversion to some risks and their indifference to others. Fifteen risk characteristics were identified, but many of them proved to be highly correlated to one another, so they were consolidated via factor analysis into two factors called "dread risk" and "unknown risk" (Slovic, Fischhoff, Lichenstein 1980; Slovic 1987). As the names suggest, characteristics describing the extent to which the consequences of a risk are dreadful comprise one factor, while characteristics capturing the unknown nature of a risk comprise the other.

Slovic's 1987 research showed that the dread and unknown factors increased or decreased for risks with the characteristics in Table 5.1. Furthermore, research at the time suggested that risks with a high unknown factor were perceived as riskier. Figure 5.4 shows a mapping of the cognitive perceptions of Slovic's subjects. Though it is not a universal mapping, it does provide some insight into the nature of risk perception when combined with the information in Table 5.1.

Since that groundbreaking research, a number of other outrage factors have been found to affect both the perception and acceptability of risks. Some of them are effects on children, the manifestation of effects, trust in institutions, media attention, accident history, benefits associated with the risk, reversibility of effects, origin (natural risks are more acceptable than human-made risks), memorability, moral relevance, and the responsiveness of the risk management process. The riskier a situation "feels" based on these kinds of characteristics, the less acceptable or the more unacceptable it is in people's perceptions.

TABLE 5.1

Factors that Increase or Decrease Dread and Unknown Aspects of Risk Consequences

Increases Dread	Decreases Dread
Uncontrollable	Controllable
Dread	No dread
Global catastrophic	Not global catastrophic
Fatal consequences	Nonfatal consequences
Not equitable	Equitable
Catastrophic	Individual
High risk to future generations	Low risk to future generations
Not easily reduced	Easily reduced
Risk increasing	Risk decreasing
Involuntary	Voluntary
Increases Unknown	**Decreases Unknown**
Not observable	Observable
Unknown to those exposed	Known to those exposed
Delayed effect	Immediate effect
New risk	Old risk
Risk unknown to science	Risk known to science

Source: Slovic (1987).

5.4.1.3 Know and Engage Your Audience

"Audience" is a tricky word; it suggests a one-way communication in its common usage. Here it is used to identify a particular group of the public that will be the target of a risk communication activity. I will use "public" to mean the collection of all audiences.

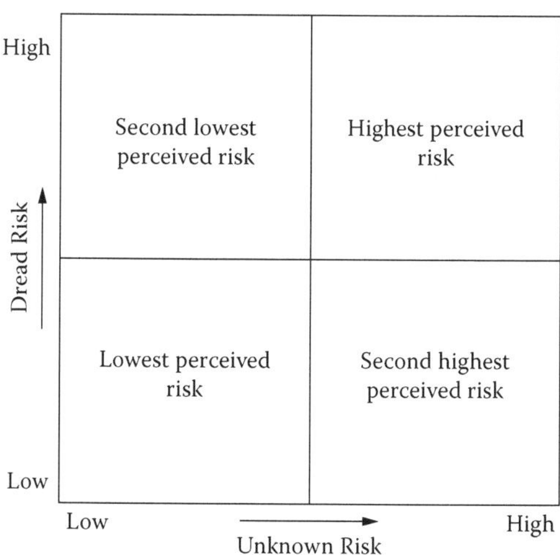

FIGURE 5.4 The effect of dread and the unknown on risk perceptions.

There are many different kinds of audiences, and some will be more difficult to persuade or communicate with than others. Take care to avoid the mistake of thinking you have one monolithic audience. Government, industry, academia and research institutions, media, consumers and consumer organizations, and the general public might comprise your audiences. Within the general public are many audience subpopulations that vary based on such things as family situations, locations, education, professions, physical differences, cultural differences, generational differences, language differences, social status, past experience with the risk, prior knowledge of the topic, attitudes toward the responsible organization, and so on. Covello and Cohrssen (1989) offered seven rules for engaging the audience that have stood the test of time. They are summarized here to help guide your risk communication efforts.

5.4.1.3.1 Rule 1: Accept and Involve the Public as a Legitimate Partner

People expect the opportunity to participate in decisions that affect their lives in a democracy. Risk communicators must demonstrate respect for the public; they are going to hold you accountable. Do not attempt to diffuse the public's concern or to preempt any action they may be inclined to take. Instead, aim to develop an involved, interested, reasonable, thoughtful, solution-oriented, and collaborative public-involvement program. Involve people early and in meaningful ways. For example, there may be an important role for the public when preparing the problems-and-opportunities statement, the objectives-and-constraints statement, and the list of questions risk assessment is to answer. If so, that involvement must be planned from the beginning of the risk management activity. This is work that would be done as part of the second task in the internal risk communication process (see Figure 5.2).

5.4.1.3.2 Rule 2: Plan Carefully and Evaluate Your Efforts

You need different communication strategies for different audiences and situations; these strategies must be carefully planned. Begin your communication planning with clear and explicit objectives. Be sure to evaluate your information ahead of time; know its weaknesses as well as its strengths.

Use an effective spokesperson with good presentation and interaction skills. Prepare two or three talking points, word them simply, and learn them cold. Pretest your message whenever it is possible to do so. Pretest with typical people, not activists or community leaders. Always pretest your message before going on television. Then carefully evaluate your efforts and learn from your mistakes.

5.4.1.3.3 Rule 3: Listen to Your Audience

Listen to the audience if you expect them to listen to you. Identify their concerns. They are often more concerned with fairness, trust, credibility, competence, control, caring, and voluntariness than they are with the details of your risk assessment.

Don't assume what people know, think, or want done about the risks. Find out what they are thinking. Never walk into a meeting with no preparation. Use interviews and focus groups to learn. Arrive early to meetings and mingle to find out what people are thinking. Let all interested people be heard. People come to the table with prior life experience, beliefs, personal knowledge, and values. They can be a valuable source of information, especially about social values. Recognize the public's emotions and let them know you have heard them and understand their concerns.

5.4.1.3.4 Rule 4: Be Honest, Frank, and Open

Trust and credibility are your most important assets. If you lose them, they are difficult to regain. State your credentials, but do not expect them to validate you. If you do not know the answer to a question, do not fake it. Admit you do not know and get back to them with an answer.

Give people risk information as soon as possible. Do not speculate about or distort the level of risk. Admit mistakes when you make them. Be sure to discuss uncertainties and the strengths and weaknesses of your data. If you must err, err on the side of sharing too much information rather than too little.

5.4.1.3.5 Rule 5: Coordinate and Collaborate with Other Credible Sources

Avoid conflicts with other credible sources of information. Allies can make the risk communication task easier. Develop relationships with other sources of risk information, preferably in advance of a crisis. Coordinate your messages so the public hears a consistent interpretation of the situation. Determine who is best able to answer questions about risk and let them speak.

Never be blindsided by new information. Monitor the public media on your issue as well as your technical sources. Avoid public disagreements, but acknowledge uncertainty when it leads to different interpretations. If others do not coordinate their message with yours, don't argue; be respectful to the other party, but state your position clearly as well as your reasons for it.

5.4.1.3.6 Rule 6: Meet the Needs of the Media

The media are a major channel for disseminating risk information. They are essential to your ability to tell your story and to get information out to your audiences. The media are usually not out to get you; they are out to get a story, so don't be the story. Be accessible to the media and understand their needs for simplicity, conflict, and a "hook" for stories. Supply a hook the media can use. Prepare media materials in advance and tailor them to the specific type of media you use. They should be sufficient for a reporter to tell the whole story in print, video, or audio.

It is wise to establish long-term relationships with media representatives well in advance of a crisis. If a reporter uses you as a reliable source when they need one, they're more likely to come to you when you need to get word out.

5.4.1.3.7 Rule 7: Speak Clearly and with Compassion

Risk assessment is science based, but communication is not. Avoid technical language, jargon, and acronyms. Use simple, nontechnical language and be sensitive to local norms and expectations about speech and dress. Use concrete, relevant, and simple examples. Vivid metaphors and effective risk comparisons can help to put risks in perspective. People respond better to stories than to theories or a recitation of facts. Tell stories, but be consistent with your message.

Be sure to respond to emotions that people express, e.g., fear, anger, helplessness, outrage. When responding to emotional outbursts and histrionics, never cut someone off. Speak with them gently. Convey empathy for the person's response while at the

SOURCES OF STAKEHOLDER ANGER

Fear
Threat to self
Threat to family
Frustration
Feeling powerless
Feeling disrespected
Feeling ignored

Source: **Risk Analysis 101 USDA, APHIS, PPQ.**

same time expressing skepticism over inaccurate things that may have been said. This is not the time to challenge core community attitudes and beliefs!

Never restate a problem in objective terms without the emotional content. Regain control of the discussion by restating the concerns expressed. Watch your body language; it is the greater part of communication.

5.4.1.4 Psychographic Information

Psychographics is the use of demographics to study and measure attitudes, values, lifestyles, interests, beliefs, and opinions, usually for marketing purposes. Psychographics can also help you deal with your different audiences and to construct messages. For example, some psychographic measures with significance for risk communication are self-esteem, involvement, anxiety, fear, and trust.

Self-esteem embodies our feelings of self-worth and the effectiveness of our own actions. Risks that deal with our health or well-being will be perceived through the lens of our self-esteem. Self-esteem, through its self-efficacy dimension, can affect our perceptions of risk management options like weight loss, exercise, and changes in personal behavior. Groups or individuals with low self-esteem present unique risk communication challenges.

The theory of vested interest (Crano and Burgoon 2001) identifies four levels of involvement that reflect the degree of concern an audience has regarding a risk. These are:

- Value-relevant involvement
- Outcome-relevant involvement
- Impression-relevant involvement
- Ego-relevant involvement

When people are involved in an issue because it is relevant to their value system, they can be hard to persuade if the situation challenges their values, especially highly engrained ones. To succeed in communication with these groups, your message must reflect their values.

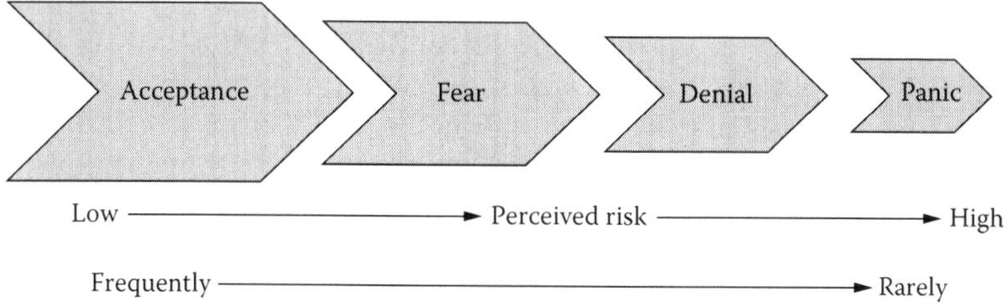

FIGURE 5.5 Progression of our reactions to a perceived risk. (*Source*: Sandman and Lanard 2003, 2005.)

Those involved in an issue because of the personal consequences (i.e., outcome) of the issue can be persuaded if they believe what you propose is in their best interest. The key of the risk communication message is to show them how the topic affects their personal interest and is in their best interest.

Impression-relevant involvement stems from behaviors that serve to create or maintain a specific image of the individual. This self-image tends to inhibit change in general. Effective risk communication must ensure the audience that the actions you want them to take are not silly and that people will not think less of them because you advocated them.

When a person's involvement is motivated by ego, that person can be difficult to persuade. Messages that threaten the ego evoke defensive reactions, and defensiveness causes people to disparage the source of the message. Consequently, it is important to avoid insulting people.

When the risk is perceived as high and efficacy is perceived as low, anxiety results. The good news is that anxious people are motivated to seek information. The bad news is that anxiety interferes with our ability to process information.

Fear and trust are adaptive survival processes. Fear rises rapidly and is slow to cease. It is easily reestablished. Trust, on the other hand, is slowly acquired and easily destroyed. Once destroyed, it is very difficult to reestablish.

Sandman and Lanard (2003, 2005) suggest that reactions change with the perception of risk, as shown in Figure 5.5, and that humans usually adapt well to risk. As the perception of the risk increases, our reactions progress from acceptance through fear, denial, and finally panic. Panic, however, is a rare response.

Fear is an adjustment reaction that is natural in a crisis. Sandman and Lanard (2005) say that fear:

- Is automatic
- Comes early
- Is temporary
- Is a small overreaction
- May need guidance
- Serves as a rehearsal
- Reduces later overreaction

Smart risk communicators know this and encourage, legitimize, ally with, and guide these adjustment reactions.

Overreacting to a risk is a natural first reaction when it is new and potentially serious. Typically, we will pause, become hypervigilant, personalize the risk, and take extra precautions that are at worst unnecessary and at best premature.

When this fear grows it leads to denial, which is less common than fear but more dangerous because it keeps people from taking precautions. Risk communication can reduce denial by legitimizing the fear, taking action by doing something, and empowering people to decide how to respond by providing them with a range of actions they can take.

Panic is a sudden strong feeling of fear that prevents us from reasonable thought or action. Panicky feelings are not unusual, but actual panic is quite rare. We often worry that providing people with unfavorable information or that presenting them with a dire scenario will result in "panic." This can lead communicators to withhold information or to overassure people. The orderly evacuation of the World Trade Center Towers on September 11, 2001, and the January 15, 2009 emergency ditching of a jetliner in the Hudson River, provide vivid examples of how rare panic really is. Most people can cope with and manage their fear. Risk communicators can help mitigate the fear and anxiety by empowering people with information that builds self-efficacy—"This is what you can do..."—and that assures them that response will work. Fearful people need information they can process easily; that means nothing complicated. Keep the message sensitive and simple. Give anxious people specific instructions. Repeat the message as often as possible.

5.4.1.5 Risk, Stress, and the Communication Model

Communicating to an emotional and possibly untrusting or, worse, distrustful audience is one of the most difficult tasks you may ever face. Do it well and life may not be simple, but it will be a whole lot easier than if you do it poorly. The basic communication model includes the following components:

- Sender: communicator
- Receiver: public, partners, stakeholders
- Channel: medium used to convey information
- Message: content presented
- Feedback: receiver's response message
- Noise: barriers that may interfere with reception (physical, receiver's stress level)
- Environment: time and place

During normal risk communications situations when stress is low, trust in the communicator is based on that person's level of competence and expertise. Covello's research (2002) indicates that as much as 85% of trust may be based on these credentials (see Figure 5.6).

When risks are perceived as high or in crisis situations, communication takes place in high-stress circumstances. Trust factors change rather dramatically during times of high stress, as seen in Figure 5.7. Competence and expertise become far less important, while listening, caring, and empathy become the primary factors for establishing trust. Honesty and openness also appear as important factors. These trust factors are

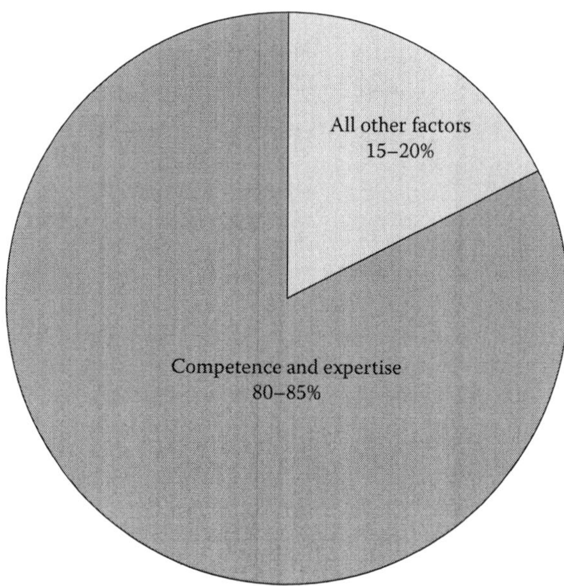

FIGURE 5.6 Trust factors in low-stress situations. (*Source*: Covello 2002.)

routinely assessed within the first 30 seconds of communication. There is no second chance during stressful circumstances.

The basic communication model changes during high-stress conditions, as summarized in Table 5.2. The effectiveness of the sender now depends on credibility and trust. The receiver's ability to process complex information is reduced, so

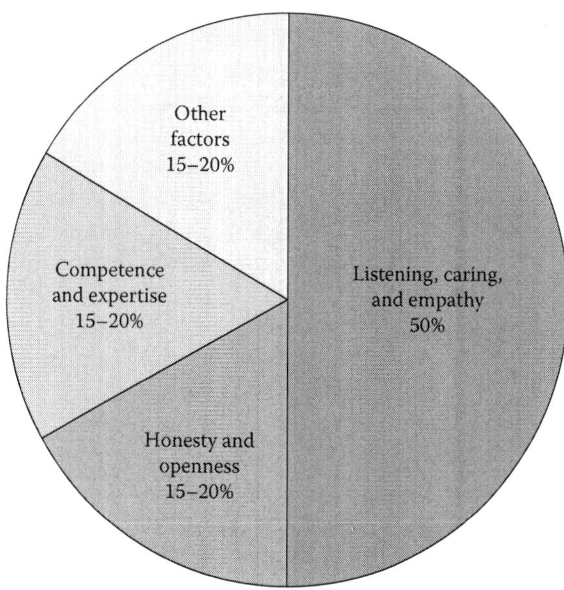

FIGURE 5.7 Trust factors in high-stress situations. (*Source*: Covello 2002.)

TABLE 5.2

Communication Shifts in Low- to High-Stress Situations

Low Stress	High Stress
Process an average of seven messages	Process an average of three messages
Information processed linearly (1, 2, 3)	Information processed in primacy (1, 3, 2) or recency order (3, 2, 1)
Information processed at average grade level	Information processed at 4 levels below average grade
Focus on competence, expertise, knowledge	Focus on listening, caring, empathy, compassion

Source: Covello (2002).

the messages must be simplified. Feedback is essential to gauging the public's response.

People process information very differently during high-stress situations. We can handle fewer bits of information at a time, and we process it in a different order. Newspapers tend to write for an eighth-grade reading level. If that is the average grade, then during high-stress situations, information is processed at a fourth-grade level.

Mental noise caused by the stress of fear impedes the receiver's ability to accurately process information. To counteract these changes, risk communicators should:

- Simplify the message (reading level)
- Reduce the number of message points to a maximum of three points
- Use short sentences
- Use numbers carefully
- Use pictures or graphics to present ideas

5.4.1.6 Three M's of Risk Communication

Given the potential role of stress in risk communication, there are three categories of tools to consider in risk communication. Referred to as the *three M's of risk communication* (Eisenberg and Silverberg 2001), they are:

- Message: What to say

TIME LIMITS

- 20 minutes combined for all speakers at a public meeting
- 2 minutes to answer a question at a public meeting
- 8–10-second sound bites for answering media questions during an interview
- 3 minutes for interacting with the public for every minute you speak

***Source*: Eisenberg and Silverberg (2001).**

- Messenger: Who to say it
- Media: How it should be presented

5.4.1.6.1 Message

There can be many different purposes for a risk communication (Fulton and Martinez n.d.), including:

- Raising awareness
- Educating/informing
- Achieving consensus
- Changing behavior
- Changing perception
- Receiving input

The message needs to be consistent with your purpose.

During the initial stages of message development, there are three helpful questions (Eisenberg and Silverberg 2001) to consider:

1. What are the three most important things you would like your audience to know?
2. What three things would your audience most like to know?
3. What are the three points your audience is most likely to get wrong unless they are emphasized and explained?

Avoid messages that convey only technical facts. Convey empathy, caring, honesty, openness, dedication, and commitment in your verbal and nonverbal messages. To maximize the information your audience hears, understands, and remembers:

- Structure and organize your message
- Limit your information to three key messages
- Keep your messages short
- Present each message in 7–12 words followed by two to four supporting facts
- Repeat your key messages: Tell them what you are going to tell them; then tell them; and finally tell them what you told them

5.4.1.6.2 Messenger

The best spokesperson is not always the topic expert. You need someone who can show empathy, stay organized, understand the audience, and speak clearly—someone with credibility and expertise. Credible speakers have the requisite expertise, but they are also trustworthy and likeable. They are similar to the audience and communicate well nonverbally. Recall that credibility is based on the mix of trust factors identified previously and in the following paragraphs. Expert speakers have advanced knowledge and/or degrees in the area they are speaking about, and they speak with authority, assured in their knowledge. However, the best messenger in a low-stress situation may not be the best messenger in a high-stress situation.

AVOID

Humor
Negative terms
Guarantees and absolutes
Complex language
Jargon
Personal beliefs
Attack
Worst-case speculation
Numerical details

***Source*: Eisenberg and Silverberg (2001).**

People need to know that you care before they will care about what you know. Active listening skills including paraphrasing, providing active feedback, and controlling nonverbal cues are an important part of being perceived as a trustworthy speaker.

To be trustworthy, be balanced. Focus on a specific issue. Pay attention to what the audience already knows and be respectful in tone, recognizing the legitimacy of people's feelings and thoughts. Be honest about the limits of scientific knowledge.

If you make a promise or a commitment, keep it. Come early and stay late. Engage people one on one. Provide a phone number and an e-mail address where you can be reached.

Limit your use of notes and show a high level of organization and logic. Dress professionally; avoid over- or underdressing. Be assertive and avoid hedging. Make sure the audience knows your credentials.

5.4.1.6.3 Media

How will you be presenting the information? Media comprise vehicles, channels, and applications for your message appropriate to your audience. Communication vehicles may be written, oral, visual, interactive, computer based, experiential, or technology assisted. Channels include media, advertising, public meetings, one-on-one

BODY LANGUAGE

- Make eye contact while slowly sweeping the room.
- Avoid darting eyes or staring.
- Keep your hands open at about waist level.
- Don't cross your arms, make a fist, clasp your hands, put your hands in your pockets, or make large waving hand movements.
- Lean slightly forward from the waist.
- Avoid slouching and standing or sitting rigidly.

***Source*: Eisenberg and Silverberg (2001).**

opportunities, Internet, word of mouth, speaker bureaus, and the like. Examples of applications include such things as fact sheets, pamphlets, reports, news releases, newsletters, Web pages, wiki spaces, public notices, flyers, posters, exhibits, videos, journal articles, fact sheets, and so on.

Message media are chosen on the basis of impact and influence. Impact refers to how widespread the impact of your message will be, while influence refers to the kind of persuasive influence (e.g., credibility) the channel has. The choice of channel goes back to knowing your audience. Where are they? How do you reach them? Where do they get their information? Do they read newspapers, listen to radio and watch TV, or do they text, tweet, instant message, e-mail, cruise the Internet, and rely on reference groups for information?

5.4.1.7 Critical Differences in Crisis Communication

Risk communication has been defined in various ways, but most definitions include some version of a two-way exchange of information and opinion about risks. We have seen previously that when hazard and outrage are both high, we are often engaging in a crisis communication strategy. Glik (2007) defines crisis communication more narrowly as "the exchange of risk-relevant and safety information during an emergency situation." The primary purpose of crisis communication is to motivate the audience to action.

The communication goals are different for more routine risks than for crises. Risk communication addresses what could go wrong and how it could happen. Crisis communication deals with what is happening right now. A crisis is a dynamic, usually unexpected event that involves a significant threat, ongoing uncertainty, and greater intensity than longer-term risk situations (Sellnow et al. 2009). There is not time for many of the best-practice techniques described in the risk communication literature. The coordination, collaboration, consensus building, issue resolution, and public-involvement interactions often prescribed will not likely be possible.

Whereas risk communication can be planned, tested, and strategic, crisis communication is spontaneous. Risk communication usually takes place before an event occurs, while crisis communication is post event. The risk communication model is multidirectional, proactive, and relatively certain. In a crisis, communication is unidirectional, reactive, and far more equivocal. Seeger and Ulmer (2003) summarize some other potential differences between the two strategies, as shown in the Table 5.3.

CONSEQUENCES OF POOR CRISIS COMMUNICATION

- People may not make good choices or may make them too late.
- Public frustration (or outrage) may develop, and once the public reaches an "outrage" state, it is very difficult to go back.
- Messages may be misinterpreted or misunderstood, causing bad feelings.
- The public may start to mistrust the organization.

***Source*: Risk Analysis 101 USDA, APHIS, PPQ**

TABLE 5.3

Differences between Risk Communication and Crisis Communication

Risk Communication	Crisis Communication
Risk-centered: focuses on harm or risk occurring in the future	Event-centered: focuses on a specific event that has occurred and produced harm
Messages may include known probabilities of negative consequences and how they may be reduced	Messages address current state or conditions: Magnitude, immediacy, duration, control/remediation, cause, blame, consequences
Based on what is currently known	Based on what is known and what is not known
Long term (precrisis stage)	Short term (crisis stage)
Message preparation possible (campaign)	Less preparation (responsive)
Personal scope	Community or regional scope
Mediated: commercials, ads, brochures, pamphlets	Mediated: press conferences, press releases, speeches, Web sites
Controlled and structured	Spontaneous and reactive

Source: Seeger (2002).

The person who is a good public relations communicator or even a good risk communicator may not be the best crisis communicator. The outrage can be expected to be greater, but the three M's still apply.

5.4.1.8 Explaining Risk to Nonexperts

Explaining risk data is not the primary purpose of risk communication, but sometimes it is necessary. Risk assessors have to explain the risk to risk managers. Nonexpert stakeholders and the general public are also going to need scientific and technical information from time to time. There are three principles (Sandman 1987) for accomplishing this difficult task: simplify, personalize, and use risk comparisons.

5.4.1.8.1 Simplify

The challenge is to make hard ideas clear. The best way to do this is to simplify the language rather than the content. You cannot tell the public everything you know, so we need some guidelines for deciding what to say and what to leave out. Sandman and Lanard (2003) suggest three rules of thumb for deciding what gets included and what gets left out.

First, tell people what they need to know. Answer their questions. Provide instructions for coping with a crisis. Stress these things.

Second, tell people what they have to know to both understand and **feel** that they understand the information they are given. The trick here is to know what the audience might get wrong and provide the information that prevents that error. Testing messages is especially useful in this task.

Third, help people understand that there is more than what you are telling them so that additional information at a later date won't make them feel misled. You are building a framework to support an evolving understanding of the problem.

Explaining risk is difficult because people prefer hearing about things that are safe or dangerous. The public is more comfortable with these extremes. To avoid them, risk trade-offs and risk comparisons may be useful.

Although the nature of the risk may, itself, be complex and uncertain, people can understand risk trade-offs, risk comparisons, and risk probabilities when they are carefully explained. Because of the way risk is perceived, the public can be expected to be resistant to the idea that their risk is modest when they are outraged or that it is substantial when they are not. In the long run, effective risk communication relies more on effective ways of addressing the anger, fear, powerlessness, optimism, and overconfidence of the public than it does on finding clever ways to simplify complex information.

The risk information you do prepare is most likely to reach the public through the mass media. Consequently, you are often simplifying risk information for journalists. Journalists are going to simplify the information for their readers. You are more likely to get a better result if you simplify complex information for them than you will if you give them the complex information to simplify. This is especially true for broadcast media that rely on short sound bites.

The greater concern then becomes to avoid oversimplifying the information and misleading the audience. Both your integrity and the public's trust are at stake. The key is to be prepared in advance of this communication with mass media. Know precisely what it is you want the journalist and his audience to take away from your message. In a crisis, you may have little time to prepare, but take the time you have and use it to prepare. If a journalist takes a different approach to the story, you can then quickly answer the less relevant question and follow with the longer prepared answer to the question that "should" have been asked. Alternatively, you are prepared to suggest a focus for the story or interview.

Make fact sheets part of your preparation. This can clarify your most important points without oversimplification. The greatest challenge is when reporters are demanding more information than you have and everyone is hurrying to respond to the crisis. In that case, be honest and warn reporters that you are in a rush and qualify your remarks by acknowledging that you are simplifying the response; then provide an overview of the sorts of information you lack or are omitting for the sake of simplification.

Think about the things people are most likely to get wrong about your message and then provide information to prevent this mistake or discuss the mistake directly. As you and the public find out more about the situation, you want to make sure your simplification holds up as solid and accurate rather than as misleading. There is nothing wrong with being incomplete. Being incorrect is another matter.

5.4.1.8.2 Personalize

Make it personal, and the public is much more likely to understand the risk. Experts tend to gravitate toward the big (societal) picture and policy issues, while the audience for your risk information is interested in the smaller (personal) picture and their own options. Individual voluntary decisions are very different from social policy decisions. Ordering the city of Galveston, Texas, to evacuate is a fundamentally different kind of decision than deciding that you want to leave the island.

PERSONALIZE

"Persons not heeding evacuation orders in single family, one or two story homes will face certain death."

National Weather Service
Hurricane Ike Warning for Galveston, September 2008

"The best way to guard against the flu is to get vaccinated, which helps to protect you, your loved ones, and your community."

CDC official
Seasonal flu vaccination, September 2006

It helps to understand the reporter's or the audience's viewpoint. Personalizing the issue brings it to life. It makes the abstract concrete. A focus on real people making real decisions is the best way to personalize a risk.

The personal judgments of the experts are often a powerful indicator for the public. It is important to separate the big-picture policy decisions from the small-picture personal ones for your audience. You may be more concerned about the societal risk, but you should be prepared to talk about both.

Sometimes the scientist must go against her instinct for the sake of good risk communication. Science is devoted to abstraction and deriving principles and theories from data. The public, on the other hand, wants concrete, specific, and personal information, especially when it comes to novel risks. Examples, anecdotes, and images help to personalize a risk.

Compare a leaking landfill to coffee grounds, a flood to a bathtub overflow, finding the source of a food-borne disease outbreak to finding a needle in a haystack. Good communication relies on vivid and memorable examples and images.

RISK COMPARISONS CAN HELP WHEN...

1. The source of the comparison has high credibility and is more or less neutral.
2. The situation is not heavily laden with emotion.
3. The comparison includes some acknowledgment that factors other than relative risk are relevant, i.e., the comparison does not dispose of the issue.
4. The comparison aims at clarifying the issue, not at minimizing or dismissing it.

***Source*: Covello and Allen (1988).**

5.4.1.8.3 Risk Comparisons

Risk comparisons are controversial. Some experts like to avoid them; others embrace them as a useful tool for explaining risk to nonexperts. The alternative—providing the risk estimate details—is often not practical. In general, the public does not understand the scale or the units of measurement. What is 7×10^{-7}? Who knows what a picocurie is or what cfs or CFU mean? For that matter, who knows if a milliliter is a little or a lot? Worse, the consequences and endpoints are often intimidating, threatening, or unattractive. Flesh-eating bacteria, increased lifetime risk of cancer, lost life expectancy, disease, habitat destruction, and so on, are hard to understand and unpleasant to contemplate.

The challenge is to find a middle ground between safe and dangerous and to present scientific facts that are comprehensible to the audience. Risk comparisons are an option. They help make risk numbers more meaningful and put risks into perspective by comparing this risk to other risks. Covello, Sandman, and Slovic (1988) have developed a taxonomy of risk comparisons that provides a useful guide to this option:

- The most acceptable risk comparisons
    - Comparisons of the same risk at two different times
    - Comparisons with a standard
    - Comparisons with different estimates of the same risk
- Less desirable risk comparisons
    - Comparisons of the risk of doing something versus not doing it
    - Comparisons of alternative solutions to the same problem
    - Comparisons with the same risk as experienced in other places
- Even less desirable risk comparisons
    - Comparisons of average risk with peak risk at a particular time or location
    - Comparisons of the risk from one source of a particular adverse effect with the risk from all sources of that same adverse effect
- Marginally acceptable risk comparisons
    - Comparisons of risk with cost, or of one cost/risk ratio with another cost/risk ratio
    - Comparisons of risk with benefit
    - Comparisons of occupational risks with environmental risks
    - Comparisons with other risks from the same source
    - Comparisons with other specific causes of the same disease, illness, or injury
- Rarely acceptable risk comparisons: Use with extreme caution!
    - Comparisons of two or more completely unrelated risks
    - Comparison of an unfamiliar risk to a familiar risk

Concrete examples of these can be found at http://www.psandman.com/articles/cma-4.htm.

SCALE COMPARISONS

One-in-a-million: One drop of gasoline in a full-size-car's tankful of gas
One-in-a-billion: One four-inch hamburger in a chain of hamburgers circling the earth at the equator two and one-half times
One-in-a-trillion: One drop of detergent in enough dishwater to fill a string of railroad tank cars 10 miles long
One-in-a-quadrillion: One human hair out of all the hair on all the heads of all the people in the world

***Source*: Covello et al. 1988.**

Comparisons must be relevant to your audience. Those rarely acceptable risk comparisons often turn out to be false arguments, based on a flawed premise. Risks have certain contextual characteristics for the audience, as we saw in the discussion of risk perceptions. The comparisons should not violate any of the important risk characteristics. In other words, do not compare a familiar risk to an unfamiliar one or a voluntary risk to an involuntary one, and so on. The risk comparisons must be appropriate and truly comparable in the eyes of the audience.

5.4.1.9 Explaining Uncertainty

Sometimes risk assessors are the risk communicators. There will always be uncertainty in risk assessment. When explaining uncertainty to risk managers, the job falls to the assessors. When explaining uncertainty to the public, professional risk communicators may be involved. In either case there are a number of simple rules of thumb (Sandman 2010a, 2010c) that will make the task easier.

Acknowledge uncertainty from the outset. Do not wait for someone else to discover what you do not know. Bound your uncertainty with a range of possibilities that are credible.

Clarify that you are more certain about some things than others. Tell people:

- What you know for sure
- What you think is almost but not quite certain
- What you think is probable
- What you think is a toss-up
- What you think is possible but unlikely
- What you think is almost inconceivable

Tell people what has been done and what you continue or plan to do to reduce the uncertainty. If you are going to be unable to reduce the uncertainty further, say so. Report everyone's estimates of critical uncertain values, not just your own. Never hide behind uncertainty. If the existence of a problem is uncertain but likely, say so. Neither should you perpetuate uncertainty. If there are things you can do to answer the uncertain questions, do them.

TWELVE TIPS FOR EXPRESSING UNCERTAINTY

1. Ride the risk communication seesaw.
2. Try to replicate in your audience your own level of uncertainty.
3. Avoid explicit claims of confidence.
4. Convert expert disagreement into garden-variety uncertainty.
5. Make your content more tentative than your tone.
6. Show your distress at having to be tentative and acknowledge ours.
7. Explain what you have done or are doing to reduce the uncertainty.
8. Don't equate uncertainty with safety—or with danger.
9. Explain how uncertainty affects precaution taking.
10. Don't hide behind uncertainty.
11. Expect some criticism for your lack of confidence.
12. Don't go too far.

***Source*: Sandman (2004).**

No evidence of an effect is not evidence of no effect. Be especially careful not to say there is no evidence of a particular effect if you have not looked for the evidence.

Let people know when finding out for sure is less important than taking appropriate precautions now. Acknowledge that people disagree about how to respond to uncertainty and that different people may do different things. "Based on the information available I have decided I will not get the swine flu shot, but my wife is going to." Help people become involved in reducing uncertainty for themselves. Give them ways to learn about their own vulnerability. Tell them how to learn about their flood risk; let them know how to get up-to-the-minute information about the neighborhoods that may be affected by the spill. Show them how to find the batch number and date on the peanut butter, and so on.

Research shows that acknowledging uncertainty diminishes the perception of your competence while it increases people's judgment of your trustworthiness. Rarely do we say things like

> "I have no idea if the side effects of the swine flu vaccine are more or less dangerous than the risks of going unvaccinated. Because of the speed with which the vaccine was developed, no one has any data to estimate that yet."
>
> "The source of this latest outbreak of salmonellosis is not yet known; the evidence is mixed and very confusing."

Experts rarely say they do not know something, and they probably should do so far more often. No one likes to sound ignorant, so we often focus on what we do know, inadvertently leaving out important or useful information about what we do not know.

Indeed, part of the risk assessor/communicator job is to be more precise about uncertainty and the level of confidence in their results and in what they are saying. Consequently, experts may sound more certain than they really are. If you sound certain and turn out to be wrong, credibility and trust can be grievously wounded. The best way to ensure that the media do not make you sound more certain than you are is to proclaim your uncertainty.

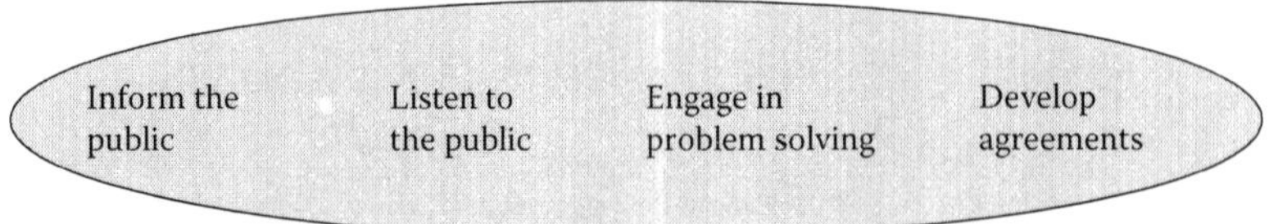

FIGURE 5.8 Varying intensity of different public-involvement activities. (*Source*: Creighton 2005.)

Another human trait is to be biased toward providing too much reassurance. Optimism is one of our fundamental biases. Uncertainty is not symmetrical: We tend to underplay negative outcomes and overplay positive ones.

5.4.2 Public involvement

As noted earlier, public/stakeholder involvement is sometimes considered part of risk communication and sometimes seen as a separate process. Its basic purpose is often to increase awareness or to build public support for a course of action. Public involvement is not going to be required for all risk management activities, whereas some amount of risk communication will be. Consequently, we'll restrict our consideration of the topic to a brief review of some public-involvement practices.

The reasons for involving the public in a risk management activity vary, as shown in the public-involvement continuum in Figure 5.8 (Creighton 2005). The extent of the participation process varies with its purpose. Informing the public is far less intensive than is partnering and developing agreements.

Commonly stated goals for a public-involvement program include the following:

- Incorporate public values into decisions
- Improve the substantive quality of decisions
- Resolve conflict among competing interests
- Build trust in institutions
- Educate and inform the public

That the public should have a say in decisions that affect their lives is one of the core values of public involvement. The risk analysis team needs to seek out and involve those potentially affected by the subject risks and their management decisions. In a good public-involvement program, the public's contributions will help define the problems and the risk management objectives. They will have the opportunity to exchange information and opinions and influence decisions. A good public-involvement program communicates how that will happen. The program should convey the interest of the decision makers while it simultaneously meets the process needs of participants. The best programs let participants help define how they will participate.

Involving the public improves the quality of decisions and, through consensus building, it also minimizes cost and delay that can result from processes that exclude the public and leave them no option for participation other than adversarial ones. Public involvement builds trust and helps an organization maintain its credibility and legitimacy. A program that anticipates public concerns and attitudes is easier to implement.

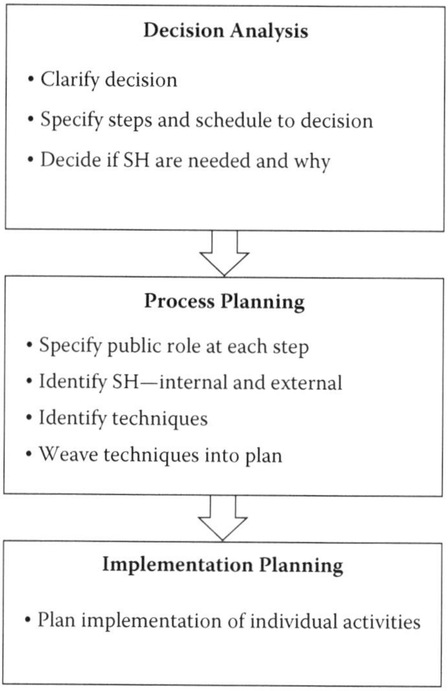

Decision Analysis

- Clarify decision
- Specify steps and schedule to decision
- Decide if SH are needed and why

Process Planning

- Specify public role at each step
- Identify SH—internal and external
- Identify techniques
- Weave techniques into plan

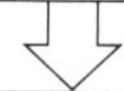

Implementation Planning

- Plan implementation of individual activities

FIGURE 5.9 Blueprint for developing a public involvement. (*Source*: Creighton 2005.)

5.4.2.1 Planning Stakeholder Involvement

Creighton (2005) offers a blueprint for developing a public-involvement program that is summarized in Figure 5.9. The first step is the analysis of the decision context, a subject that was covered in Chapter 3, "Risk Management." To adapt this step to developing a participation program, a few more questions might be added, such as:

- Who needs to be involved in the decision analysis?
- Who is the decision maker?
- What is the decision being made or problem being addressed?
- What are the steps in the decision-making process?
- When will they occur?
- What institutional constraints or special circumstances could influence a stakeholder's participation process?
- Is stakeholder involvement needed?
- If so, what level of participation?
- Will the decision be controversial?
- Will the decision require trade-offs of one value for another?

A good process begins by being clear about why you want public involvement. Do you want a better-informed public? Is it to fulfill legal requirements or to give the public a voice before a decision is made? Do you need support or informed consent? Must you have buy-in for success? Are you trying to change behaviors or save lives?

Planning the participation process can also be guided by some questions:

- Who needs to be on the planning team?
- What are the issues?
- Who are the audiences/publics?
- What is the level of controversy and how do we prepare for it?
- What do we want to accomplish at each step?
- What are our stakeholder-involvement objectives?
- What do stakeholders need to know to participate effectively?
- What do we need to learn from stakeholders?
- Do special circumstances affect our techniques?
- What techniques are best?
- What will we include in the plan?

Identifying stakeholders is especially important to ensure that no one is left out and to reach all parties interested in the issue. It is helpful to consider who benefits or loses in a given situation and who uses the relevant and affected resources. Different people and groups can be expected to participate in your process to varying extents. You need effective techniques to involve them all. Information exchange remains the goal of these activities.

You may need to explain the nature of the risk management activity and the decision process. Or you may want to know how different groups see the problem, who sees themselves as affected, as well as how the problem affects them. One key to good public involvement is to not let the issue slip from view. Although you cannot expect less active participants to sustain interest over a long time, bear in mind that suspicion often grows when an issue disappears. Use a variety of techniques to keep people productively involved in your process. Participation is especially important in the weeks leading up to decision points, and there may be many decision points in a risk management activity.

Implementing a good participation plan will take a mix of skills. You'll need a spokesperson, technical experts, and facilitators. You may also need people with a wide array of communications skills. Do your homework and take care not to surprise

LEVELS OF PARTICIPATION

1. Unsurprised apathetics: do not participate, have little interest
2. Observers: keep abreast and generally do not participate
3. Commenters: very interested, will attend meeting or send letter
4. Technical reviewers: other agencies, peers
5. Active participants: commit time and energy in order to influence decisions
6. Co-decision makers: those who will make or veto a decision

***Source*: Creighton (2005).**

TABLE 5.4
Traditional and Internet to-and-from Communication Techniques

Traditional To	Internet To	Traditional From	Internet From
Briefings	Data, models, reports	Advisory group or task force	Web conferencing
Exhibits and displays	Twitter	Charette	Wiki spaces
Feature stories	Hotline	Coffee Klatch	Virtual communication
Repositories	Up-to-the-minute information	Computer simulation	Interactive Web sites
Mailings	Chatroom, discussion boards	Consensus simulation	Interactive Web sites
Media interviews	Multimedia	Field trip	Shared spaces
Media kits	Interactive	Focus groups	
Talk shows	Downloads	Hotlines	
News conferences	Distance learning	Interviews	
Newsletters	Publish information about events	MCDA	
News releases	Podcasts	Shared vision planning	
Newspaper inserts and advertisements	Instant messaging	Large groups/small group meetings	
Panels			
Presentations			
PSAs			
Symposia			

elected officials or other community leaders and opinion makers. See to the needs of the media. If you are meeting with the public, visit the site in advance. Be sure to back up your technology, have a Plan B, and never outnumber the public.

There are many effective ways to communicate, and technology is growing the list of possibilities all the time. Even though communication is a two-way process, it is convenient to think of "to" and "from" techniques as seen in Table 5.4.

Although one must remain aware of the digital divide that can separate different audience segments, it is exciting to consider the new possibilities for communication and participation that the Internet provides. Online learning classrooms, real-time chat, Twitter, instant messaging, podcasts, wiki spaces, collaborative working environments, Web conferencing, YouTube videos, interactive learning tools, data visualization techniques, Google, and all manner of emerging social networking techniques and tools make this an exciting time to be interested and involved in public involvement.

The way people are working is changing. More and more collaborative work environments are popping up. More and more work is becoming defined by Don Tapscott's (2006) four organizing principles:

- Open: all are welcome
- Peered: no one is in charge
- Shared: communal ownership
- Global: worldwide

PRINCIPLES OF CONSENSUS COMMUNICATION

- Ensure stakeholder or audience participation early and throughout the risk analysis process.
- Listen to and honestly address the public's specific concerns.
- Convey the same information to all segments of your audience.
- When possible, allow stakeholders to participate in risk management decisions.
- Ensure that there are effective feedback mechanisms between the communicators and stakeholders.
- Plan how you will balance the interests of various stakeholders.
- Address uncertainty.

***Source*: Neeley (n.d.), USDA APHIS PPQ.**

Now is a great time to experiment, innovate, and collaborate. Imagine asking the world to help you solve your problem—and getting an answer! Think about using wikis so affected citizens can participate in a more active way. Spread your wings and fly, experiment with new technologies, vary your approach.

5.4.3 Conflict Resolution

There will be times when external risk communication includes public involvement, and public involvement will require conflict resolution. Conflict resolution or consensus communication is often used to bring a number of parties to consensus on how to manage a risk. It is an effort to get people on "the same page." It is most useful for addressing particularly contentious, controversial, or divisive issues (Neeley n.d.). Although there are many reasons for conflict, three are almost inevitable (Deep and Sussman 1997) in a risk management activity. Ours is a world of increasing complexity and rapid change. A consequence of that is growing diversity. Three television networks once exhausted the broadcast options for the United States. The number of options now numbers in the hundreds, and each of these options has viewers in the many thousands or even millions. The first inevitability is that different people want different things. There are few risk management solutions that will satisfy everyone.

Second, risk management activities involve and affect people. To affect people is inevitably to experience conflict. People will miscommunicate, misunderstand, jump to conclusions, suffer bruised egos, hold incompatible beliefs, and have incompatible needs. Humanity breeds conflict.

Third, limited resources mean even the winners in decision processes rarely get exactly what they want. Instead we are often "satisficing," i.e., trying to get the best situation possible given the available options and constraints. For the losers in a decision process, the situation is direr, so conflict flares readily. Given the inescapable nature of conflict, conflict management may at times be one of the risk manager's most needed skills and one of risk communication's critical tasks.

Deep and Sussman (1997) offer a treasure trove of practical lists for managing conflict productively. Eleven different lists comprising 105 different ideas are included in their conflict management chapter. It is a great place for the novice to begin. There is also a rich professional literature on conflict management (see, for example, Burton 1968, Kelman and Fisher 2003, Kriesberg 1998), and three forms of conflict management are addressed in the literature. These are conflict settlement, conflict resolution, and conflict transformation. Conflict settlement includes any conflict strategy that aims at a definite end of the conflict without necessarily addressing its basic causes (Reimann 2004). Conflict-resolution approaches include strategies that can be used to find an exit from the conflict's damaging dynamic that aim at reaching a satisfactory solution for all parties involved. Galtung (2000), Lederach (1995), and others have suggested that the conflict context, its structure, the parties involved, and the general conflict issues may at times be transformed into a more agreeable situation.

Like the topic of public involvement, conflict resolution is too complex and too well developed elsewhere to address it in detail here. It is, however, important to understand that conflict resolution may be considered part of the risk communication program in the broadest constructions of the risk communication component.

5.5 Summary and Look Forward

Risk communication has both internal and external tasks. Internally, the coordination between risk managers and assessors is essential to the success of the risk analysis process. Not too many years ago, many thought managers and assessors may need to be separated almost to the point of sequestering the assessors so their objective work would not be tarnished by the subjective concerns of managers. Now we know better and recognize the importance of collaboration, coordination, and cooperation between managers and assessors.

The external communication task usually receives most of the emphasis in discussions of risk communication, and the extent of this task varies from one context to another. The narrower view of the risk communication component focuses on specific risk and crisis communications. An increasingly more common, broader view of this component includes those communications, but also may include public-involvement and issue-resolution responsibilities as well.

The hazard and outrage dimensions of risk necessitate at least four types of risk communication strategies. These dimensions can affect the perception of risk, which is an important consideration for both risk managers and risk communicators. There are many unique challenges to effective risk communication, not the least of which is understanding the special challenges of communicating with people who are stressed and fearful. The three M's of risk communication—message, messenger, and media—are an important focus for any risk communication process.

Because the risk analysis process is for making decisions under uncertainty, risk communicators must develop skill at explaining risk and critical uncertainties to nonexperts. This is a task that is aided by simplifying, personalizing, and using risk comparisons. Learning to express uncertainty effectively and developing more effective techniques for communicating complex scientific information and its attendant uncertainty remains a challenge for all risk communicators.

With the three risk analysis components now described, the next chapter begins to consider the application of these principles to the risk management task of problem identification.

REFERENCES

Burton, John W. 1968. *Systems, states, diplomacy and rules.* Cambridge, U.K.: Cambridge University Press.

Chess, C., and B. J. Hance. 1994. *Communicating with the public: Ten questions environmental managers should ask.* New Brunswick, NJ: Center for Environmental Communication.

Chess, C., B. J. Hance, and P. M. Sandman. 1989. *Planning dialogue with communities: A risk communication workbook.* New Brunswick, NJ: Rutgers University, Cook College, Environmental Communication Research Program.

Covello, Vincent. 2002. Message mapping, risk and crisis communication. Invited paper presented at the World Health Organization Conference on Bioterrorism and Risk Communication, Geneva, Switzerland. http://www.orau.gov/cdcynergy/erc/Content/activeinformation/resources/Covello_message_mapping.pdf.

Covello, Vincent T., and Frederick H. Allen. 1988. Seven cardinal rules of risk communication. OPA-87-020. Washington, DC: Environmental Protection Agency.

Covello, Vincent T., and John J. Cohrssen. 1989. *Risk analysis: A guide to principles and methods for analyzing health and environmental risks.* Washington, DC: Council on Environmental Quality.

Covello, Vincent T., David B. McCallum, and Maria Pavlova. 1989. Principles and guidelines for improving risk communication. In *Effective risk communication: The role and responsibility of government and non-government organizations,* ed. Vincent T. Covello, David B. McCallum, and Maria Pavlova, 3–19. New York: Plenum Press.

Covello, Vincent T., and Miley W. Merkhofer. 1993. *Risk assessment methods: Approaches for assessing health and environmental risks.* New York: Plenum Press.

Covello, Vincent T., and Jeryl Mumpower. 1985. Risk analysis and risk management: An historical perspective. *Risk Analysis* 5 (2): 103–120.

Covello, Vincent T., Richard Peters, Joseph Wojtecki, and Richard Hyde. 2001. Risk communication, the West Nile Virus epidemic, and bioterrorism: Responding to the communication challenges posed by the intentional or unintentional release of a pathogen in an urban setting. *Journal of Urban Health.* 78 (2): 382–391.

Covello, Vincent T., and Peter M. Sandman. 2001. Risk communication: Evolution and revolution. In *Solutions to an environment in peril,* ed. Anthony B. Wolbarst, 164–178. Baltimore, MD: Johns Hopkins University Press.

Covello, Vincent T., Peter M. Sandman, and Paul Slovic. 1988. *Risk communication, risk statistics, and risk comparisons: A manual for plant managers.* Washington, DC: Chemical Manufacturers Association.

Crano, W. D., and M. Burgoon. 2001. Vested interest theory and AIDS: Self-interest, social influence, and disease prevention. In *Social influence in social reality: Promoting individual and social change,* ed. Fabrizio Butera and Gabriel Mugny, 277–289. Seattle, WA: Hogrefe & Huber.

Creighton, James L. 2005. *The Public participation handbook: Making better decisions through citizen involvement.* San Francisco: Jossey-Bass.

Deep, Sam, and Lyle Sussman. 1997. *Smart moves: 140 checklists to bring out the best in you and your team.* Rev. ed. Cambridge, MA: Perseus Publishing. http://www.questia.com/library/book/smart-moves-140-checklists-to-bring-out-the-best-in-you-and-your-team-by-sam-deep-lyle-sussman.jsp.

Eisenberg, Norman A., and Beverly R. Silverberg. 2001. *Food safety communication primer: A guide for conveying controversial or sensitive food safety information to concerned audiences.* College Park, MD: Joint Institute for Food Safety and Applied Nutrition.

Fischhoff, B. 1986. Helping the public make health risk decisions. In *Effective risk communication: The role and responsibility of government and non-government organizations*, ed. Vincent T. Covello, David B. McCallum, and Maria Pavlova, 111–116. New York: Plenum Press.

Fischhoff, B. 1995. Risk perception and communication unplugged: Twenty years of progress. *Risk Analysis* 15 (2): 137–145.

Fischoff, Baruch, Paul Slovic, Sarah Lichtenstein, Stephen Read, and Barbara Combs. 1978. How safe is safe enough? A psychometric study of attitudes towards technological risks and benefits. *Policy Sciences* 9:127–152.

Food Insight. 2010. Risk communicator training for food defense preparedness, response and recovery: Trainer's overview. http://www.foodinsight.org/Resources/Detail.aspx?topic=Risk_Communicator_Training_for_Food_Defense_Preparedness_Response_Recovery.

Fulton, Keith, and Sandy Martinez. n.d. *Risk communication primer: A guide for communicating with any stakeholder on any issue that impacts your mission.* Houston, TX: Fulton Communications.

Galtung, Johan. 2000. *Conflict transformation by peaceful means (The Transcend Method), participants and trainers manual.* New York: United Nations.

Glik, Deborah, 2007. Risk communication for public health emergencies. *Annual Review of Public Health* 28:33–54.

Hance, B. J., C. Chess, and P. M. Sandman.1990. *Industry risk communication manual.* Boca Raton, FL: CRC Press/Lewis Publishers.

Heath, Robert L., and H. Dan O'Hair, eds. 2009. *Handbook of risk and crisis communication.* New York: Routledge.

Johnson, B. B., and V. Covello, eds. 1987. *The social and cultural construction of risk: Essays on risk selection and perception.* Dordrecht, Netherlands: D. Reidel Publishing.

Kelman, Herbert C., and Ronald J. Fisher. 2003. Conflict analysis and resolution. In *Oxford handbook of political psychology*, ed. David O. Sears, Leonie Huddy, and Robert Jervis, 315–357. Oxford, U.K.: Oxford University Press.

Kriesberg, Louis. 1998. *Constructive conflicts. From escalation to resolution.* Lanham, MD: Rowman & Littlefield.

Krimsky, S., and A. Plough. 1988. *Environmental hazards: Communicating risks as a social process.* Dover, MA: Auburn House.

Lederach, John Paul. 1995. *Preparing for peace: Conflict transformation across cultures.* Syracuse, NY: Syracuse University Press.

Lundgren, Regina E., and Andrea H. McMakin. 2009. *Risk communication: A handbook for communicating environmental, safety and health risks.* 4th ed. New York: Wiley.

Neeley, Alison. n.d. Risk communication applications and case studies. Slide presentation for Risk Analysis 101, Raleigh, NC, USDA APHIS PPQ.

Reimann, Cordula. 2004. Assessing the state of the art in conflict transformation. In *Berghof handbook for conflict transformation.* Berghof Research Center for Constructive Conflict Management. http://www.berghof-handbook.net/articles/.

Sandman, Peter M. 1987. Explaining risk to non-experts: A communications challenge. *Emergency Preparedness Digest* (October–December): 25–29.

Sandman, P. M. 1989. Hazard versus outrage in the public perception of risk. In *Effective risk communication: The role and responsibility of government and non-government organizations*, ed. Vincent T. Covello, David B. McCallum, and Maria Pavlova, 45–49. New York: Plenum Press.

Sandman, Peter M. 1999. Risk = Hazard + Outrage: Coping with controversy about utility risks. *Engineering News-Record*, October 4.

Sandman, Peter M. 2010a. Dealing with uncertainty. Peter M. Sandman Risk Communication Web Site. http://psandman.com/handouts/sand13.pdf.

Sandman, Peter M. 2010b. Four kinds of risk communication. Peter M. Sandman Risk Communication Web Site. http://www.psandman.com/handouts/sand17.pdf.

Sandman, Peter M. 2010c. Acknowledging uncertainty. Peter M. Sandman Risk Communication Web Site. http://www.psandman.com/col/uncertin.htm.

Sandman, Peter M., and J. Lanard. 2003. Fear of fear: The role of fear in preparedness…and why it terrifies officials. Peter M. Sandman Risk Communication Web Site. http://www.psandman.com/col/fear.htm.

Sandman, Peter M., and J. Lanard. 2005. Adjustment reactions: The teachable moment in crisis communication. Peter M. Sandman Risk Communication Web Site. http://www.psandman.com/col/teachable.htm.

Seeger, M. W. 2002. Chaos and crisis: Propositions for a general theory of crisis communication. *Public Relations Review* 28 (4): 329–337.

Seeger, M., and R. R. Ulmer. 2003. Explaining Enron: Communication and responsible leadership. *Management Communication Quarterly* 17 (1): 58–85.

Sellnow, Timothy L., Robert R. Ulmer, Matthew W. Seeger, and Robert Littlefield. 2009. *Effective risk communication: A message-centered approach*. New York: Springer.

Slovic, Paul. 1987. Perception of risk. *Science* 236 (4799): 280–285.

Slovic, Paul, Baruch Fischoff, and Sarah Lichenstein. 1980. Facts and fears: Understanding perceived risk. In *Societal risk assessment: How safe is enough?* ed. C. Schwing and W. A. Albers, 124–181. New York: Plenum Press.

Slovic, Paul, Nancy Krauss, and Vincent T. Covello. 1990. What should we know about making risk comparisons? *Risk Analysis* 10 (13): 389–392.

Tapscott, Don, and Anthony D. Williams. 2006. *Wikinomics: How mass collaboration changes everything*. New York: Penguin Group.

United Nations. 1998. Food and Agricultural Organization and World Health Organization. *The application of risk communication to food standards and safety matters*. Rome, Italy: FAO.

University of Minnesota. 2006. Terrorism, pandemics, and natural disasters: Food supply chain preparedness, response and recovery. Symposium summary, November 1, 2006. http://foodindustrycenter.umn.edu/prod/groups/cfans/@pub/@cfans/@tfic/documents/asset/cfans_asset_250263.pdf.

Yoe, Charles, and Kenneth Orth. 1996. Planning Manual. IWR Report 96-R-21. Alexandria, VA: Institute for Water Resources.

6

Problem Identification for Risk Management

6.1 Introduction

There is a great chapter in Morgan D. Jones's (1998) book *The Thinker's Toolkit*. It is called "Thinking about Thinking," and its primary thesis is that the human mind is not analytical by nature. He explores the fallibility of human reasoning and suggests that the best remedy for the mind's ineffectiveness is to impose some structure on the way we think. This is what risk analysis does well. Jones argues, persuasively in my opinion, that structuring our analyses is at odds with the way our minds work naturally.

He argues that, left to our own devices, we often begin a problem analysis by formulating our conclusions, thus beginning where we ought to hope to end our analysis. With an intuitive preference for one solution, often the first one that seems satisfactory, we tend to give insufficient consideration to alternative solutions. We tend to confuse the gathering and analysis of data with the process of thinking about a problem.

Jones identifies seven traits that get in the way of our ability to analyze and solve problems. Risk analysis attempts to influence our thinking by making it more analytical. This can simultaneously limit the "damage" our fallible human reasoning can inadvertently do to decision making. The first of these traits is the emotional dimension to our thoughts and decisions. Emotion can hijack our ability to reason, as anyone who has thrown a broken item across a room has discovered. Serious mistakes can be made if decisions are made when emotions are hot. Risk analysis avoids emotional decisions.

Our minds are constantly taking shortcuts. We do not teach our minds how to work. They work as they work, and most of us rely on such shortcuts as stereotyping, personal bias, prejudice, hunches, jumping to conclusions, intuition, and the like. Risk analysis strives to replace shortcuts with science.

A third trait is that we are driven to see patterns in the world around us. We are conditioned to fill in missing information and to edit out what does not fit the pattern. The mind is not objective; it can construe random events as nonrandom or, when convenient, reverse the process and see no pattern where there is one. This compulsion to see patterns can mislead us when we analyze problems. Risk analysis requires evidence to support patterns.

We rely on our biases and mind-sets for much of our decision making. Bias is the unconscious belief that conditions, governs, and compels our thinking and our behavior. It is instinctive and beyond the reach of our conscious mind. Our biases ensure that we give high value to information that reinforces our biases and assumptions, and they encourage us to discount or reject inconsistent information. We are unwitting slaves to our biases and the perspectives they create. They constrain and bound what we think and see. Consequently, our minds often operate analogically, and our logic

DO YOU SEE A PATTERN?

The numbers 9/11 (9 plus 1 plus 1) equal 11
American Airlines Flight 11 was the 1st plane to hit the World Trade Center
There were 92 people on board (9 plus 2)
Sep. 11 is the 254th day of year (2 plus 5 plus 4)
There are 11 letters each in "Afghanistan," "New York City," "the Pentagon," and "George W. Bush"
The WTC buildings themselves form the number 11

Then what of the second flight that hit the WTC, United Airlines Flight 175, or American Airlines Flight 77 that hit the Pentagon, or United Flight 93 that crashed in Pennsylvania? You can see patterns if you cherry-pick the facts you examine.

is largely superficial. New experiences are interpreted in light of old ones, and our inferences are based on similarity. When we rely on bias, our reasoning only needs to be plausible to satisfy us. Good risk analysis avoids bias.

Fifth, we need to provide explanations for this uncertain world, regardless of the accuracy of those explanations. Jones calls us an explaining species. An eclipse is a dragon eating the moon! We seek cause and effect even when there is none. We want our perceptions to mean something because knowing is pleasing and not knowing is not. When an event has no particular meaning, we find one anyway. Subconsciously, we do not care if it is valid; it enables us to move on, and that is often sufficient. We are quite adept at coming up with explanations that do not fit the evidence very well, and our stunning indifference to the validity of our explanations is profoundly telling. Explanations need not be true to satisfy our need to know and explain. Risk analysis seeks sound, evidence-based explanations for what we see and what we do about what we see.

We tend to seek evidence that confirms our beliefs while devaluing what does not. In other words, we naturally prefer what confirms our existing views. When that causes us to (subconsciously) seek out evidence that supports our beliefs and judgments while eschewing information that does not, decision making can suffer. We focus by nature. In fact, our very survival is aided by our ability to focus. The problem often is that we tend to see in a body of evidence what we expect to see, i.e., what we are looking for, and we do not see what we do not expect or do not want to see. The mind is very good at reconfiguring evidence to make it consistent with our expectation. This advocacy position is the basis of our political and judicial systems. Advocates are trained to take a position and marshal supporting evidence to defend that position while doing their best to weaken all other perspectives. Our systems reward subjective argument more often than objective analysis. When the advocate is also the decision maker, advocacy can be destructive. Risk analysis does honor values and the explanations they can lead us to, but it also seeks a scientific explanation whether it conforms to social values or not.

Finally, Jones asserts that we can stubbornly cling to untrue beliefs. In fact, we often treat beliefs like prized possessions. They make us feel good, so we cherish and protect them, but many of our most cherished beliefs are untrue because we often prefer to believe what we prefer to be true. Our willingness to embrace untrue beliefs can have disastrous effects on our ability to analyze and resolve problems. Risk analysis challenges beliefs. It separates what we know from what we do not know and then identifies key uncertainties and addresses them. It identifies assumptions and challenges them. It forces us to consider different perspectives and challenges the way we think. Risk analysis is analytical decision making, and it asks us to change the way we think about things.

Thus, this is where people often struggle. We humans are prone to reacting to what we think the problem is instead of seeking to understand what the problem really is. As a consequence, we often treat symptoms and ripple effects of problems instead of the problems themselves. You've got to get the problem right. Then you have to write the problem down. Risk analysis is oriented toward finding the right problem. If you solve the wrong problem, well, you've still got problems. Remember: risk comes in two "flavors": loss and opportunity. A successful process depends on your ability to find the right problem or the right opportunity.

You're the risk manager and your boss has just said, "We have to do something about…." How do you begin to do that? You begin by asking, "Why?" Why are we initiating this activity? If your answer involves the words authority, regulation, law, previous studies, the boss, and the like, you may be technically correct, but you're missing the point. There is a reason for every risk management activity. You face problems or have opportunities. Risk management begins by identifying and clearly articulating those problems and opportunities. This comes before budgets, schedules, milestones, profits, stakeholders, shareholders, politics, or any of the other pressing duties you might have with a risk management activity.

Assemble as many of the risk analysis team members as you can identify, sit down together, and begin where you are. "What are the problems here?" "What opportunities do we have to improve conditions here?" In some cases you may have to rely on the vague and generic wording of a higher authority that set you in motion. In other cases you may have three full hard drives and a bank of file cabinets full of information about a situation that has been studied off and on for the last 30 years. At times, the problem may kick down your door unexpectedly. Wherever you are, this is where you begin.

Take a piece of paper. At the top of your paper write "Problems and Opportunities." Date the page because this is going to change many times before it is finalized. Risk analysis is an iterative process, and so is the identification of problems and opportunities. Your first entry is, "Waterfowl are disappearing in the Babylon Ranch area of the Voodoo River." In time you will learn what kind of waterfowl are disappearing, and you will understand why they are disappearing. But today you are beginning, and you work with what is available to you.

It's easier to identify problems than opportunities. Human nature finds it easier to point out negatives than positives. But on day one, while you have whatever your team looks like assembled in one place, you also need to begin to think about the opportunities you might have to improve conditions. Write them down, too. Your second entry might be, "Preserve wildlife at Babylon Ranch on the Voodoo River." Never mind that you might not have the authority to do this. Never mind that you are not sure anyone "out there" wants to do this. You need a place to start, and this is it. Now you have the first draft of your "Problems and Opportunities Statement."

When people ask why you are doing this risk management activity, you can say, "Because waterfowl are disappearing from the Babylon Ranch area of the Voodoo River and because we are exploring the possibility of preserving wildlife there." The takeaway point here is that you must have a clear understanding of why you are doing what you are doing. At this point it is better to be clear than to be right. It is day one.

There will be time to better understand the problems and the opportunities. In time you can be both clear and right. Risk management is always best when its purpose is clear. In an iterative process, that purpose will and should change and evolve as uncertainty is reduced. That is okay. That is a good risk management process.

If it is day two of your activity and you do not yet have the first draft of your problems and opportunities statement, you are not managing as well as you could. That piece of paper is the "why" of what you are doing. Do not neglect it.

Problem identification was defined earlier as a three-part process of problem recognition, problem acceptance, and problem definition as seen in Figure 6.1, Problem identification is the "seemingly" simple act of recognizing that a problem exists and figuring out what it is. It is also the subject of this chapter.

Following this introduction, we spend some time describing and differentiating problems and opportunities. Next we consider how people become aware of problems and opportunities by focusing on the notion of triggers and inputs. The chapter presents a number of different techniques for identifying problems and opportunities and concludes by considering the problems and opportunities (P&O) statement.

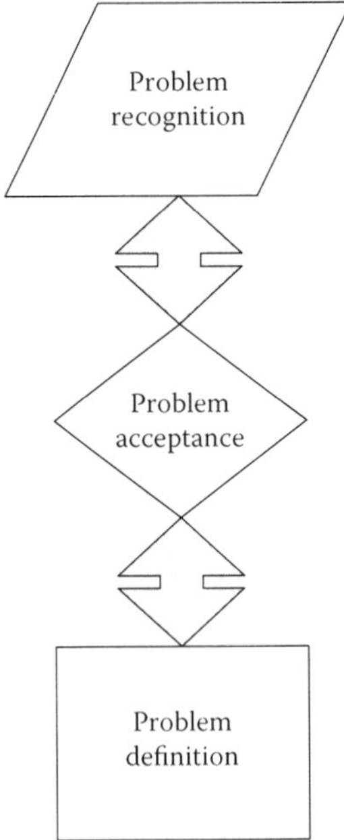

FIGURE 6.1 Problem identification elements.

6.2 What's a Problem? What's an Opportunity?

Merriam-Webster's New Collegiate Dictionary defines a problem as follows:

> prob•lem
> **1 a :** a question raised for inquiry, consideration or solution **b :** a proposition in mathematics or physics stating something to be done
> **2 a :** an intricate unsettled question **b :** a source of perplexity or vexation
> **synonyms** see MYSTERY

In risk management, we might informally think of a problem as a situation that we would like to address and solve. Most of us are pretty good at recognizing problems. A risky problem is going to involve a potential loss and usually a hazard, while a risky opportunity is going to involve the potential for an uncertain gain.

The most common kinds of problems are well-structured problems. Facts play a larger role than judgment with these problems. With well-structured problems the existing state and the desired state are both clear, and the methods needed to reach the desired state are fairly obvious. Textbook problems are archetypical examples of well-structured problems that simply require you to apply a set of principles to reach a solution. Another example would be when a piece of equipment fails that we know how to fix. We can figure out how many people we need, how long it will take, and what the costs will be. Thankfully many of our problems are well-structured. In an uncertain world, however, a good many problems are not so well-structured.

Risk analysis is well suited to these poorly structured problems, which tend to be situational and emergent dilemmas. Judgment often plays a bigger role than facts when solving these problems. As the amount of uncertainty increases, the role of judgment increases. Poorly structured problems have existing and desired states that are unclear, making it more difficult to confidently identify the method(s) for reaching the desired state. They are poorly structured because one or more of the problem elements are unknown or not known with any degree of confidence (Wood 1983).

These problems may require the integration of numerous knowledge domains, and there are often no explicit means for determining an appropriate action. The parameters of the problem may not lend themselves to ready manipulation. Understanding them may require people to express their personal opinions or beliefs about the problem (Jonassen 1997). Decision makers must make judgments about these problems and then defend them. Their solutions may require uniquely human interpersonal activities (Meacham and Emont 1989). Complex problems with values in conflict are much more difficult to solve. Understanding the nature of the ill-structured problem is, then, necessarily an essential starting point for problem solving.

We are less adept at recognizing opportunities. An opportunity is defined by Webster as:

> op•por•tu•ni•ty
> **1 :** a favorable juncture of circumstances
> **2 :** a good chance for advancement or progress

TABLE 6.1

Differences between Problems and Opportunities

Characteristic	Problem	Opportunity
Focus	Existing undesirable condition; a description of what is or might be	Future desirable condition; a description of what could or should be
Core message	Negative—an objection	Positive—a desire
Occurrence	Past: usually occurred	Past: usually did not occur
	Existing: usually occurs	Existing: may or may not occur
	Future "without action": usually expected to occur	Future "without action": may or may not be expected to occur
Relationship to other assets	An existing condition that may adversely affect other assets	Existing condition does not affect other assets
Implicit objectives of action	Return to a past condition that was not considered objectionable	Create a future condition considered to be desirable
	Create a future condition that would not be objectionable	Return to a previous condition considered to be desirable
Consequences of doing nothing	Usually direct, immediate, and adverse	Usually indirect and long term due to benefits foregone

Source: Yoe and Orth (1996).

Informally, we might define an opportunity as a gain or situation we would like to realize or a target we'd like to hit. It is useful to clearly distinguish the terms *opportunity* and *problem* so that we can intentionally identify both when initiating a risk management activity. See Table 6.1 for some differences between problems and opportunities.

A problem is usually an existing undesirable condition perceived as negative or objectionable. People, resources, or other assets of value are usually adversely affected under existing or expected future conditions. The implicit objective of responding to a problem is to return to a problem-free circumstance or to create a future condition that is not objectionable. In contrast, an opportunity is usually a positive condition that does not yet exist and will not likely exist unless we take intentional steps to make it happen. Under existing conditions, the absence of an opportunity does not adversely affect other assets, even though their full potential may not be realized. Management's implicit objective with respect to opportunities is to create a future condition that will realize those opportunities.

There are also similarities between problems and opportunities, as seen in Table 6.2. They are both often multiple in number, identified by people and through analysis, and stated in practical specific terms. They are measurable in the sense that we can at least tell the difference between better and worse conditions. They are achievable, and ideal situations can sometimes be described for both of them.

A problem statement need not be elaborate. It can be as simple as the example in Table 6.3. The representation or definition of these problems will take considerably more explanation. Each problem needs to be thoroughly developed by the risk managers. It is important that risk managers and assessors understand exactly what problems they are addressing and that other interested parties can also see a clear statement of the problems under consideration.

TABLE 6.2

Similarities between Problems and Opportunities

Characteristic	Similarity
Number	Variable, few to many
How stated	In practical, meaningful, operational terms in a single statement
Source	Developed from people, observation, analysis, and documentation
Specificity	Specific, narrow; essentially limited
Specific subject	Usually limited to a specific asset
Specific location	Usually found in a particular place or locale (example: "study area")
Specific measurability	Moderate to high; usually measurable or easy-to-recognize change that would result in a "better" or "worse" condition
Ability to achieve	High; problems can be solved, opportunities can be realized
"Ideal"	An "ideal" usually exists and can be identified
	The "ideal" is not the same as the existing condition
	The "ideal" is not the same as the long-term "without" condition

Source: Yoe and Orth (1996).

TABLE 6.3

Simple Problem Statement Examples

Redlands Creek Basin Problem Statement

The problems in the Redlands Creek Basin are:

1. Loss of fish habitat in Redlands Creek due to urbanization
2. Flood damages in the industrial section of Central City
3. Stream bank erosion along Campus Park
4. Saltwater intrusion in the Redlands Bay estuary
5. Loss of coastal wetlands along the South Ditch section of Redlands Bay

Food Safety Agency Problem Statement

The problems of concern to the regulatory agency are:

1. Increasing resistance of *Campylobacter* in chicken to fluoroquinolones
2. Declining efficacy in the use of fluoroquinolones for the treatment of campylobacteriosis in humans

Opportunities should, likewise, be clearly stated. An opportunity statement need not be elaborate. It can be as simple as the example shown in Table 6.4.

6.3 Becoming Aware of Problems and Opportunities

Some problems come bursting through your door at 4:30 on Friday afternoon and set up camp on your desk. Other problems are more subtle and do their best to avoid detection. Some are as familiar as the faces of our parents or children, while others we've never seen before nor will we see them again. Some problems stay only a brief while and others refuse to leave. Among these extremes are infinite varieties of ways for problems to be discovered. Although the precise manner in which a problem is identified will vary from one organization to the next, there are some useful

TABLE 6.4

Simple Opportunity Statement Examples

Redlands Creek Basin Opportunity Statement

There are opportunities in the Redlands Creek Basin to:

1. Increase wildlife habitat along Campus Park
2. Restore indigenous fish species in the upper basin
3. Provide increased recreational opportunities along the waterfront

Food Safety Agency Opportunity Statement

There is an opportunity for the regulatory agency to:

1. Reduce incidence of all campylobacteriosis in humans due to consumption of chicken
2. Provide a healthier work environment for on-farm employees
3. Reduce disease in flocks

generalizations that can be made about the origins of problems and how we initially become aware of them. Likewise, the same is true for opportunities. To avoid the tedium of constantly repeating "and opportunities" after each mention of a problem, I will stop coupling the two terms. The techniques presented here for problem identification are equally adaptable for identifying opportunities.

6.3.1 Triggers and Inputs

Some events trigger or initiate a process or reaction that identifies a problem. The *Deepwater Horizon* explosion of 2010, melamine in milk products, or a controlled flight of an aircraft into terrain are examples of problems that force themselves upon us and cannot be ignored, although they can be misidentified! Passive problem identification is sufficient for the more aggressive issues that come to you. Other problems only become apparent as inputs (information, reactions, events, opinions, and the like) accumulate over time; you must look for these kinds of problems.

Triggers and inputs are indicators of problems or opportunities that may require a risk management decision. A trigger may be a single event (e.g., a decision made by a higher authority, a natural disaster, infrastructure failure, terrorist attack, disease outbreak) or an accumulation of inputs into the program manager's continuous program assessment and evaluation. When triggers or early warning signs indicate that a risk cannot be addressed within the current program capabilities, it is reaching a point of importance and urgency that may require a risk management response.

Triggers come from a variety of sources and can be thought of in several broad types (FDA 2003). They are:

- Authority triggers
- Crisis triggers
- Science and technology triggers
- Emerging or "on the horizon" triggers
- Strategic plan triggers

EXAMPLES OF INPUT SOURCES

Academia
Professional literature
Environmental monitoring
Toxicity testing
Disease surveillance, epidemiological studies
Lack of compliance with standards
Permit applications
Inspection
Community reaction
Legal action
Media or interest group reporting
What other nations do
Research
Staff feedback
Strategic planning

Authority triggers are "pulled" by higher authorities in an organization. Someone with decision-making authority directs what we consider a situation. Examples include Congressional authorizations and appropriations, new policies, and decisions by the boss.

A real or perceived crisis can trigger a risk management activity. Examples include a message from the media alerting you to an issue, a public outcry and subsequent media involvement, natural or manmade events, and adverse comments by stakeholders, critics, or others. Crisis triggers can also result from growing public awareness of an apparent risk, the concerns of susceptible subpopulations, decreased consumer confidence in your organization and what it does, geopolitical events, natural disasters, accidents and the like.

Science or improved scientific methodology can act as a trigger. Newly acquired scientific knowledge that reveals previously unknown risks of specific hazards or a technological improvement that presents new opportunities may act as triggers. Examples include emerging concerns like climate change, new information that indicates that a problem or hazard is of greater concern than previously thought such as the "discovery" of acrylamide in fried foods, and new technology that makes new solutions viable, for example, nanotechnology and genome mapping.

A well-planned, forward-looking forecasting practice can identify emerging or on-the-horizon triggers. This entails scanning your organization's risk landscape to see how emerging events might affect you now and in the future. How might recent or future earthquakes, storms, floods, or civil unrest in the world affect your programs and opportunities? What trends and global events now on the horizon could affect your organization in time? Strategic plan triggers are the things we choose to address in order to realize the organization's goals and, therefore, its vision.

Among the inputs that may accumulate and steer the risk management process are some recurring elements. Some potentially important inputs include:

- Science (including research findings). We need to understand the facts contributing to a problem. Many organizations intentionally collect data and/or evidence and develop or assemble an understanding of the hazards, opportunities, or other risk factors of the situations or issues that may negatively or positively impact their organizational mission.
- Policy/precedent. The institutional history of an organization is an important source of information. Standard operating procedures (SOPs), formal policy, and informal ways of doing things can be important sources of input. When SOPs become insufficient, for example, that is useful information that is beginning to accumulate.
- Legal issues. The underlying legislative history, authorities, laws, precedents, regulations, programs and such that support the potential responses of your organization can be important. Furthermore, they can change.
- Economics. Costs, benefits, and cost/risk trade-offs are practical elements of any risk management activity that must be considered.
- Political issues. Awareness and knowledge of key state, local, congressional, media, consumer, and international geopolitical concerns with respect to a presenting issue are essential inputs to consider in a risk management response.
- Cooperation and collaboration. Working with knowledgeable and expert stakeholders, partners, academicians, industry, other governments, and international expert(s) to ensure that a shared vision of the problems and their solutions is developed is an essential input source.

An accumulation of inputs from any combination of these areas can surface a problem or opportunity and trigger a risk management activity.

When a problem first comes to your attention it is likely to be poorly structured. Once it reaches your radar screen, the next task is to decide whether you will accept the problem or not. In the public sector this decision may be based on authorities, and in the private sector resources or an organization's mission may provide the basis for accepting or rejecting a problem. Acceptance is usually the most straightforward of the three problem identification tasks, and it is not considered further here.

Think of problem recognition as articulating the initial understanding or description of the problem. Once an identified problem is accepted as an appropriate issue to pursue, a more detailed problem definition can follow. Problem identification, getting the problem right, is one of the most important steps in the risk management task. To aid you in accomplishing this task, a number of problem identification techniques are offered. These may be applied in the problem recognition or problem description stages of problem identification.

6.4 Problem and Opportunity Identification Techniques

Focus on the major things, circumstances, events, and conditions that cause something to happen or not happen. These are the factors that cause issues to arise, and therein lie your problems and opportunities. Focus on the factors that drive situations. Do not waste time on subtleties, as they are not significant. When a subtlety becomes important, it is no longer a subtlety (Jones 1998).

What is not as it should be? Be specific. Are you sure you have identified the actual problem and not just its symptoms? Some problems are relatively routine and quite familiar. They require little problem identification effort. When the needle on your fuel gauge is at or below E and the car engine chokes and dies, there is little uncertainty about what the problem is or how to solve it. Other problems are a little harder to solve; there is a lack of information and some general uncertainty about what the problem really is. Often we can modify existing problem frameworks to accommodate these. The most difficult problems are those that are complex and ambiguous, the ill-formed problems spoken of earlier. They are accompanied by little information and a lot of uncertainty. These poorly structured problems require the most effort to identify.

When a diverse group of people, however large or small, is expected to come to a common understanding of anything, process is important. Problem identification requires at least the risk managers and their assessment team to have a common view of the problem(s). In the cases where the risk management activity is being undertaken by a government organization in some stewardship role (think post-Katrina New Orleans, H1N1 flu, food safety, airport security), it can involve a great many more people.

In my experience, many adults have little use for a formal process, especially well-educated adults. We are all adults, some with advanced degrees; certainly we can identify a problem, right? Maybe we can't, at least not easily. Communication is still our species' greatest challenge, and a formal process can greatly aid communication.

The techniques presented in this section will help you better understand complicated and difficult situations. Having a structured and methodical way to accomplish this all-important first step for risk managers keeps problems from seeming huge and overwhelming. If we cannot provide certain structure to the problem itself, we can at least provide structure to the search for that problem. Techniques often help provide the laser-like focus needed for a successful problem identification process. (See Chapter 3 for a description of this process.)

This section presents 11 examples of techniques that can be helpful in identifying problems and opportunities. There are many more than that identified in the literature (see VanGundy 1988). Use these techniques. Invent your own. Be systematic in identifying problems and you will never regret it. Haphazard approaches are often regretted. Look through the summaries of the techniques presented here. Find a couple you like and give them a try the next time you need to identify a problem.

6.4.1 Appreciation

This technique (Mind Tools 2007) is used to extract maximum information from the sparse facts that are available. Simply start by stating a fact about your problem or opportunity and ask the question, "So what?" What are the implications of this fact as far as the problems it may present? Keep on asking the question until the team has drawn all the inferences possible from the fact. Turn to the next fact and repeat the process. The inferences and facts can then be used to provide an initial description of the problem. This technique is useful when little information is available about a problem.

This technique is especially useful for unanticipated events like the explosion at the *Deepwater Horizon* oil-drilling rig in 2010. There has been a fire and explosion aboard the platform. So what? So lives may be lost. So what? The answer to this question can lead down many paths. Another response may be that we do not yet know its cause. So what? We do not know the risks to which we may still be exposed. So what? And so it goes.

6.4.2 Be a Reporter

Investigative reporters identify and describe problems and opportunities all the time. You can do a lot worse than to ask some what, where, how, why, when, and who questions to help you identify problems. What makes you think there is a problem? Where is it happening? How is this problem occurring? Why is there a problem, i.e., what is causing it? When is it happening? Who is involved with this problem? Be as specific in answering these questions as you can be.

When you have answers to these or similar questions, write the lead to the newspaper story describing this problem. If you do not have a unique "hook" for the story, consider one of these. "This is what is happening and it should not be happening." Or, "This is what is not happening that should be happening."

Do not leap to fixing blame for a problem. Understand the facts of the problem first. Once BP was blamed for the Gulf oil spill in 2010, our attention was divided between stopping the oil spill and blaming BP at a time when perhaps addressing the oil spill was the greater of the two concerns. When a person's temperature spikes into a dangerous fever, it is more important to bring the fever down than to pinpoint the cause.

PITFALLS IN PROBLEM DEFINITION TO AVOID

1. No focus: Definition too vague or broad
   Example: Lack of biodiversity in the watershed.
2. Focus is misdirected: Definition is too narrow
   Example: How can we improve conditions for the mottled duck?
3. Statement is assumption-driven
   Example: How can we stop harmful human disturbances?
4. Statement is solution driven
   Example: Mallow Marsh needs a water control structure.

Often the solution is more important than the cause. Nonetheless, there are times when understanding the cause is essential to solving the problem. So provide the most specific answers to these questions that you can. Specifics make a better story and they define the problem more precisely.

Good reporters follow a good story. You should feel free to continue to ask additional questions once you have the basic facts. If the problem involves an asset or resource, for example, we might zero in on an aspect of the story and ask: What happened or will happen to the resource/asset? What is or will be that event's impact? Who or what does that impact affect? When did or will that impact happen? How did it or will the impact occur? Why did it or will it occur? What could be done about what happens, its impact, who or what it affects, and when, how or why it occurs?

Ask and answer good questions. Write the lead to the story. Follow the story. These three simple steps will give you a good start on identifying a problem.

6.4.3 Utopia

Problems may affect an organization, a system, a process, a population, or any other number of targets. A problem may be an event, a situation, or an ongoing issue. For the next several techniques it will help to develop a little shorthand for all these possibilities. So, when you hear the word "situation," feel free to think any or all of the causes of problems discussed here. Likewise, the decision context is generalized to an organization to keep the language simple.

Look at your organization's situation. Create an idealized or utopian situation for your organization. What do you want to see happen? What do you want to ensure does not happen? What does an ideal future look like? Now compare the existing situation to this utopian ideal. What are the differences? Why do these differences exist? What do the differences suggest to you about the problems and opportunities in your situation? Use the insights gained through a comparison of your situation to a utopian situation to articulate your problems.

6.4.4 Benchmarking

Xerox is credited with developing this technique for comparing practices among businesses. It can be easily adapted to any situation. Who is the very best at what they do in your situation? Who sets the standard for performance? What is that standard? In benchmarking you identify a situation similar to your own in significant resources, size, and other important aspects that is considered the best there is. Then you compare your situation to the benchmark.

Just as with the utopia technique, you are looking for deviations, i.e., the gap between where you are now and where the benchmark situation is. These deviations should be described as specifically as possible. What are the deviations? Where are they occurring? When do they occur? How large are the deviations? Can you explain the deviations? What factors explain them? What are the possible explanations for the most important deviations?

Make each possible cause stand on its own or in combination with others to satisfactorily explain the deviation. Now choose the most likely cause(s) of the deviations and you have an initial problem identification.

This technique differs from utopia in that it uses an actual situation as the benchmark for comparison rather than a utopian ideal. Utopia is used when there is no benchmark for your organization or situation.

6.4.5 Checklists

Examine your situation with checklists in hand. The best checklists provide a comprehensive set of questions to ask, conditions to monitor, factors to consider, and the like. Deep and Sussman (1997) provide a set of 140 lists to aid in decision making. *That's a Great Idea* by Tony Husch and Linda Foust (1986) has several situation analysis checklists. Many of these lists are designed to help you identify problems and opportunities. Arthur B. VanGundy's (1988) *Techniques of Structured Problem Solving*, though out of print, is well worth the search; it offers 105 different creative problem-solving techniques.

The best checklists may be the ones you develop specifically for your own situation. Many food processors, for example, have adopted the Hazard Analysis and Critical Control Point (HACCP) plan approach to food production. This approach breaks the production process down into its basic elements and identifies those points at which things can go wrong. Such an analytical process helps identify classes of problems associated with the production process. The medical profession provides many excellent examples of checklists useful in diagnosing health problems. You may have to develop your own checklists initially, but once developed they can prove to be an invaluable aid in identifying recurring kinds of problems.

6.4.6 Inverse Brainstorming

Inverse brainstorming unleashes the destructive energy in us that we have been suppressing since childhood! James T. Higgins (2005) offers this technique and 100 others in his excellent book, *101 Creative Problem Solving Techniques*. In this technique, you begin by considering your situation satisfactory. Then you nitpick it. See how many things you can find wrong with it. These potential obstacles and shortcomings become the bases for your problem identification.

Inverse brainstorming can be especially useful in dealing with opportunity risks. With this technique you look at an opportunity and brainstorm all the things that might prevent it from being fully realized. Why can't ecosystem habitat be increased? The technique can be particularly good at exposing the implicit assumptions that lie in our unconscious minds. When only one person or when like-minded people look for what can go wrong, beliefs and values may limit our ability to see the reality of a situation. These beliefs can often be exposed by expanding the inverse brainstorming circle to include those outside your organization or the immediate situation.

Little gets people's creative juices flowing faster than deconstructive criticism. Ask people to find ways to create problems with your situation. Do a premortem on your situation and think about what can go wrong. Ask people how we can decrease

habitat, increase cases of food-borne illness, introduce radon into our homes, cause the structure to fail, and so on. Get into a destructive problem-causing mode. Think like a bad guy or at least a severe pessimist. The insights gained from this process can help you identify the things that can go wrong as well as the ways they can happen.

6.4.7 Bitching

Closely related to the inverse brainstorming technique is bitching. This gives full reign to the negative energy of humans. It differs in that while it can consider obstacles to opportunities, it may be more useful for problems. It simply invites people to let it rip and tell us what is wrong with our situation.

As a practical matter, you gather as many people as possible who know the situation and brainstorm (see Chapter 7) their lists of complaints. As the number of complaints begins to grow, they can be organized into subject areas. The subject areas are in turn grouped as necessary to get to a definition of problems.

6.4.8 Draw a Picture of the Problem

Draw a picture of the problem to make sure you are identifying the real problem. Drawing pictures taps into the creative side of the brain. Seeing the problem in a picture is a useful way to gain insights into the problem and its linkages with other ideas. The picture can be a flow diagram or a crude symbolic drawing; do not let a lack of artistic sense stop you from drawing the problem. Many problems often lend themselves nicely to visualization. There are no further rules to this technique. In one version of this technique, each member of the group is invited to draw a picture of the problem. The group then develops a consensus picture from the various individual versions. In another version, the group is tasked with developing a drawing jointly from scratch. The pictures are explained to everyone's satisfaction, and this explanation becomes the basis for the problem description.

6.4.9 Mind Maps

Mind maps are a specific kind of problem picture that is growing in popularity and ease of use. Mind maps are diagrams used to represent many ideas and their linkage to a central concept. They are extremely useful for visualizing, structuring, and classifying ideas. They are also useful for analyzing, comprehending, synthesizing, recalling, and generating new ideas as well as for illustrating problems. Mind maps can be done by hand. The basic steps for creating a mind map are:

- Write the key problem word or phrase in the middle of a blank sheet of paper. Draw a circle around it.
- Think of as many related subtopics as you can. Write them down and connect each of them to the center with a line.
- Treat each subtopic as if it is the central idea in its own mind map. Identify facts (or, for a complex model, another tier of subtopics) for your subtopics and repeat the process. Generate as many of these lower-level subtopics as you see fit. Connect each one to its corresponding topic.

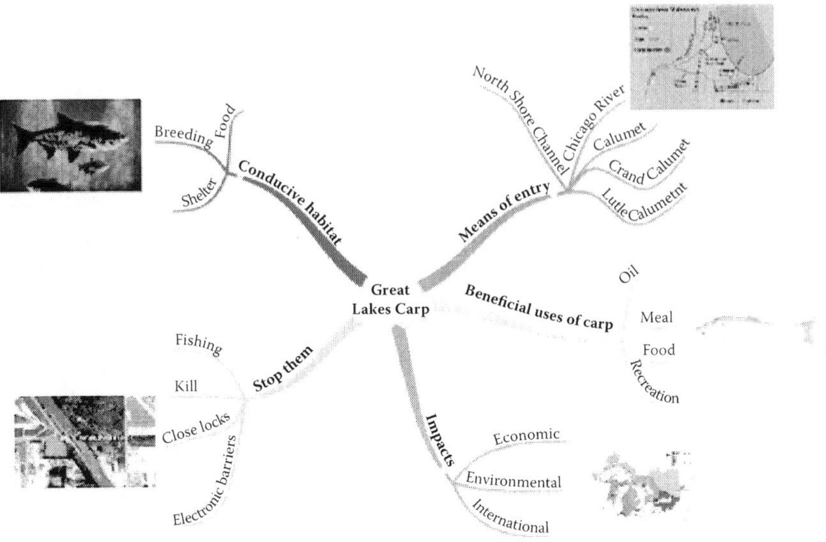

FIGURE 6.2 Mind map example for Asian carp entering Lake Michigan.

- Be as visual as you can be. Use colors, photographs, stick-figure drawings, sketches, and symbols generously.
- Limit your words. Keep your topics, facts, and phrases as short as possible. One word is good; a picture is better.
- Engage your brain; use variety. Different-size print or script, colors, thickness, lengths, curvature, alignments, and so on should be used. They encourage our creativity.

Figure 6.2 shows how a mind map aids thinking about the problem of the Asian carp reaching Lake Michigan in the United States. There is concern that this invasive species may become established in the Great Lakes. Once a map like this is drawn, it becomes easier to see the complexity of the problem and to develop a well-focused problem statement from it.

The example here is kept intentionally simple. The map suggests ways to frame the problem in terms of pathways, impacts, science, or even solutions. The map shows entry pathways, impacted areas, structures on the waterways, and even the species itself. The pictures are often more effective than words in stimulating thought. The eventual problem description can be developed by integrating the key insights from the mind map.

6.4.10 Why-Why Diagram

The why-why diagram identifies possible causes of a problem, and when done well it quickly sorts ideas into useful categories. It is sometimes called a cause-and-effect diagram, fishbone diagram, or Ishikawa diagram. This technique is most useful when causes are important to know. It helps the team think through the causes of a problem in a thorough manner. It pushes the team to consider all potential causes of a problem

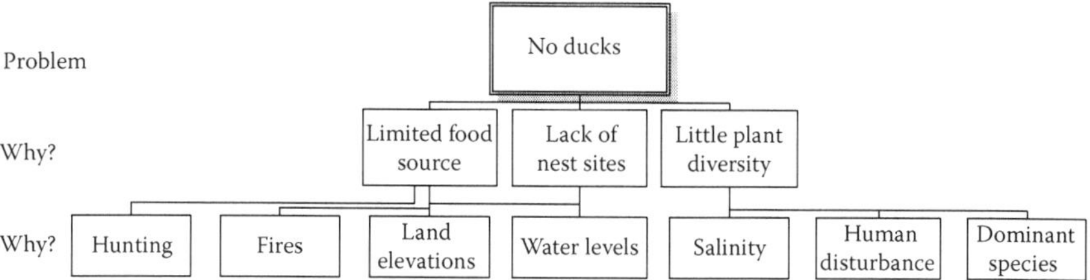

FIGURE 6.3 Why-why example for mottled duck habitat.

rather than focusing too quickly on a single cause, a symptom of the problem, or on affixing blame. The why-why approach is a useful tool for moving from an early recognition of the problem to a more mature description of the problem.

It can be done as simply as this: Divide a piece of paper into at least three horizontal sections. Write the problem in one section of the paper. In the middle section ask why this is a problem, ask why it happens, and list as many reasons for the problem as possible. The third section is the second level of "whys." Here you write the reasons for the factors identified in the middle section. Add as many subsequent why sections as you need to sketch and understand the problem. An example, explaining the absence of ducks at a marsh site, is shown in Figure 6.3.

Convenient and easy-to-use software tools have been developed to assist more sophisticated why-why models. Such tools can be easily found on the Internet by using the search term "fishbone diagram."

6.4.11 Restatement

Once you have begun to define a problem or opportunity, restate it in as many ways as you can. Look at and imagine your situation from as many different perspectives as possible. The basic idea is to see your problem through many different sets of eyes to broaden your perspective of a problem.

The obvious place to begin is with your situation's different stakeholders. Consider the problem of public elementary education in the United States. How would school administrators phrase the problem? What would politicians, students, teachers, administrators, janitorial services, the local newspaper, textbook makers, neighbors, crossing guards, and so on say?

Once you have exhausted the list of stakeholders, it can be helpful to think about how different professions and demographic groups would state the problem? What would engineers say? What about medical doctors, police officers, African-Americans, Asians, liberals, conservatives, the uneducated, the wealthy?

Imagine how different media would characterize your problem. What would CNN say? What about Fox News or ESPN? How would the magazines *Time*, *All About Beer*, *Vanity Fair*, *Popular Mechanics*, *Mad*, or *Home and Garden* state your problem? Finally, you might consider how specific individuals, real or not, might state the problem. What would the mayor or governor say? How about the president of the company or your fourth-grade teacher? What would JFK, Aristotle, Bart Simpson, Popeye, or Superman say about the problem?

This technique can be a lot of fun when an honest effort is made to explore the different perspectives on a problem. It is especially helpful for identifying biases and

TABLE 6.5

Techniques for Problem Restatement

Every problem can be viewed from multiple perspectives. Restating a problem as many ways as possible is a good way to explore these perspectives and the dimensions of a problem. To obtain a good definition of the problem, restate it in as many ways as possible. Here are a few ways to restate a problem that have proven effective. Restating should take no more than 5 or 10 quality minutes when you start with a basic statement of a problem.

Initial statement: Freshwater fish and invertebrate species are being eliminated.

1. Paraphrase: Restate the problem using different words without losing the original meaning.
   Paraphrased: How can we preserve and restore freshwater fish and invertebrates?
2. 180 degrees: Turn the problem on its head.
   180 degreed: How can we eliminate freshwater fish and invertebrate species?
3. Broaden the focus: Restate the problem in a larger context.
   Broadened focus: How can we achieve greater biodiversity?
4. Redirect the focus: Boldly, consciously change the focus.
   Redirected focus: Saltwater fish species are on the increase.
5. Use the why-why approach: Ask "why" of the initial problem statement. Then formulate a new problem statement based on the answer. Then ask "why" again and restate the problem based on the answer. Repeat this process a number of times until the essence of the "real" problem emerges.
   Why? Because salinity levels are changing in the marsh.
   Restatement: Salinity levels are changing in the marsh.
   Why? Because salt water has entered this formerly freshwater marsh.
   Restatement: Salinities in the freshwater marsh are increasing.
   Why? Because the navigation channel has introduced a saltwater wedge to the river that feeds the marsh.
   Restatement: Dredging has increased salinities in the marsh, eliminating freshwater fish and invertebrates.

---

assumptions when your original problem description is compared to those you identify. The goal is simple: Identify as many different ways of seeing and expressing the problem as possible. Write them down. Do you see the problem any differently? Do you understand it better? How can you best consolidate these views into a cohesive statement? Table 6.5 offers a slightly different approach to the restatement technique.

6.5 The P&O Statement

Your problems and opportunities statement isn't real until it is written down. Your risk management activity should never be without a written list of current problems and opportunities. Once your list starts to take shape, it takes very little time, e.g., in a team meeting, to ask if there are any revisions or evolutionary changes to be made to the P&O statement. Keep it current and up to date. Make sure every team member and all the significant stakeholders always have access to a current copy.

Publish it on your Web page. Tell people that these are the problems as you see them. Ask them, "Do you agree? Did we get them right? Have we missed anything? Is there anything here that does not belong?" If your risk management activity is not in the public sector, ask your risk analysis team these same questions. Vet your work. It takes some time, but it takes less time to do it right than to do it over.

What should the P&O statement look like? It is short, a page or two at most if you have multiple problems and opportunities. Each problem and opportunity is succinctly stated in a sentence or two at most. They are numbered for convenience. Eventually, you may develop an expanded definition and description of each item, but you need not do so for the P&O statement.

None of this brevity is intended to reduce you to sound bites and clichés. It is intended to focus the risk management activity and crystallize the attention of managers, assessors, and stakeholders. The statement enables you to say why you are conducting a given risk management activity. The risk assessment can still weave a story and explain the problems and opportunities in as much detail as necessary. Your P&O statement needs to be clear, concise, and complete. It is the reason for the risk management activity. In this respect, it is like a mission statement for the activity. Continuing the analogy, the objectives and constraints statement, then, is like the vision statement for the activity.

An optional step between the P&O statement and the risk assessment's full definition of the problem is to prepare a fact sheet or P&O profile, not to be confused with the risk profile. For example, a profile might identify the source that first recognized the problem, the public or extra-organizational concerns about the problem, as well as a few technical details to help flesh out the problem a little more. A simple template for a profile is shown in Table 6.6.

How do you know you have a good P&O statement? When people understand it. If you show it to someone and they understand why you are doing an assessment, you have succeeded. How do you know if they understood it? You must vet it, publish it, communicate it, and seek feedback on it. How do you know when it is final? When people agree with it or at least accept it. It is final when you no longer find a reason to change it.

Don't wait until you have a good P&O statement to begin using it. A good statement takes time to develop, but a useful statement is the one you have. There are two things to do with a useful P&O Statement: You can improve it, or you can use it.

6.6 Summary and Look Forward

Problems are undesirable conditions we want to avoid. They usually require risk-avoidance strategies. Opportunities are desirable conditions we seek to capitalize upon. They usually require risk-taking strategies. It is critically important that problems and opportunities be as clearly and carefully identified from the outset of a risk management activity as possible. One of the most common and most avoidable weaknesses of a poor risk management activity is solving the wrong problem or failing to realize an opportunity.

Identifying problems and opportunities is important. Risk managers would be well served to know and use one or more problem-identification techniques. Several techniques to help you prepare an effective problems and opportunities statement have been offered in this chapter.

TABLE 6.6

Problem and Opportunity Profiling Template

Problem/Opportunity Profile

1. Source: What source first identified the problem or opportunity? Examples: higher authority, media, stakeholders, conversation with elected officials, experts, field observations.
2. Public concerns
    a. Advocate: Who is the spokesperson for the problem or opportunity? Identify specific groups, agencies, and individuals.
    b. Basis: What is the advocate's basis for the problem or opportunity? Examples: homeowners who have experienced flooding, state agency legally mandated to oversee wildlife resources.
    c. Background: In the advocate's view, what is the problem or opportunity, and what are the causes and effects?
    d. Other stakeholders: Who else believes the problem or opportunity does or does not exist? Why or why not? Identify specific groups, agencies, and individuals.
3. Technical analysis
    a. Subject: State the problem or opportunity.
    b. Location: Describe the location of the problem or opportunity; map it if possible.
    c. Measurement: Identify one (or more) measurable indicator that is used to measure change in the problem or opportunity.
    d. Conditions: Describe past, present, and future conditions related to the problem or opportunity:
        (i) Historic condition
        (ii) Existing condition
        (iii) Future "without project" condition
    e. Decision criteria: Identify any standard, target, or other criteria that may be used to define the magnitude of a problem or opportunity. For example: state water quality standards, design specifications, legislated targets, strategic plan goals, profits, costs, and so on as appropriate.

    Causes and effects: Describe causes and effects of the problem or opportunity.

    Conclusion: Does the technical analysis support the public concerns about the problem or opportunity? Why or why not? How do the historic, existing, and future conditions compare to any applicable decision criteria?
4. Problem/opportunity statement: Write a clear and brief description of the problem or opportunity.
5. Information

    Sources: List sources of information about the problem or opportunity.

    Information needed: Briefly describe the types of additional studies needed to address the problem or opportunity.

Source: Yoe and Orth (1996).

Many professionals are wary of techniques; they tend to regard them as gimmicky and unnecessary. My experience says some parts of the risk management process can be vastly improved with a good technique or two. Problem and opportunity identification is one such instance. Another instance is at the points in the risk management process where divergent thinking is needed. This is where brainstorming, another useful tool for the risk manager's toolbox, can be very helpful. Brainstorming is the topic of the next chapter.

REFERENCES

Deep, Sam, and Lyle Sussman. 1997. *Smart moves: 140 checklists to bring out the best in you and your team*. Rev. ed. Cambridge, MA: Perseus. http://www.questia.com/.

Food and Drug Administration. 2003. *CFSAN's risk management framework*. College Park, MD: Center for Food Safety and Applied Nutrition.

Higgins, James. M. 2005. *101 creative problem solving techniques: The handbook of new ideas for business*. Rev. ed. Winter Park, FL: New Management.

Husch, Tony, and Linda Foust. 1986. *That's a great idea*. Oakland, CA: Gravity.

Jonassen, David H. 1997. Instructional design models for well-structured and ill-structured problem-solving learning outcomes. *Educational Technology Research and Development* 45 (1): 1042–1629.

Jones, Morgan D. 1998. *The thinker's toolkit*. New York: Three Rivers Press.

Meacham, J. A., and N. C. Emont. 1989. "The interpersonal basis of everyday problem solving." In *Everyday problem solving: Theory and applications*, ed. J. D. Sinnott, 7–23. New York: Praeger.

Mind Tools Limited. 2007. *The Mind Tools e-book*. 5th ed. London: Mind Tools Limited. http://mathamlin.com/files/MindToolsEbook.pdf.

VanGundy, Arthur, Jr. 1988. *Techniques of structured problems*. New York: Van Nostrand Reinhold.

Wood, P. K. 1983. Inquiring systems and problem structures: Implications for cognitive development. *Human Development* 26:249–265.

Yoe, Charles, and Kenneth Orth. 1996. *Planning Manual. IWR Report 96-R-21*. Alexandria: Institute for Water Resources.

7

Brainstorming

7.1 Introduction

Good risk management begins with divergent thinking: lots of ideas and lots of different perspectives. Too often we are prisoners of our own beliefs, biases, mind-sets, organizational cultures, and authorities. We have difficulty seeing and accepting other perspectives on a problem and its solutions. We see more obstacles to opportunities than we see possibilities.

Every organization faces limitations on what it can and can't do, but no organization should face a limitation on what it can think. If the best ideas are beyond your authority, new authorities may be obtainable. If they are really good ideas, someone else may be willing to implement them on their own. We can never be sure we have the best idea unless we have selected it from among many other good ideas.

Brainstorming is a proven effective methodology for generating ideas. In my experience, it is rarely used and when it is, it is not always used effectively. Follow-through and follow-up are often missing. Virtually everyone has heard of or used brainstorming techniques at one time or another, and most people have an opinion about brainstorming. With no research behind me and only my own gut instinct, I again offer my own pet theory. Adults, especially well-educated adults, do not like to use techniques. They seem gimmicky and unnecessary for smart people. Serious people often have difficulty thinking in an uninhibited fashion. So, despite its proven benefits, many people do not seem terribly fond of using group techniques like brainstorming.

There seems to be at least some support for this idea in the literature. Isaksen (1998) reports a National Science Foundation official at a conference saying, "We all know that brainstorming is nothing more than executive entertainment." De Bono (1992) said of brainstorming, "It was designed for use in the advertising industry.... Novelty and gimmicky does attract attention.... I find that people who have a brainstorming background tend to perform rather poorly...they are always looking for the way out, exotic idea and often miss the simple, practical idea which is at hand."

That is an unfortunate experience, if literally so, because I have used brainstorming consistently and always with good results. I have no solutions to help those uncomfortable with brainstorming feel more comfortable with it, beyond trying it a few times. It can be an invigorating and productive experience.

This chapter begins by considering some of the things you can brainstorm. The four cornerstones of brainstorming are identified, followed by four pitfalls to avoid. Suggestions for things to do both before and during the session are offered to help avoid the pitfalls and other problems during the session. Once this background is completed, we'll consider a range of oral and nonoral brainstorming techniques and follow this by carefully considering my favorite technique. This is followed by a description of the colored-dots evaluation process, which works well with large groups. The chapter concludes by considering whether you need a professional facilitator or not.

7.2 What Can You Brainstorm?

Brainstorming can help us find solutions to problems, and risk managers often need to find creative solutions to problems. Actually, brainstorming can be used for a wide variety of purposes in a risk management activity. It may be helpful to use brainstorming techniques to frame and represent the problems and opportunities. Once the problem is articulated, brainstorming can obviously be used to formulate risk management options that could solve the problem.

Brainstorming can also be used to identify hazards of concern, opportunities for improvement, and stakeholders for an issue, or you can use brainstorming to figure out what the appropriate decision criteria or risk metrics might be. Brainstorming is a flexible tool that does not require a highly organized process. Once you master a few techniques, you'll find many opportunities to apply them.

7.3 Background

Alex Faickney Osborn (1963), an advertising manager, is credited with at least popularizing, if not introducing, the brainstorming methodology in his book, *Applied Imagination: Principles and Procedures of Creative Problem Solving.* In the third revised edition of his book, Osborn offered four basic rules for brainstorming that are still valid today. They are presented here as the cornerstone concepts for a good brainstorming session. They are followed by four barriers to a good brainstorming process.

7.3.1 No Evaluation

Brainstorming is for generating ideas, not for evaluating them. Fear of criticism or of looking foolish is one of the major reasons people hold back their thoughts and ideas in open forums. To provide a safe and fun environment for generating ideas, there is no criticism of any idea during the brainstorming session. Neither is there praise for an idea. Divergent thinking is the goal of a brainstorming session, and ensuring that people feel comfortable with the notion of generating unusual and incomplete ideas is essential to achieving that goal. In time, the ideas will need to be evaluated, but that does not occur during the idea-generation stage of brainstorming.

Related to the idea of no evaluation is the no censoring rule. Not only should individuals not evaluate the ideas of others, pro or con, they should not evaluate their own ideas. Do not censor your thinking; make an effort not to exclude or keep an idea to yourself. If it pops into your head and expressing it aloud will not land you in jail, then by all means share the idea.

7.3.2 Unusual Ideas

Brainstorming is not just after ideas; it is especially interested in unusual ideas. Creating an enthusiastic and fun atmosphere that encourages participation by all team members will loosen the creative juices. Unusual ideas are welcomed for a couple of reasons. First, the seed of a solution is often found in an unusual idea. Some of our best solutions come from new ways of looking at the world. Second, unusual ideas often inspire others to generate unusual ideas of their own or to build on another's ideas. Unusual ideas are contagious. Third, it is easier to back off from the weird ideas than it is to step up the tame ideas. Encourage everyone to exercise their creativity. Encourage people to consider wild ideas.

Work is serious. But work can be fun, too. The fun atmosphere is an important element of successful brainstorming, in my experience. In a serious atmosphere, wild ideas are just stupid, a waste of time. A bad experience with brainstorming is likely to sour one on the potential for this technique. So do not overlook the importance of tending to the relaxed, fun aspect of brainstorming.

7.3.3 Quantity Counts

Another goal of brainstorming is generating the greatest number of ideas in the least amount of time possible. The more ideas generated, the more likely we are to have some really good ones. To get more ideas "out there" in a given period of time, don't worry about the details. A germ of an idea is sufficient. A word is often enough. Brainstorming is not the time for details. The best techniques are those that generate the longest list of ideas.

7.3.4 Combine and Improve Ideas

Stealing and plagiarizing are encouraged. Take someone else's idea and tweak it. Take two ideas and combine them. Build on the ideas of others and add to their thoughts. Look to other ideas for your own inspiration. Hear a silly idea? Out-silly it. Keep the process going; feed on the creativity of the group. Adapting and improving an idea is every bit as valuable as generating an original idea.

7.3.5 Four Pitfalls

In addition to my own pet theory about an aversion to seemingly gimmicky techniques, the literature (Beasley and Jenkins 2003) identifies four common hurdles to brainstorming's effectiveness. These are group domination, free riding, "groupthink," and "groupshift."

Because the purpose of brainstorming is to involve the entire group in generating and sharing as many ideas as possible, group domination is one of the most deleterious problems that can arise. It occurs when one or a small number of participants from the group dominate the process and squash the creative energy of the group. This obviously reduces the likelihood that the effort will produce results of value.

The more hierarchical the structure of the organization, the more susceptible a group is to this pitfall. Junior, especially newer, employees are often inclined to defer to senior employees or supervisors. Intimidation is still the style of senior staff in some organizations. Some group members may defer to those with more responsibility for the effort, believing they have had an opportunity to think in more detail about the issue and are better qualified to speak to the issues. In addition to the hurdles of hierarchy and responsibility, there are also individuals with, shall we say diplomatically, great confidence in their ability and a determination to present their views. The benefits of brainstorming may never materialize if any of these conditions are present. Fortunately, brainstorming techniques like the "3× Yeah" process described later in this chapter can be designed to minimize the opportunities for group domination.

Free riding, sometimes called social loafing, can also limit the success of a brainstorming activity. It occurs when people fail to engage in or disengage from the process. Free riders expect others to do the heavy lifting in the process. This phenomenon is more likely when a group is large and it is easy to hide or get lost in the crowd. People may disengage for a number of reasons, however, including the absence of a motive to participate, the lack of incentives for active participation, lack of buy-in to the team, and feelings of futility possibly even related to group dominance.

One of the more insidious reasons for free riding is fear of evaluation. There is often a strong negative reaction to this sort of group exercise, especially among those who fear they could lose dignity, standing, or credibility. These kinds of fears may prevent people from offering creative or even goofy ideas. They can also make people less likely to voice an unpopular idea or opinion. Some people may feel they lack the experience or qualifications to speak about a topic, or they may be afraid to do so in front of a more experienced group. Self-censuring and censoring can cause free riding in a group process. Prior notice of a brainstorming session and its agenda, accompanied by clear ground rules, an encouraging environment, and small groups can often offset a tendency to free ride.

Groupthink can arise from the snowball effect when a group is anxious to reach a consensus for any reason at all. It is another impediment to the creativity of a good brainstorm. Groupthink limits the number of ideas that are generated or, if the brainstorm includes an evaluation step, the number of ideas evaluated.

Any number of group dynamics may limit the members' willingness to offer new or additional ideas. Some groups may be inclined to align too quickly with a single viewpoint on an issue. When this happens, it is easy for new ideas to reinforce the central idea and give rise to that snowball effect. Others, perhaps anxious to limit the pain of a brainstorming session, may quickly abandon their perspective in favor of the group's emerging consensus. Groups are particularly susceptible to this effect when the individuals have not bought into the process and are anxious to get back to their "real" work. This dynamic can create a rush to judgment that counters the benefits of brainstorming.

Polarization or groupshift is a fourth pitfall to a successful group process. Consensus building is often a desired outcome of a brainstorming process. We would like to think that reaching a conclusion on an issue is a desirable thing. Groups can at times harden their positions on an issue. Those with a certain view of a problem, upon hearing the quite different views of others, may dig in their heels and harden their positions. Conservatives on an issue become more conservative and liberals on an issue become more liberal. Extreme views become more extreme, and the benefits of brainstorming are limited. Keeping evaluation out of the idea-generating portion of a brainstorming process is the best way to avoid groupshift.

7.4 Avoid Problems in Your Process

The best way to avoid the problems described in the previous section is with a well-defined process. That would be a process based on Osborn's four principles and one that intentionally seeks to minimize the manifestation of the four pitfalls described in the preceding discussion. A number of useful ideas can be gleaned from the literature (Beasley and Jenkins 2003, Panitz 1998, Sutton 2006) and from practical experience with the process. Some simple preparations will help with virtually any brainstorming technique.

There are a few things to do prior to the brainstorming session. First, know why you want to hold a session. You need a very specific topic, so carefully define the focus of your session. Someone or some group must identify the problem that needs to be solved by the brainstorming session. Make sure everyone is clear on the focus of your session.

Second, decide what kind of session you will run, who will run it, and whom to invite to it. There are any number of brainstorming techniques, as a quick Google search will confirm. A few specific techniques I have found success with follow. Every session needs a leader, and that person needs to be familiar with Osborn's principles, the things that can go wrong, and the technique being used. We'll return to the idea of a facilitator near the end of the chapter when we discuss whether you need a professional facilitator or not.

Make sure you have the right people at your session. A small group is usually six or fewer (Aiken et al. 1994); a large group is more than six. A small group tends to complete tasks more quickly. It also minimizes the potential for group domination and free riding. A larger group tends to provide better solutions to problems and more ideas because there are more people thinking about the topic. The downside of a large group is that it tends to deter some people from participating. Although there are some specific large-group brainstorming techniques, a common compromise when faced with the need or desire to involve many people is to divide the large group into subgroups and to work at the subgroup level. You can have several small groups working at once.

Third, assign homework. Let everyone involved know what the problem is well in advance of the scheduled brainstorming session so that they can begin to think about it. Providing a context for the brainstorming session is important. This is also a good first opportunity to allay fears and to begin to create a comfortable environment for all to work in. The session leader should encourage all group members to use their different experiences and knowledge to enhance the process. Stress the value of fresh eyes from new employees and address other easily anticipated obstacles to openness at the time the session is announced. Also let people know that this session will not be time-consuming.

Fourth are the logistic concerns. Secure a meeting space and prepare any necessary materials. These requirements will change with the type and size of your session. It is usually desirable to schedule the brainstorming session as a stand-alone meeting rather than as part of a larger meeting. Once the presession preparation is done, you can turn your attention to what needs to be done during the session. Six suggestions are offered.

First among the session tasks is to establish ground rules. The importance of this task is difficult to exaggerate. Establishing the ground rules is an essential part of providing a comfortable environment for the sessions, which, in turn, is the best way to combat the pitfalls of brainstorming. Each group member should know what is expected of herself and others. All participants should feel that their input is valued and that their voices will be heard. To that end, participants should understand how the session will proceed and how the ideas generated during the session will be used by risk managers.

Second, set the tone for the session. Value each individual and all ideas. Not everyone enters the session with the same expertise or commitment to the process, so it is vitally important to establish that comfortable environment. The leader must establish a tone that invites and encourages all members to contribute to the session. An open mind is to be valued over conformity during the session.

Third, establish a "zero tolerance" on all criticism, defensiveness, and commentary during the idea-generation stages of the session. Nothing can shut down the productivity of a brainstorming session faster than a laughing supervisor, rolling eyes, or a stern rebuke. Criticism can quickly put an end to one's willingness to participate in the process. Comments and cross talk can also divert the session from its purpose to other topics, so they are best avoided. The facilitator needs to make it clear that no criticism of any idea will be tolerated during the session, without doing so in a heavy-handed way. One of the best ways to ensure this is to establish and enforce rules barring evaluation and cross talk during the oral brainstorming session.

Fourth, always encourage more ideas, not less. The group should make an effort to produce the largest number of ideas possible. A skilled facilitator may find it useful to reframe an issue to spur a growing number of ideas.

Fifth, credit the group, not individuals. Brainstorming is a group activity, and a good facilitator will recognize the group process and group ownership of ideas. Stressing the group process is more likely to increase the group's interest in and commitment to the process than does recognizing individual contributions to the process.

Sixth, follow the rules or "it ain't brainstorming!" The facilitator cannot remain silent if the group begins to lose its focus or if rules are violated. One of the reasons people are not fond of brainstorming is a prior experience in a poorly run session. We have all been asked to brainstorm when there were no rules or when the few rules that were in place were not followed.

7.5 A Few Good Techniques

The variety of brainstorming techniques have been covered too well in too many places to compete with here. What I would like to do is distinguish between oral brainstorming and nonoral brainstorming techniques, then offer a more detailed consideration of a technique I have found easy and effective to use. Some, like

Aiken et al. (1997), consider brainstorming to be oral idea generation, and they would call nonoral idea generation brainwriting. This distinction works well enough to adopt here.

7.5.1 Brainstorming

Open brainstorming (Beasley and Jenkins 2003) is a process with few rules and procedures. It is characterized by its unstructured technique. Think of the leader standing at the front of a room writing down ideas as they come to mind in a free-for-all kind of gestalt. That is open brainstorming. It is frequently the first experience many of us have with brainstorming, and it often comes about when a meeting bogs down, someone says, "Let's brainstorm," grabs a marker, and steps to the flipchart. Open brainstorming is prone to ignorance of Osborn's simple tenets as well as to the pitfalls discussed previously (and many others not mentioned) that result from poor preparation. In general, open brainstorming is not going to be terribly effective.

SCAMPER

Alex Osborn (1963) posed questions to help spur brainstorming creativity that was adapted by Bob Eberle into an easy-to-remember acronym.

Substitute. What if I used a different ingredient, material, person, process, power source, place, approach, or tone of voice?

Combine. Could I use a blend, an alloy, an assortment, or an ensemble? Could I combine units, purposes, appeals, or ideas?

Adapt. What else is like this? What other idea does this suggest? Does the past offer a parallel? What could I copy? Whom could I emulate?

Modify. Could I add a new twist? Could I change the meaning, color, motion, sound, odor, form, or shape? What other changes could I make?

Magnify. Could I add something, such as more time or greater frequency? Could I make it stronger, higher, longer, or thicker? Could I add more ingredients? Could I duplicate, multiply, or exaggerate it?

Minify. Could I subtract something? Could I make a condensed or miniature version? Could I make it lower, smaller, shorter, or lighter? What if I omit something, or streamline, divide, or understate it?

Put to other uses. Are there new ways to use this as it is? Does it have other uses if I modify it?

Eliminate. Could I remove anything? Could I disaggregate or decompose it?

Rearrange. Could I interchange components? Could I use a different pattern, layout, or sequence? What if I transpose the cause and effect? What if I change the pace or the schedule?

Reverse. Could I transpose positive and negative? What if I tried the opposite or reversed roles? What if I turn it backward, upside down, or inside out?

Round-robin brainstorming is a somewhat more structured technique that provides each group member with a turn to present his or her ideas. Turns do help minimize group dominance problems, but they can also stifle creativity and spontaneity. Waiting for one's turn can lessen feelings of connectedness to the group, and it leads to productivity gaps as people wait for their turn. A common variation of the round robin is to ask for one idea at a time, going around the group until the supply of ideas is exhausted. Those who are out of ideas simply pass.

These techniques encourage social interaction and a high level of group cohesion. The disadvantages are that: People must take turns, a significant concern with large groups; ideas must often be summarized, and content is lost; the process lacks anonymity, so some people may not contribute or say what they really think.

7.5.2 Brainwriting

Brainwriting has been devised as a group of techniques that help overcome some of the problems with oral brainstorming. Brainwriting can be done in a face-to-face setting or in nominal group settings. Electronic brainwriting, using digital technology, has developed as a nominal group brainwriting technique in the last two decades. Brainwriting is characterized by individual and silent idea generation followed by written communication of these ideas. The greatest advantages of this technique are: No one needs to await a turn, so the group may be more productive; all ideas are recorded; anonymity can be preserved for all group members.

Brainwriting is less likely to require a professional facilitator, and it can work with people who have little experience with brainstorming. It is also useful when group domination is a real concern or when there is conflict among individuals in the group. Its principal limitation is that it may not meet the social interaction needs of group members, which could cause groupshift and the attendant hardening of polarized opinions (Aiken et al. 1997). There are probably as many brainwriting techniques as there are brainstorming techniques, so we will focus on two broad categories of techniques: poolwriting and gallery writing.

7.5.2.1 Poolwriting

Poolwriting pools ideas. Although the details of this technique vary widely, the basic steps are some version of those described by Geschka, von Reibnitz, and Storvik (1981). Assuming the preparation has been done, poolwriting consists of:

1. Each individual silently writes down his ideas about the topic on a sheet of paper.
2. The papers are all placed in the center of the table (or some other pool).
3. Each individual takes a sheet of paper from the pool, reads the ideas on the new sheet, and uses them to stimulate new ideas, which are added to the sheet.
4. Once new ideas are added to the sheet, it is exchanged for another sheet from the pool. Often there may be a rule that at least one new idea must be added to a sheet before it can be returned to the pool.

5. Each individual continues to write down ideas and exchange sheets for three or more iterations.
6. An optional step would include reading aloud the ideas and evaluating them.

Anonymity is preserved by not signing the uniform sheets of paper. Everyone can work at the same time, so no time is spent waiting for a turn or listening to others. There is a permanent record of each idea generated. The downside of poolwriting is that individuals do not get to see all of the ideas generated during a meeting. Different people are likely to see different things. Some of these limitations have been overcome, and some advantages have been enhanced, by electronic poolwriting.

7.5.2.2 Gallery Writing

An interesting innovation of the poolwriting technique is to make the process a bit more open in a technique called gallery writing (Aiken et al. 1997). This also helps overcome the problem of limited exposure to the ideas generated in brainwriting. It involves some version of the following steps:

1. Flip chart papers are attached to the walls of a room, or the flip charts themselves can be stood around the sides of a room.
2. Each individual silently writes down her ideas about the topic on a sheet of paper.
3. Individuals wander around the room and record their ideas on the flipcharts.
4. Individuals then read others' ideas and add new ideas to the sheets of paper.

This technique moves people around the room instead of moving paper like poolwriting does. Gallery writing enables people to write and view comments simultaneously. This is believed to increase group cohesion, but at some cost of anonymity, as others can see what an individual is writing.

7.5.2.3 Electronic Brainwriting

Brainstorming techniques are being combined with software technology. Brainwriting techniques have been readily adapted to digital technologies in what has come to be called electronic brainwriting. Electronic brainwriting improves over the manual version by increasing anonymity. No one will recognize your handwriting, and no one needs to see what you write. This helps to remove organizational politics and personalities from the brainstorming process. Flexible scheduling and/or extended time periods can be offered for the session. This can enable better-developed contributions if participants have time to collect their own thoughts, read those of others, and reflect more on the topic after hearing new perspectives on an issue.

Some software systems may offer creative incentives to participants. Free-rider problems can be overcome, for example, by offering points based on the quantity and quality of ideas that enable participants to share in real (e.g., cash, compensatory time) or symbolic (e.g., plaques, candy bars) rewards. Incentives have been shown to be effective in improving idea generation outcomes (Toubia 2006). Electronic techniques are well suited to larger groups. For example, it is possible to open a process for an extended period of time, say two weeks, that enables participants to post ideas anonymously and continuously from anywhere in the world. They can work on their own schedule whenever an idea strikes them. Internet searches for electronic brainwriting software and electronic meeting systems will introduce you to some of the available options.

7.6 3× Yeah

The available brainstorming techniques are numerous. I learned my favorite one from Ken Orth, a planner and friend. It closely follows Olson's method, and although the use of a professional facilitator would no doubt improve the process, I do not believe that one is required. What is required, however, is a champion of the technique familiar with the approach who will be able to make the participants feel safe enough to get a little creative. The process, called "3× Yeah" (as in the Beatles's "Yeah, yeah, yeah"), is designed to add a little fun to it. It works like this.

- Provide materials
- Identify the question
- Explain the process
- Silent idea generation (brainwriting)
- Group idea generation (brainstorming)
- Preliminary evaluation
- Award prizes
- Follow up

You will need a workspace that is flexible enough to provide individuals with a private writing surface and enough room to break out into small groups of six or less. Index cards and pencils are needed for private idea generation and a flip chart with markers (or some other communal form of recording, like a digital file) is needed for each group. Post-it flipcharts or some means to display the group results (e.g., LCD projector) are also needed.

The issue you are brainstorming needs to be clearly stated and well focused. People need a specific task, a single question to which to respond. Brainstorming is not so good for ill-defined problems and general tasks. Every process begins with a well-defined question on which the group will focus.

For the purposes of better illustrating the process, we'll use a trivial example I have used in many classes: pizza tongue burn. We have all scalded our tongue or the roof of our mouth on a hot piece of pizza. We will call this problem "pizza tongue burn." Our task is simple: Identify as many ways to manage the risk of pizza tongue burn as possible.

The process itself is simple to explain. It has the following simple rules:

- No evaluation
- No judgment
- Quantity counts
- Time is limited
- Follow the process
- Generation of ideas is separate from judgment of ideas

Explain the importance of not evaluating or judging ideas. Stress the importance of not censoring one's thoughts. If something comes to mind, write it down; do not worry if it is a good idea or a bad one, a complete one or a germ of one. Capture it. The goal is generating as many ideas as possible that are related to the problem we are solving. Let people know that this is a time-limited process. The entire process should take well under one hour. It is important to follow the process and to give it a chance to work. Let everyone know they will get a chance to evaluate the ideas when they are done generating them.

Group dynamics, as we now know, are a common impediment to a brainstorming process. When bosses, opinion leaders, or other strong personalities are in the room they can intimidate the other members, sometimes intentionally, sometimes inadvertently. To try to avoid this effect, the process begins with each individual silently generating a list of ideas. This is essentially a brainwriting task.

Make sure that everyone has several index cards or a piece of paper and something to write with. Ask them to write legibly because their cards will be collected. Begin the process by telling them they have three minutes to list ten ideas for managing pizza tongue burn. During this time they are to work silently. If they can go beyond ten ideas, please do so. After three minutes have passed, ask if anyone got more than ten ideas. If anyone did, ask how many ideas they got. Then ask if anyone got ten, nine, and so on until most of the group has raised their hands. If you'd like to inject a little fun into the process, toss the person with the most ideas a token prize of a candy bar or some other trinket that would be regarded as fun without crossing over into competitiveness or jealousy.

Congratulate folks on their efforts and results. Tell everyone to draw a line beneath their last idea and explain that the first time through we are getting all the obvious stuff, the easy ideas. Give them three more minutes and ask for ten new ideas. Let them continue adding the ideas to their cards. After three minutes repeat the query of your group, asking if anyone got more than ten. Before moving on, ask if anyone failed to get any new ideas. It will be rare that someone did not; be gentle with anyone who comes up blank. Take the process seriously, but have fun with it.

An experienced facilitator can decide if it is time to move into groups or to try one more round of silent idea generation. The basis for this decision will be the number of new ideas generated in the second go-round. If there are people getting five or more new ideas, it may be worth mining the group one more time. This time give them one minute and ask for three ideas. You never want to ask people to generate their own ideas more than three times.

The next step is to form small groups for some oral brainstorming. Use a random process of some sort to do this (see text box). Counting off is a simple way to break the

COUNTING OFF

Instead of counting off by threes or fives to form small groups, try this instead. Ask someone for their favorite letter of the alphabet, TV show, city to visit, color, beer, or any random thing you can imagine. Write their response down for all to see. Then ask your next goofy question of another person. When you have as many answers as groups you want to form, have them count off by the answers, e.g., A, Leave it to Beaver, Berlin, Green. These become your group names, and the exercise usually throws enough people that it can provide a few laughs along the way.

Warning: Never attempt humor if you are humor impaired.

chosen seating patterns and to form groups. You want no fewer than three in a group and no more than six, but these are suggestions and not requirements.

The group's charge is for people to take their index cards with them and, as a group, to make the longest list of unique ideas they can, no repeats allowed. These ideas will be numbered and recorded by the group on flipchart paper, a computer, or some other medium that others will be able to see. The usual process is for everyone to offer one idea from their list going around the group until no one has any new ideas left. A freewheeling style is an acceptable alternative in this step of the process.

Everyone is encouraged to add new ideas to their list as they listen to others. If an idea gives them a new idea, add it to the list. If you can combine two or more ideas in a novel way, write it down. Offering the group with the most ideas a prize—ice cream on a stick, candy bars, an hour off with pay—can spur some friendly competition among the groups. Usually twenty minutes will be sufficient for this task. If you need more time, take it.

When the group idea generation is completed, ask each group to select their best idea, their worst idea, and their wildest idea. For a variation and to get people to read the ideas of others, have the groups exchange lists and vote these "honors" to another group's list. Have the group report back on its selections, collect the index cards and flip chart paper, award prizes if you use them, and call it a productive hour.

The critical question now is what are you going to do with all of this information? If there is no productive follow-up, then you have just wasted an hour of everyone's time. What you have likely done, if you follow this process, is generated a lot of ideas from a lot of people in less than an hour.

Compile the suggestions and feed them back to all participants and interested others as swiftly as you can. Now it is time for convergent thinking, going from the many ideas to weeding out the weaker ideas to get to the best ideas. There are many creative problem solving (CPS) techniques that can get you from your long list of ideas to a short list. You can even adapt some of the qualitative risk assessment approaches of Chapter 9 to the task. Two good resources for this evaluation task include VanGundy's (1988) out-of-print (but well worth searching for) book on structured problem solving and Mind Tools Limited's (2007) e-book. Two less modest recommendations include Yoe and Orth's (1996) chapters on evaluation, comparison, and selection of plans and Yoe's (2002) introduction to multicriteria decision analysis.

7.7 Group Evaluations with Colored Dots

Group decision making is a complex topic supported by its own rich literature, a literature I choose to overlook here in favor of a simpler technique that has proven quite useful for group consensus finding. It is often useful to learn how a group of stakeholders, the risk analysis team, or decision makers seem to think about an issue with alternative courses of action before more-formal action is taken. One of the simplest ways to gauge the relative importance or desirability of a number of alternatives is to rank them.

The problem is there are a great many more-or-less sophisticated ways to do that. One of the more useful ones is to let everyone whose opinion is valued vote the strength of their preference. So imagine that our 3× Yeah process has generated a great many ways to end the scourge of pizza tongue burn. How would the restaurant owners in a given chain decide which of these ideas should be implemented? Colored-dot voting is a simple visual way to quickly develop an idea as to whether consensus exists or not.

Ideally, the options would be arrayed on flip chart paper on the walls of a room. Each participant would be allocated a limited number of colored dots with which to vote. For simplicity, suppose each person got three different-colored dots: blue, green, and red. Assign points to each color. Let blue = 3, green = 2, and red = 1. Allow the group time to roam about the room to examine the various alternatives that have been proposed. Ideally, this process might follow a reading of the ideas accompanied by a chance to clarify the meaning of each.

Each stakeholder has six points with which to vote. They may cast all their votes for a single alternative if they feel strongly enough about it, or they can distribute their votes across three different alternatives. When the voting reveals a clear preponderance of dots for a relatively few ideas, it is easy to quickly see the group's consensus. An absence of dots for an alternative is an equally compelling piece of evidence about the group's consensus.

The technique is neither sophisticated nor foolproof. But it can be compelling. If the only votes my favorite idea got were my own, I can quickly gauge its popularity. If it is subsequently eliminated from consideration, I may be unhappy about that fact, but in a democratic decision-making environment I can readily see that this was going to happen and that it is not the arbitrary whim of faceless risk managers.

7.8 Do You Need a Facilitator?

Osborn's original description of the brainstorming process was designed with a facilitator as an essential part of the process. So you might wonder whether you need a professional facilitator or not. There is literature that suggests that a group facilitator is necessary to get the most from a brainstorming session. No doubt this is true. Managing and facilitating the group's behavior is important if we hope to get the group's best effort. Oxley, Dzindolet, and Paulus (1996) found that groups with facilitators generated more ideas than those without. However, Isaksen (1998) reviewed related literature that showed facilitators were used relatively infrequently, in 7 of 50 reviewed studies, to no ill effect.

Someone has to champion the process and be familiar with the brainstorming technique you are using. You are better off using a different problem-solving technique than using a poorly planned brainstorming session. If your idea of brainstorming is to go around the room and have everyone give you ideas, you'll probably be better off with a facilitator. But brainstorming is not brain surgery, and you can get pretty good at it by doing it a few times and having it add value to your decision-making process. Planning and organizing the sessions, keeping it on subject and on schedule while trying to steer it toward some practical solutions is basically the job description for a session leader.

A facilitator becomes most valuable when a process bogs down or fails to find its energy. Being able to unstick a process is a skill an experienced facilitator can bring to a process. Keep in mind that each office or team has its own dynamic as well. Sometimes a group will respond differently to an expert than they will to you, even though you and the expert may do exactly the same things.

If the stakes are high, a facilitator can be useful. On the other hand, there is much to be said for building the capacity for brainstorming within your organization. Between VanGundy and the Internet you should have no problem finding a wide variety of techniques and tricks to keep a brainstorming session productive.

7.9 Addendum

I have used the pizza tongue-burn issue as a classroom exercise many times. One of these was during an online food safety risk analysis course. Although a discussion board is far from an optimal setting for a brainstorming session, participants offered over 200 distinct solutions to the problem. When you have 200 ideas, a few of them have to be good, even if only by accident. The list was long and multicultural, but it was undeniably creative. I have culled 25 of the more "interesting" ideas, unedited, to give you a flavor for the results of a brainstorming process.

1. Remove tongue to avoid pizza "tongue." (This is to echo a Chinese saying: "Chopping off your toes to avoid being bitten by sand-worms.")
2. Consume pizza without touching the tongue by feeding blended pizza through a stomach tube.
3. Have a pizza fool try the pizza first to ensure the pizza is not burning the tongue.
4. Make a wish or pray that the pizza will not burn your tongue, then eat the pizza with faith that the wish is granted or the prayer is heard.
5. Convey warning messages on package or on display such as, "Be careful to eat it when you are an old person or a child!"
6. Serving pizza on a cold plate.
7. [Let's be serious for a moment.] Affix a label saying "Hot Pizza."
8. Develop a pizza cheese additive that will color the cheese according to how hot it is (traffic-light colors could be used for temperature ranges).
9. Engineer pizza ingredients (cheese, ham, etc.) that cannot be heated beyond a certain temperature.

10. Develop a disposable heat-absorbing wrap/film that could be used to cover the pizza after it has been heated.
11. Develop a paper band that can be put on top of a pizza and that will change colors, similar to what is used to measure pH values of liquids.
12. Ban pizza consumption.
13. Develop consumer user-friendly temperature probes and educate consumer to measure pizza temperature prior to consumption.
14. Pizza burn crisis center with hotline.
15. To make children practice eating hot food at school lunch. We get a strong stimulus repeatedly, we can adjust ourselves to it. Likewise, if we practice eating hot food from childhood, we will be able to eat hot food more easily.
16. While you are eating pizza, asking someone to whip you and cause you pain in the hopes that it will get you forget the pain in your tongue.
17. Add "how to prevent pizza burn" to school curriculum.
18. Be strong: Pizza burn is nothing.
19. Sponsor a company to research and produce a cheese with a biothermal marker in it, so the pizza changes color as it cools.
20. Place pizza in strong plastic envelope and sit on it. Start counting each second. If you can reach 50 without burning yourself, it is okay to eat. If you have to stand up rapidly, then stop and get ready to repeat. Wait two minutes and repeat. This is known as "bottoming out" the pizza.
21. Obtain postcard with picture of President Obama shaking hands with President Putin. On the other side of this card in your own writing make a forgery of their signatures with a sentence giving you best wishes from them both. Go into pizza restaurant. Show the card to the manager and say: "If you serve me a pizza that burns my tongue; then I have to tell you as a consequence of the burn one of these two friends of mine will destroy your restaurant."
22. Train an elephant to place its trunk through a window. Cook many pizzas at different temperatures and offer them to the elephant. Train the elephant to waggle its ears when the temperature is correct. Take the trained elephant to a pizza restaurant and park it outside near a window so that its trunk can reach your table. Only eat pizza when you see the elephant waggling its ears.
23. Give the delivery guy the wrong address.
24. Apply Orajel (or other numbing agent, e.g., Novocain) to tongue before eating. Assemble napkins or towels nearby to clean up the St. Bernard–like drool/mess that will inevitably result.
25. "Pizza burn" is an unheard of phenomenon for a Chinese family from Guangdong.

7.10 Summary and Look Forward

Brainstorming is a simple, convenient and effective tool for promoting the kinds of divergent thinking risk managers need at different times in the risk management process. There are more variations of brainstorming techniques than you will ever need in a lifetime. The four sacrosanct elements of a good process are: quantity counts, no evaluation of ideas, unusual ideas are welcomed, and combining ideas to continue the process is desirable. Find a useful technique or two that combines these elements and use them. One good brainwriting technique and one good brainstorming technique should serve risk managers well.

Continuing our focus on risk management topics, the next chapter turns to economics. Someone is always going to care about costs, no matter what your other risk management objectives are. It is essential that every risk management professional understand a little bit about the economic realities of choice: We can't do everything, and every choice costs us the opportunity to have done something differently. They also need to know a little something about how economic motivations influence the response to risk management decisions.

REFERENCES

Aiken, M., J. Krosp, A. Shirani, and J. Martin. 1994. Electronic brainstorming in small and large groups. *Information and Management* 27:141–149.

Aiken, M., H. Sloan, J. Paolillo, and L. Motiwalla. 1997. The use of two electronic idea generation techniques in strategy planning meetings. *Journal of Business Communication* 34 (4): 370–382.

Beasley, Mark S., and J. Gregory Jenkins. 2003. A primer for brainstorming fraud risks: There are good and bad ways to conduct brainstorming sessions. *Journal of Accountancy* 196 (6). http://www.allbusiness.com/finance/insurance-risk-management/715064-1.html.

de Bono, E. 1998. Serious creativity. *Journal for Quality and Participation* 11 (3). http://www.debonogroup.com/serious_creativity.htm.

Geschka, H., U. von Reibnitz, and K. Storvik. 1981. *Idea generation methods: Creative solutions to business and technical problems*. Columbus, OH: Battelle Memorial Institute.

Isaksen, S. G. 1998. A review of brainstorming research: Six critical issues for research. Buffalo, NY: The Creative Problem Solving Group, Inc. http://www.cpsb.com/resources/downloads/public/302-Brainstorm.pdf.

Mind Tools Limited. 2007. *The Mind Tools e-book*. 5th ed. London: Mind Tools Limited. http://mathamlin.com/files/MindToolsEbook.pdf.

Osborn, Alex F. 1963. *Applied imagination: Principles and procedures of creative problem solving*. 3rd ed. New York: Charles Scribner's Sons.

Oxley, N. L., M. T. Dzindolet, and P. B. Paulus. 1996. The effects of facilitators on the performance of brain-storming groups. *Journal of Social Behavior and Personality* 11 (4): 633–646.

Panitz, Beth. 1998. Brain storms. *ASEE Prism* 7:24–29.

Sutton, Robert. 2006. Eight tips for better brainstorming. *Business Week,* July 28.

"The History of S.C.A.M.P.E.R." fuzz2buzz, accessed July 1, 2011, http://www.fuzz2buzz.com/en/group/creative-thinking/discuss/history-scamper.

Toubia, Olivier. 2006. Idea generation, creativity, and incentives. *Marketing Science* 25 (5): 411–425.

VanGundy, Arthur, Jr. 1988. *Techniques of structured problems*. New York: Van Nostrand Reinhold.

Yoe, Charles. 2002. *Trade-off analysis planning and procedures guidebook. IWR report 02-R-2*. Alexandria, VA: Institute for Water Resources.

Yoe, Charles, and Kenneth Orth. 1996. *Planning manual. IWR Report 96-R-21*. Alexandria, VA: Institute for Water Resources.

8

Opportunity Costs and Trade-Offs

8.1 Introduction

No matter what the risk management issue is, someone is always going to care about costs. Whether the problem falls in the realm of public health, public safety, profitability, or anything else, costs matter. This chapter* explains why costs, or opportunity costs as economists tend to think about them, should matter to everyone. Because costs matter, it is essential that all risk managers have some basic understanding of a few economic principles. These include scarcity, opportunity cost, valuation, and marginal analysis. Each is addressed in this chapter. The chapter concludes by considering the difficult and subjective business of identifying and making trade-offs and the economic analysis techniques used to support those decisions.

8.2 Economics for Risk Managers

Economics is the study of how society establishes the institutions and social norms that result in the allocation and distribution of its scarce resources. It includes individual behavior and such things as how people decide what and how much to buy, how to spend their time, how much to work, how much to save, and so on. It includes firms' behaviors, including what and how much to produce, as well as how to produce it. It helps to explain how firms will choose from the many ways they have to respond to a government regulation, compliance being only one of them. Economics includes how society, often through government, decides to divide its resources among national defense, public health, public safety, public works, consumer goods, protecting the environment, and other needs. Economics can help guide the choice in determining how much and what kinds of risk management are best for society.

Decision making is at the heart of economics. Decision making under uncertainty is at the heart of risk management. It stands to reason that the two disciplines are closely intertwined. Mankiw (2009) offers 10 economic principles for understanding how

* I would like to gratefully acknowledge the many helpful comments and contributions of Dr. Richard Williams to this chapter. His knowledge of and experience in the world of public choice and regulatory economics are, if rivaled at all, rivaled by very few. The discussion of rent-seeking behavior and government failure are due to Richard, as is a significant portion of the first trade-off discussion. There is no part of this chapter that does not bear some part of his fingerprints.

RESOURCES

Resources are often grouped into categories such as land, labor, and capital. Land includes all natural, environmental, and agricultural resources. This includes everything that is in, on, under, and moving over the land as well as all that springs forth from it.

Labor includes all forms of human and animal work, intellect, and productivity. Capital resources are basically everything else. This includes all means of production and everything that has been produced for the benefit of society.

Resources are real things. Do not confuse them with financial resources. The two are similar only to the extent that financial resources are useful only insofar as they enable one to command real things.

people make decisions, how they interact, and how economies as a whole function. They are:

1. People face trade-offs.
2. The cost of any action is measured in terms of forgone opportunities.
3. Rational people make decisions by comparing marginal costs and marginal benefits.
4. People respond to incentives.
5. Trade can be mutually beneficial.
6. Markets are usually a good way of coordinating trade.
7. Government can potentially improve market outcomes if there is a market failure or if the market outcome is inequitable.
8. Productivity is the ultimate source of living standards.
9. Money growth is the ultimate source of inflation.
10. Society faces a short-run trade-off between inflation and unemployment.

These principles provide effective touchstones for the topics presented in this chapter.

8.3 Economics and Decision Making

Individuals, businesses, government, and society at large all make decisions. Economics offers several insights into the principles of decision making that risk managers cannot afford to ignore. Four of these principles are offered here.

8.3.1 Trade-Offs

Scarcity is a fundamental fact of life. Resources are limited in relation to society's wants and needs. We simply do not have enough resources to do everything we want or need to do. This simple fact of life holds for the individual, the firm, government, and society at every imaginable level.

Because of scarcity, we cannot do everything, so we have to make choices. By their nature, choices require that we do one thing and not do another. Why did we choose a baked potato instead of french fries? Why did we watch one television program and not another? We are not always able to articulate why we choose to do certain things. Most of our choices are context-specific, i.e., they change depending upon the context in which they are made. The fact is, whether we can articulate why we do what we do or not, we always are choosing between at least two options, and choices require trade-offs.

It is important to distinguish between private choices, such as the examples just given, and public choices. Public choices are those made in the public sphere, and they often involve consequences for many people. The choice made by a risk manager may have personal consequences (e.g., a philosophical desire for a particular outcome, job promotion, or a desire to reward one's allies), but the choice also has consequences for those who benefit from and those who bear the costs of the choice. In some cases, such as local political issues, those who are impacted by the choices may affect the outcome directly through voting. In other cases, such as the appointed head of a federal regulatory agency, those affected have little control over their fate.

Whether acknowledged or not, public choices require trade-offs of multiple effects. For example, increased public health protection increases costs that must be paid for by consumers and may result in job losses in industry. A public choice to allow speed limits to increase (the benefit is reducing travel time) may increase the risk of death and injury. Some public trade-offs are obvious, such as paying more for increased security, health, or safety, but others are indirect and require study to understand.

An important recurring social trade-off is that of efficiency vs. equity. Efficiency means society is getting the most from its scarce resources, i.e., what is sometimes called "getting the biggest bang for the buck." Suppose there are two ways to prevent people from crashing at a local intersection and both have the same probability of reducing crashes. One is a stoplight that costs $300 per year and another is to have a full-time police officer directing traffic for $30,000 per year. It would be inefficient to hire the police officer. In fact, by buying the stoplight you could, if you wished, get 99 more stoplights that might save many more lives at the same cost as one police officer.

Equity refers to how resources are distributed, not how efficiently they are used. If in the previous example, a public decision maker decided that the police officer deserved the job to achieve some sort of social justice, then that would be a decision based on equity. When risk managers choose, for example, to protect a small minority of consumers who are at high risk (either highly sensitive or highly exposed), those decisions generally are more driven by equity than efficiency. One could argue that money spent on the Transportation Security Agency's air travel security rather than on feeding the hungry, reducing highway deaths, or providing health care is an equity decision rather than an efficiency one. Most decisions in government involve elements of efficiency and equity, and they are often in opposition to one another. Those who are stewards of some public trust must always choose, that is, make trade-offs between the two.

As long as there are choices to be made—and every decision problem has at least an action/no action choice—there will be trade-offs. Decision making is best served by making the nature of the trade-offs explicit. Benefit-cost analysis, an economic tool, has been developed to try to help decision makers identify the trade-offs with a common metric (dollars). It is one tool by which trade-offs can be addressed. We'll return to this topic later in the chapter.

8.3.2 Opportunity Cost

If you think of cost as the money price of obtaining something, think again. Economists wish to challenge that notion. The cost of something is what you must give up to get it. Cost is measured in real things sacrificed for a choice. Money, the most common measure of cost, is simply a shorthand means of communicating what economists call opportunity cost. When we choose one thing we have simultaneously forgone the opportunity to have made another choice. This is the essence of opportunity cost.

Any one risk management option (RMO) will cost us the opportunity to have chosen other RMOs and other real outcomes. Opportunity cost is one of the most important concepts to grasp because it is not well served by simple accounting measures. This is in part because of the differences between explicit and implicit costs.

Imagine the Rolling Stones on tour yet again. Face value on a ticket is $300 and you have one for row three. As you walk into the arena you see an aging hippie in ponytail and red Converse All-Stars holding a sign announcing he will pay $1,000 for a seat in the first three rows. You smile and walk into the concert. What did it cost you?

Your explicit costs were $300. Think of explicit costs as those costs you must pay from your income or savings. But there was also an implicit cost to your going to the concert. When you saw the sign you had a choice: Sell your ticket for $1,000 or see the concert. Had you sold the ticket, you could have reimbursed yourself $300 for the explicit cost of your ticket and made a tidy $700 profit. When you decided to keep your ticket, you were saying no to a $700 gain. This $700 is an implicit cost of your decision. Think of an implicit cost as an opportunity for gain that you declined. Opportunity cost is explicit cost plus implicit cost, in this case $300 + $700. The concert cost you $1,000. Note that this simple example conveniently ignores other associated costs, some explicit (travel and hotels perhaps) and some implicit (the value of your time spent going to the concert).

It is very easy for a decision maker to take an incomplete view of costs. It is important for someone to carefully consider the opportunity costs of your risk management decisions.

OPPORTUNITY COST EXAMPLE

If you go to the vending machine and come back with a bag of M&M's and I ask what that cost you, you are likely to reply $1, or whatever the money cost was.

But if you stood before the machine with your only dollar, weighing the choice of a Twix (candy with a cookie crunch) or M&M's (melts in your mouth not in your hands), and then chose the M&M's, then that choice would have cost you the opportunity to have a Twix. So the economist's answer is that the M&M's cost you a Twix. It is a real thing forgone.

Others may have chosen a bag of chips, a peppermint, pretzels, and so on. It becomes unwieldy to list all the opportunities forgone by a single purchase with so many different people and preferences, so money prices serve a wonderfully efficient shorthand means of communication. When we say it costs $1, that stands for everything real you could have done with that $1.

There are several notions of opportunity costs that are important in the public sphere for risk choices. The first is that individuals make private choices to reduce their own risks all of the time. Buying a Mercedes rather than a Toyota Tercel may at least in part be because they are safer cars. Having regular medical and dental checkups reduce risks. Smoke detectors are an excellent investment in risk reduction. When government spends to reduce risk, one opportunity cost is that the people who pay taxes to support government programs have less money to make their own private expenditures to reduce risk.

The next notion is that there is an almost limitless menu of public risks that we ask government to address. Workplace safety, food safety, drug safety, airline safety, terrorism, crime, mental health, and literally every sphere of life is affected by public-sector decisions. If we think of risk as, for example, risk of death, it becomes a reasonably easy job to compare how much it costs to reduce various risks associated with death. We can compare the costs of reducing one death from cancer to the costs of reducing one death from a car accident, for instance. When we do, we discover that there is an extremely broad distribution of costs of lives saved. With limited resources, the opportunity cost of addressing one risk rather than another becomes a significant concern. Any risk manager who has been faced with the need to set risk management priorities—as is done routinely in food safety, health care, international trade, and many other areas of responsibility—knows this only too well.

Tengs et al. (1995), in a fascinating but now somewhat dated study, identified the cost per year of life saved for 500 different measures. The costs ranged from negligible for such measures as smoking advice for pregnant women, compression stockings to prevent venous thromboembolism, and automatic seat belts in cars to much higher amounts. Table 8.1 shows some sample values selected from the appendix of Teng's paper.

TABLE 8.1

Selected Lifesaving Interventions and Their Cost Effectiveness

Risk Reduction Measure	Cost per Year of Life Saved
Mandatory seat belt use and child restraint law	$98
Flammability standard for upholstered furniture	$300
Mandatory motorcycle helmet laws	$2,000
Dual master cylinder braking system in cars	$13,000
Ban asbestos in brake blocks	$29,000
Kidney transplant from cadaver	$29,000
Ban asbestos in specialty paper	$80,000
Annual mammography for women age 40–49	$190,000
Benzene emission control at pharmaceutical manufacturing plants	$460,000
Annual cervical cancer screening for women age 20 and above	$1,500,000
Ban asbestos in thread, yarn, etc.	$34,000,000
Trichloroethylene standard of 2.7 (vs. 11) microgram/L in drinking water	$34,000,000
Control of benzene equipment leaks	$98,000,000
Sickle cell screening for non-Black low-risk newborns	$34,000,000,000
Chloroform private well emission standard at 48 pulp mills	$99,000,000,000

Source: Tengs et al. (1995).

Imagine yourself as the risk management czar with the power to enact RMOs that can impose costs of up to $100 billion on society. If your objective is to preserve life, what measures will you enact? You could impose a chloroform private well emission standard at 48 pulp mills and save one year of life and be done with the task. Or you could begin by imposing a mandatory seat belt use and child restraint law that would only cost $98 for every year of life saved. The National Highway Transportation Safety Administration estimates that seat belts save 11,000 lives annually. If, for convenience, we assume an average additional life expectancy of 35 years, that is 385,000 years of life saved at $98 per year for a total cost of about $38,000,000—a much better deal than the chloroform standard. Note that this is not the intended use of Teng's data so much as a convenient gimmick to enforce a point.

Note, also, the relative costs of the asbestos bans and benzene controls. These suggest that total bans on the use of substances may not always be the most efficient risk management strategies. Recall that your $100 billion of costs imposed on society will also affect the ability of taxpaying citizens and industries to pay for private risk reduction decisions. The opportunity costs of RMOs often involve some risk-risk trade-offs somewhere else along the line.

8.3.3 Marginal Analysis

Rational people are systematic and purposeful in trying to do the best they can to achieve their objectives. In economics, this means we make decisions by weighing, if only qualitatively, the costs and benefits of our actions when we act rationally. Acting rationally is taken as a minimum requirement for risk managers. If they do not decide rationally, they should. Likewise, people affected by an RMO will respond to it in a way that is rational for them. If risk managers do not seriously consider the rational responses to their decisions, the desired risk management objectives may not be realized.

A rational person will consider marginal changes, i.e., incremental adjustments, to an existing plan. Imagine we have a public health risk issue and we are trying to decide how much regulation is appropriate. There is a hazard associated with a food, and it can be reduced to varying extents at the costs shown in Table 8.2. Note that there are two costs shown. Total cost indicates it would cost $0.50 to reduce the pathogen to 100 ppm but $10 to get it down to 50 ppm.

In addition, there is the cost of each additional increment of pathogen reduction via increasing regulation level. This is marginal cost (MC), the cost of one more unit of

TABLE 8.2

Costs of Different Regulation Levels for a Pathogen

Regulation Level	Total Social Cost	Marginal Cost [MC]
100 ppm	$0.50	NA
90 ppm	$1.25	$0.08
80 ppm	$3.00	$0.18
70 ppm	$5.00	$0.20
60 ppm	$7.25	$0.23
50 ppm	$10.00	$0.28

TABLE 8.3
Benefits of Different Regulation Levels for a Pathogen

Regulation Level	Total Social Benefit	Marginal Social Benefit
100 ppm	$4.00	NA
90 ppm	$7.00	$0.30
80 ppm	$9.00	$0.20
70 ppm	$10.00	$0.10
60 ppm	$10.50	$0.05
50 ppm	$10.75	$0.03

hazard reduction. Think of marginal cost as the change in cost required to produce one more unit of output; in this case, output is the regulation level or hazard reduction. It is defined as the change in total cost divided by the change in output. Going from 100 ppm to 90 ppm, costs change by $0.75 while the hazard reduction changes by 10 ppm. Thus, $0.75/10 yields the marginal cost of $0.075, say $0.08, for each additional part reduced. At some point, marginal cost tends to exhibit the increasing trend shown in Table 8.2.

Hazard reduction is expected to have benefits for society, and these benefits are shown in Table 8.3. Total benefits range from $4 for a reduction to 100 ppm to $10.75 for a reduction to 50 ppm. What we gain from each increment of hazard reduction is the difference in the total social benefit divided by the difference in the hazard reduction. This is called the marginal social benefit in the table. In going from 100 to 90 ppm, benefits rise from $4 to $7, a change of $3. The hazard level also changes by 10 ppm, so $3/10 = $0.30, as shown.

We see that marginal benefits (MB) are decreasing. This is a common pattern that results from the economic principle of diminishing marginal utility. It means that, usually, the more you have of something (including risk-reducing regulation), the less valuable additional units of that thing become.

Given these facts, the question for risk managers now becomes what is the most appropriate level of hazard reduction in this instance. Risk management decisions can be based on science and social values, so any answer is in fact possible. However, one of the social values that people will always care about, especially with public decision making, is economic efficiency. Economic efficiency relies on marginal analysis, and this is not intuitive for many risk managers and stakeholders.

Should society pay $10 to reduce the hazard to 50 ppm? Its value to society is, after all, $10.75. Would you trade $10 for $10.75? Most people would say yes. Table 8.4 shows a net profit to society for every level of hazard reduction through regulation.

However, the answer changes rather dramatically when we use marginal analysis to choose the optimal level of hazard reduction. Note, for example, that net social benefits rise to a maximum of $6 at a hazard level of 80 ppm, then decline to $0.75 at 50 ppm. In a world of scarcity, where we cannot do everything, we do not want to waste resources. Choosing 50 ppm wastes resources, as the marginal analysis in Table 8.5 shows.

Table 8.5 reproduces all of the information from Tables 8.2–8.4 and adds two new columns. Note that the table shows a positive level of net social benefits for every level of hazard reduction, as noted previously. Now look at the marginal cost and marginal

TABLE 8.4

Total Net Social Benefits for Different Regulation Levels for a Pathogen

Regulation Level	Total Social Cost	Total Social Benefit	Net Social Benefit
100 ppm	$0.50	$4.00	$3.50
90 ppm	$1.25	$7.00	$5.75
80 ppm	$3.00	$9.00	$6.00
70 ppm	$5.00	$10.00	$5.00
60 ppm	$7.25	$10.50	$3.25
50 ppm	$10.00	$10.75	$0.75

TABLE 8.5

Marginal Analysis of Different Regulation Levels for a Pathogen

Regulation Level	Total Social Cost	Total Social Benefit	Net Social Benefit	Marginal Cost [MC]	Marginal Social Benefit	Marginal Net Benefit	Do It?
100 ppm	$0.50	$4.00	$3.50	$0.05	$0.40	$0.35	Yes
90 ppm	$1.25	$7.00	$5.75	$0.08	$0.30	$0.23	Yes
80 ppm	$3.00	$9.00	$6.00	$0.18	$0.20	$0.03	Yes
70 ppm	$5.00	$10.00	$5.00	$0.20	$0.10	–$0.10	No
60 ppm	$7.25	$10.50	$3.25	$0.23	$0.05	–$0.18	No
50 ppm	$10.00	$10.75	$0.75	$0.28	$0.03	–$0.25	No

benefits. If you held the marginal cost in your hand, would you trade that sum of money for the corresponding amount of marginal benefit? That is, in its essence, the choice the risk manager is making on society's behalf. Would you trade $0.05 for. $0.40? Of course you would. The net gain is $0.35 for every additional ppm reduced, that is a good deal for you and it would be a good deal for society as well. Would you spend $0.08 for a $0.30 gain? Yes. Would you incur marginal costs of $0.18 for a chance to gain $0.20? Sure. Would it be wise to spend $0.20 to gain $0.10? No. We would not make this deal knowingly. So we would stop at 80 ppm because it is the last level of hazard reduction that paid for itself. Not coincidentally, this is also where net social benefits are maximized.

The rule for rational marginal behavior is to undertake any change for which MB ≥ MC and avoid any change where MB < MC. Applying this rule will also maximize total net benefits to society. Looking at marginal net benefits, we see they exceed zero for 80 ppm but become negative for any additional reduction. In a world where it is not possible to do everything, it is wise to avoid doing things that require costs in excess of their value to society. Risk managers should take care to request information in a marginal analytical framework so they can better see the trade-offs between benefits and costs.

8.3.4 Incentives

People respond to incentives. Incentives are rewards or punishments that induce people to act in specific ways. When gasoline prices rise, people drive less and they buy more hybrids and fewer SUVs. Likewise, with risk management, people respond to incentives. It is important to understand both the intended and unintended incentives associated with a risk management option.

Consider a hypothetical contaminant in a food. Suppose a baseline study of ten businesses shows that all ten test for the contaminant, but five do not test adequately; that is, they use neither a sensitive enough method to detect the contaminant down to the lowest levels that matter nor do they test frequently enough to actually reduce risk. Imagine an RMO that requires mandatory testing for the contaminant at the level that the regulator believes is adequate, based on marginal analysis. Food-safety regulators may expect they have solved the problem by requiring testing.

This requirement has an incentive effect on businesses, but it may not be the expected one. Changes in requirements present people with choices that necessitate changes in behavior. So imagine that two of the firms not testing adequately meet the requirements, one goes out of business, one chooses to stop testing altogether to lower costs and hopes not to be detected by inspectors, and another moves the company overseas where there are no testing requirements. For the last two companies, risks may have actually increased. The RMO provides everyone with a choice of how to respond. An RMO may present some with a reward and others with a punishment. Those already testing are rewarded by the increased cost of operation imposed on their competition by the RMO. Companies not already testing are punished by higher costs of operation. Responses to these incentives should be anticipated by risk managers.

It can be useful, at times, to incentivize an RMO to ensure faster and more complete implementation of the planned measures by those responsible for implementing the measures. Subsidies, rebates, tax breaks, coupons, and such can be effective financial incentives. Helping affected parties understand and then realize the benefits of the measures can also be helpful.

8.3.5 Rent Seeking

Rent seeking is a special kind of incentive, worthy of its own discussion. It is easy to imagine that all choices made by public managers should be made with the best interests of people at heart, whether that means being more efficient or more equitable. In reality, it is not always or even often done this way. The study of "public choice" is dedicated to understanding how choices are actually made in the public arena, instead of the way we would like to think they are made. Whenever government has the power to make choices for others, and some will benefit and some will lose, people are willing to expend resources to try to influence decisions in a way that benefits them. That behavior is called "rent seeking." The term *benefit* is defined here to mean that they either gain financially from the decision or they seek to gain an outcome from the decision that matches their personal philosophy.

In the testing example considered above, consider that some firms gain from having their rivals spend more to increase testing. Such a requirement would be worth trying to influence in the hopes that the regulator decides in that way. If you are the firm that is not testing sufficiently, you might invest in lobbying to avoid it. Some, in the advocacy sphere, may spend to advocate for testing because they believe the product is unsafe.

Those who are both best organized and most committed tend to lobby for or against the regulation to realize the most gain from the decision. Those who bear the costs of the decision, usually widely distributed among consumers, generally will lose in this situation. Individual consumers have so little to gain or lose from the decision that such groups rarely unite to protect their collective (often large) interests. In many cases, there are politicians and bureaucrats who also may gain by rewarding rent seekers.

This system of rent seeking, rent avoidance, and political and bureaucratic rent distribution is well entrenched in the 150 regulatory agency system of the United States. It is just as important to understand why certain decisions get made and how we can avoid the worst excesses of this system as it is to study rational ways to address risks.

8.4 Economic Basis for Interactions among People

8.4.1 Trade

Trade includes both interstate commerce and international trade. A great deal of the recent rising interest in risk analysis can be traced to international trade issues. As used here, the term *trade* means that rather than being self-sufficient, people can specialize in their careers and firms can specialize in producing one good or service and exchange it for other goods. Individuals as well as countries benefit from specializing and then trading.

Trade provides for a greater variety of goods from which consumers can choose. It also frequently offers a better price abroad for goods that domestic firms produce. Conversely, we can often buy other goods more cheaply from abroad than can be produced at home. The mutual benefits of trade ensure that this will remain an important dimension for many risk management decisions.

SPS AGREEMENT

The Agreement on Sanitary and Phytosanitary Measures (SPS) grew out of the Uruguay round on multilateral trade negotiations from 1986 to 1994. The SPS Agreement recognizes the right of governments to protect the health of their people from hazards that may be introduced with imported food by imposing sanitary measures, even if this means trade restrictions. It obliges governments to base such sanitary measures on risk assessment to prevent disguised trade protection measures.

Sanitary measures need to be nondiscriminatory, not more trade-restrictive than necessary, and based on sufficient scientific evidence. To harmonize these measures as widely as possible, governments should base their sanitary measures on international standards where they exist.

Codex Alimentarius standards are explicitly recognized as the standards considered consistent with SPS provisions.

8.4.2 Markets

A market is a group of willing buyers and sellers. Thanks to international trade and technology, they no longer need to be in a single location. Markets are used to organize economic activity in most countries around the world. Planned economies, the primary alternative to market economies, have waned with the fall of Communism, and even nominally Communist countries today have mixed economies that rely on markets to varying extents. International trade is almost exclusively a market activity.

Markets are used to decide what goods to produce, how to produce them, how much of each to produce, how much to charge for them, and who gets them. Market economies answer these questions through the decentralized decisions of many consumers and firms as they interact in markets.

Supply (willing producers and sellers) and demand (willing consumers and buyers) interact in a market and determine both prices and quantities of goods and services that change hands. Prices function to allocate society's scarce resources. Price simultaneously reflects the good's value to buyers and the supplier's cost of producing the good. Market-determined prices guide self-interested consumers and firms to make decisions that, in most cases, maximize society's economic well-being.

Market operations sometimes transform risks. Risk management options can sometimes have unanticipated effects on markets and market prices. A desire to protect infants in airplanes might result in a requirement for parents to purchase a seat and use a safety restraint. The increased cost of family travel may cause more families to choose to drive, which could result in more infant deaths in highway accidents. Thus, an effort to reduce the risk of infant deaths could result in an increase in infant deaths due to consumer responses to market incentives.

Prices provide incentives to consumers and producers. Risk managers are well advised to consider the price/incentive effects of any RMO they are considering.

8.4.3 Market Failure

When markets work, they can work very well in allocating resources. Sometimes markets fail to allocate society's resources efficiently for a variety of reasons. When this happens, government intervention has the potential to alter and improve market outcomes.

Common causes of market failure include:

- Asymmetric information
- Market power
- Externalities
- Nature of certain goods

Asymmetric information occurs when one party knows more about a product than another. If exporters sell a product with melamine in it to give the impression of a higher protein content, this is asymmetric information. If a peanut product company continues to sell its products after they have tested positive for *Salmonella*, that is asymmetric information. Buyers, unaware of the facts about the quality of the product, may overestimate the value of that product and unwittingly pay a price that is greater than the true value of that product to them.

MARKET FAILURE = MARKET OPPORTUNITY

It should be noted that every market failure is also a market opportunity. If consumers have less information than producers, there is a market to produce and sell that information. If consumers care about air pollution (another form of market failure), there is a market to produce and advertise goods that do not pollute the air.

Asymmetric information can be a two-way street. Buyers with inside information can benefit if a large entertainment company buys land in a rural area with the intention of building an entertainment park or with the knowledge that there will be a zoning change, they may pay a price substantially under its true market value. Risk communication can be an effective tool for correcting certain asymmetries of information.

When a single buyer or seller can exert significant influence over prices or output, we can observe a failure of market power. Monopolies are the best example. Here a single firm can set the price for their good at any level they desire. Although monopolies provide the best known example of market power, any imperfectly competitive situation can distort the market outcome. Market power enables a producer to charge more and produce less than the social optimum would dictate in a market.

Risk management options that reduce competition or help consolidate market power can contribute to market failures. On the other hand, government RMOs may be designed to reduce market power and to promote efficiency in markets as well. Just as RMOs can affect market operations, market operations can sometimes make effective RMOs.

An externality is an economic side effect, and it occurs when an economic activity (consumption or production) affects a bystander to that activity. These side effects are never fully reflected in the prices of goods. Externalities don't enter the cost or benefit decisions of either buyers or sellers; that is why they are called external. They can be harmful (negative externalities) or beneficial (positive externalities), but the harmful ones seem to attract more attention.

Pollution risks may be the classic example of a negative externality. Food-borne illness in the United States was estimated to cost $152 billion in health-related costs (Scharff 2010). These are serious public health externalities, as are the additional costs to industry from reputation externalities and recalls. It is easily argued that RMOs have the potential for significant positive externalities to the extent that they can reduce some of these effects. Addressing negative externalities has long been a primary motivation for government regulation of industry.

The final cause of market failure is related to the nature of the good itself. What firm, for example, can produce public health or public safety? There are some things that can only be consumed by everybody (relatively speaking) or nobody at all. These are called public goods, and they have distinct characteristics. They are:

- *Nonrival*: One person's consumption of them does not stop another person from consuming them, i.e., the supply of the good is not reduced when a person consumes it.

- *Nonexcludable*: If one person can consume them, then anyone can consume them, i.e., it is impossible to stop another person from consuming them.

A subset of these public goods are also nonrejectable. People can't choose not to consume them even if they want to. National defense, homeland security, and public health are examples of nonrejectable public goods. Risk management options can increase the supply of public goods.

Some risks threaten common goods. Nations that continue whaling provide such an example. Common goods are goods that are nonexcludable but are rival. Many environmental resources are common goods. Government regulations can sometimes effectively reduce the risk to common resources.

8.4.4 Government Failure

Market failure should always be compared to the possibility of government failure. Even when we can identify a market failure, we cannot guarantee that government intervention will improve on the situation. The previous section on rent seeking shows that governments, when influenced by special interests, can make decisions that are not "rational" in an economic sense.

If there is such a thing as a true public servant not beholden to any outside interests, there is still one problem that is not solvable. This was identified by Nobel Prize winner Friederich Hayek (1945) years ago in what he termed "the knowledge problem." Any modern economy is founded on a great sea of private know-how that is dispersed among many specialized participants. No one, not even a state agency, can amass all the knowledge that each participant "on the spot" inevitably acquires. This suggests that government is always making decisions under uncertainty and would be unable to be fully rational even if that was its intent.

8.5 Principles of the Economy as a Whole

8.5.1 Living Standards and Productivity

There is a significant variation in standards of living across countries and over time. According to the Central Intelligence Agency's World Factbook (accessed March 15, 2010), the gross domestic product per capita for the richest nations in the world can be well over 100 times greater than for the poorest nations.

Economists agree that productivity, the amount of goods and services produced per unit of labor, is one of the most important determinants of a country's standard of living. Productivity depends on the equipment and technology available to workers as well as their education and skills. Studies consistently show that other factors like labor unions and competition from abroad have far less impact on a country's living standard. The best risk managers will consider the potential effects of their RMOs on the productivity of their affected populations. In many cases, however, because of the way we organize our regulatory agencies, these kinds of considerations are not taken into account because of the single-minded nature of the statutory authority mind-set.

8.5.2 Inflation and Unemployment

Inflation is an increase in the general level of prices. Inflation is usually caused by excessive growth in the quantity of money. This causes the value of money to fall. Simplistically speaking, in the long run, the faster the government creates money, the greater is the inflation rate. In the shorter run, say two years or less, economic policies tend to push inflation and unemployment in opposite directions. For example, government spending tends to reduce unemployment and exacerbate inflation, while decreasing the money supply can reduce inflation and increase unemployment. The rare risk management issue large enough to affect the economy as a whole needs to consider the potential effects of the RMOs on inflation and unemployment.

8.6 Making Trade-Offs

The concepts introduced here argue, we hope persuasively, for considering the economic effects of any risk management decision. At the same time, we recognize that not everyone will share this opinion, and other values may supersede economic considerations at times. So we return one more time to the topic of trade-offs, this time to consider a few more notions that arise when making them. Additional discussions on trade-offs are found in Chapter 3 under multicriteria decision analysis (section 3.5.3.1).

In the broadest sense, a trade-off is giving up one thing to get another. Some choices entail trade-offs while others do not. Optimization choices, for example, do not always acknowledge trade-offs. An optimization choice is usually minimizing or maximizing some single objective. If we are maximizing net national economic development or minimizing deaths, the decision rule is simple.

That does not mean the choice is trivial; it simply means that decisions based on such choices do not use trade-off analysis or multicriteria decision-making models. There may be a great deal of analysis required to get the information upon which the choice will be based, but in such choice settings the solution is imbued in the model or the decision rule we use. Once the decision model is chosen, there is no further choice to be made. The model makes the decision, and trade-offs may not be explicitly considered.

Not all decisions are so easy. Reservoir storage reallocation studies provide a good example of trade-offs. Will storage be allocated to hydropower or flood control? Storage filled with water to drink can't be left empty to hold potential floodwaters. An acre-foot of water can be used for withdrawal purposes (irrigation, water supply) or in-stream purposes (navigation, water quality, habitat). More of one usually means less of another.

Hadari (1988) offers a formal definition of a value trade-off that I have modified slightly here. A value trade-off exists if a risk manager must choose a course of action whose implementation involves at least two values, V_a and V_b, both held as positive values. Two conditions must hold. First, the alternatives available would each necessarily entail sacrificing, at least to some degree, either V_a to V_b or the opposite. To use more technical language, past some point, the values to be upheld are divergent. Second, no common unit of measurement applies to both V_a and V_b, i.e., the values are incommensurable.

There is no formal value trade-off if these two elements do not hold. If there is no divergence, then we do not have to give up one value to gain another. If the values are commensurable, then in theory this decision could be made using optimization analysis. We need trade-off analysis because of conflicts among values and a lack of a common unit to measure relative gains and losses in implementing RMOs that reflect a variety of values.

Some trade-offs are explicit. One more acre-foot of irrigation water means one less acre-foot of water supply. When, for example, a given land or water resource has competing and mutually exclusive uses, the trade-off is an explicit one, and the terms of the trade-off may be fixed by the laws of our physical universe.

An acre of forest can be forestland or it can be an acre of mall parking. This is an explicit trade-off defined by this obvious one-to-one correspondence in outputs. Not all explicit trade-offs will be so easy to define. More agriculture in a watershed means more fertilizers and pesticides used on crops that can degrade drinking water quality. An increase/decrease in the use of pesticides and fertilizers means a decrease/increase in drinking water quality. The trade-off is explicit, and the laws of our physical universe fix the terms of the trade-off even though we may be unable to ascertain them in a precise manner.

In contrast to these explicit trade-offs stand subjective trade-offs. The terms of a subjective trade-off are fixed by the value systems and preferences of risk managers, decision makers, and the public. There is no explicit trade-off for jobs lost and illnesses prevented. That trade-off is subjective because its terms of trade are based on something other than the laws of the physical universe. Explicit trade-offs can sometimes be easier to quantify and measure than subjective trade-offs.

As mentioned previously, many risk management decisions involve trading off values. Consider, for example, two RMOs alike in all respects except that one has 100 acres of wooded urban recreation and the other uses those 100 acres to create wetlands inaccessible to the public. The trade-off is explicit. But weighing these two

A FEW RESOURCES

The topics in this chapter are treated in a very superficial manner. For additional information about the techniques in this sections consult the following:

Cost Effectiveness Analysis Methods and Applications, 2nd ed. By Henry Levin and Patrick McEwan, Sage Publications, Thousand Oaks, CA, 2000, ISBN 0-7619-1934-1.

Cost-Benefit Analysis Concepts and Practice, 2nd ed. By Anthony E. Boardman, David H. Greenberg, Aidan R. Vining, and David L. Weimer, Prentice Hall, Upper Saddle River, NJ, 2001, ISBN 0-13-087178-8.

Risk-Benefit Analysis. By Richard Wilson and Edmund A. C. Couch, Harvard University Press, 2001, ISBN 0-674-00529-5.

Regional Economic Development: Analysis and Planning Strategy. By Robert J. Stimson, Roger R. Stough, and Brian H. Roberts, Springer, Berlin, 2006, ISBN 10-3-540-34826-3.

alternatives will still require a subjective trading off of the risk manager's values. In any true trade-off, someone must at some point attach subjective weights to the values being traded off.

A good trade-off analysis is transparent, and a transparent trade-off analysis makes the subjective nature of a trade-off of values as explicit as possible. This usually happens by specifying the weights used to make the trade-offs.

If there are no value trade-offs, then there is rarely a problem selecting the best risk management measure. Without trade-offs we are simply identifying the optimum plan. Only when we have to give up something to get something incommensurable do we face a problem in the decision-making process. The optimization paradigm does not often work for risk management decisions. It almost invariably involves analysis that must take multiple criteria into account. One reason for this is that someone is always going to care about costs, and more risk reduction invariably means more cost.

8.7 Economic Analysis

Economics is the study of resource allocation. While we can rely on markets to produce the "right" amount of shoes, pens, and potato chips, markets may fail to capture the value of reductions in risks being managed by the public sector. In fact, many risks are allocated perfectly well by the market, although not everyone is comfortable with the outcome. Two examples might be smoking and obesity. The risks of each of these are widely understood so that, in a sense, the market works. There are, however, many risks that society has chosen to allow government to manage, instead of leaving it to people in their own private choices. They are allocated by risk managers and decision makers higher up in the decision-making pyramid.

Risk management decisions will always affect the allocation of resources; as a result, economic analysis is always potentially relevant to risk managers. In a world of scarce resources, economists maintain that only RMOs that result in positive net benefits are economically efficient. If risk managers enact measures with negative net benefits, they must do so on some basis other than economic efficiency.

Economists have many methods for conducting their analyses. Five that are most germane to evaluating risk management options and aiding risk management decisions are introduced here. The introductions are brief and spare on details. The purpose of this chapter is to raise awareness of the availability of these techniques rather than to explain their usage.

8.7.1 Cost-Effectiveness Analysis

Cost-effectiveness analysis is concerned with achieving a fixed objective with the least expenditure of resources. In a risk management context, it might mean finding the least costly way to obtain a given level of risk reduction. For example, what is the least expensive way of cutting the risk in half? Alternatively, it may be looking for the RMO that provides the greatest risk reduction for a given expenditure.

Cost effectiveness relies on a measurable output like the number of illnesses reduced, days without accidents, habitat units created, increased tons of commerce, loans repaid, quality-adjusted life years (QALY), and the like. It also requires estimates of the costs of obtaining different levels of output. Cost-effectiveness analysis is most useful when the outcomes of an RMO cannot be readily or reliably monetized but can be quantified.

8.7.2 Incremental Cost Analysis

When risk managers are not considering a fixed level of risk reduction or a fixed expenditure on risk reduction as they do with cost effectiveness, incremental cost analysis is an effective technique. Incremental cost is the cost of a little more (i.e., an increment) of an output. When the increment is one additional unit of output or an arbitrarily small increment of output, incremental cost is sometimes called marginal cost. It can be used to compare different resource allocation options in like terms. The health-care profession has used cost effectiveness and incremental cost analysis to guide health-care budget allocation decisions.

Consider the risk of children being poisoned by eating vitamin pills like candy. Imagine the following independently implementable risk management measures have the costs and impacts shown in Table 8.6. The incremental cost is defined as the change in total cost divided by the change in output. It is a synonym for marginal cost and is often the preferred term for changes in output larger than a unit. In this example the change in output is poisonings prevented. The incremental costs show that redesigning the container is the best buy, i.e., it has the lowest incremental cost of poisonings prevented for $15 million, followed by individually bubble wrapping each pill at a cost of $30 million, for a cumulative RMO cost of $45 million.

When incremental costs take an order-of-magnitude jump, as they do for the education option, this break point often signals when enough has been spent. The analysis here helps risk managers allocate a limited budget on risk management measures. Incremental cost analysis helps risk managers obtain the biggest bang for the buck. Or it may suggest that it is reasonably cost effective to prevent 2,000 poisonings at a cost of $55 million, but it is not worth spending an extra $20 million for only 100 more reductions. That money may best be spent elsewhere.

TABLE 8.6

Incremental Cost of Poison-Prevention Options

Risk Management Measure	Additional Cost	Poisonings Prevented	Incremental Cost of Poisonings Prevented
Labeling	$10 million	250	$40,000
Bubble wrap	$30 million	1100	$27,300
Redesigned container	$15 million	650	$23,100
Education	$20 million	100	$200,000

BCA

Benefit-cost analysis is simply rational decision making. People use it every day, and it is older than written history.

Our natural grasp of costs and benefits is sometimes inadequate, however, when the alternatives are complex or the data uncertain. [sic] Then we need formal techniques to keep our thinking clear, systematic, and rational. These techniques constitute a model for doing benefit-cost analysis. They include a variety of methods:

- identifying alternatives;
- defining alternatives in a way that allows fair comparison;
- adjusting for occurrence of costs and benefits at different times;
- calculating dollar values for things that are not usually expressed in dollars;
- coping with uncertainty in the data; and
- summing up a complex pattern of costs and benefits to guide decision-making.

It is important to keep in mind that techniques are only tools. They are not the essence. The essence is the clarity of the analyst's understanding of the options.

***Source*: Treasury Board of Canada Secretariat (1998).**

8.7.3 Benefit-Cost Analysis

When risk managers seek the option that will do the most good given the choices available, benefit-cost analysis can be a useful tool. Benefit-cost analysis (BCA), also called cost-benefit analysis, analyzes the advantages and drawbacks of a risk management option. All the advantages and drawbacks are estimated in monetary terms. If the advantages (benefits) measured in, say, dollars exceed the drawbacks (costs) in dollars, the option is said to be economically efficient. This means its value to society exceeds its costs to society. The most desirable RMO from a BCA perspective is the one with maximum positive net benefits, and this is not necessarily the option with the highest benefit-cost ratio.

8.7.4 Risk-Benefit Analysis

We all accept risks to realize the benefits associated with the risky behavior. Risk-benefit analysis is another comparison method available to risk managers. It compares the risks of a situation to its related benefits. A tolerable level of risk may be judged to be warranted by the fact that substantial benefits accrue in relation to the risk. A risk-benefit analysis is to some extent the reverse image of cost effectiveness. In cost-effectiveness we are unable to express the advantages in monetized terms, whereas the costs of risk management are more easily estimated. In risk-benefit analysis the risks (costs) remain quantified in risk-related terms while the benefits that accompany that risk are often monetized. Another application of risk-benefit analysis is found in pharmaceuticals, where statistically rigorous methods are used to demonstrate the level of risk patients and other decision makers are willing to accept to achieve the benefits provided by a new drug or health-care product.

Institutional review boards—bodies that review research proposals for organizations—have developed their own jargon and guidance on risk-benefit analysis. Risks to research subjects posed by participation in research should be justified by the anticipated benefits to the subjects or society. This requirement is clearly stated in all codes of research ethics, and it is central to U.S. federal regulations. The risks to which research subjects may be exposed have been classified as physical, psychological, social, and economic (Levine 1986).

8.7.5 Economic Impact Analysis

Economic impact analysis (EIA) studies how the direct benefits and costs of an RMO affect the local, regional, or national economy. The economic impacts of RMOs usually include effects on jobs, incomes, prices, taxes, and possibly measures of economic welfare like consumer and producer surplus or QALYs. Thus, economic impact analysis is intended to measure these types of economic effects associated either with the status quo or with particular RMOs that may be implemented. BCA measures direct benefits and costs of an RMO. It typically does not convert these direct effects into their indirect effects on the economy, such as changes in employment, wages, business sales, or land use. This is the role of EIA.

The most common forms of economic impact analysis trace spending through an economy and measure the cumulative effects of that spending in the impact region. The impacts forecast the number of jobs created or lost by a risky event or an RMO. Many economic impact models also predict impacts on personal income, business production, sales, profits, and tax collections. It is not unusual for an economic impact analysis to show impacts on hundreds of different sectors of a region's economy.

8.8 Summary and Look Forward

Resources, i.e., real things, are always scarce, and so we can't do everything. We have to make choices. Choices cost us the opportunity to do other things, so they entail trade-offs. These trade-offs are best examined at the margin, i.e., marginal analysis. Risk managers are always well advised to consider the incentive effects of their decisions, especially when the response to a decision is essential to the success of a risk management strategy.

We are ready to turn now from the risk management task to risk assessment. This is the evidence-based analytical process that is designed to reduce uncertainty to the greatest extent possible while providing risk managers with the information they need to achieve their objectives and solve problems.

Risk assessment does not have to be quantitative. It does not have to take years or months, and it does not have to be expensive. Qualitative risk assessment, the subject of the next chapter, is often sufficient to provide the evidence risk managers need to make decisions. When uncertainty is great, it is sometimes the only option for assessing risks. Qualitative risk assessment is often the first iteration in a quantitative risk assessment. Qualitative methods are especially useful tools for establishing risk management priorities. So it is appropriate that our focus on risk assessment should begin by considering some qualitative risk assessment techniques.

REFERENCES

Boardman, Anthony E., David H. Greenberg, Aidan R. Vining, and David L. Weimer. 2001. *Cost-benefit analysis: Concepts and practice*. 2nd ed. Upper Saddle River, NJ: Prentice Hall.

Hadari, Saguiv A. 1988. Value trade-off. *Journal of Politics* 50 (3): 655–676.

Hayek, Friedrich A. 1945. The use of knowledge in society. *American Economic Review* 35 (4): 519–530. http://www.econlib.org/library/Essays/hykKnw1.html.

Levin, Henry, and Patrick McEwan. 2000. *Cost effectiveness analysis methods and applications*. 2nd ed. Thousand Oaks, CA: Sage Publications.

Levine, Robert J. 1986. *Ethics and regulation of clinical research*. 2nd ed. Baltimore, MD: Urban and Schwarzenberg.

Mankiw, N. Gregory. 2009. *Principles of microeconomics*. 5th ed. Mason, OH: South-Western Cengage Learning.

Scharff, Robert L. 2010. Health-related costs from foodborne illness in the United States. Produce Safety Project. http://www.producesafetyproject.org/admin/assets/files/Health-Related-Foodborne-Illness-Costs-Report.pdf-1.pdf.

Stimson, Robert J., Roger R. Stough, and Brian H. Roberts. 2006. *Regional economic development: Analysis and planning strategy*. Berlin: Springer.

Tengs, Tammy O., Miriam E. Adams, Joseph S. Pliskin, Dana Gelb Safran, Joanna E. Siegel, Milton C. Weinstein, and John D. Graham. 1995. Five-hundred life-saving interventions and their cost-effectiveness. *Risk Analysis* 15 (3): 369–390.

Treasury Board of Canada Secretariat. 1998. *Benefit-cost analysis guide*. Ottawa, ON: Treasury Board of Canada Secretariat.

Wilson, Richard, and Edmund A. C. Couch. 2001. *Risk-benefit analysis*. Boston: Harvard University Press.

9

Qualitative Risk Assessment

9.1 Introduction

Risk assessment provides the scientific and other evidence required to support risk management decision making under conditions of uncertainty. It is often divided into two broad types: qualitative and quantitative. Qualitative risk assessment is distinguished primarily by its lack of reliance on numerical expressions of risk. That means that qualitative risk assessment depends on risk descriptions, narratives, and relative values often obtained by ranking or separating risks into descriptive categories like high, medium, low, and no risk. When the relative values are numeric but nominal or ordinal in character, such as when index numbers are used, the risk estimate is said to be semiquantitative, but they remain more qualitative than quantitative in character. Quantitative risk assessment relies on numerical expressions of risk.

The qualitative risk assessment process is identical to that used for quantitative risk assessment. It is not a lesser kind of assessment; it is a different kind of assessment. The difference is in the details. When the details are not needed or when the details are not available for risk management decision making, qualitative risk assessment may be the most appropriate choice.

Qualitative risk assessment is often used for screening or separating risks to determine which risks merit risk management's attention or, perhaps, a more detailed quantitative assessment. Qualitative assessments can also be used, however, to provide all the information needed for risk management decision making. Qualitative assessment provides an effective means of compiling, combining, and presenting evidence to support a statement about risk sufficient for decision making. A formal qualitative process provides consistency and transparency in decision making. It is an organized, reproducible method based on science, sound evidence, and the four generic risk assessment steps presented in Chapter 4. Done well, qualitative assessment is flexible and consistent, easy to explain to others, and supports risk management decision making.

Because of these attributes, qualitative assessment has proven especially useful when theory, data, time, or expertise is limited. When uncertainty is great, a qualitative risk assessment may be the best available option. It can also be useful for dealing with broadly defined problems where quantitative risk assessment is impractical. This is especially true for problems intended to establish risk management priorities.

Qualitative risk assessments are used in regulatory guidance and requirements documents for international trade, food safety, and health risk assessment. They are also widely applied internally by organizations to support the kinds of risk-based decision making that may fall short of the full risk analysis model described in this book. The qualitative risk assessment toolbox is growing with the increasing application of qualitative assessment methods. Some of these tools, which are the subject of this chapter, include:

- A generic process
- Increase or decrease risk
- Risk narratives
- Evidence mapping
- Screening
- Ratings
- Rankings
- Enhanced criteria ranking
- Operational risk management (risk matrix)
- Qualitative assessment models
- Multicriteria decision analysis

Regardless of the methodology or tool chosen, there are a few tasks that are necessary preparation for any kind of risk assessment. These are the basic risk management tasks that prompt a risk assessment:

- Identify the problem
- Identify the objectives
- Identify the questions to be answered

With this necessary preparation in mind, we begin by considering the basis for many qualitative risk assessment methods and tools.

9.2 Risk = Probability × Consequence

Risk was defined as the product of probability and consequence back in Chapter 1. This is a handy conceptual model for understanding risk more than a formula for calculating it. If either factor is zero, there is no risk. In a very general sense, qualitative risk assessment methods describe, explain, and narrate these two factors using the best available evidence in nonquantitative ways. Most qualitative methods either explicitly split the risk into these two factors for consideration or implicitly consider them. It is helpful to understand this as you move into the substance of this chapter.

When the probability and consequences of a risk cannot be estimated quantitatively, it can be informative and useful to think carefully about the nature of the probability and the nature of the consequences. Gathering and presenting sound science, evidence, and logical arguments about the nature of these factors often enables risk assessors to describe risks in ways that support decision making or to separate risks into meaningful categories for further subsequent consideration.

The generic process presented in the following section is based explicitly on a decomposition of these two factors. Operational risk management, also known as the risk matrix, is based on identifying and defining discrete categories of each of these two factors to obtain an overall risk rating. Separating methods (screening, rating, and ranking) are useful for reducing a long list of potential risks or hazards to a prioritized shorter list for further consideration. The best of these methods develop science-based criteria for separating that are usually related to either the probability or the consequence of the risk. Many techniques will address both factors to some extent, but it is not unusual for a method to focus on one or the other. The best qualitative methods address both the probability and the consequence of the risks under consideration.

The simplest tools—increase or decrease risk and risk narratives—are the least structured. So let me suggest that you structure these methods around these two risk factors as well. There are a number of tools in the qualitative risk assessment toolbox that are not necessarily intended to provide for a complete risk assessment. The evidence map provides a good example of such a tool. Although there is nothing to prevent its use as a risk assessment tool, it is most useful in addressing significant matters of uncertainty encountered in a risk management activity. The more complex methods include qualitative risk assessment methods developed for specific applications and more generic tools like multicriteria decision analysis. These methods can be expected to address both the probability and the consequence of a risk.

The methodology and, frequently, its application do not always make these fundamental factors evident. You can read some risk assessments and never see any of the four generic assessment steps or hear the terms *probability* or *consequence* or much of the other terminology developed in this book or the risk literature. Learn to look for them; they are often there but unlabeled. When they are not found, that is a cause to carefully evaluate the utility of the qualitative assessment in question. We'll begin our exploration of qualitative assessment techniques by considering a generic approach that can be made to work for virtually any kind of risk problem.

The following sections provide descriptions of a number of qualitative risk assessment tools and methods. I am not yet sure that the descriptive language for some of these techniques has coalesced to the point that all would agree with the language I use. So, with apologies for all of the invented language found below, let us begin.

9.3 A Generic Process

A generic qualitative risk assessment process* is presented here and is offered as a first approach to addressing a risk qualitatively when a better starting point is not apparent. The process is conceptually quite simple. It begins with the familiar conceptual model of risk presented below. Call this tier one of the process.

$$\text{Risk} = \text{Probability} \times \text{Consequence}$$

Each of these two factors is taken individually and decomposed into the critical elements that explain the probability and the consequence for a specific issue. The elements that comprise the probability may be multiplicative when a series of

* A debt of "inspiration" is owed to Richard Orr, Susan Cohen, and Robert Griffin (1993), all of APHIS at the time, who developed a generic risk assessment process for the USDA Animal Plant Health Inspection Service (see text box). The process presented here mimics and expands their basic approach.

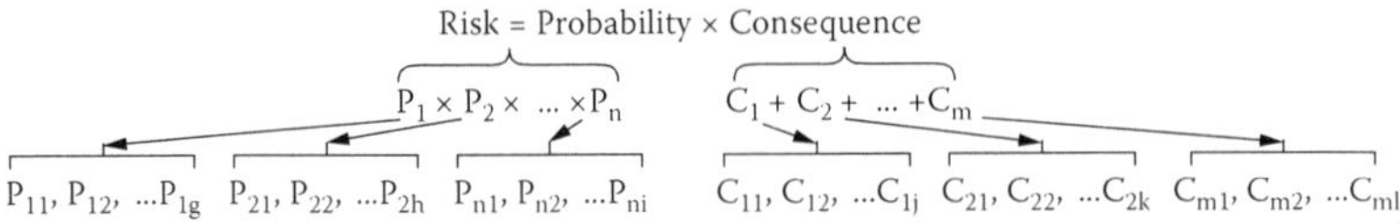

FIGURE 9.1 Decomposition of risk for a generic qualitative risk assessment process.

independent elements must all be present for a nonzero probability to exist. They could be additive elements if, for example, they represent separate pathways. The elements of the consequences could likewise be additive (±) if multiple consequences are relevant. It is conceptually possible that consequence elements could be multiplicative, but this is likely to be a rare instance. Call these critical elements tier two of the process.

Each tier-two element is then supported by a variable number of facts or bits of evidence. Call this evidence tier three. This three-tiered method is shown in Figure 9.1. Note that the index letters vary to indicate a potentially asymmetric numbers of facts in support of each separate component in the process.

The simple idea underlying this generic process is that if you can identify the risk(s) of interest and then think about the probability and consequence of each risk, it is

APHIS'S GENERIC PROCESS

APHIS uses a systematic approach to pest risk assessment based on the notion of a generic process. They are concerned with the risk of establishment of a plant pest as the result of importing plants and plant products from another nation.

The probability of establishment has four components:

1. *Presence potential*: probability that organism is on, with, in pathway
2. *Survival potential*: probability that organism survives transit
3. *Colonization potential*: probability that organism colonizes and maintains population
4. *Spread potential*: probability that organism spreads

Scientific evidence of presence, survival, etc., are put forward to support the rating given to the individual component.

Consequences of establishment are:

1. Economic damage potential
2. Environmental damage potential
3. Perceived social, political, and other damage

Ratings for these components are likewise supported by factual elements that determine ratings for each kind of consequence.

The APHIS process has continued to evolve beyond this original approach, which provided the inspiration for the generic process described in the text.

desirable to identify elements of that probability and consequence to better understand the risk. This approach has additional benefits for risk managers if any of the tier-two elements could be managed to zero or reduced in order to lower the overall risk.

The process begins by identifying all the potential risks relevant to the decision context. The generic process would be completed for each individual risk. In fact, if desired, a generic process could be completed for each independent pathway for a risk. Clearly articulating the probability and consequence of interest for a risk is the starting point. For example, if the risk is that connected to an invasive species, then we'll be concerned with the probability of establishment and the consequences of establishment. For levee systems, we may be concerned with the probability of overtopping and the consequences of overtopping along with the probability of catastrophic failure and the consequence of catastrophic failure.

Armed with a clear identification of the specific probability and consequence, we decompose these tier-one factors into their tier-two elements. Think about the probability factor. What sequence of events is necessary for the risk to occur? Identify those things that must happen for a nonzero probability to result. If any one of these components is missing, then the probability of the risk occurring goes to zero. Thus, we seek to expand the probability as follows:

$$P = P_1 \times P_2 \times \ldots \times P_n$$

This is conceptually multiplicative; therefore, if any one of the P_n components is zero, the entire probability of the risk occurring goes to zero. However, when there are multiple independent exposure pathways, the probability could be represented in an additive fashion.

$$P = P_1 + P_2 + \ldots + P_n$$

Combinations of multiplicative and additive elements are also conceivable. So the task is to represent these elements in as realistic a mathematical form as possible. Each individual probability element is, in turn, decomposed into a sequence of events, facts, arguments, or bits of evidence as appropriate. To minimize the redundancy of the discussion, it will proceed assuming a multiplicative definition of the probability.

In a similar manner, at tier two, we identify the positive and negative consequence components that could result from this risk as follows:

$$C = \pm C_1 \pm C_2 \pm \ldots \pm C_m$$

The consequences are more logically additive. If any one component goes to zero, there are still other potential consequence components that could prevent this overall term from going to zero. For example, we may be concerned with environmental, economic, and legal consequences. If we subsequently learn that there are no legal consequences, there may still be environmental and economic ones. Also note that the uncertain consequence can be a loss or a potential gain; hence its sign can be positive or negative.

The third tier of this process is to now take each tier-two element and decompose it. For example, take C_i and break it down into its events, facts, arguments, or bits of evidence, hereafter called facts, for the purposes of the qualitative assessment.

Up until this point, the process is a systematic decomposition of the basic risk, first into two factors, then into the elements of these factors, and finally into the supporting facts about those elements. Ideally, there will be quantitative and/or qualitative evidence available at the fact level to support qualitative judgments about the risk potential. These facts about the elements serve as the basis for assigning categorical potential risk ratings for the P_i and C_i elements. Thus, each P_i and C_i element will individually, based on the evidence, be rated a "high," "medium," "low," or "no" risk potential. Although the ratings themselves are qualitative, they are to be based on the best available evidence about the elements of the probability (exposure) and consequence (harm or opportunity) as represented by the P_i and C_i.

The ratings for the individual elements are combined into overall ratings for the probability and consequence factors. These factor ratings are in turn combined to determine an overall potential risk rating. This result is then used as an input to decision making.

This generic process simplifies the risk assessment by reducing the required inputs and calculations to a manageable set of facts and judgments based on these facts. Meanwhile, the rating logic is transparent and easy to apply. The value of the process lies not so much in the qualitative judgments of the actual ratings as in the thought process that goes into defining components and elements and supporting one's rating reasoning with evidence. The process also provides a handy basis for documentation. The simple categorizations produced by this generic process and the arrayed facts are relatively easy to communicate to risk managers and stakeholders.

To summarize the process after facts are assembled, it goes like this:

- Use tier-three facts to make risk ratings for tier-two components. Rating categories must be carefully defined or explained to assure they are mutually exclusive, collectively exhaustive, and based on science and sound logic.
- Use tier-two element ratings (high, medium, low, no) to define ratings for the probability and consequence factors.
- Use tier-one factor ratings to define the overall risk potential (high, medium, low).

The manner in which these last three steps are executed is somewhat flexible and contingent on the nature of the risk. Conceptually, it may be easier to explain by starting with the highest order rating, i.e., tier one, of the risk and working down to the elements. Sample overall risk ratings follow. They should be considered an example of what could be done rather than a prescription for what must be done.

Beginning with the overall risk potential and working backwards, the first task is to identify the number of risk categories and to define each of them. We'll use three: high, medium, and low. We will also continue in a generic fashion for now. The potential risk ratings for an identified risk are defined below assuming a risk of loss.

High: The risk potential is unacceptable based on the evidence; risk management action or additional assessment is indicated.

Medium: The risk potential may or may not be tolerable; depending on the offsetting benefits of the risky activity, risk management may be indicated.

TABLE 9.1

Determination of Overall Risk Rating Based on Factor Ratings

	Consequence Factor			
Probability Factor	**High**	**Medium**	**Low**	**No**
High	High	High	Medium	No
Medium	High	Medium	Low	No
Low	Medium	Low	Low	No
No	No	No	No	No

Low: The risk potential is tolerable or acceptable, and no risk management is indicated.

The overall risk potential could be determined as shown in Table 9.1. These ratings are subjective and are not amenable to proof, but the ratings and the scientific basis for them are subject to review and question. Let's continue to work down the model and consider how the factor ratings in Table 9.1 might be determined.

The probability dimension rating in the table could be determined in the following manner. Each probability component (P_i) at tier two would receive a high, medium, low, or no rating based on the facts obtained for it. The underlying evidence and facts, not the ratings, are the key for this process. Choosing an H, M, L, or N rating is subjective, but the facts used to make that judgment make the process transparent for review. In fact, the H, M, L, N ratings usually cannot be precisely defined or measured. If they could, a quantitative assessment would be preferable; so, they remain judgmental. A sample set of descriptions for probability component ratings follows.

High: Probability of undesirable result is unacceptable based on (tier three) evidence.

Medium: Probability of undesirable result is borderline tolerable/unacceptable based on (tier three) evidence.

Low: Probability of undesirable result is tolerable or acceptable based on (tier three) evidence.

No: Probability of undesirable results is zero or so low as to be effectively treated as a zero based on (tier three) evidence.

For any set of multiplicative probability elements $P_1 \times P_2 \times \ldots \times P_n$, the overall probability rating takes the rating of the element with the lowest probability rating. The total probability rating is, therefore, somewhat conservatively biased to overestimate the probability of the risk occurring. For example, a probability with four components rated HHHL has an overall rating of L. In fact the overall rating will always be lower than the lowest probability component rating.

To see the conservative nature of this rule, consider a numerical example with the following probabilities, $.9 \times .9 \times .9 \times .1$. This would yield a probability of .07, which is smaller than any of the component probabilities. Thus, using the lowest rated element ensures that the probability is not qualitatively underestimated because of the rating rule.

The HMLN rating of an individual probability element is determined based on the facts that can be gathered about the element. These are the facts that describe the basis of a tier-two element rating. The rating for each element, P_i, relies on a number of facts or bits of evidence. Thus, the rating for P_1 is based on specific facts P_{11}, P_{12}, and so on. Likewise the P_2 rating is defined by evidence P_{21}, P_{22}, and so on. Each fact or bit of evidence, P_{ij}, is a science-based or objective attribute or "explainer" of the probability element. The HMLN rating for any P_i is a subjective judgment based on the sum total of the evidence available in the facts P_{i1}, P_{i2}, ..., P_{im}. However, providing transparent definitions of these category ratings would be a valuable addition to any generic process.

This is an important point to emphasize. There are qualitative assessment techniques that emphasize assigning a HMLN rating without requiring explicit evidence or providing any explicit rational for the rating. Many rely on the reputation of the assessment team to sell the assessment. These are not useful techniques. Only the evidence can sell the assessment, and we want the evidence not only to speak in this generic process, but to be made available for review as well. One scenario for doing this is that the assessors consider all the evidence, assign a rating, and cite the evidence as the basis for the rating. There will be statements like, "P_1 is rated high based on the following elements of evidence: P_{11}, P_{12}, etc." A preferred scenario is that the assessors define what a high rating is for probability element P_i, then define all other category ratings used. Then the evidence is assembled and compared to the rating definitions in order to assign an HMLN rating.

The facts used to rate the relevant elements must be made available to support the HMLN rating for the probability component. In addition to providing the evidence about the rating elements, the assessors should also include the sources of the facts used as the basis for the rating. Is this fact a matter of general knowledge, is it from the literature, an extrapolation, or is it professional judgment? All are valid, and the source of the evidence for each element should be identified.

When the element rating for P_i is provided, the assessor should attach a confidence rating for each such judgment based on the quality of the evidence and the remaining uncertainty upon which the rating was based. It can be a simple confidence rating. Adapting the confidence scale originally used by APHIS, the following ratings provide an example: very confident (VC), reasonably confident (RC), moderately confident (MC), reasonably diffident (RD), and very diffident (VD). These can be defined in any manner that makes sense for the assessors. But they also need to be defined to be meaningful to risk managers and others.

Consequence ratings work in a manner similar to that described for probability ratings. Working from the top down, Table 9.2 shows how a hypothetical consequence rating might be obtained. The consequences will vary in type and number from risk to risk. Table 9.2 demonstrates how their judgments about consequence element ratings are combined into an overall consequence rating. The scheme shown in the table is essentially saying that economic and environmental consequences are equally important and more important, respectively, than other consequences. Note that the overall consequence rating is based on the rating for the higher of these two consequences. That is an arbitrary decision rule that you should feel free to modify to suit your needs.

TABLE 9.2

Determination of Overall Consequence Factor Rating Based on Consequence Element Ratings

Economic Consequences	Environmental Consequences	Other Consequences	Overall Consequence Rating
H	H,M,L,N	H,M,L,N	H
H,M,L,N	H	H,M,L,N	H
M	M,L,N	H,M,L,N	M
M,L,N	M	H,M,L,N	M
L	L,N	H,M	M
L,N	L	H,M	M
L	L,N	L,N	L
L,N	L	L,N	L
N	N	N	N

Descriptions of risk potential consequence element ratings follow.

High: Consequence of undesirable result is unacceptable based on (tier three) evidence.

Medium: Consequence of undesirable result is borderline tolerable/unacceptable based on (tier three) evidence.

Low: Consequence of undesirable result is tolerable/acceptable based on (tier three) evidence.

No: There are no undesirable consequences based on (tier three) evidence.

The individual tier-two consequence element, C_i, ratings are determined based on the facts and evidence available to support the rating judgments. Thus, in a manner analogous to that described for probabilities, the assessor might say element C_i, environmental consequences, is rated medium based on the following elements of evidence C_{i1}, C_{i2}, and so on. The sources of the evidence should be identified in the documentation of this process. The consequence component rating should also have a confidence rating attached to it. When the overall consequence and probability ratings are produced, the overall confidence rating becomes the least confident rating for each component. The overall confidence rating for the risk rating as shown in Table 9.1 is the lower of the overall probability and consequence confidence ratings.

9.3.1 Example Setup

Asian carp are a nonindigenous fish species that is now present and plentiful in the Illinois River. Many people and groups are concerned that the fish may soon invade Lake Michigan and possibly become established there, outcompeting native species and then possibly moving on into Canadian rivers and waterways. So let's consider the risk of interest to be the establishment of a breeding population of Asian carp in Lake Michigan and test-drive this generic process a little.

The basic risk of establishment is represented as:

$$\text{Risk} = P_{est} \times C_{est}$$

The tier-two elements of the probability and consequence are given as

$$P_{est} = P_{arrive} \times P_{survive} \times P_{colonize} \times P_{spread}$$

The logic here suggests the carp must arrive at the lake, survive once they get there, and establish a reproducing colony that is capable of spreading throughout the lakes and into the rivers of Canada. It is shown as a multiplicative process because if any one of those probabilities is a zero, then the overall probability of establishment is zero and there is no risk.

The level-two consequence elements might be as follows.

$$C_{est} = -C_{econ} - C_{environ} - C_{other} + C_{econ} + C_{other}$$

Note that economic consequences can be positive. A new and abundant fishery can be a source of fish oil, fish meal, food, and other beneficial activities. It can also have negative economic consequences on existing fisheries. Consequences are additive because, although any one of them may be a zero, that would not preclude other nonzero effects.

To establish HMLN ratings for each of these components we go to tier three, which includes facts and evidence. For example, to estimate the probability that the fish will arrive at the lakes, we would list the various pathways by which the fish could be introduced to the lake as well as any known barriers to any of these pathways. At that point it is time to gather evidence on the likelihood that the carp could reach the lake through a given pathway. Based on this accumulated evidence, the individual probability elements shown here would be rated to obtain an overall probability of arrival. The assessor would then say, "The probability of arrival is rated H, M, L, or N based on the following facts," while providing the source for each fact and an overall confidence rating for the probability of arrival.

In a similar way, the probability of survival is broken down into its most important elements, which might include various life requisites like a food source, predators, cover, breeding grounds, and so on. Then factual evidence on these life requisites for the carp and the lake environment are gathered to produce an H, M, L, or N rating for this probability component. The same would be done for each remaining probability element.

Consequence elements are rated the same way. Environmental consequences are broken down into specific elements, which might include their voracious appetites that can extinguish native snails and mussels, their direct competition with other fish for food resources, and the physical danger they pose to boaters because of their leaping ability. The facts about these elements are investigated and, on the basis of the weight of the evidence, a subjective environmental consequence rating is proffered. The various elements are rated and combined upward for an overall consequence rating, consistent with a new table like that in Table 9.2. The probability and consequence (tier one) factor ratings are combined to provide an overall risk rating of HML consistent with Table 9.1 or one like it. Once again, it is emphasized that the subjective ratings are less important than the evidence and thought process that are marshaled with this methodology.

The elements and, subsequently, the facts and evidence provide a veritable array of flips and switches that risk managers might be able to manipulate to reduce the probability of establishment or the consequences. Feel free to adopt or adapt any part of this generic process as well as any process or tool that follows.

9.4 More or Less Risk

When risks have been ignored, even paying the slightest attention to them can be a step in the right direction. For some simple problems, it may be enough to know if things are getting more or less risky. What has the snowstorm done to the roofs in the community? How has the earthquake affected food safety in the damage area? Being able to say that risk has increased or decreased and to present the evidence or rationale for why we think so may be the simplest form of a qualitative risk assessment.

Do we now have more or less risk than we had before? It can be useful to think separately about the probability and consequence of the risk. What has happened to make the risk more or less likely to occur? Has anything happened to make the consequences of the risk more or less severe? When the impacts of events are cumulatively aligned, this technique can be a useful simple tool. If we have reasons and evidence to support the notion that a risk is more likely to occur and the consequences may be more severe, it is easy to conclude the situation is more risky. Our evidence and rationale once again become the support for this judgment. Simply identifying the direction of change in a risk and the specific reasons for that change can be a positive step forward. Take care to identify the elements of your judgment that are uncertain.

Unfortunately, such a simple method is not much good for netting out changes in risk factors. When some circumstances tend to increase a risk while others tend to decrease a risk, this technique will be of no value. The next logical step up the ladder of qualitative sophistication is the risk narrative.

9.5 Risk Narratives

Earlier, risk assessment was described as the work necessary to answer four simple questions about a given situation.

1. What can go wrong?
2. How can it happen?
3. What are the consequences?
4. How likely is it?

Using simple narratives that answer these questions honestly and directly can be a very effective form of qualitative risk assessment. Describe the risky situation and support your description with the facts that are available while honestly communicating what you do not know. Often what we lack is an understanding of the risk, and a simple risk narrative provides a hypothesis for how things can go wrong. Answering each of those four questions and summarizing the evidence upon which the answer is based is a helpful organizing technique. Think of this narrative as a risk hypothesis. Be sure to identify the uncertain aspects of your hypothesis.

Tell the story of the whole of the risk if you can. Even without a quantitative estimate of the risk, it can help to tell the story of the existing risk. The best risk narratives will go beyond that, however. When possible, tell the story of the risk management options' effectiveness, i.e., the risk reduction story. Then tell the story of the residual, transferred, or transformed risks. In some less complex situations, risk narratives may fully meet the needs of many risk management activities. They should accompany every risk estimate, qualitative or quantitative. These narratives will always aid the risk communication task.

9.6 Evidence Maps

Summarizing scientific evidence about a potential hazard is a fundamental task in risk assessment. In an increasing number of instances, the existing data may be incomplete, inconsistent, or even contradictory on some significant matters of uncertainty. What is the likelihood and magnitude of sea level rise this century, or will it happen at all? Do mobile phones cause cancer? What will normalization of relations between the United States and Cuba mean for the two countries? Evidence maps help say what is certain about hazards, what is uncertain, and why.

Evidence maps have been proposed by Schütz, Wiedemann, and Spangenberg (2008) as a tool for summarizing the scientific data on these uncertain issues in an easily accessible form. The maps provide a transparent, consistent, and reasonable characterization of the available evidence. The purpose of an evidence map is to illustrate the logic that leads experts to their conclusions about the existence of a potential hazard or risk. It is a risk evaluation technique that identifies and communicates how experts evaluate current scientific evidence on chosen topics subject to uncertainty.

There are three core elements to an evidence map. These are:

- The *evidence basis*—the number and quality of relevant scientific studies
- The *pro and con arguments*, with supporting and attenuating arguments
- The *conclusions* about the existence of a hazard, with remaining uncertainties identified

This evidence framework can be used to identify the conclusions experts have reached regarding the risk potential of an uncertainty or a hazard. It identifies the specific evidence/arguments used to justify these conclusions and does so in a way that also identifies the consensus/disagreement that exists and the uncertainties that remain.

The basic format for an evidence map is shown in Figure 9.2. Experts select the studies from the available literature that are suitable for a risk evaluation. From these studies they extract the arguments for a risk (pro-argument) and the arguments against a risk (contra-argument). They carefully document any evidence that attenuates or supports the arguments pro and con and draw some tentative conclusions while noting the remaining uncertainties surrounding the issue. The pro and con arguments along with the uncertainties handily summarize what is and is not known about a hazard or topic being mapped. Such a concise summary cannot include all the relevant detail, so the maps are often accompanied by a narrative report. The paper by Schütz et al.

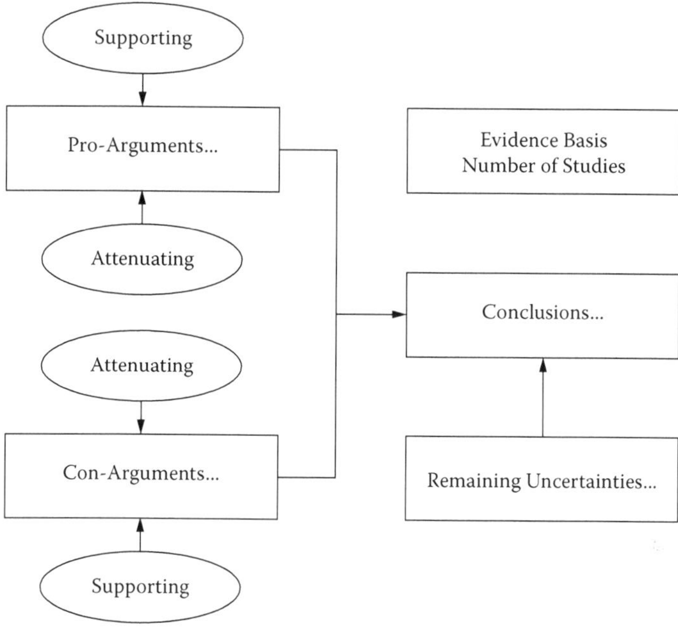

FIGURE 9.2 Template for an evidence map. (Adapted from Schütz, Wiedemann, and Spangenberg 2008.)

(2008) provides several explicit examples of this promising technique for organizing evidence. It is a potentially useful qualitative risk assessment tool.

9.7 Ordering Techniques

Putting things in order is an instinctive process we use all the time to aid in decision making. We look down a menu and make a selection. We surf the channels of our television and stop at the program we want. We scan a newspaper and choose what to read. It is all instinctive. However, if pressed to say why we chose as we did, we would be able to give reasons that we chose what we did and why we did not chose other things. When pressed for reasons, we can say what we like and dislike and why. Likewise, in risk assessment we can often put hazards and opportunities in some sort of order. When we make such decisions jointly, for an organization, or as stewards of some public trust, it is important to make the reasons explicit. They comprise the evidence upon which decisions will be based.

Qualitative risk assessment techniques can be especially useful for separating things and establishing priorities. Ordering techniques, which include chronologies, screening, rating, and ranking tools, can be useful for going from a long list of things to a short list of things. So if you are trying to figure out what hazards, pathways, mitigation measures, potential risks, and the like are most risky, most important, or of most interest to decision makers, these are useful tools to consider.

COX'S CAVEAT

Qualitative risk rating systems tend to make two types of errors: (1) *Reversed rankings*, i.e., assigning higher qualitative risk ratings to situations that have lower quantitative risks; and (2) *Uninformative ratings*, e.g., frequently assigning the most severe qualitative risk label (such as "high") to situations with arbitrarily small quantitative risks. This results in assigning the same ratings to risks that differ by many orders of magnitude.

Consequently, despite their consensus-building appeal, flexibility, and the appearance of thoughtful process in input requirements, qualitative rating systems do not always provide enough information to support accurate discrimination between quantitatively small and quantitatively large risks. As a consequence, the value of information (VOI) they provide to risk managers can be low to zero if most risks are small but with a few large risks, as qualitative ratings are sometimes unable to confidently distinguish the two risks. This suggests that quantitative risk assessment methods should be considered when qualitative ones may prove unreliable.

***Source*: Cox et al. (2009).**

9.7.1 Chronology

One of the most elementary techniques for ordering information is the chronology. Chronological ordering is sometimes not only a handy way to present information, it can be important as well. Showing the sequence and timing of events can sometimes reveal cause-and-effect relationships. Chronologies better enable us to see patterns, identify important events, and recognize significant gaps in our understanding of cause-and-effect relationships.

A good chronology begins with the earliest significant date or event. Make a list of all the events relevant to the risk of interest. Assign the appropriate time or date to each, and then arrange them in chronological order. A vertical timeline is an effective display. Whether the display is vertical or horizontal, always put the date or time first, followed by the description. This preserves the chronology.

For a variation that may provide additional insight, put your chronology in context by adding additional chronologies. Thus, in addition to the chronology for the risk itself, you might add a chronology with a matching timeline to show what was going on at the same time in your organization, with your competitors, the nation, the stock market, weather conditions, or any other contextual environment that may be useful to understand. The juxtaposed detailed sequence(s) of events often enable us to draw inferences about the risk.

9.7.2 Screening

Screening or sorting is the most basic structuring technique. It is the process of separating elements into categories of interest and no interest through a systematic evidence-based process. It would be used to identify hazards of potential concern or of no concern. It could be used to decide which risk initiatives will be funded this year and which will not, or to decide which risk management options are worthy of consideration as a potential solution for a given problem. It is not the tool to be used to find the best item among the "piles" of elements; it is the tool to use to create the piles.

Screening requires items to be screened and carefully defined categories (or piles) into which they are to be separated. In general, there are usually two kinds of screening categories, each of which are some version of in or out. Screening criteria can be chosen to either screen items onto the short list of interest (in) or to screen items off of the long list (out). You will need evidence-based criteria to use for separating your items into categories. For each item you will need some evidence of how that item scored on each criterion. Finally, you will need a synthesis algorithm for using the criteria and your measurements to separate your long list of items into discrete and separate categories of items.

There are several common screening algorithms for separating items into the desired categories. A domination procedure requires an item to be better or worse on all criteria than all other items. This could be used to separate the best or worst from the rest of a population of items. A conjunctive procedure requires an item to meet all predetermined criteria thresholds for inclusion in a category. A disjunctive procedure requires an item to meet at least one criterion threshold.

Elimination by aspects is sometimes used. This begins by identifying the most important criterion from your set of criteria. A cutoff value is set for it, and all items that do not meet the cutoff value are eliminated or screened out. Then you identify the next most important criterion, set a cutoff for it, and eliminate all items that do not meet it. This process continues until you have the desired subset of screened-in items. Weighting rules can also be used. First you rank the criteria against one another. Then you rank the items against the criteria. Weighting rules tend to be more useful for ranking than screening, but they can be used for screening. An example using weighting rules is found in section 9.7.4.2.

9.7.3 Rating

Rating is an activity that individually scores or rates each item of interest to the decision maker. It goes further than simple screening by systematically separating elements into multiple categories of varying degrees of interest. Items with like ratings are gathered into like groups, where the groups usually, but not always, have an ordinal logic to them. Think of the Motion Picture Association of America rating system for movies: G, PG, PG-13, R, and NC-17. Movies are also rated on a five-star basis. Hazards may be rated high, medium, or low. Probabilities of risks may be rated from rare to regular. Consequences could range from negligible to catastrophic. The requirements for a rating system are essentially the same as those for a screening system, i.e., items to be rated, rating categories, criteria for determining the ratings, evidence to measure the item against the criteria, and a synthesis algorithm. Rating, as compared to screening, will usually require more categories and often more criteria, hence more evidence.

9.7.4 Ranking

Ranking is an activity that distinguishes differences among individual items and assigns a position to one thing relative to other things. It is a systematic process used to put items in an ordinal sequence when used in a qualitative setting. Ranking can also be a cardinal or scalar ranking when data are available. Ranking requires essentially the same elements as a screening or rating process, but it may also include weighting the importance of the various criteria. Ranking is a simple process when objective criteria measures are available. A ranking process is described in the next section.

9.7.4.1 Enhanced Criteria-Based Ranking

There are countless ordering techniques in usage. You should feel free to invent your own techniques as long as they legitimately meet the science-based standards of risk assessment and the needs of decision makers. The example offered here uses a technique informally called enhanced criteria-based ranking. The ranking is based on criteria grounded in evidence. The technique is enhanced by its structure and by inclusion of a built-in "does it pass the red face (embarrassment) test" step.

This technique follows eight systematic steps:

1. Criteria
2. Evidence-based ratings
3. All possible combinations of ratings
4. Ranking
5. Evaluate reasonableness of ranking
6. Add criteria
7. New combined rating
8. New ranking

Like all good risk assessments, it begins with a clearly focused question. Like all good ordering techniques it requires a list of hazards, pathways, mitigation measures, or other items to be ranked.

Example of Enhanced Criteria-Based Ranking

Preliminaries. In this example we want to identify nematodes from Mexico that are of quarantine risk concern to the United States. This example is based on an actual ranking done by USDA (n.d.).* The nematodes of potential concern were:

- *Globodera rostochiensis*
- *Meloidogyne chitwoodi*
- *Punctodera chalcoensis*

* Plant Epidemiology and Risk Analysis Laboratory. Undated. "A Demonstration Exercise Using Enhanced Hazard Identification for Criteria-Based Ranking of Pest Risk." Prepared for Risk Analysis 101, Raleigh, USDA APHIS PPQ.

- *Cactodera/Heterodera* spp.
- *Ditylenchus destructor*
- *Bursaphelenchus cocophilus*
- *Ditylenchus dipsaci*
- *Radopholus similis* and *R. citrophilus*

Step 1: Criteria. The first step is to identify a few science-based criteria that will enable you to answer the question. Design your criteria to reflect the most important aspects of the risk you are evaluating. Bearing in mind the "probability × consequence" definition of risk, it is often useful to develop a criterion or two for each of these risk factors.

In general, the number of criteria is limited. Five or less is a good rule of thumb for this methodology. If you really need more criteria, you may also need a more sophisticated tool. Once the criteria are identified, you must decide whether the criteria will receive different weights or not. For simplicity, equally weighted criteria are used for this example.

The most critical task in this first step after identifying the criteria is defining mutually exclusive and collectively exhaustive scenarios for each criterion. The general practice for this particular technique is to carefully define what constitutes a high, medium, or low risk potential in such a way that the ranking can be based on evidence. Carefully documented criteria and their high, medium, and low scenarios provide a substantial and transparent science-based core for documenting the process and its resulting judgments. The criteria and scenarios for this example follow.

Criterion 1: Survival potential

H = High cyst former, survives with or without host
M = Medium survives in soil without host
L = Low survives only with host

Criterion 2: Host range

H = High many commercial hosts
M = Medium 2–4 commercial hosts
L = Low 0–1 commercial host

Criterion 3: Distribution

H = High wide distribution in Mexico
M = Medium limited distribution in Mexico
L = Low found in United States

Numerical values are avoided in the ratings, i.e., H, M, and L are used rather than 1, 2, and 3, because numbers suggest scalar content to most people. Letters are preferred because they avoid that potential error and they better serve the relative comparisons we are making. We have been trained to do arithmetic with numbers and tend to do so naturally.

That is not appropriate with a qualitative approach like this, so letters are preferred when possible.

The criteria chosen must enable the assessor to discriminate among the hazards or items to be ranked. If there is an important criterion for which all items receive the same rating, then it is not a useful criterion and should be dropped. It will be sufficient to note that it was considered but was not used in the assessment because it did not help to separate or rank the items.

Step 2: Evidence-Based Ratings. With a list of hazards and the criteria for ranking them identified, it is time to gather the evidence needed to rate each hazard for each criterion. This is the most critical step in establishing the value and credibility of any qualitative assessment: gathering science-based evidence to support the qualitative judgments. Expert judgment may be used to critically evaluate the available information and develop estimates for each of the nematodes against the three criteria defined previously. However, that judgment must be documented as carefully as any other evidence.

Ratings for this example, based on the life history and experiential knowledge about the nematodes, are shown in Table 9.3. The facts and evidence used to make these ratings would be referenced and included in the documentation of the process.

Step 3: All Possible Combinations of Ratings. This step determines how the various ratings are combined and weighted to produce rankings. Prior to ranking the hazards, you must determine the hierarchy of risk potential. When different weights are used for the criteria, the importance of this step becomes more evident. For the equal weights assumed in this example, it is a straightforward process. All the possible combinations of ratings are shown as follows:

Greatest risk	HHH
	HHM, HMH, MHH
	HHL, HLH, LHH, HMM, MMH, MHM
	HLM, MHL, HML, LMH, MLH, MMM, LHM
	HLL, LHL, LLH, MML, LMM, MLM
	MLL, LML, LLM
Least risk	LLL

TABLE 9.3

Nematode Ratings for Each Criterion

Nematode	Criterion 1	Criterion 2	Criterion 3
P. chalcoensis	H	L	M
G. rostochiensis	H	M	M
M. chitwoodi	M	H	L
D. destructor	M	H	L
R. similes	L	L	L
Heterodera & Cactodera	H	M	L
B. cocophilus	L	M	H
D. dipsaci	L	H	L

TABLE 9.4

Subject Ranking Clusters of Nematodes with Three Criteria

Nematode	Rating	Ranking
G. rostochiensis	HMM	Greatest risk
Heterodera & Cactodera	HML	Moderate risk
M. chitwoodi	MHL	
P. chalcoensis	HLM	
D. destructor	MHL	
B. cocophilus	LMH	
D. dipsaci	LHL	
R. similes	LLL	Least risk

Note that if, for the moment, we let H = 3, M = 2, and L = 1, the first row sums to nine, the second to eight, etc. If in the first step we had decided that criterion 1 were twice as important as criterion 2, an HMM would be a higher risk than MHM or MMH, and the groupings would look different with different weights. The rows show the combinations of ratings that receive the same ranking. Finer distinctions can be obtained, if desired, by using weights or more criteria.

Step 4: Ranking. Using the results of steps 2 and 3, rank the nematodes according to descending relative risk. At this point it may help to group them into subjective clusters of relative risk or concern. The nematode rankings are shown in Table 9.4. Note that the order of similarly rated items is arbitrary.

At this point in the process the greatest and least risk among the candidate hazards are objectively trivial to identify. The distinctions among the six nematodes ranked as risks of moderate potential were not sufficient to separate them further in the judgment of the analysts.

Step 5: Evaluate Reasonableness of Ranking. Take a look at the rankings. Do they make sense? If not, why not? What is missing? This step was intentionally inserted into the process to minimize the likelihood of an error based on overlooking some important criterion. Step 5 asks analysts to evaluate the thought process to this point in time before finalizing the rankings.

In this example, the team thought *P. chalcoensis* would be ranked highest. This was the nematode that stakeholders and scientists were initially concerned about. Why did it show up as a moderate risk? The answer lies in the nature of the criteria.

Step 6: Add Criteria. The initial criteria did a good job of presenting evidence in support of the probability of a nematode being introduced into the United States. They did not do as well addressing the consequences of that introduction. The original criteria overlooked the economic consequences of the risk of introduction. Although the number of hosts potentially affected does capture some aspect of consequence (as well as probability), the economic importance of the host is not addressed in the original criteria. A fourth criterion addressing the economic importance of the hosts was added to more accurately reflect the risk.

Criterion 4: Economic value of the hosts

H =	High	affecting major U.S. crop(s)
M =	Medium	affecting U.S. crop(s), but not major U.S. crop(s)
L =	Low	not affecting crops of strong economic significance

With this criterion, the process will better reflect the potential risk posed by the nematodes.

This step is sometimes misunderstood. Its purpose is not to manipulate the analysis to get the answer you want. That sort of manipulation could be easily hidden from the outset if that was the assessor's intent. This enhancement step is intended to reveal and document the evolutionary thinking about the question at hand for the sake of transparency.

A new criterion will not always be added. Sometimes criteria will be dropped as redundant or they will be replaced by better criteria. If the criteria change in any way, steps 2 and 3 are repeated. If no new criteria are added, the process is complete at this point.

Step 7: New Combined Rating. Each nematode is rated against the evidence for the new criterion to obtain a new combined rating for each nematode as shown in Table 9.5. It will also be necessary to prepare a new set of all possible combinations that reflect the weights of the new and old criteria. That task is left to the reader. All that remains after that is to establish a new ranking.

Step 8: New Ranking. Using the order established in the revised list of all possible combinations, a new ranking is established. Once that is done, the nematodes are again grouped into subjective clusters of designated risk potential as shown in Table 9.6. Note that where the lines are drawn for the subjective groupings is arbitrary. *Punctodera chalcoensis* has now been designated as one of the greatest risks.

The tables presented here provide an effective documentation of the decision process. The criteria, ratings, and rankings are all transparently clear. Others may challenge any part of the evidence or judgment and present alternative evidence for doing so. Focusing on the evidence that underlies the rankings is a strength of the process.

TABLE 9.5

Nematode Ratings with Four Criteria

Nematode	Criterion 4 Rating	New Combined Rating
G. rostochiensis	H	HMMH
Heterodera & Cactodera	L	HMLL
M. chitwoodi	H	MHLH
P. chalcoensis	H	HLMH
D. destructor	M	MHLM
B. cocophilus	M	LMHM
D. dipsaci	H	LHLH
R. similes	H	LLLH

TABLE 9.6

Final Ranking of Nematodes

Nematode	Rating	Ranking
G. rostochiensis	HMMH	Greatest risk
M. chitwoodi	MHLH	
P. chalcoensis	HLMH	
Heterodera & Cactodera	HMLL	Moderate risk
D. destructor	MHLM	
B. cocophilus	LMHM	
D. dipsaci	LHLH	
R. similes	LLLH	Least risk

9.7.4.2 Paired Ranking

Let's consider another ranking problem through the lens of a different technique, paired ranking. This is an alternative form of weighted ranking. Following the method of Jones (1998), there are nine steps to this technique. They are:

1. List all criteria for ranking
2. Rank the criteria pair-wise
3. Assign percentiles to the criteria you will use
4. Construct a ranking matrix
5. Rank all items pair-wise for each criterion
6. Calculate weighted criteria ratings
7. Sum the criteria ratings
8. Establish the rankings
9. Conduct a sanity check

Imagine that a food producer is going to offer a new product line and is trying to decide which of four suppliers (W, X, Y, Z) to choose for a critical ingredient. Suppose the producer has decided that the criteria for making this choice will be the cost of the ingredient (C), the supplier's quality control measures (QC), and the future reliability (R) of the supply. Let's apply the paired-ranking technique to this problem assuming that the decision maker wants to minimize the risk of any problems associated with the supply of the ingredient over the next three years.

Example of Paired Ranking

Step 1: List all criteria for ranking. The first step is to identify all the relevant criteria you can think of using to rank your items.

Step 2: Rank the criteria pair-wise. The second step narrows that list to the most important criteria through pair-ranking. For simplicity, this example uses only three criteria, but any number of criteria can be used. The caveat remains the same as before. If you need a lot of criteria to provide the ranking, you may need a more sophisticated assessment method.

Pair-wise ranking works like this. Given three criteria—C, QC, and R—we ask which is more important: C or QC, C or R, QC or R. These choices are presented for all possible pairings of criteria. Given the risk of concern, assume that the risk managers have a clear transitive ranking QC > R > C. Thus, QC would receive two most important "votes," R would receive one, and C would receive none.

When you are dealing with a larger number of criteria, the preferences may not always be so obviously transitive. With a large number of criteria, the process can also grow tedious quickly. Nonetheless, pair-wise ranking can be a useful way to identify the most important criteria from among the list of candidates. What lends an air of credibility and confidence to pair-wise ranking, however, is when the assessors carefully record the reasons and evidence used to support their rankings.

In the present case, QC is more important than reliability because it is essential that the company has complete confidence in the quality and safety of the supplier's product. Laboratory certificates, a total quality management plan, and a hazard analysis and critical control point (HACCP) plan may be important features of QC. Reliability is more important than cost because the success of the new product line rests on a dependable supply of the ingredient over a long term. Cost is not a trivial consideration; it is just the least important of the three.

Step 3: Assign percentiles to the criteria you will use. This is a subjective judgment step. There is no simple deterministic set of values to use. You should not use the votes a criterion received to calculate these weights. The weights should reflect the risk manager's preferences. If a tool like this is used for making public policy, additional care will need to be taken with this step. For this example, assume that the weights are QC = .5, R = .3, and C = .2. The weights are usually easier to use and for others to understand if they sum to 1 (or 100). Note that equally weighted criteria is not an option with this methodology.

Step 4: Construct a ranking matrix. The risk matrix* is a simple table with the criteria in columns and the items to be ranked in rows. A completed matrix is shown at the end of this example.

Step 5: Rank all items pair-wise for each criterion. This step in the process is another round of pair-wise comparisons that is based on evidence. Taking the first criterion, cost, we ask which supplier is better on cost: W or X, W or Y, etc. Table 9.7 summarizes the evidence available for this decision problem.

Step 6: Calculate weighted criteria ratings. Using the data from Table 9.7 for the pair-wise rankings, Table 9.8 shows the votes received by each supplier for each criterion. The relative weights appear in the column headings. Weighted criteria rankings are calculated by weighting the number of votes by the criteria weights to obtain the ratings shown in Table 9.9.

Steps 7 and 8: Sum the criteria weightings and establish the rankings. The ratings are summed and the suppliers are ranked according to their

* Do not confuse this use of risk matrix with the subsequent usage under the discussion of operational risk management.

TABLE 9.7

Evidence for Criteria Used in Ranking Matrix

Supplier	Cost (C)	Quality Control (QC)	Reliability
W	Most costly	Second best controls	Most reliable
X	Third most costly	Third best controls	Second most reliable
Y	Least costly	Least controls	Third most reliable
Z	Second most costly	Best controls	Least reliable

TABLE 9.8

Votes Received in Pair-Wise Comparison of Suppliers for Each Criterion

Supplier	Cost (C) .2	Quality Control (QC) .5	Reliability .3
W	0	2	3
X	2	1	2
Y	3	0	1
Z	1	3	0

TABLE 9.9

Calculation of Weighted Ratings for Each Criterion

Supplier	Cost (C) .2	Quality Control (QC) .5	Reliability .3
W	0 × .2 = 0	2 × .5 = 1	3 × .3 = .9
X	2 × .2 = .4	1 × .5 = .5	2 × .3 = .6
Y	3 × .2 = .6	0 × .5 = 0	1 × .3 = .3
Z	1 × .2 = .2	3 × .5 = 1.5	0 × .3 = 0

rating in Table 9.10, which shows a completed ranking matrix. Supplier W is the best choice.

Step 9: Conduct a sanity check. This technique can be used to rank items in a simple and straightforward way. It represents the judgments of the people that established the weights and made the pair-wise comparisons—and no one else's. These judgments are only as good as the evidence they are based upon. When this technique is used to make private decisions about risks, it is a perfectly valid technique to aid decision making.

TABLE 9.10

Completed Risk-Ranking Matrix

Supplier	Cost (C) .2	Quality Control (QC) .5	Reliability .3	Total Score	Rank
W	0	1	.9	1.9	First
Z	.2	1.5	0	1.7	Second
X	.4	.5	.6	1.5	Third
Y	.6	0	.3	.9	Fourth

9.8 The Risk Matrix

The risk matrix, sometimes known as operational risk management (ORM), is another qualitative technique based on the simple equation, risk = probability × consequence. It is commonly used to help organizations rank their program risks and occasionally, as for the military (USDOD 2000), it is used to help manage risks. Like any qualitative technique, this one assumes that the relevant risks have been identified through a rigorous risk management process before the assessment is begun.

The idea of a risk matrix is simple. The probability dimension of a risk is envisioned as a continuum from 0 to 1. This continuum is broken into segments or categories. Although the segments could be defined quantitatively, we are dealing with qualitative methods in this chapter, and so the categories are expected to be qualitatively defined, such as improbable, remote, occasional, probable, and frequent (USDOD 2000). Likewise, a number of quantitative or qualitative categories are defined for the consequences as well. An example is negligible, marginal, critical, and catastrophic. We will soon return to consider how these categories are defined, but for now let's look at the matrix and the ranking of risks.

Using the categories developed, a matrix is constructed as shown in Figure 9.3. The cell in the lower left is clearly the least risky combination of probability and consequence that can be obtained. The cell in the upper right is unambiguously the riskiest. Risk clearly increases as we move along the northeast axis from low risk to high risk. Because the example matrix is not square, that is the strongest and most definitive statement we can make.

Within any row we know that moving east (right) increases the risk, i.e., *q* is riskier than *p*. Within a column moving north (up) increases risk, so that *c* is riskier than *f*. Once we attempt to compare cells in different rows and columns, it becomes more difficult to judge the relative risk. For example, which is riskier, *l* or *p*? The probability of one is remote with negligible consequences; the other is improbable with marginal consequences. If these two combinations described the risk of fire to your home, which would you prefer? We face the same conundrum at the risky end of the matrix. Cell *c* is a risk that is frequent and critical, while *g* is probable and catastrophic. Which is the greater risk?

Certainly these are questions that many individuals can answer for themselves, but when it comes to group decision making or public stewardship responsibilities, agreement becomes more of a problem. Thus the risk matrix has a fundamental subjectivity

Frequent	a	b	c	
Probable	d	e	f	g
Occasional	h	i	j	k
Remote	l	m	n	o
Improbable		p	q	r
	Negligible	Marginal	Critical	Catastrophic

FIGURE 9.3 Risk matrix example.

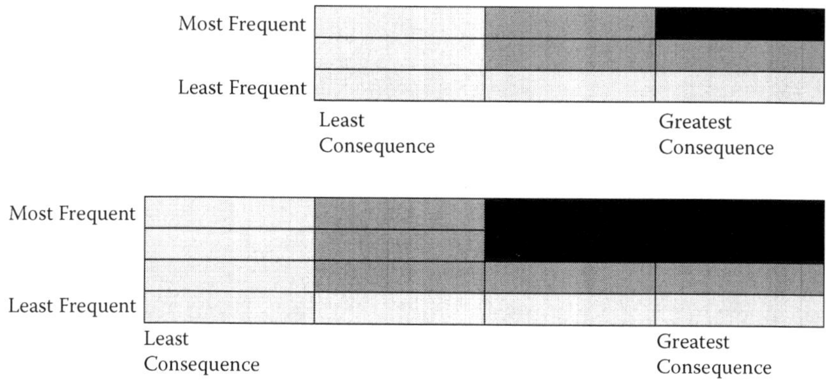

FIGURE 9.4 Consistent color-coding for a 3 × 3 and a 4 × 4 risk matrix.

about it that can present a problem if it is not explicitly recognized by its users. If you do an Internet search on the phrase "risk matrix" and look at the resulting images, you will see that people usually address this issue by filling the cells in with green, amber, and red colors in a bewildering array of patterns.

It is relatively simple to make distinctions between cells separated by some distance from one another. No one would argue that *g* is less risky than *i*; both factors are risky. It is when trade-offs have to be made—when a risk is greater in one dimension and less in another—that the subjectivity of this technique cannot be avoided.

In actual practice, these issues most often are addressed through subjective judgments. Cox (2008a) has proposed that following his three axioms—weak consistency, betweenness, and consistent coloring—small square risk matrices should be grouped as shown in Figure 9.4 if one follows the convention of grouping risks as dark (stop), medium (caution), and light (go).

In the extreme case, assessors sometimes try to put each of the matrix cells in its ordinal sequence of increasing risk. Thus a 4 × 4 matrix would have 16 categories of risk from first (most risky) through sixteenth (least risky). This is the approach taken in the mishap risk guidance developed for the Department of Defense (USDOD 2000), where they ranked 20 different cells. However, this was done not so much to rank the risk relative to one another as to group the cells and identify the appropriate risk manager (by rank or position) responsible for making decisions for the various risk groupings. For example, risks ranked 1 to 5 were classified high risks, where the component acquisition executive was the designated risk manager/decision maker. So, you see, there are many inventive and possible uses for the risk matrix.

The risk matrix is easy to use and, consequently, easy to abuse. There are numerous junk applications of this technique that you need to avoid. The most egregious misuse of this technique occurs when participants in the process, who may or may not have any degree of expertise in the relevant subject matter, are asked to rank a consequence (or a probability) from 1 to 10. Only slightly less offensive is the technique that creates consequence or probability categories and asks participants to assign each risk to a consequence and a probability category without regard to producing any evidence for doing so. If this technique is not data driven, it may, as Hubbard (2009) suggests, be less than worthless. Cox (2008a, 2008b, 2009) has gone on to a far more sophisticated

ANOTHER COX CAVEAT

In an examination of risk matrices, Cox (2008a) found the following limitations:

Poor resolution: Risk matrices correctly and unambiguously compare maybe less than 10% of randomly selected pairs of hazards. They can assign identical ratings to quantitatively very different risks, a problem called "range compression."

Errors: Risk matrices have assigned higher qualitative ratings to quantitatively smaller risks, sometimes leading to worse-than-random decisions.

Suboptimal resource allocation: Allocating resources decisions cannot be based on the categories provided by risk matrices.

Ambiguous inputs and outputs: It is difficult to provide objective categories of consequences when there are significant uncertainties.

Risk matrix inputs, i.e., probability and consequence categorizations, and resulting risk rating outputs require subjective interpretation. Different users can obtain opposite ratings of the same quantitative risks. These limitations suggest that risk matrices should be used with caution, and only with careful explanations of embedded judgments.

critique of the manner in which the resulting matrix is used to rank risks. Hubbard and Cox have been scathing in their justified criticism of this technique in particular. Cox points out that the matrices are not usually subjected to rigorous validation studies, and there are potential pitfalls as described previously and in the text boxes of this chapter.

Now let us return to the matter of identifying the categories of probability and consequence. Assessors must decide how fine the distinctions they will need for decision making must be. In general, one might expect square matrices, despite the 5 × 4 example above. Of course, if the assessors have a good reason for greater refinement in one dimension of the risk than in another, there is nothing that prevents them from reflecting that in their matrix. However if you feel you need a sizeable number of categories for each risk dimension and have the evidence to support that, you may want to consider doing a more quantitative risk assessment. This technique is limited in the real range of distinctions it can reliably depict.

The key to a useful risk matrix is the extent to which your categories of probabilities and consequences are effectively defined based on criteria that can be observed, measured, or judged objectively. There can be no escaping the need for judgment with a risk matrix, so the key is to explain the evidence upon which those judgments will be based. This means that category definitions (not simply rank 1 to 10 or rate as frequent or infrequent) should reflect the proffered evidence as transparently as possible for a qualitative assessment process.

The USDOD (2000) defines catastrophic consequences as those that "could result in death, permanent total disability, loss exceeding one million dollars, or irreversible severe environmental damage that violates law or regulation." A rating of catastrophic consequences for a risk would then be based upon tangible (even if narrative) evidence of one or more of these conditions. A frequent probability has two different definitions: one for a specific item or individual, and another for a fleet or inventory. The first definition says the risk is "likely to occur often in the life of an item, with a probability of occurrence greater than 0.1 in that life." The inventory definition says the risk is "continuously experienced."

Alternative formulations of the frequency definitions may specify a frequency based on an interval of time, such as happens daily, weekly, monthly, annually, and so on. As with a consequence rating, the integrity (and hence the utility) of a probability rating stands and falls on the quality of the evidence used to provide and support it.

When the categories have been defined, the evidence gathered, and the ratings completed, the risks can be ranked. Sometimes the risks are ranked according to priorities. For example, risk managers may decide that all "red" risks must be managed immediately, amber risks should be monitored, and green risks ignored. In other applications, risks are rated and dropped into a cell in the risk matrix. The cells are then ranked in order of priority that meets the needs of risk managers.

As noted throughout this discussion, risk matrices are a matter of some controversy. Some argue that they are effective tools and that they have indeed been adopted by numerous organizations and applied to a wide range of risk problems. However, if the matrices are not well constructed with ratings supported by evidence, this can lead to risk management problems, including the misallocation of scarce resources.

9.9 Qualitative Risk Assessment Models

Many qualitative risk assessment techniques have been developed by and for a variety of practitioner communities. Qualitative risk assessment models (QRAM) are based on a qualitative rendering of an existing risk assessment model. In this book I have offered four application-free steps in risk assessment:

1. Hazard or opportunity identification
2. Consequence assessment
3. Likelihood assessment
4. Risk characterization

Many practitioner communities have developed their own specific steps for conducting a risk assessment. For example, in the food-safety community, there are a variety of risk assessment models. Some are summarized in Table 9.11. Read down the columns to see the risk assessment steps.

A QRAM simply completes the established steps using qualitative data. Of the models presented in Table 9.11, only the antimicrobial resistance risk assessment is routinely conducted as a qualitative risk assessment (U.S. FDA 2003). This qualitative CVM model shown in Figure 9.5 relies on the now-familiar evidence-based ratings of high, medium, or low for each of the major steps in the risk assessment. These are subsequently combined to obtain an overall risk estimate.

QRAMs like this one are adaptations of more general notions about risk to a specific application. At the outset of this chapter, the generic process was built around

TABLE 9.11

Steps from Selected Food Safety Risk Assessment Models

Microbial Risk Assessment	Pesticide Risk Assessment	Pest Risk Assessment	Antimicrobial Resistance Risk Assessment	Nutrient Risk Assessment
Hazard identification	Identify residue of interest	Pest categorization	Release assessment	Problem formulation
Hazard characterization	Toxicity studies	Assessment of the probability of introduction and spread	Exposure assessment	Hazard identification
Exposure assessment	Determine NOAEL[a]	Assessment of potential economic consequences	Consequence assessment	Hazard characterization
Risk characterization	Select safety factor	Degree of uncertainty	Overall risk estimate	Dietary intake assessment
	Determine ADI[b] Identify MR[c]	Conclusion of the pest risk assessment stage		Risk characterization
	Exposure assessment (intake)			
	Risk characterization (ADI/intake)			

[a] No Observed Adverse Effect Level.
[b] Acceptable Daily Intake.
[c] Maximum Residue Limit.

the risk = probability × consequence notion. An alternative approach is to adapt this sort of qualitative method to other models of risk, like the CVM antimicrobial risk assessment of Figure 9.5.

In Chapter 4, risk assessment is described as the four-step process shown in Figure 9.6. A general approach to developing a QRAM is to choose a conceptual model such as that one or one from Table 9.11, for example, and then adapt it to produce qualitative risk estimates. For example, in the present instance, assessors might define HML ratings for each of the first three steps presented in Figure 9.6, using science-based evidence as the basis for the judgment. The risk characterization rating (step 4) would be the overall risk estimate, and it would be based upon the ratings of the three previous steps.

For example, the first step requires an estimate of the probability that the hazard exists and is present. This probability could be ranked as an H, M, or L on the basis of the best available evidence once these likelihoods have been defined as objectively as possible. Next, the consequence of an exposure must be estimated as H, M, or L. Once again, these and all subsequent ratings must be objectively defined so they can be supported with evidence. We then would estimate the probability of these

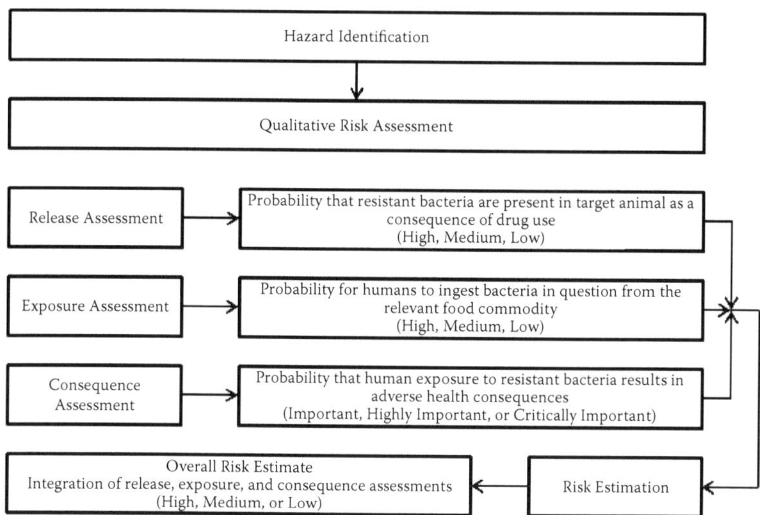

FIGURE 9.5 Components of a qualitative antimicrobial resistance model. (Adapted from FDA 2003.)

consequences in the likelihood assessment step as an H, M, or L. This qualitative assessment, based on the best available evidence, would result in a triplet of ratings, like MHM or some such combination. It remains then for the assessor to construct a table of all possible combinations of ratings and the overall risk characterization rank that each warrants—something akin to Table 9.1. A confidence code, as discussed earlier in the chapter, could accompany the overall risk characterization.

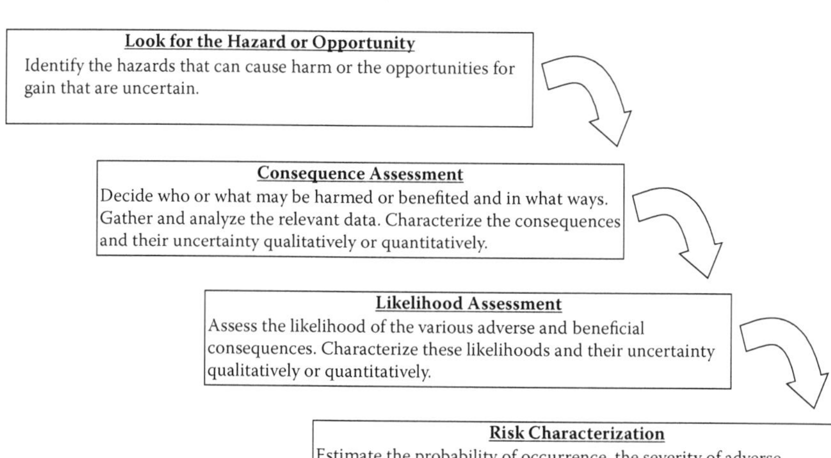

FIGURE 9.6 Application-free risk assessment model.

9.10 MCDA

Multi-Criteria Decision Analysis (MCDA), first encountered in Chapter 3 and revisited in Chapter 8, is a well-established technique for making trade-offs of quantitative and/or qualitative information that involves the personal preferences of decision makers. MCDA offers a great deal of promise to risk assessors. Let's consider an example.

MCDA begins with a basic question we are trying to answer. Suppose that question is which lock gates along a waterway present the greatest potential risk to health and safety and therefore should be repaired first? MCDA requires alternatives to analyze, so we will use five hypothetical locks numbered one to five.

MCDA, in a risk-based context, also requires evidence-based criteria for evaluating the alternatives and answering the question. It will also need subjective weights for defining the relative importance of the criteria and, therefore, trade-offs. Finally, it needs an algorithm for pulling all of this information together. MCDA methods are often distinguished based on the algorithm used to complete the analysis. For this example, Criterium DecisionPlus* Software has been utilized. Criterium DecisionPlus provides the capability to use the Analytical Hierarchy Process (AHP) or Simple Multi-Attribute Rating Technique (SMART) to assist the decision process. SMART originates from multi-attribute utility theory and was used for this example.

The hierarchical model used for this qualitative risk assessment is shown in Figure 9.7. The purpose of the risk assessment, shown on the left at the top of the hierarchy, is to identify the riskiest lock gates. To the right are the evidence-based criteria that will be used to answer this question. They are frequency of use, age, consequence of failure, and maintenance record. All but the consequence criterion relate to the probability of a lock gate failure. To the right of the criteria are the alternatives being evaluated, i.e., the locks whose gates present a potential risk to health and safety.

As with other criteria-based screening tools, assessors must decide which, if any, of the criteria are most important for decision-making purposes. This will usually be a risk management decision rather than a risk assessment one, although it is wise to reserve the possibility that objective weights may exist for some problems. The weights used in this instance are shown in Figure 9.8. These are subjective weights entered based on a 100-point (maximum importance) scale. The consequence of failure is judged to be twice as important as the age of the gates, for example.

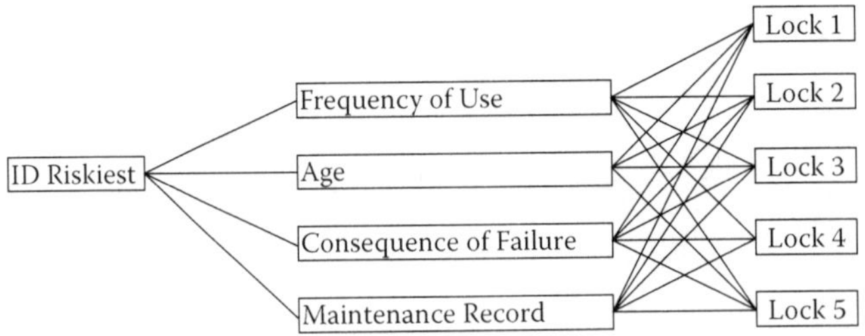

FIGURE 9.7 Hierarchical MCDA model for ranking the potential risk of lock gates.

* Criterium DecisionPlus is a trademark of InforHarvest Inc.

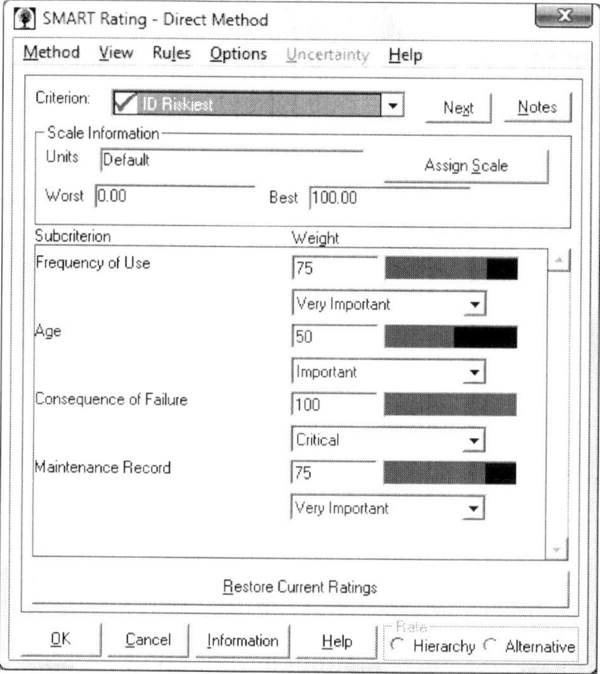

FIGURE 9.8 Decision criteria weights for MCDA analysis of the potential risk of lock gates.

Evidence for each of these criteria must be gathered, assessed, and entered into the model. An example of how this might be done is shown in Figure 9.9. In this example, age is the only empirical data available. It is measured in years. The other criteria were rated qualitatively. The consequence of failure ranged from trivial to critical. Frequency of use was ranked from 1 (most frequent usage) to 5 (least frequent usage). The maintenance record was based on a quality scale from unsatisfactory to the finest. It is critically important to understand that these qualitative ratings must be carefully defined and then based on the best available evidence supplemented by expert knowledge and professional opinion, all of which becomes part of the documentation of the rating process.

Different data-entry techniques were chosen to demonstrate the potential of the method. A wide variety of data entry options are available. Note that data may be entered as a distribution to reflect the uncertainty in any criterion measurement. For simplicity, those capabilities of the method are not developed here.

The model and the criteria definitions are such that a high score indicates the least risky lock gates and a low score shows the gates with the greatest risk potential. Figure 9.10 shows the results of the synthesis of the inputs using the SMART algorithm. Lock 3, with a low score of 0.33, has been ranked the riskiest set of lock gates based on this qualitative assessment. Locks 1 and 2 have the least risky lock gates. MCDA provides the capability of using qualitative data to produce semiquantitative estimates of risk.

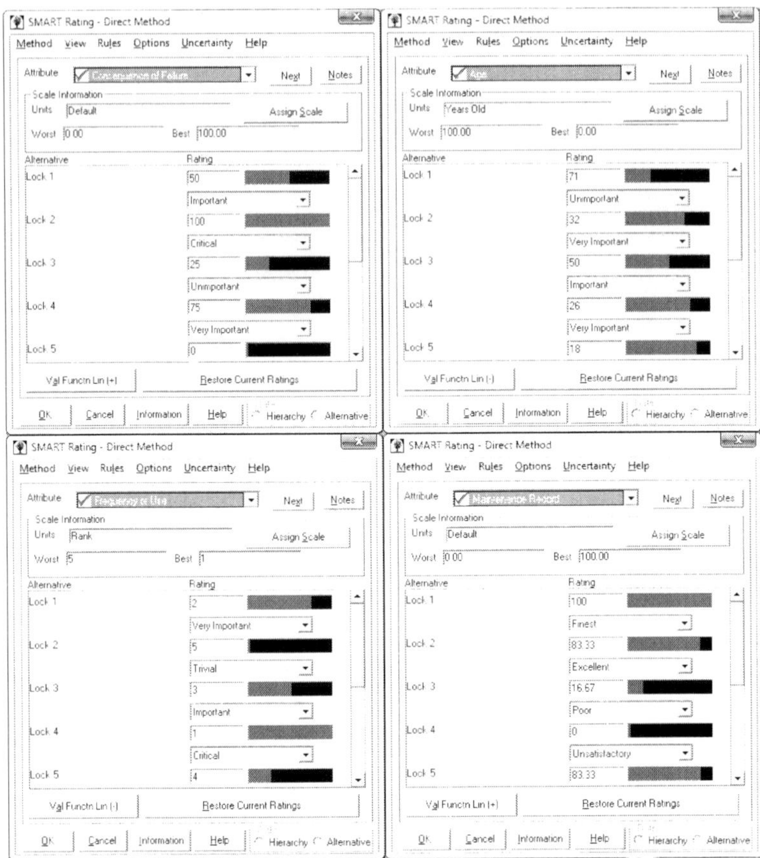

FIGURE 9.9 MCDA input data for a qualitative risk assessment.

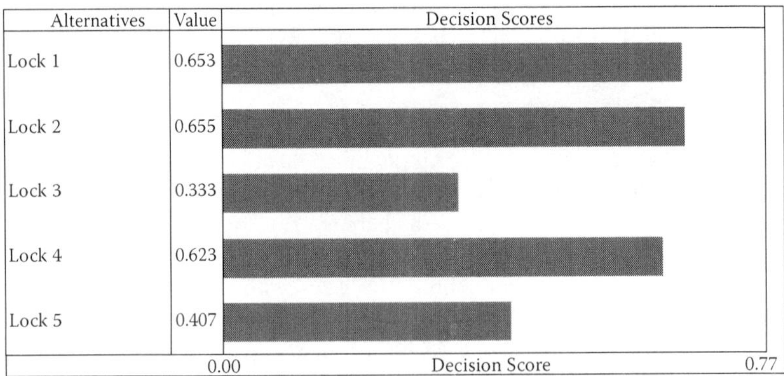

FIGURE 9.10 SMART decision scores for each alternative set of lock gates.

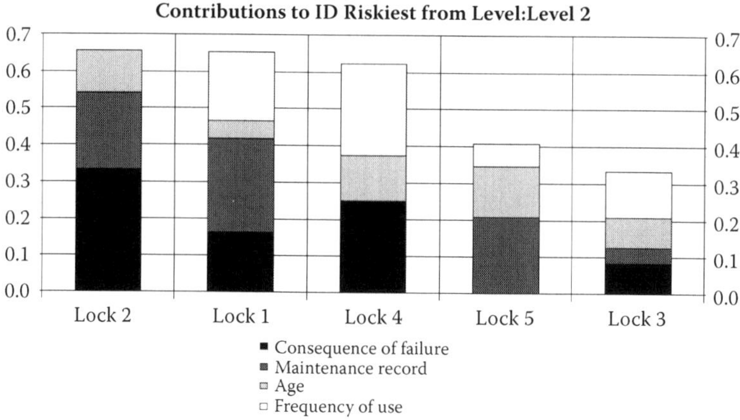

FIGURE 9.11 Sensitivity of alternative score to the contributions of each criterion.

The software packages available for MCDA vary in their features, but most provide some capability for sensitivity analysis. As we will learn in Chapter 16, this may be the most robust tool for exploring the uncertainty in a decision context. Figure 9.11 provides an example of a sensitivity analysis technique.

The heights of the bars show relative safety, so shorter bars represent riskier locks. Because of the way the criteria were defined and entered, this figure is somewhat counterintuitive. The smaller the color bar, the greater the risk; the larger the bar, the less the risk. The colors show the positive (safe) contributions of that lock to the individual criteria. Lock 2, for example, shows no positive contributions from frequency of use, so this is a major risk factor for that lock.

Lock 2 does have a large bar for consequence of failure. That means the potential consequences of a failure would be relatively small. Lock 5, by contrast, has no bar for consequence of failure, indicating that this is a large component of the risk. A quick look at this graphic indicates the nature of the risk at each alternative lock. The absence of a particular colored bar or the small size of a color indicates a greater risk factor for that lock.

MCDA helps risk managers sort through a prespecified set of alternatives to identify those alternatives with the characteristic of interest, i.e., greatest risk potential, greatest risk reduction potential, and so on. The quality of information used in MCDA can be scientifically derived hard data or it can accommodate subjective interpretations of ratings for one or more criteria. MCDA, like risk assessment, does not produce decisions; it produces information upon which informed decisions can be based.

9.11 A Semiquantitative Risk Assessment Example

Qualitative methods that make limited use of numerical estimates of risk are sometimes called semi-quantitative assessments. An example is offered here to illustrate the idea. Hurricanes Katrina and Rita focused national attention on levee safety. A national levee safety program was implemented to meet three goals:

1. Reduce risk and increase public safety through an informed public, empowered to take responsibility for its safety.
2. Develop a clear national levee safety policy and standards.
3. Maintain a sustainable flood damage reduction system that meets public safety needs.

One part of this program includes a levee inventory and a technical risk assessment of each levee. With thousands of levees to consider, it may be useful to develop tools for screening the levees to identify those with the greatest potential risk to humans and property.

This example presents a technique that could be adapted to such a purpose. It is not the method being used in the National Levee Safety Program. The example begins with an abbreviated version of the three pieces of paper that a good risk management process should produce: a problems and opportunities statement, an objectives and constraints statement, and a list of questions to be answered by the risk assessors. These are the most essential elements for defining a decision context. A sample risk assessment capable of answering the risk manager's questions is presented. The manner in which the uncertainty encountered was handled is summarized.

Example of Semiquantitative Risk Assessment

Problems:

1. Locally constructed levees are prone to underperformance or failure resulting in property damage and risk to life.
2. There is a complete lack of data on many of the levees to be evaluated in a national levee safety program.

Objectives:

1. Identify those levees of greatest potential risk to the populations they are to protect.
2. Protect life, health, and safety.
3. Reduce property damage.
4. Identify data gaps.

Questions:

1. Which levees in the region present the greatest potential risk to life and property?
2. Which levees should be subjected to a technical risk assessment first?

Figure 9.12 shows the risk ranger model. Using the basic conceptual model—risk = probability × consequence—to rank the potential risk of the region's levees in a qualitative manner, the challenge is to identify criteria that aid in assessment of the probability of an unsatisfactory performance as well as an assessment of the consequences. The criteria developed for this qualitative risk

This levee is rated a 89

Probability

A. How old is the levee? 5
1. Unknown
2. 10 years or under
3. Over 10 and up to 25 years
4. Over 25 and up to 50 years
5. Over 50 years

B. Who owns the levee? 3
1. Unknown
2. More than one owner
3. Private levee
4. State or local government

C. How well is it maintained? 5
1. Unknown
2. Regular mainetnance by known authority
3. Periodic maintenance by known authority
4. Irregular maintenance
5. No maintenance

D. Construction quality? 4
1. Unknown
2. State-of-the-art engineering design and construction
3. Standard engineering design and construction
4. Substandard design and construction

E. Number of flows confined in last ten years? 4
1. Unknown
2. None
3. One
4. Two or more

F. Any known problems? 2
1. Unknown
2. Yes
3. No

Consequence

G. How vulnerable is the population? 2
1. Unknown
2. Highly vulnerable (low income, elderly, low education, minority)
3. Moderately vulnerable (housing close to levee)
4. Low vulnerability (housing removed from the levee)

H. How large is the population at risk ? 6
1. Unknown
2. Less than 1000
3. 1000 to 10000
4. 10000 to 100000
5. 100000 to 1000000
6. Over 1000000

FIGURE 9.12 Levee risk ranger: a qualitative risk assessment tool.

assessment are shown below. We begin with criteria to assess the probability of an unsatisfactory performance of the levee.

A. How old is the levee?
   1. Unknown
   2. 10 years or under
   3. Over 10 and up to 25 years
   4. Over 25 and up to 50 years
B. Who owns the levee?
   1. Unknown
   2. More than one owner
   3. Private levee
   4. State or local ownership
   5. Federal ownership
C. How well is it maintained?
   1. Unknown
   2. Regular maintenance by known authority
   3. Periodic maintenance by known authority
   4. Irregular maintenance
   5. No maintenance
D. Construction quality?
   1. Unknown
   2. State-of-the-art engineering design and construction
   3. Standard engineering design and construction
   4. Substandard design and construction
E. Number of flows confined in the last ten years?
   1. Unknown
   2. None
   3. One
   4. Two or more
F. Any known problems?
   1. Unknown
   2. Yes
   3. No

There is a virtual absence of information on many of the nation's levees. Thus, for a screening tool to be useful it must rely on reasonably available data rather than on the higher quality engineering data that will ultimately be required for a technical risk assessment. The logic reflected here is that this tool uses reasonably available data that provide clues to the potential risk associated with the levee. It is presumed that older, private structures that are not built to exacting standards, are poorly maintained, have passed flood flows and may have known problems are more likely to perform unsatisfactorily or even to fail. Consequence criteria follow.

G. How vulnerable is the population?
   1. Unknown
   2. Highly vulnerable (low income, elderly, low education, minority)

3. Moderately vulnerable (housing close to levee, much housing in flood plain)
4. Low vulnerability (housing removed from the levee, less housing in flood plain)

H. How large is the population at risk?
1. Unknown
2. Less than 1,000
3. 1,000 to 10,000
4. 10,000 to 100,000
5. 100,000 to 1,000,000
6. Over 1,000,000

The size and social vulnerability of the population were considered to be the two criteria of most importance in the early screening stages when the focus is on protecting lives.

Evidence is gathered to rate each levee against each criterion. If any entry is unknown, that assessment is flagged as a "data gap" and no rating is provided. Otherwise the choices for each criterion are converted to an order of magnitude. The "riskiest" choice is rated a 1, the second riskiest 0.1, the third riskiest 0.01, etc. An unknown entry is rated a 0. The product of all eight entries is calculated. The largest and smallest possible products are calculated, and the range between these two values is normalized over the [0,100] interval. The normalized value is the levee's rating. This is a semiquantitative method. Although the rating is numerical, it remains qualitative in information content. The numerical ratings have only ordinal-level information content.

All the levees in the region are assessed, and their qualitative ratings enable assessors to answer the risk manager's questions. The levees with the highest numerical ratings have the greatest risk potential. It is understood that this assessment proceeds under conditions of considerable uncertainty. When the very rudimentary data of this tool are not available, the levee cannot even be ranked. Presumably such an assessment would highlight the need for additional data at some sites while enabling risk managers to identify those levees that should first be subjected to a more rigorous technical risk assessment.

9.12 Summary and Look Forward

Qualitative risk assessment is growing in credibility and acceptance when it is based on techniques and methodologies grounded in science and sound evidence that account for uncertainty in reasonable ways. The qualitative risk assessment toolbox is growing in terms of the number of available options and their applications and use. Still, there are legitimate concerns that some of these tools can be easily abused.

Many of the best techniques find some way to address the "risk = probability × consequence" conceptual model, marshaling the evidence to understand the probability of a risk and its consequences. A generic approach based on this model was presented in this chapter. Criteria-based ranking techniques and the risk matrix remain two popular qualitative tools. If the evidence is not clearly presented and documented—and if the uncertainty is not adequately addressed—you should be wary of any risk assessment technique.

Unless a qualitative risk assessment provides the information risk managers need for decision making, quantitative risk assessment is usually preferred. The next eight chapters are devoted to quantitative risk assessment topics. We begin with modeling. Quantitative risk assessment requires models. Not all of them need to be complex, but they are all models. Chapter 10 describes the practice and art of model building. It provides an introduction to the types of models that are used in risk assessment and presents a 12-step model-building process to help you. It concludes with a look at the technical and craft skills a modeler uses. Additional details on model building are sprinkled throughout the chapters on quantitative risk assessment.

REFERENCES

Cox, L. A., Jr. 2008a. What's wrong with risk matrices? *Risk Analysis* 28 (2): 497–512.

———. 2008b. Some limitations of "Risk = Threat × Vulnerability × Consequence" for risk analysis of terrorist attacks. *Risk Analysis* 28 (6): 1749–1761.

———. 2009. What's wrong with hazard-ranking systems? An expository note. *Risk Analysis* 29 (7): 940–948.

Hubbard, Douglas W. 2009. *The failure of risk management: Why it's broken and how to fix it.* Hoboken, NJ: John Wiley and Sons.

Jones, Morgan D. 1998. *The thinker's toolkit.* New York: Three Rivers Press.

Orr, Richard L., Susan D. Cohen, and Robert L. Griffin. 1993. Generic non-indigenous pest risk assessment process: The generic process for estimating pest risk associated with the introduction of non-indigenous organisms. Unpublished in-house document. U.S. Department of Agriculture.

Plant Epidemiology and Risk Analysis Laboratory. Undated. A demonstration exercise using enhanced hazard identification for criteria-based ranking of pest risk. Prepared for Risk Analysis 101, Raleigh, USDA APHIS PPQ.

Schütz, H., P. M. Wiedemann, and A. Spangenberg. 2008. Evidence maps—a tool for summarizing and communicating evidence in risk assessment. In *The role of evidence in risk characterization: Making sense of conflicting data*, ed. P. Wiedemann and H. Schütz, 151–160. Weinheim, Germany: Wiley-VCH.

U.S. Department of Defense. 2000. Mishap risk: DOD standard practice for system safety. Washington, DC: DOD.

U.S. Food and Drug Administration. 2003. Guidance for industry evaluating the safety of antimicrobial new animal drugs with regard to their microbiological effects on bacteria of human health concern #152. Rockville, MD: Center for Veterinary Medicine. http://www.fda.gov/downloads/AnimalVeterinary/GuidanceComplianceEnforcement/GuidanceforIndustry/UCM052519.pdf.

10

The Art and Practice of Risk Assessment Modeling

10.1 Introduction

Quantitative risk assessment relies on models. A model is an abstraction of reality used to gain clarity about a problem or its solutions by reducing the variety and complexity of a situation to a level we can understand. This simplification enables us to gain insight into systems, processes, events, scenarios, problems and their possible solutions. These insights can help us understand or control some aspect of a problem or opportunity. Models help us to address complex phenomena, force us to synthesize knowledge, and enable us to solve problems, test hypotheses, examine strategies, set priorities, and communicate all of this to others in a cost-effective and time-efficient manner.

It is not always easy to understand people mean when they use the word *model*. Sometimes there are models within models and the language can be confusing. A risk assessment model, for example, may actually comprise a consequence model, a likelihood model, and a risk characterization model along with numerous computational components. So a model could be the entirety of all these parts or any one of the component parts. "Probability model," right or wrong, is a phrase often used to describe a probability distribution used in a Monte Carlo simulation. So In this chapter, *model* refers to the overarching representation of a process, i.e., the whole of the risk assessment model rather than its component parts. The parts that comprise a model are called modules or components.

Risk assessment models have many practical uses and yield a number of important benefits in addition to those above. A valid model accurately represents the relevant characteristics of the decision problem. Valid models are a relatively inexpensive way to learn about the potential effects of uncertainty. Uncertainty often ensures that important data will not become available until some point in the future. Risk assessment is designed to address such uncertainty, and its models help bridge the gap between what we know and what we do not know in a way that aids timely decision making. These models enable us to explore more or less desirable futures from the present. They also help us explore the effectiveness of things that would be impossible to do in reality. Crash-test dummies are human models that enable us to design safer cars. We use them to understand failure and improvement scenarios. Assessing the risk of risk management option (RMO) failure modes (levee failures, nuclear accidents, financial collapses, oil spills, and other catastrophic incidents) affords us the opportunity to anticipate future risks. Models empower us to examine the likely efficacy of these measures before they are actually implemented.

A wide variety of model types are available for use by risk assessors, and this chapter begins by discussing some of them. The chapter then narrows its focus to mathematical simulation models built in a spreadsheet environment. This is one of the easiest model building environments in which one can work. A 13-step model-building process is offered to help guide the efforts of novice model builders. The remainder of the chapter is devoted to a discussion of model-building skills. These skills are broken into technical skills and craft skills. Technical skills might be likened to the science of modeling while craft skills are more like the art and practice of modeling.

10.2 Types of Models

There are many ways to discuss and categorize models. The most common models are mental, visual, physical, mathematical, and spreadsheet models (Powell and Baker 2007). Mental models are conceptual notions and ideas about how things work in reality. They are internal representations of external realities, i.e., mental images translated to verbal or written descriptions. Mental models are abstract models. The risk management and risk assessment models of Chapters 3 and 4 are good examples of mental models. Many of these models tend to be informal or conceptual frameworks, although well-worked-out formal theories can be mental models as well.

Visual models include maps, figures, graphics, and charts that show how things work in reality or how ideas are related to one another. When coaches draw plays on the sidelines, they are using visual models. When someone gets up at a meeting and goes to the whiteboard and begins to sketch the ideas being discussed, they are using visual models.

Physical models are usually analog or iconic models. Analog models look like the reality they represent. Cockpit simulators are used to train airline pilots, bridge simulators to train ship captains and pilots. Working in replicas of the planes and ships they will command, pilots can practice a wide variety of situations and circumstances. Analog simulators of all types are becoming more common. Engines used in shop classes and human body parts used in doctors' offices are examples of other kinds of analog models.

Iconic models are scaled-down replicas of the object, system, or process under study. They have been used to design cars, buildings, and new pieces of equipment. Master planners rely on iconic models to communicate their visions. Beach cross sections can be designed and tested in a wave tank before beach nourishment projects are built. The plastic models we built as children are additional examples. In the not-too-distant past, iconic models of watersheds were quite common.

There is also a class of models intended to represent a critical or interesting physical dimension of reality or to give a physical dimension to an abstract concept without replicating it or scaling it down. Examples include models of DNA molecules and models of three-dimensional mathematical functions like the utility functions of economics.

There are many kinds of mathematical models, and they are used in every area of scientific endeavor. Mathematical models generally rely on functional relationships among dependent and independent variables. Every quantitative risk assessment model is a mathematical model. Their basic characteristic is that they use mathematics to describe a system, situation, object, or problem. Mathematical models can be categorized as prescriptive, predictive, or descriptive models (Ragsdale 2001).

A prescriptive model produces the best value for a dependent variable. Examples of prescriptive models include things like linear, integer, and nonlinear programming models, networks, and so on, that can be optimized. They prescribe the most unambiguous course of action for the risk manager when they are available.

Predictive models predict the value of a dependent variable based on the specific values of the independent variables. When the functional form of the relationship between the dependent and independent variables is known, we often have prescriptive models. When that functional form is unknown and has to be estimated, we have predictive models. Many risk assessment models are predictive models. Such models are used to route flood waters and to anticipate hurricane tracks, to estimate pathogen growth, as well as to predict the performance of markets and other complex systems. Examples of predictive models include regression and time-series analysis.

The third category of models, descriptive models, may be the most common in risk assessment. These models are characterized by their uncertainty. That uncertainty may include uncertainty about the functional form of a relationship or, when very precise functions are known, they may include uncertainty about the exact values of one or more independent variables. These kinds of models describe the range of outcomes or behaviors that are possible in a system that is plagued by uncertainty. Simulations, queuing, and inventory models are examples of descriptive models.

Mathematical models can also be described in accordance with the family tree of system models seen in Figure 10.1. A deterministic model has no stochastic, i.e., random, components. Thus, the next state of the system is always determined by predictable actions built into the model. In a static model, time is not a significant variable, i.e., the data do not "age." The passage of time is a significant consideration in a dynamic model. Dynamic models often offer an explicit representation of the sequence in which events occur. Continuous dynamic models involve states that evolve continuously. They often rely on differential equations. In contrast, discrete dynamic models produce states that are a piecewise constant function of time. A discrete model may evolve a new state every 15 minutes, every day, or annually, whereas

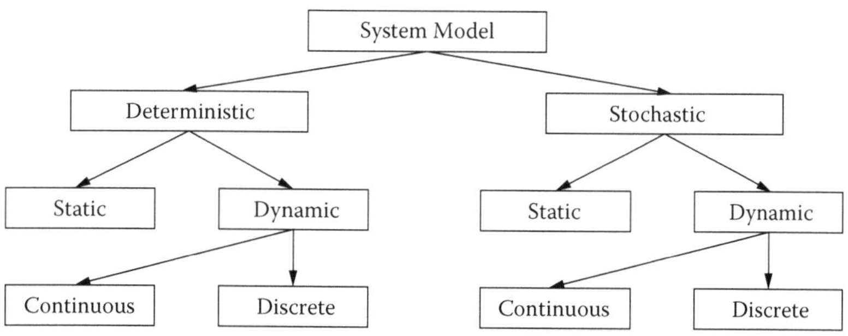

FIGURE 10.1 A taxonomy of model types.

a continuous model is continuously evolving new states. The same kinds of models are found in the stochastic universe, where risk assessors deal with random components in the systems they model.

Simulation models, the focus of this chapter, are a subset of mathematical models that can exhibit any of the characteristics of the family tree of system models in Figure 10.1. Monte Carlo simulation models built in the spreadsheet environment enable us to explore the stochastic branch of models and will occupy most of our attention, but there are many more-complex system simulation models in use that are worth a brief discussion. Evans and Olson (2002) identify four such simulation modeling approaches.

Many of these system models involve the flow of some entity or object through a system. The entity might be an attribute (purity, resistance, strength), a condition (new, fresh, complete), a piece of information (message, order, profile), or a physical object (an egg, a towboat, a human being, a pathogen). System simulation models reproduce the most important activities and dependencies that control the flow of one or more entities through a system over time. These sorts of models tend to be too complex for a spreadsheet.

Activity-scanning simulations describe the activities that occur during fixed time intervals. The model is incremented by some fixed time period and it describes the activities of interest that occurred during that time period. For example, the model will describe what happens in the first time period and then advance to the next time period and continue the process. If we model the movement of towboats on a waterway, this would include all the activities that occur within a given time increment such as distances traveled and positions, traffic maneuvers, and dockside operations. If our model follows eggs from laying to consumption, it might represent what activities occur in the first 24 hours, followed by the activities in the second 24 hours, and so on. If we are monitoring bacteria in an egg, it might describe all the physical activities that occur within the egg at fixed time intervals.

Process-driven simulation models focus on the process an entity must move through in a system. So, for example, if we're concerned about how an egg gets from the farm to a store, the process could be described and modeled as follows:

> Egg forms → laid → transported to collection → collected → transported to processing → washed → graded → separated → transported to packaging → placed in carton → palletized → shipped to retail

If we think in terms of discrete modules, this sort of model would use modules corresponding to the parts of the process rather than modules defined as increments of time. So we could examine what happens in the washing process or the shipment to retail outlets.

Another type of simulation model is called event-driven. Events are occurrences in a system at which point changes in the system (or entity under consideration) occur. These events take place at a moment in time. Events are processed in chronological order, and time is advanced from one event to the next. These models focus on events that change the system, and so they are often computationally efficient. This kind of model describes changes in the system that occur at the instant an event occurs. For our egg example, contamination of the egg in utero, breakdown of the yolk membrane, and cooling on the pallet are events of potential interest that can change the egg entity from risk free to risky when assessing the risk of Salmonella in shell eggs. If we are modeling towboats, the risky events of interest might include traffic maneuvers when meeting, passing, or overtaking other vessels in the waterway or perhaps when navigating difficult turns.

When variables of interest change continuously over time, continuous simulation is appropriate. The amount of money moving through an economy, water in a stream, oil in a pipeline, or milk in a production run are all examples of continuous variables. These continuous variables are frequently called state variables. Continuous simulations rely on equations defining relationships among state variables that enable dynamic system behavior to be studied. System dynamics is a popular form of continuous simulation.

Simulation is a legitimate technique for solving problems when analytical solutions are not possible. Whereas analytical models represent reality, simulation models imitate it in a simplified manner. Simulation models allow us to conduct controlled experimentation that is otherwise impossible. Simulations can produce information that reveals new facts about the problems we work on. They are suitable for a broad range of applications and can be effective training tools.

Building models generates insight that enables us to make better decisions. Modeling improves our thinking about and understanding of a problem, and it enables us to experiment and learn about it as well. As this discussion suggests, a wide range of simulation modeling options is available. Spreadsheet models are an especially useful platform for building risk assessment models because they require no special programming language skills or abilities. These spreadsheet models are the focus of the remainder of this chapter.

10.3 A Model-Building Process

There is no one way to build a mathematical spreadsheet model for risk assessment. Modeling may be the most idiosyncratic part of the risk assessment process. Nonetheless, it can be useful to have a process in mind. The model-building process presented in this section has served me reasonably well over the years; feel free to adopt or adapt it. The 13 steps in this process include:

1. Get the question right
2. Know the uses of your model
3. Build a conceptual model
4. Specify the model
5. Build a computational model
6. Verify the model
7. Validate the model
8. Design simulation experiments
9. Make production runs
10. Analyze simulation results

11. Organize and present results
12. Answer the question
13. Document the model and results

10.3.1 Get the Question Right

Know what information the model needs to produce. If the risk management questions are not clear or if they are not the right questions, then nothing that follows in the model-building process will make any difference at all.

Different questions can lead to very different models. Consider the following questions that relate to a food-safety concern with *Vibrio parahaemolyticus* (*Vp*) and oysters. How many people get sick from *Vp* in oysters? What is the probability of getting sick from eating a raw oyster? How effective would refrigeration of oysters within 60 minutes of harvest be in reducing illness from *Vp* in oysters? What is the risk of eating raw oysters to people with liver disease?

Each of these questions will require different data and a different model structure. Get the right question, then get the question right. The first step in developing any model is understanding what question(s) the model needs to be able to answer. Be sure to discuss and clarify the meaning and intent of each and every question before you begin to build your model.

10.3.2 Know the Uses of Your Model

Understand what the model is expected to do. Is the model identifying research needs, developing a baseline risk estimate, attributing risk to different hazards? Will it be used to evaluate risk management options? What kinds of options might they be? Is it a learning tool? Will it be shared with others or used again? Who will use it? Will it be added to over time, or is it to be completed once and for all? Is it the basis for a regulation?

Understand what the model cannot do. Risk managers, in particular, must be made aware of the limitations of the model and its outputs. The seemingly innocent request to evaluate an unanticipated RMO after a model is built may be impossible because the necessary data were not collected or the model structure does not support that analysis.

Anticipate as many potential uses of the model before you begin to build it as possible. This can only be done through close collaboration and clear communication between risk managers and risk assessors.

10.3.3 Build a Conceptual Model

This is where your model-building effort is most likely to succeed or fail. Models fail less often due to data and parameter value issues than they do for faulty conceptualization of the problem to be represented. This step proceeds from the abstraction of ideas and notions to the cold, hard reality of details. The best modelers will include the important processes and exclude unimportant ones. This step calls on both science and art.

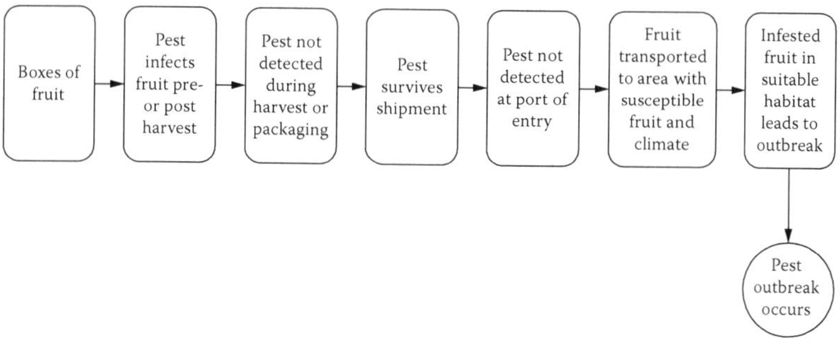

FIGURE 10.2 Conceptual model for fruit import pest infestation risk.

You now have specific questions to answer. Work out your risk hypothesis for each question before you begin to build a model. Develop your risk narrative. That narrative should answer the four questions:

1. What can go wrong?
2. How can it happen?
3. What are the consequences?
4. How likely is it?

Use the qualitative generic process if it fits your problem better than the four questions do. Be practical, not bound to any one approach. This is the step where you work out the major ideas for your model. What are the state variables? How are they related to one another? The conceptual model is a mental model. A sample for a pest outbreak associated with a fruit import is shown in Figure 10.2.

10.3.4 Specify the Model

Once the conceptual model has been developed it is time to convert it into a specification model. This means moving from concepts to develop the relationships, equations, and algorithms that will describe the ways the various components of your conceptual model will work together. This is the model construction step that usually stops short of actually building the final computer program. It may help to think of it as the paper and pencil exercise of figuring out the calculations that will be needed in your model. It is more than this, but that provides a good mental image of the essence of this task.

The functional form of all relationships as well as the model logic are built in this step. Placeholders and dummy values can be used in lieu of data. The specification model defines the calculations and other inner workings that will make the model run. It often includes a detailed sketch of the spreadsheet model. In the current example, this will concern such things as the mathematical process that determines how many boxes of fruit are imported, how we identify which are infected with pests, how we calculate the percentage of the pests that avoid detection and survive transit, and so on.

PARAMETER SPECIFICATION

From minimal to maximal precision:

- Guesstimates ("hand waving")
- Expert opinion
- Bounded estimate
- Survey-based data
- Data-based parameters (averages, etc.)
- Curve-fit parameters (trade-offs)
- Parameter sensitivity testing
- Entire parameter sets by optimization

***Source*: Risk Analysis 101, USDA, APHIS, PPQ.**

10.3.5 Build a Computational Model

If we think of the specification model as the skeleton, then the computational model puts flesh on the bones. In this step you complete the computer program needed to run the model you have specified and collect the data you will need to make the model operational. The output of this step is a fully functional spreadsheet model.

Placeholders are replaced with real data. Dummy values may be replaced by probability distributions. At this point you have a working risk assessment model. You will learn how to choose probability distributions to represent knowledge uncertainty and natural variability in Chapter 12. For now we focus on the steps required to build a model rather than on the details.

10.3.6 Verify the Model

Verification is an extremely important separate and formal step in the model-building process. You want to get the equations, calculations, cell references, and all the details just right. This is when you ensure that your computational model is consistent with the specification model and that the specification model is correct. You want to ensure that you have built your model correctly. So you debug, review, and test your model to ensure that the conceptual model and its solution are implemented correctly and that the model works as it was intended to work.

Many model builders verify their own work. The obvious advantage of this approach is their intimate knowledge of the model's structure. The disadvantage is that they have often been working with the model for so long they can no longer see potential problems. Having a model verified by someone other than its builder(s) can be time consuming and more costly, but it is often more effective in finding flaws. It is certainly more effective in increasing the credibility of the model and confidence in its outputs.

10.3.7 Validate the Model

Once your model is verified to ensure that it does what you want it to do in the way you want it done, it is time to validate your model. Is it a good representation of reality? Have you built a model that represents reality closely enough to provide information to support decision making? The weakest form of model validation is a model that accurately describes the system being modeled. Next, we ask if the model produces plausible results by considering whether the model can reproduce independent results like statistics or real data.

There are basically three ways to validate your model. You can validate model outputs, model inputs, or your modeling process. Historical data validity uses historical inputs to see if your model reproduces historical outputs. If weather conditions that always produced rain in reality produce sunny days in your model, you have a problem. There are at least two practical issues with this form of validation.

The first, and most pressing for risk assessment, is the lack of historical data. You must hold back some of your data and not use it in the model-building process in order to use it to validate your model. A common technique is to divide the available data. Part of the data, say half, is used to build the model and develop all the requisite relationships in the model. The other half is used to test the model. If actual data are input into the model, we expect to get predictions that closely match the actual output data from a valid model. Risk analysis is a decision-making paradigm designed for conditions of uncertainty. Consequently, data gaps and insufficient information frequently make historical data validation impossible. Frequently there are not sufficient data for confidently building a model, much less enough to hold some back.

A second problem is how closely must a model replicate historical conditions? If you are fortunate enough to have the data, how many rainy days must your model replicate for us to declare it valid? The answer to that question is most often a pragmatic one rather than a statistically rigorous one.

A second validation approach is data validity. When the outputs cannot be validated, it may help to validate the inputs. Data validity means you have clean, correct, and useful data. Furthermore, it ensures that all input data and probability distributions are truly representative of the system modeled. When simulating situations that have never existed, it is often impossible to assure data validity.

When neither the inputs nor the outputs can be validated we have few options but to try to validate the reasonableness of the process. Face validity, the third option, asks experts to examine the model or its results to determine if they are reasonable. A model has face validity when it looks like it will do what it is supposed to do in a way that accurately represents reality. One face validity test is to present knowledgeable experts with real-world data and model output data. If they cannot distinguish the two or find no significant divergence between the two, the model that produced the data may be considered valid on the face of its output data. However, a model that looks like it should work is quite different from a model that has been shown to work.

Validation can be checked at several levels. For instance, each module in a model should be validated. In a microbiological risk assessment, for example, it may be possible to validate a pathogen growth model or a dose-response curve even though the entire model cannot be validated. It may be advisable to validate components of some of the modules. A probability distribution chosen to represent a variable value within a module might be validated, for example. Of course, the whole-system level, i.e., the entire model, should be validated as should any benchmark cases. Graphical comparisons, confidence intervals, statistical tests, and hypothesis tests can be useful validation tools to help determine whether model predictions are within an acceptable range of precision.

Verification and validation, together, provide the basis for the assessors' degree of confidence in the model. Risk managers should always inquire after the specific steps taken to verify and validate the model. Risk assessors should always inform risk managers about the verification and validation effort and its results. Some organizations use a model certification process to address questions of model verification and validity for models to be adopted for use throughout the organization.

10.3.8 Design Simulation Experiments

Characterizing the range of outcomes in an uncertain world is not a simple thing to do. There are often many scenarios to investigate. Assessors may be asked to explore historical conditions, existing conditions, a benchmarked baseline condition, future conditions, improved conditions, worst case, best case, different geographic locales, different seasons, and so on. Consequently, the assessor must carefully plan the simulation experiments that must be run to answer the risk manager's questions and to fulfill the intended uses of the model. This means deciding what parameters and inputs will vary, what assumptions will be used for which experiments, and so on.

If you sit down with your model and just start making runs, it is likely that, sooner or later, you will get what you need. It is also likely you'll waste a lot of time making runs you did not need. Carefully identifying the various conditions for the simulations you want to run is essential to an efficient risk assessment process. Arrange your series of experiments efficiently. It may make sense to do them in a specific order if significant adjustments must be made to your model for different conditions. Write down the runs you will do in the order you will do them. Take care to verify all alterations to a model and save significantly different versions of your model as separate files.

10.3.9 Make Production Runs

This could easily be an extension of the preceding step. Once you have identified the range of experiments you want to run, the next step is to run them. It is very easy to fall prey to the excitement of running your model as soon as it is built. Many assessors skip over the verification, validation, and experiment design steps. You have likely spent a good bit of time and effort to get to the point of a computational model, and you want to use it. Unfortunately, few if any models are error free. If you make a slew of runs with your model and then find that decimal point error in the formula, everything must be repeated. Resist the temptation to jump directly to production runs.

One of the most frustrating, common, and avoidable problems is failure to document a production run. Every time you run your model you should carefully record the nature and purpose of the run (e.g., existing risk estimate to establish a baseline measure of the risk) and make note of your model's initial conditions, input parameters, outputs, date, the analysts, and so on. Enter this information into a log you keep on a separate worksheet in your model. Keep it up to date.

It can also help to be systematic in making your runs. For example, using our current example, if we are investigating pest establishment risks for several fruits, several countries of origin, and several pests, it makes a lot of common sense to work smart! If setting up the model for a specific commodity takes time, it may make sense to do all the runs for avocados before moving to the peaches. On the other hand, if it takes more effort to set up the pest or country parameters, it may make sense to sequence the runs in another fashion. This is why you design your experiments: so you can run them efficiently.

It is especially important to take great care when making runs that reflect the presence of a new risk management option. These runs need to be carefully documented. It may also be important to record any fixed seeds* that were used to initiate your simulation process in case there is a desire to reproduce the simulation at a future date.

Take special care to save all outputs from a production run and to carefully identify them. Unless you are absolutely sure about the outputs you will and will not need to complete your risk assessment, save all of your simulation outputs if possible. It is far better to save outputs you will never need than to need outputs you never saved.

10.3.10 Analyze Simulation Results

This is the all-important process of getting useful information from data. As the runs are completed, the assessors need to analyze the results and learn what the simulations have taught us about the problem. This analysis is statistical in nature, and it needs to account for the uncertainty in the inputs and to carefully convey that and the resulting uncertainty in the outputs to the risk manager. It is through the analysis of your simulation results that you will craft answers to the risk manager's questions, always taking care to characterize the remaining uncertainty in useful and informative ways.

10.3.11 Organize and Present Results

The information assessors glean from the simulations needs to be organized and presented to decision makers in a form that is useful. At this stage of the modeling process, useful information is information that supports decision making and carefully addresses the significant uncertainties encountered. Presenting useful information for decision making is the topic of Chapter 17.

10.3.12 Answer the Question(s)

The entire reason for doing a risk assessment as well as for building an assessment model is to provide risk managers with the specific information they need to make a decision. Be sure you answer the questions. Do it in a question-and-answer format. Do not assume that your well-organized and well-presented report narrative and results will accomplish this purpose. Answer each question specifically. To the extent

* Seeds are explained in Chapter 14 and Appendix B.

that lingering uncertainty affects those answers, this must be carefully portrayed as well. Once you have adequately answered the questions, feel free to summarize the insights you have gained, offer specific observations, and even to conjecture in a responsible way.

10.3.13 Document Model and Results

Many risk assessments are going to be documented by some kind of report. Printed risk assessments are as common as crows. Careful documentation of your results is assumed to be an essential part of your model-building process. Few would disagree with that statement. However, there is more that needs to be done.

You need to document your model. That means explaining the structure of the model, including relevant descriptions of the preceding steps, your conceptual and specification models, the source and quality of your data, the results of the verification and validation efforts, as well as the history of production runs. And that is not all of it! You also need to provide the equivalent of a user's manual in your documentation.

You have likely spent so much time on this risk assessment model that your in-box is piled high with tasks that all had to be done yesterday. There may be no time to document your model, but you intend to do so the first chance you get. Alternatively, you are sick of living with the model; you know it inside and out; and you can't stand to spend another minute dealing with it. Both of these scenarios pretty much ensure that your model will never get documented. That could be a problem.

As familiar as you are with the model today, in six months time there is a good chance it will look like someone else's work to you. Inevitably there will be questions you are asked to answer, and it will take 10 times longer than it would have had you documented your model.

Then consider what happens if the model builder gets hit by a bus or changes jobs? With the builder goes all of the organization's model expertise unless someone has carefully documented the model and how it works. As much of a pain that it is to document your model, you cannot afford not to.

10.4 Simulation Models

Simulation is the process of building a model of a system or a decision problem and experimenting with the model to obtain insight in to the system's behavior or to assist in solving the decision problem. This can be done with physical models. Do you recall the coin-operated mechanical horse of your youth that sat outside the mall? That was a physical simulation model. Physical simulation models have grown increasingly sophisticated and are now used for all sorts of training and testing.

Simulation models are often used in training exercises. Food companies hold simulated recall exercises; first responders use simulated emergency exercises for training. Everyone has participated at one point in time in the fire drill, another simulation model exercise. These simulations are valuable training tools in situations where it is prohibitively expensive or too dangerous to allow trainees to use the real equipment in real situations. Tabletop exercises provide another form of simulation training that enables decision makers to learn valuable lessons in a safe virtual environment.

While risk analysis can make effective use of these kinds of simulation models, the most common simulation models and the focus of the remainder of this chapter are computer simulation models. Computer simulations are used to model actual or hypothetical situations on a computer so that they can be studied to see how the system works. By changing variables and other model inputs, predictions may be made about how the system will behave under a wide variety of circumstances/scenarios.

Simulation models have been used for modeling the function of natural systems (flood, drought, hurricane tracking, ecosystems) and human systems (traffic flow, transportation systems, engineering, economics, the social sciences). They are used in all the natural sciences and have been used to model processes of all kinds. Clearly, analytical solutions to problems are preferred and should be sought when they exist. Computer simulations arose as an adjunct to, or substitute for, modeling systems for which closed-form analytic solutions do not exist. Simulation models are able to generate a number of representative scenarios describing the possible behavior of a system when it is either impossible or impractical to consider the entire range of possible scenarios.

Simulation models are most helpful when problems exhibit significant uncertainty, a situation commonly characterizing most risk assessment problems. Monte Carlo simulation tends to be used in static models, while systems simulation (described previously) are used more often in dynamic models. The Monte Carlo process, a sampling experiment designed to estimate the distributions of outcome variables that depend on one or more probabilistic input variables, is described in detail in Chapter 14.

Simulation is a legitimate technique for solving problems when analytical solutions are not possible. They have a broad range of applications and, in risk assessment, are often constructed in a patchwork style, i.e., a hazard module may be built first, followed by an exposure/likelihood module and a hazard characterization/consequence module. These may all be pulled together subsequently in a risk characterization module. Readily available commercial software makes simulation modeling incredibly easy to do.

Simulations are not without their downside. The larger models can be costly or time consuming to build and run. Powerful computers may be needed for the largest simulations. Model results are very sensitive to model formulation, and there is no guarantee of an optimal solution being identified from the results.

Perhaps the greatest disadvantage of simulation modeling is that it has become so easy to do. It is tempting, when faced with a problem, to sit before the computer and start to build a simulation model. That is one reason why it is important to have a model-building process. Too often this is easier than thinking the problem through and possibly finding an analytical solution to a problem. The ease of simulation modeling can encourage overlooking other techniques.

10.5 Required Skill Sets

Building a good risk assessment model requires a special skill set. Powell and Baker (2007) have proposed breaking this skill set into technical skills and craft skills. Their distinction is a useful one and is one I will build on in the remainder of this chapter. Technical skills are necessary for getting the right answer. Getting the right answer is important, in case you had not guessed that! Craft skills are extremely useful for simplifying complex problems and for modeling in new and poorly structured situations. Craft skills represent the creative side of modeling. Both skill sets are invaluable.

The discussion that follows is generally applicable, but there is a wide range in the formality with which risk analysis is conducted. In a regulatory setting, the risk analysis process and the risk assessment in particular may be quite formal. In a large organization where risk analysis has penetrated the culture, it may be practiced less formally in branches, sections, teams, and even by individuals. Some readers will find some of the ideas that follow useful; others may find them less so. For example, if you address the knowledge uncertainty and natural variability in a model used only by you because it makes sense to do so, you may not feel it necessary to design your model to communicate effectively with others. Feel free to take what is useful from the sections that follow and leave the rest. Modeling is an art, and ultimately each artist must develop her own style.

10.5.1 Technical Skills

Technical modeling skills comprise the knowledge and proficiencies required to build a specific risk assessment model. These are the skills that get you to a right answer. No one person may possess all the necessary technical skills for a complex risk assessment, but they all must be present on the assessment team.

First, technical skills require the modeler to know or to be able to understand the relevant science represented in the model. This does not mean the modeler must be a microbiologist to build a food-safety model or a civil engineer to build an engineering-reliability model. It does mean that discipline-specific knowledge must be present and available and that the modeler must be capable of translating that knowledge into a workable and reasonable model structure.

Second, there is the technology knowledge that is required. Unix, Lynux, Java, C+++, Perl, MySQL, Microsoft C, and other language skills may be required for some models. In this book, we will focus on spreadsheet models, so it is sufficient to have knowledge of Microsoft Excel or a similar spreadsheet program. If additional software tools, e.g., Palisade's @RISK, used in this text, are used, the modeler must be proficient in their use as well. (An introduction to the use of several modules of Palisade's DecisionTools Suite is found in Appendix B.)

Third, there is a wide variety of narrow, well-defined tasks the modeler must be able to carry out. There are techniques and methodologies for all sorts of calculations that may not belong to any one discipline. These include basic math and language skills, for example, as well as more specialized skills. For economic and financial models, this could include handling present-value calculations. For engineering models, it might include working with a reliability beta index model, a hazard function, or a fragility curve. For food-safety models, this task might be constructing or programming a dose-response curve. These skills also tend to be discipline based.

A good risk assessment team will have all these technical skills available. Rarely will they all be present in a single individual. It is not uncommon for a risk assessment model builder to come from a quantitative disciplinary background that may be quite different from that of the problem's context. The technical skills of risk assessment model building remain one of the more scarce resources in risk analysis. When the requisite skills are spread across many people, frequent team interaction and effective communication are essential.

10.5.2 Craft Skills

This discussion of craft skills has some general applicability, but it is intended primarily for those who work in the spreadsheet environment. Craft skills are separated into two topical areas for presentation here: the art of modeling and the practice of modeling, as this chapter title suggests. The discussion that follows attempts to flesh out the 13-step process presented previously with some art and practice tips.

10.5.2.1 The Art of Modeling

Powell and Baker (2007) offer eight modeling heuristics that are simply too good for anyone to ignore. They form the basis for this discussion of the art of modeling as supplemented by my own experiences. To see these heuristics in their original form, the work of Powell and Baker is not to be overlooked. The text box introduces the example followed in the remainder of this section.

10.5.2.1.1 Identify the Question(s) and Simplify the Problem

We will assume that the modeler begins with a well-defined decision context and a list of questions the risk managers would like to have answered. Let's also assume that the risk assessment team has considered these questions, answered those that could be answered with the available information and knowledge, and now seeks to build a model to predict answers to the remaining questions. This is where we would like the art of modeling to begin. The truth is that few risk assessment modelers ever find

THE EXAMPLE

To illustrate the heuristics described here we will use the U.S. FDA *Vibrio parahaemolyticus* Risk Assessment, July 19, 2005, "Quantitative Risk Assessment on the Public Health Impact of Pathogenic *Vibrio parahaemolyticus* in Raw Oysters" available at http://www.fda.gov/Food/ScienceResearch/ResearchAreas/RiskAssessmentSafetyAssessment/ucm050421.htm (FDA 2005).

The risk assessment model, Appendix 3, can be obtained from the following link: http://www.fda.gov/Food/ScienceResearch/ResearchAreas/RiskAssessmentSafetyAssessment/ucm185244.htm.

Vibrio in raw oysters represent a serious health risk. This risk assessment considered different geographic sources of oysters and different seasons of the year. To simplify the exposition, we'll use only the Gulf Coast of Louisiana in the summer as the single geographic region and season. In addition, we will consider only one of several questions posed to risk assessors: "What reductions in risk can be anticipated with different potential intervention strategies?"

themselves in such a situation. So let's back up and begin a bit more realistically with the risk manager who has never read Chapter 3.

You, the modeler, simply must know what question(s) you are trying to answer with your risk assessment model. There is no way around this if you are to build a useful model. Ideally, you will be the beneficiary of a good risk management approach, such as the one described in Chapter 3 of this book. If you do not know what the risk manager's questions are, sit down with your risk managers and ask them, "What do you need to know to solve the problem you are facing?" Write down their responses. Ask if there is more they need to know. Get them to understand what they have asked for and what you will provide and not provide.

Albert Einstein has been quoted as saying, "Any intelligent fool can make things bigger, more complex, and more violent. It takes a touch of genius—and a lot of courage—to move in the opposite direction." This brings us to the starting point for modeling: It is time to begin to simplify.

Build the simplest model you can. Power it with test-case data. Does it answer the question(s) risk managers have asked you? If yes, stop model building and go get real data. If not, revise the model and repeat the process. Make your model as simple as you can, so long as it meets the needs of your risk managers. It is always easier to add to a simple model than it is to simplify a complex one.

Sanchez (2007) formalizes this advice in four very concise suggestions for building a model:

1. *Start small.* Begin with the simplest possible model you can. It should cut away all the complexity that is not essential while still capturing the essence of the system. This can sometimes best be done by focusing on the questions you are trying to answer and the outcomes you'll need to provide those answers.

   Simplifying anything requires us to make assumptions. Document your assumptions as you make them. Remember that you are most dangerous as a model builder when you begin making assumptions without recognizing that you are doing so. Peer review is one of the best ways to guard against this danger.
2. *Improve your model incrementally.* It is easy to add features to a basic working model. You can often improve a model by relaxing the assumptions with a more complete representation of reality. Even this should be done simply. Prioritize the changes you could make. Make one small change and see if it is adequate.
3. *Test your model frequently.* You want your model to conform to reality, not to duplicate it. How do you know when the model is adequate? If it answers the questions you have been tasked with answering, it is quite likely to be adequate.
4. *Backtrack if you must.* In every model there comes a point when additional improvements are not worth it. When the latest change produces no measurable benefit to the answers to your questions, do not be afraid to go back to the earlier, simpler model when all you have added is complexity.

The question for this example (see preceding text box) is, "What reductions in risk can be anticipated with different potential intervention strategies?" This question requires risk assessors to first estimate the annual number of illnesses before it can address the risk reductions. The simplest model that does this is:

$$\text{Eat oysters} \rightarrow \text{get sick}$$

Clearly, this is not going to be sufficient, but it is a start. Perhaps there are things that happen before the oysters are eaten that are important or maybe there are alternative outcomes. Most importantly, we have begun to build our model.

10.5.2.1.2 Break Problem into Modules/Sections

One of the most effective ways to simplify a problem is to decompose it, i.e., to break it down into simpler components or modules. This decomposition is the basis for the conceptual risk assessment model discussed in the 13-step procedure. We can always break any risk assessment into a hazard/opportunity identification, a likelihood assessment, a consequence assessment, and a unifying risk characterization. Likewise, each of these components can be further decomposed.

When working on a specific decision problem, it is usually simpler to decompose the problem into components. The reason for doing this is that it is easier to think about and work with the components of the problem than it is to work with the whole problem. The components, when well chosen, provide a natural structure for the assessment model.

The question then becomes how many components do you need? Where do we draw the line? Powell and Baker (2007) suggest that a problem should be divided into components that are as independent of one another as possible. This number remains a subjective judgment. Each component of the model would then be specified and built separately, with the builder remaining cognizant of where each fits into the overall conceptual model. A model module would have its own worksheet(s) and would be physically separated from other modules.

In our example we might look at that simplistic model and decide it does not capture the key components of this model. It might be useful to break that eating-oyster problem down into several components such as harvesting, handling, storage, consumption, and health outcome. The actual risk assessment model is shown in Figure 10.3. The rows of the model that comprise a component are identified by the labels on the right. You would begin only with the names of the components and a conceptual model.

It's time for a brief caveat to the reader. The risk assessment chosen to illustrate the art of modeling is a real one that was actually used for public decision making. It is presented as a real example rather than an ideal one. Thus, you will find that it does not always conform to the practices described in this chapter.

10.5.2.1.3 Rapid-Iteration Prototyping

Word-processing software did not spring from the womb the way it looks today. It began with someone figuring out how to get software that would type words neatly arranged on a page. It began with a prototype, and then features were added to that model until it evolved into the word-processing software we use today. Echoing the "keep it simple" message discussed previously, the way to approach modeling is to build and refine it in the same way. Build a prototype

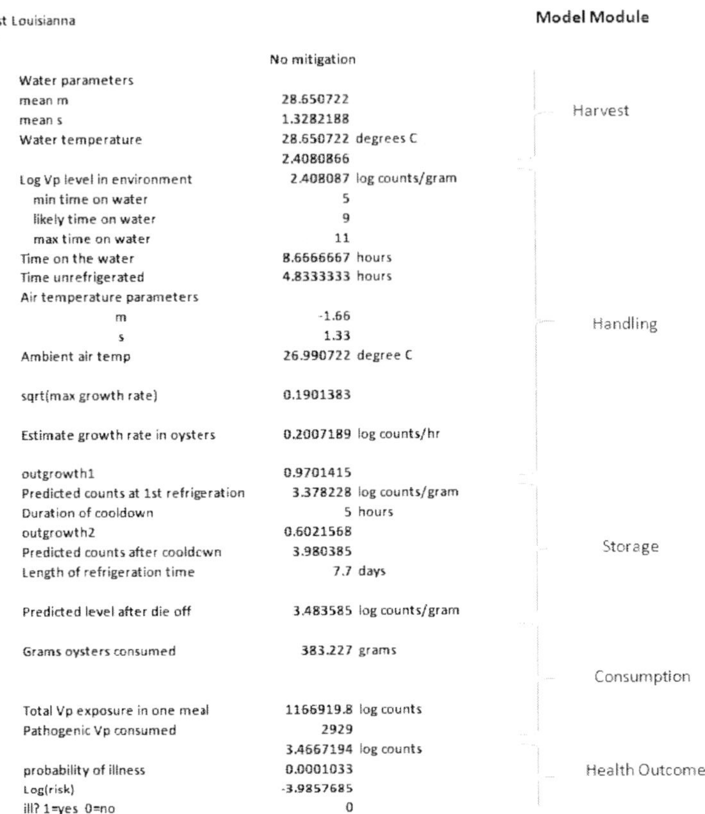

FIGURE 10.3 Key model components of FDA *Vibrio* risk assessment model.

of your model; use it; examine the gaps between what it gives you and what you want; and then refine it.

In practice, that means building a prototype of each module and testing it. Rapid-iteration prototyping means building a prototype as quickly as possible, testing it to see if it works as intended, and if it does not, moving on to the next prototype as quickly as possible. This is all done without worrying about finishing the model. The emphasis is on improving the model.

This is an especially useful approach for new model builders who struggle with wrapping their heads around a big problem. Simplify, break it into modules, and work quickly. Expect errors, insufficiencies, and problems, and this will free you to be more creative and crafty in your model building.

PROTOTYPE

A prototype is a working model suitable for testing. It takes inputs from the user and produces outputs.

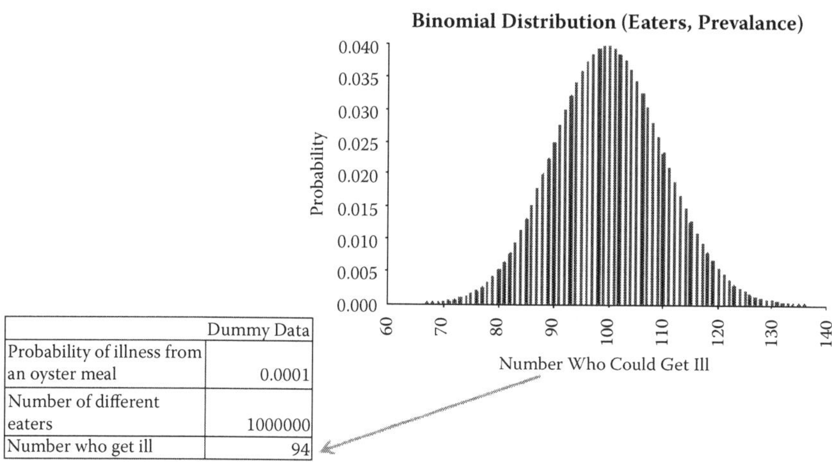

FIGURE 10.4 A simple *Vibrio* risk assessment model prototype.

Risk assessment is often required to help with ill-structured problems with lots of uncertainty. There are few things more satisfying in this situation than having a working model, no matter how crude. Beginning to bring order from chaos is psychologically very satisfying, even if the final model is many iterations off into the future. Add detail, refine your thinking, improve calculation efficiencies, and simplify the problem further. Iterate, iterate, iterate your model.

Figure 10.4 shows a simple model that could estimate the number of people who get ill annually. Note that it is using test-case data. With a probability of illness, the number of eaters, and the assumption that this is a binomial process, we actually have a working prototype of the simple "eat oysters → get sick" model described previously.

Does it give us everything we need? Probabilities could, conceptually, be estimated for different regions and seasons, but we'd have to be able to get the data. Keep in mind that our question requires us to be able to investigate the effectiveness of risk reductions associated with different RMOs. If they could be expressed as changed probabilities of illness, we might be onto something here.

As it turns out, all these data inputs are uncertain. To estimate this probability we have to decompose it into more pieces, ultimately, the pieces shown in Figure 10.3. In addition, there are other questions asked by the risk managers that this early problem model does not answer. There is a desire to model watermen's behavior and the other components identified in Figure 10.3. Nevertheless, the very first primitive model might look like this. We know we are not done because the model does not answer all the questions!

10.5.2.1.4 Graph Key Relationships in X–Y Plane

How do you get started? Express your intuition! Modeling is an abstract activity. You often begin with nothing but a weakly defined problem with no definitive structure and a crude conceptual model. From this you are to bring order and answers! It is often difficult to clarify your thoughts and to get started.

Powell and Baker (2007) point out that many people have a good intuition about the modeling challenge they face but lack the skills to represent those intuitions in a useful way. Experienced modelers often have a variety of ways to consider a problem. They may try to identify the parts of a risk assessment model. If working on a food safety problem, they may look for the hazard identification, hazard characterization, exposure assessment, and risk characterization pieces. Some may identify different components of these four steps. Still others will use analogies, draw influence diagrams, arrange objects on a table, perform experiments, develop risk hypotheses, or drink a beer.

Some will draw a picture of the problem as they understand it. Mind maps, introduced in Chapter 6, are great for fleshing out model components. Visual representations of problems can make the abstract tangible, and a problem we can look at is often easier to deal with than a mathematical or conceptual one that is stuck in our minds. Few of us are blessed with the artistic skills or confidence to think we can sketch a problem or a model. But we can all graph a freehand relationship between two variables, and that is a good place to begin.

So, assuming that we have identified the components of a conceptual model (as shown on the right side of Figure 10.3) and we have identified some critical variables in each component, we begin to ask some simple questions. What is the relationship between water temperature and the number of *Vibrio* in the water? How about the number of *Vibrio* and time on the water or the length of refrigeration? What about the probability of getting sick and the number of *Vibrio*? Even if you have no background in microbiology you may have some intuition about the nature of these relationships, and visualizing them is a good way to start.

Even simple visualizations are powerful. Consider the simple model that follows:

RMOs → Risk assessment model → Public health outcomes

It focuses our attention on three ideas. First, there are risk management options to consider. Second, they have public health outcomes, being more or less effective in lowering *Vibrio* illnesses due to oyster consumption. Third, the model (in the middle step) suggests we must identify relationships that connect risk management options to public health outcomes.

Research suggests these visualization techniques help us by externalizing the analysis. They move ideas from inside our minds to the external universe, where we and others can more readily consider them. This is especially important for risk assessment because it is usually a team effort, and externalizing ideas is essential for group work. Debate, progress, and revision are not possible until ideas are externalized and made more concrete.

Sketching the relationships between pairs of variables is also appealing because there are a limited number of relationships that are possible. A few examples are shown in Figure 10.5. As a practical matter, the linear relationship is most commonly considered. In the prototyping spirit, it makes a lot of sense to begin by ascertaining whether a relationship between two variables is positive or negative. A linear relationship assumption can always be amended as theory or increasing intuition suggests a more complex relationship.

Some relationships cannot be expressed as a simple positive or negative one. The relationship between health and a specific nutrient, for example, may look like the relationship shown in Figure 10.6. At very low levels of the nutrient, health may decline due to insufficient levels of the nutrient for peak bodily function. Across a

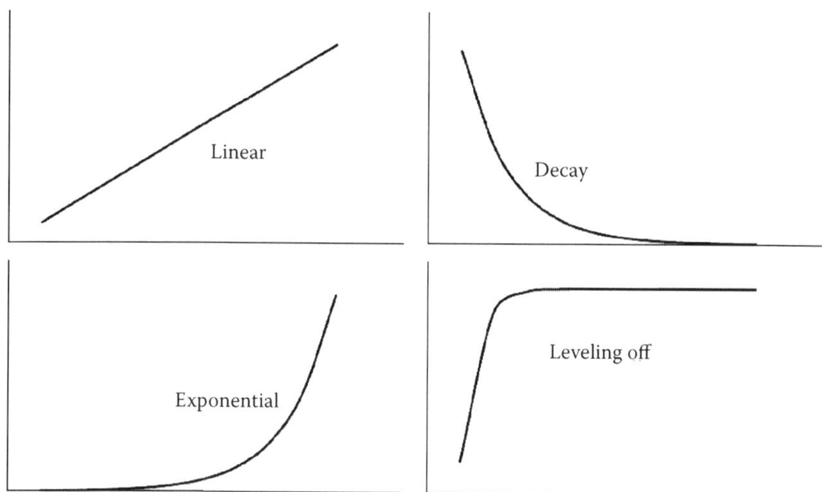

FIGURE 10.5 Sample relationships between pairs of variables.

range of levels of the same nutrient, health may increase. Once excessive amounts of the nutrient are consumed, health may once again be adversely affected, resulting in the odd negative/positive/negative slopes seen in Figure 10.6.

The point, not to be lost, is that once these ideas are reduced to paper they become more clear. This makes them easier to understand or to challenge should others disagree. Visually representing key relationships can be an effective heuristic for beginning to build a risk assessment model. Choose the most important pairs of variables in your model and sketch the relationship you think is possible. You need not be correct in the beginning. You only need to begin. So pick up a pencil and sketch.

The *Vibrio*-in-oysters model offers many possible relationships of interest. A few were mentioned previously. The rough hand-drawn sketches of Figure 10.7 could represent initial ideas about the potential natures of those relationships. V stands for the number of *Vibrio*. The first thing to capture in a relationship is the nature or direction of the relationship. Is it positive or negative? Then it may be useful to represent ideas

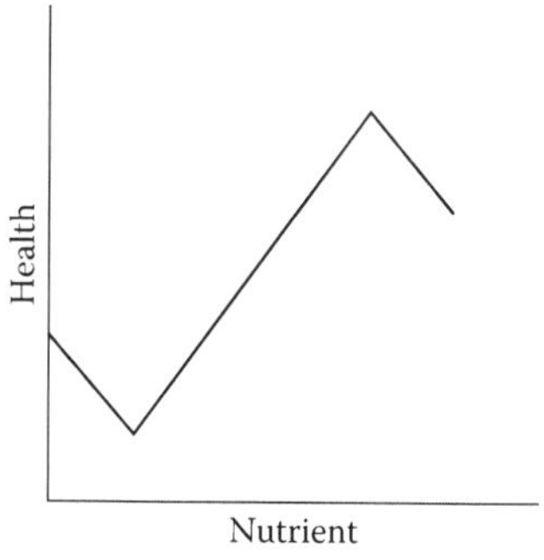

FIGURE 10.6 A nonmonotonic relationship between variables.

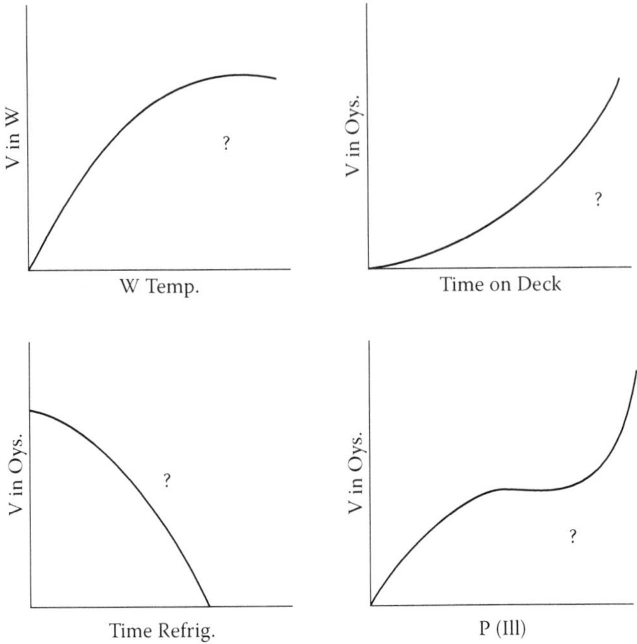

FIGURE 10.7 Hand-sketched relationship hypotheses for key assessment model variables.

you may have about that relationship, whether they are right or wrong at this point. Presenting them visually gives others a chance to review and critique your ideas.

The upper-left sketch of Figure 10.7 shows the number of *Vibrio* present in a given volume of water over a range of temperatures. This sketch suggests the number of *Vibrio* increases up to some maximum and then levels off. As the modeler is an economist, it is useful to commit these ideas to paper so that those trained in predictive microbiology might either point out the shortcomings or confirm the wisdom of the modeler's intuition. The other sketches suggest hypotheses about how other variables

FAMILIES OF CURVES

It is useful to know the functional forms of different families of curves in order to produce sketches of them. The function, its basic shape, and its mathematical form are shown as follows:

Linear functions: constant returns, $y = ax + b$
Power functions: increasing returns, $y = ax^b$ for $b > 1$, for diminishing returns $b < 1$
Exponential functions: decay, $y = ae^{-bx}$; leveling off at asymptote $y = a(1 - e^{-bx})$
S-shaped curve: rapid then slowing growth, $y = b + (a - b)[(x^2/(d + x^c)]$

***Source*: Powell and Baker (2007).**

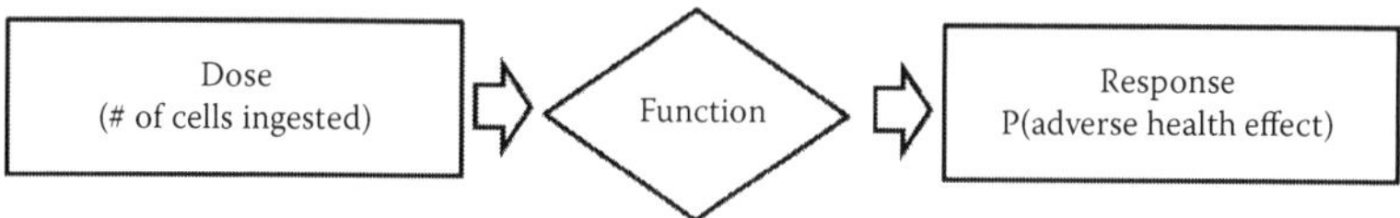

FIGURE 10.8 Dose-response links input (dose) to output (response).

are related to the number of Vibrio. They can be vetted with others before relationships are specified.

10.5.2.1.5 Identify Parameters

At some point we must move from intuition to functions and logical formulas. This is a necessary step to get from a visual representation to a mathematical formulation. In the model-building process presented earlier, this is the specification step on the way to building a computational model.

Several components or modules have been identified for the *Vibrio*-in-oysters model. Example relationships of interest were introduced in the previous section. Parameterization of a model is the step where we get specific about the model's structure and form. This heuristic may be most effectively developed and illustrated with our example.

Consider the health outcome component, specifically the relationship between the number of pathogenic *Vibrio* ingested and the probability of illness. This relationship between exposure to a hazard and probability of illness is called a dose-response relationship. A sample schematic is shown in Figure 10.8.

Dose-response data tend to have a sigmoid shape. When available data are fit to a function, it is common practice to use any one of a number of mathematical functions that exhibit this shape. One such function is the Beta-Poisson, which is of the form:

$$P_{ill} = 1 - \left(1 - \frac{Dose}{\beta}\right)^{-\alpha} \tag{10.1}$$

where P_{ill} is the probability of illness, Dose is the number of pathogenic cells ingested, and α and β are parameters to be determined based on the available data. This specifies the nature of the relationship between the dose and the health response. The function in this general form represents a family of curves. Sometimes, as in the case of the current example, it is the modeler's job to select one from among this family by choosing specific values for α and β. Identifying these parameters takes the conceptual relationship to the specified level.

FDA risk assessors determined the values for the parameters α and β to be 18,912,766 and 536,058,368, respectively. These values were determined fitting the Beta-Poisson function to the available data. The resulting dose-response curve is shown in Figure 10.9.

Probabilities of illness reach about 1% when the eater has ingested about 5.5 log of pathogenic *Vibrio* cells (about 350,000 cells). The probability of illness rises rather steadily to a 100% probability of illness at log 8.8 cells (about 630 million cells). This is just one of an infinite number of possible relationships, however. Each pair of α and β values yields a different curve. Identifying parameters, with or without uncertainty, is a critical step in the art of model building that is often well served by statistical

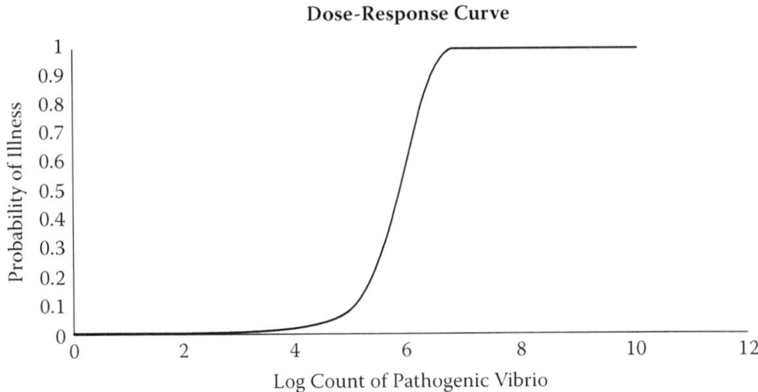

FIGURE 10.9 Dose-response curve.

techniques and judgment. When point estimates of parameters are not possible, they can be estimated by distributions.

10.5.2.1.6 Generate Ideas and Scenarios but Do Not Evaluate Them Too Soon

Best-practice modeling is a creative activity. It is not easy to be creative. The rewards for creativity are rarely obvious, while the sanctions for being wrong are often quite evident. We are often biased by our education, training, and experience to be logical and to seek the right answer. But life is often ambiguous and even maddeningly illogical. The wicked problems of risk analysis rarely have right answers; there are only answers that are better or worse than others.

Following the rules and being practical are lessons we learn at home, at play, and at work. While these are wonderful attributes, they do little to aid creativity. One of the more persistent blocks to creativity is our tendency to judge an idea prematurely. Many of our education systems do a better job teaching us how to criticize ideas than they do teaching us how to create ideas. Consequently, many of us have a strong preference for judging ideas rather than generating them, and we are quick to turn it loose against ourselves.

Model building, like much of life, requires a period of divergent thinking, followed by some intensive convergent thinking. It is terribly easy to rely on our biases, mindsets, and belief systems when we approach tasks that cry out for creativity. When trying to model a problem, begin with divergent thinking. Generate as many alternative ideas as possible. Don't evaluate. Be creative.

Brainstorming, the topic of Chapter 7, is a creative thinking technique that is both effective and easy. The only real difficulty is getting a group to commit to use the technique and to have fun with it. One of the tenets of the best brainstorming techniques is a "no evaluation/no criticism" rule. A second valuable rule is the wilder the better. Use them both early in the model-building process.

Do not fall in love with your first idea. When starting to build a model, it is important to consider as many different approaches to the problem and to answering the risk manager's questions as possible. That cannot happen in an environment where ideas are criticized and defeated before they have had a chance to fully develop. The harshest critic of most ideas is the self. We are very quick to censor our own thoughts and only slightly less quick to criticize the ideas of others.

When the critical self is in charge, the open-ended job of developing a model can seem overwhelming. The downside of every approach, the considerable problems we can anticipate, the hurdles that seem insurmountable are all very evident from the start. Modelers must learn how to quiet the critic in themselves and in others. When building a model you will make mistakes. There will be blind alleys. You may get halfway through a model and realize there is no way out. It is okay to begin again.

Begin by generating as many ideas and approaches as possible. Think of as many ways to model *Vp* and oysters as you can. Let them ferment for awhile. Refine them, explore them, flirt with them, and get to know them. Only then is it time to systematically evaluate them.

10.5.2.1.7 Work Backwards from the Desired Form of the Answer

If you've done a good job at divergent thinking, you may have many possible approaches to take with your model. Your problem now is to choose one. To do so it is often helpful to begin at the end of your model.

What do you want your answer to the risk manager's question(s) to look like? Imagine the specific form the best answer will take. Pay attention to the units of measurement your answer(s) may require. Then work backwards from that point to the best approach or the most logical starting point for your model.

Powell and Baker (2007) propose the "PowerPoint heuristic." Imagine that your answer must be summarized in a single slide. What is the essential message for that slide? Is it a figure, a graph, a table, a procedure, a number, a distribution, a recommendation? Working through this exercise helps you recognize what your critical outputs are and what the essential message of your model must be.

In the model for *Vibrio* in oysters, we want to know how many people get sick annually the way things are done now. We also want to know how those numbers of people might be reduced by a heat treatment, a freeze treatment, a rapid cooling treatment, and regulation of the harvest.* Let's stay with the estimate of the existing risk for now. What do we need to get the annual number of people who get sick? First, a rate of illness or probability of illness per eating occasion is needed along with the number of eaters. It is a simple matter to understand how these two outputs get to the essential message for us.

What do we need to get the rate of illness? We need to know how many eaters out of, say, 100,000 get sick. What do we need to get this? We are going to need a dose-response relationship and a dose. This approach suggests that we might want a model that provides the outcome for a given eater at a given meal. If that is the case, we will need a way to estimate the eater's exposure to pathogenic *Vibrio*. This is an input to the dose-response relationship.

Continuing to work backwards, to get the number of pathogenic *Vibrio*, we need to know the size of the meal and the *Vibrio* load per gram of oyster flesh. Then the *Vibrio* per gram times the number of grams will give us a log count of the pathogenic *Vibrio* the eater ingests.

* These details on the nature of the risk interventions under consideration are found in the risk assessment.

Preliminary risk assessment work may have provided insight into the nature of the available data. For example, if data on pathogen loads at the time of consumption are available, our model may have reached its beginning point. Get the data on pathogen loads and meal size and work through to the number of annual cases of illness.

In the *Vibrio* example, these data were not available, so it was necessary to use predictive microbiology to estimate it. Using the available science, the model estimates the amount of *Vibrio* present in a gram of oyster flesh at the time of harvest. That number is then allowed to grow as the oysters remain on the deck of the workboat and during the period when the oysters are being cooled down. This produces a maximum number of *Vibrio* per gram of oyster flesh. The model estimates the amount of die-off during refrigeration and uses this value to estimate the total *Vibrio* load in a meal. Along the way, the modeler learned it was easier to work with all *Vibrio* and to estimate the percentage of them that were pathogenic later in the model. Working backward from the form of the answer you want often provides the focus one needs to begin an open-ended process like model building.

10.5.2.1.8 Focus on Model Structure Not on Available Data or Data Collection Issues

"Models before data" would be a great bumper sticker for modelers. A good modeler focuses on the structure of the model and getting it to represent reality in a way that meets the information and decision-making needs of risk managers. One of the most common, most appealing, and most limiting approaches to model building is building to the data. Consequently, many modelers begin by searching for, collecting, and analyzing the data. They then build a model that uses the data they have instead of the model that should have been built.

It is tempting for a modeler to build a model that makes use of the existing data. Likewise, it is tempting to avoid including model components for which data do not yet exist. This approach is flawed from the outset. There is no automatic reason to assume the available data are accurate for your purposes. At the other extreme, if we avoid model structures that require new data, we may be failing in an opportunity to "grow the science." Research is as integral a part of risk assessment as is addressing uncertainty in our analysis.

Build the model that will solve the problem and answer the risk manager's question(s). Do not worry about the data. A good model improves the risk assessment process and often expands the body of knowledge because of new data that must be collected and organized.

The existing data are often flawed and need to be carefully screened. All of our data describe the past, and all of our risk management decisions are about the future. There is always good reason to suspect the temporal relevance of our data. Some of the more common sources of bias and errors in the data are sampling error, data collected for different purposes, masking, inappropriateness, definitional differences, geographic or cultural irrelevance, flawed experimental design, and so on. Others are found in the accompanying text box.

The best-practice method is to build the model you need. Let it tell you what data you need rather than letting the data tell you what model to build. Once the appropriate specification model is built, you can then use the available data to refine the model. This sequence may also avoid unnecessary collection of

TYPES OF BIAS

Selection biases
- Volunteer or referral bias
- Nonresponse bias

Measurement biases
- Instrument bias
- Insensitive measure bias
- Expectation bias
- Recall or memory bias
- Attention bias
- Verification or work-up bias

Intervention (exposure) biases
- Contamination bias
- Co-intervention bias
- Timing bias(es)
- Compliance bias
- Withdrawal bias
- Proficiency bias

***Source*: Hartman et al. (2002).**

additional data. If necessary, your better data-free model can be revised to accommodate the available data. Getting new data and getting better data are not always obligatory activities.

Good data do matter. But good data in a flawed model are not necessarily better than imperfect data in a good model. Good decisions are frequently driven more by good models than the data that are in them. A well-structured model can often reveal valuable insights with a rough estimate of model parameters. Frequently, insights about the problem gained by building the model are sufficient for informing the decision-making process. Data are not always necessary for informed decision making.

10.5.2.2 The Practice of Modeling

Modelers have developed many handy and helpful ideas about how to build a model in a spreadsheet environment. These are the focus of the remainder of the chapter. You'll notice a bit of overlap with the art of modeling heuristics at times. The difference is that here we assume you've tackled and solved the conceptual problems and now look to practice the art of modeling.

10.5.2.2.1 Sketch the Model—Document the Logic Flow

Once you understand the model, after your initial visualization efforts and before you begin to build your spreadsheet, sketch the logic flow or the components of the model. The visual representation will help your decide how best to organize your efforts. Figure 10.10 is taken from FDA's *Vibrio* risk assessment. It provides a very

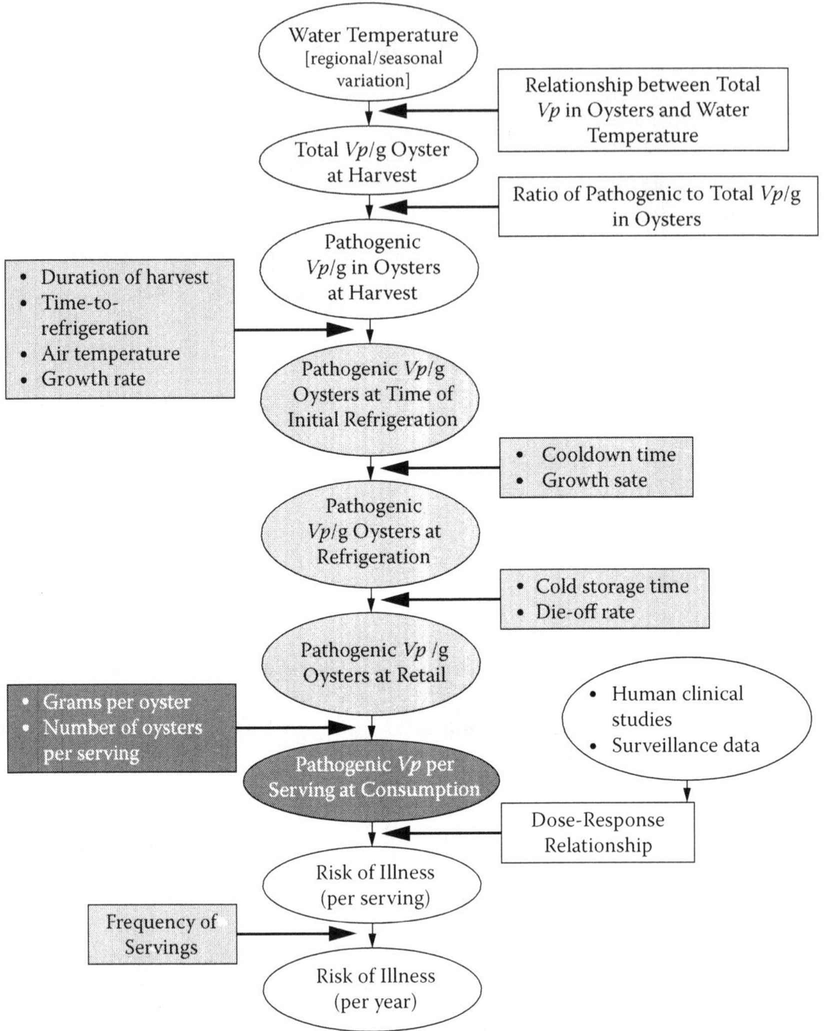

FIGURE 10.10 Schematic representation of the *Vibrio parahaemolyticus* risk assessment model. (*Source*: Food and Drug Administration 2005.)

effective visual summary of the model's construction. This diagram makes it easier to identify component modules at a finer level of detail and more intuitively than does Figure 10.3.

Make the shapes in your sketch be meaningful. Influence diagrams are a common model-sketching tool that relies on the common convention of showing a chance event as a circle, a decision as a square, a calculation as a rounded square, and a payoff as a diamond. As long as the shape and color schemes in your model sketch have meaning, their information content is enhanced.

10.5.2.2.2 Think Communication As You Design the Model

Before you begin to build your computational model, think about how you can organize the model so that it aids the risk communication process. Spreadsheet models consist of tabbed worksheets. If you use these worksheets to give your model a logical structure, it can save you and others who may use it a great deal of time in the future. Following are some suggestions for organizing a spreadsheet model.

Overview sheet. The first sheet should provide an overview to the risk assessment model. Why was it prepared and what does it do? When was it developed and by whom? Using what software? The overview sheet should be brief and it should at some point include the list of risk management questions the model was built to answer.

Instructions sheet. Documenting your model is a critically important part of the modeler's job. Spreadsheet models are not likely to be so complex as to require an actual users' manual. However, they should provide instructions sufficient to enable a future user of the model to understand how to update data and other model components as well as how to run the model and access the desired results. The second sheet should include these instructions. If the model has been verified and/or validated, the methods by which this was accomplished and the time frame should be documented here as well.

Log of use sheet. This feature is rarely included, but it can be extremely useful, especially when it is being used by a government agency for regulatory, stewardship, or other public purposes. The log of use sheet identifies who used or modified the model for what purposes and when. It can also be used to record the planned sequence of simulation experiments and the dates they were completed.

Guide to model sheets. A modular risk assessment is likely to consist of many separate worksheets. The organization of such a sheet can be difficult to discern from the short phrase on the tab. In a large model, all the tabs may not even be visible at the same time. This guide identifies the various sheets in the model and briefly describes the purpose or function of each sheet. The guide functions as a table of contents and is an indispensable aid for a large model.

Influence diagram/conceptual model. The sketch of the model needs to be presented somewhere in the model itself. It may be helpful to add a narrative to accompany the influence diagram or model layout. This may be combined with the guide to your model sheets.

Assumptions. Assumptions are critical to the construction of any risk assessment model. All explicit assumptions and as many implicit assumptions as possible should be identified and listed in an assumptions sheet. It may be helpful to organize the assumptions by model module.

Inputs. One of the most practical things you can do is create a sheet where all input variables and parameters can be entered. Separate the variables from the constants. This will make it easier for you to correct, change, revise, and update data throughout the model's lifetime.

Relationships. Adequately identifying and modeling dependencies among variables in your model is one of the modeler's greatest challenges. It is critically important to document and to convey to others the nature of these dependencies. Each interrelationship between or among variables should be described here in a narrative form. Include the formulas used to capture these dependencies when possible and indicate where in the model they can be found. In the *Vibrio* model this would include the dose-response relationship and the growth and die-off equations, for example.

Performance measures. Many models will have performance measures. These are cell values that are of interest to the risk assessor or the risk manager. They tend to be intermediate calculations rather than outputs. The number of *Vibrio* cells per gram of oyster flesh is a performance measure in our example. A model that calculates a benefit-cost ratio would have costs and benefits as performance measures. A transportation model might have travel time as a performance measure. Identify all the performance measure(s) and include the formula used to calculate them when possible. An example is presented in Table 10.1. In some instances it may be useful to include a cell reference for the performance measure (not shown in the table) so it can be easily located in the risk assessment model and observed.

Outputs. Collecting all model output cell values in a single place is extremely helpful with complex models. Models constructed in a modular fashion may have outputs of interest spread over several worksheets. In an event tree, model outputs may be scattered over many cells in a worksheet. Consolidating all output cell values on a single page is not only convenient for observing outputs, but it can also be an effective editing and debugging aid. Note that these outputs are not detailed statistical and graphical outputs of simulations. They are only images of the output cells, similar to the model values shown for performance measures in Table 10.1.

Modules. Each model component should have its own module, and each module should have its own worksheet(s). The worksheets should be labeled descriptively, and each should include a brief narrative explaining the function of the module. If a model sketch is used, it is helpful to indicate what part of the model is being represented by the module. The module sheets are the guts of the model.

Historical data used to generate model inputs. Often model inputs are derived from historical data. When this is the case, it is desirable to include these data in the risk assessment model itself if it is practical to do so. The size or proprietary nature of the data files may make it impractical. When the data cannot be included, its source and location should be noted.

TABLE 10.1

Performance Measure Examples for a Risk Assessment Model

Performance Measure	Model Value	Formula
Log *Vp* level in environment	2.0963456	−1.03+0.12*E10+RiskNormal(0,0.886)
Predicted counts at first refrigeration	2.8192952	IF((E12+E27)>6,6,E12+E27)
Predicted level after die-off	2.6691434	E31−0.0027*24*E32

Graphics page. Some models make use of dynamic graphics so that model effects and relationships can be more easily observed. If your model makes use of dynamic graphics, gather and present them all in a single worksheet when practical to do so.

10.5.2.2.3 Organize Your Model into Modules

My best advice to new and aspiring modelers is: Do not attempt to work out the entire model in your head at all once. Put things on paper. Sketch the model's major parts. Practice the art of modeling. Once you have identified the components of your model, you have the framework for a modular model design.

Repeat the model-building process for each module. Treat the module like a mini-model. Build and work on one module at a time. Sketch it unless it is simple and obvious. Identify its subcomponents. Consider making subcomponents separate modules. Don't go overboard with the decomposition, but make an effort to define modules that make sense to yourself and others.

If you are modeling a process, the modules are often easy to define. The egg production process in Figure 10.11 provides an easy identification of modules. If there is no clear process, allocate specific tasks in the model (e.g., assessment and hazard characterization) to different sheets.

Use common sense. Some models are compact even though they have several discrete components. Though it is always wise to think in terms of modules and to build in terms of modules, it is not always necessary to separate the model physically along those lines. The FDA model we have been looking at, for example, does not use a modular approach.

10.5.2.2.4 Handling Inputs and Outputs

There are reasons beyond good communication for segregating your input and output values. Efficiency and accuracy are at the top of that list. A spreadsheet model should provide a single point of entry for the values of all model inputs, i.e., variables and parameters. Isolate the data inputs to a single sheet of the model and it will be easier to input data, check for errors in inputs, avoid errors in formulas, and revise and update your inputs.

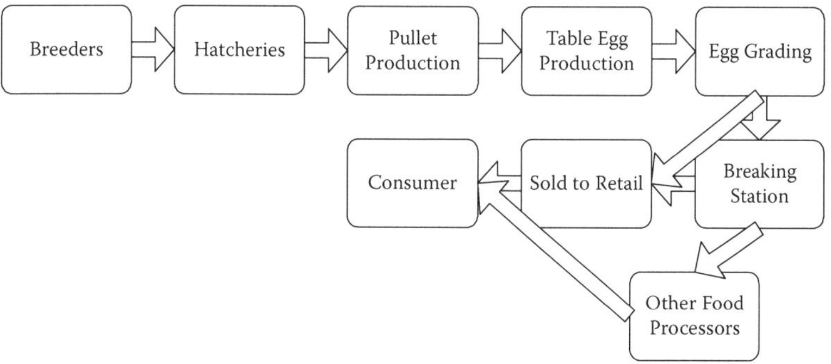

FIGURE 10.11 Process flow for shell eggs and egg products.

	B	C	D
3	Direct Entry Model		
4		Weight in pounds	Formula Syntax
5	Man 1	184	RiskNormal(182,16)
6	Man 2	172	RiskNormal(182,16)
7	Sum	357	D5+D6
8			
9	Cell Reference Model		
10		Weight in pounds	Formula Syntax
11	Mean	182	182
12	SD	16	16
13	Man 1	141	RiskNormal(D9,D10)
14	Man 2	185	RiskNormal(D9,D10)
15	Sum	326	D11+D12

FIGURE 10.12 Comparison of direct parameter and cell reference parameter entry.

Formulas in your model should not include constants of numerical values nor distribution formulas for variables. Enter all inputs into the cell of a single input sheet and then make sure all your formulas use the cell references of that cell in the input sheet in place of the actual values.

Consider a simple example of a model that calculates the combined weight of two men. Figure 10.12 shows two ways to enter the data. Notice that column and row identifiers are found in the figure's margins. The direct entry model shows the formula using the actual numerical values as arguments in the distribution. The cell reference model shows arguments for the distributions using cell references.

The advantage to using cell references and a single point of entry for input data is that if new information shows the standard deviation to be 17 pounds instead of 16, then a single change in the cell reference model at cell C12 updates the model. Using direct entry, the modeler must be sure to make the changes in every instance where the standard deviation appears in the model.

Ideally, all input values like the mean and standard deviation would be entered on a separate worksheet in a more complex model and then referenced there. The input arguments, cells D9 and D10, would be located on the inputs worksheet. On that input sheet, separate your inputs by type. Constants are numerical values that do not change for a model run. Variables are numerical values that do change in a model run.

There are two kinds of variables: deterministic and stochastic. Deterministic variables vary in a known pattern, as the hours of operation or the duration of a process might. These may well change within a model run. Decision variables and value variables are also deterministic variables, although we may be uncertain about what the most appropriate values for these variables are. In that case, different values may be chosen for multiple simulations of an otherwise identical model, or we may simply

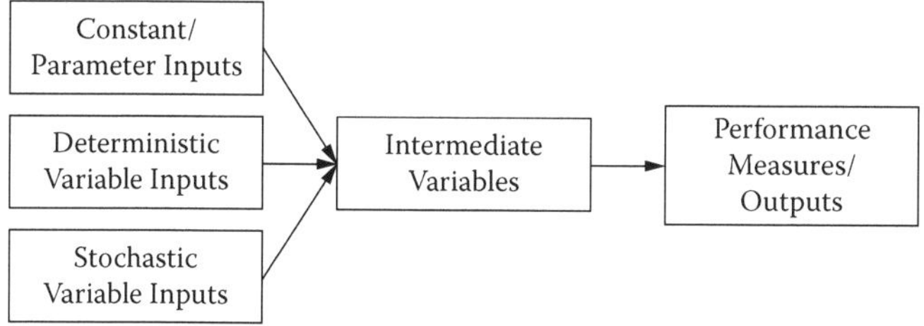

FIGURE 10.13 The role of intermediate variables in model input/output relationships.

want to do some sensitivity analysis of specific variables that have no true value. Stochastic variables have an element of randomness to them. They are usually represented by probability distributions. Segregate constants, deterministic variables (including decision and value variables), and stochastic variables in your input sheet.

You may want to segregate the intermediate variable calculations of your model in their own worksheet as well, as represented in Figure 10.13. These include the performance measures mentioned earlier. Intermediate variables link inputs and outputs through intermediate calculations. The ability to monitor intermediate values is a valuable debugging feature of a model. When intermediate variables are performance measures, there may be intrinsic value in being able to observe them as one steps through a model. Consider gathering an image of all significant intermediate variables in a single place in your model.

Gathering an image of all model outputs in a single sheet provides a convenient way to find all the values of interest for answering the risk manager's question(s) in a single place. Consider developing modeling conventions for your organization or at least for yourself. One of the simplest of these is to use color-coded fonts. Choose one color for any cell that includes an input value, another color for intermediate calculations, a third for outputs, and perhaps a fourth for data. This way the color patterns of your spreadsheet cells immediately convey useful information to anyone who knows the color code.

10.5.2.2.5 Output Reports

Each planned model run will produce outputs and performance measures of interest. These outputs should be documented in data arrays, tables, and graphics, i.e., in output reports. There should be an output report for each run of the model. A large number of model runs may necessitate storing them in a separate spreadsheet file.

10.5.2.2.6 Design for Use

Design your model to be used. We've already said, "Build the simplest model possible but no simpler." The corollary to that rule of thumb is to build your model as simply as possible. To make your model easier to use: calculate first, simulate second. Although there may be no closed-form analytical solution to your overall model, there may be analytical parts of the model. Whenever you have a choice between analyzing and simulating, choose analyzing. Keep your cell formulas simple. When you have a complex formula, break it down into several intermediate calculations. This will help avoid difficult-to-find errors, and it may speed up your model calculations.

As you determine your model's structure and begin to quantify the variables and their interrelationships, one of the issues you must resolve is the level of detail or the resolution of your model. One way to frame this issue is as a trade-off between the realism of the model and its utility.

As the terms are used here, realism means the extent to which the model faithfully represents the details of the real-world process being modeled. Utility means the extent to which the model provides useful answers to the risk manager's questions. A simple model may be sufficient for answering the risk manager's questions, but it may not be a very realistic model. On the other hand, a realistic model is not guaranteed to produce useful answers.

Realistic models tend to be larger if only because they include more details. Greater detail is rarely worthless. However, greater detail also requires significantly more data to enliven those details, and data availability is always a real concern for realistic models. Realistic models do have the advantage of getting to a tangible level of understanding, a level of resolution that is easy to envision. In a large, detailed model, no one part of the model is likely to be "too" critical. Indeed, aggregating the results of many smaller components to get an output can be more accurate than directly estimating an output. If these are important concerns, you will want a more detailed, more realistic model. Expect such a model to require a lot of data.

On the other hand, a more abstract model with less detail can be a lot smaller. Sometimes a more abstract model is a necessity and not a choice, as it may not be possible or practical to include the additional detail in a model—although you know better than to build to your data! There may be too much knowledge uncertainty about the system being modeled, or it could simply be cost prohibitive to develop a realistic model. Oftentimes, capturing the key elements of a system in your model is sufficient to support decision making.

In general, when a model provides useful answers to the risk manager's questions, it is ready to use. Design your model for use. Then, if you decide that incremental additions to the model's realism are justified by the value of the information these enhancements produce, you can add as many as you like. But first, build a model you can use.

10.5.2.2.7 Minimize Stupid Mistakes

Take extra care with modeling dependencies and relationships among variables and parts of your model. We live in a complex universe, and many phenomena are treated as independent of one another, often because it is too difficult to deal with the more complex reality. Independence is not to be an assumption of convenience for the modeler. Things must truly be at least functionally independent to be so represented in your model.

Your model's structure should prevent the generation of any unrealistic scenarios. To ensure this, you must recognize and model any interdependencies among your model's variables, calculations, and components. Things that always happen together should happen together in your model. Things that can never happen together, things that move in the same direction, things that move in opposite directions—all must be accurately represented in the structure of your model. Dependencies can be built into a model using complex logic arguments, functions, correlations, linkages, and embedded arguments, among other means. Use whatever means necessary to ensure

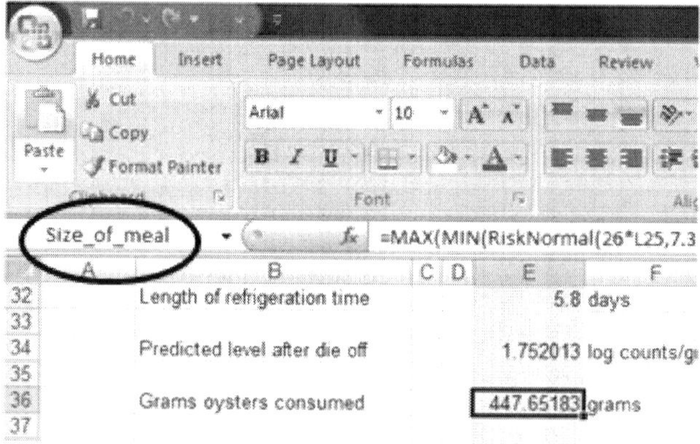

FIGURE 10.14 Use range names for key variables.

that you have carefully tended to important real-world dependencies and relationships so they are reflected accurately in the workings of your model.

Copy and paste carefully. Remember that when you paste a cell formula your software is adjusting cell references for relative changes in their new location. This often results in unanticipated errors. Always make sure cell references are as desired after pasting copied material. Remember to use dollar signs ($) when you want to preserve the absolute location of a cell when you copy and paste.

Use range names for key inputs and intermediate values. Using simple descriptive names for a cell or range of cells makes formulas that reference these cells much easier to read and understand. In Figure 10.14, cell E36 has been named "Size_of_meal" in the example shown here. This is far more informative than using "E36."

Compare the difference between the mathematically identical formulas in Table 10.2. To understand the first formula you would have to locate the cells referenced and see what the referenced values are before it is understood. The second formula uses range names and is far easier to understand. The more descriptive quality of the range names limits the likelihood of a stupid mistake like a cell reference typo where you type E35 instead of E36.

Use Comments generously in your model to document assumptions and explain things. Figure 10.15 presents an example to demonstrate the concept. This comment is not found in the original FDA model; it was created to illustrate the point made here about the value of comments. Comments can help eliminate stupid mistakes by reminding you and the naïve user of your model of things you might dangerously forget without the reminder. They are also excellent documentation aids.

Create generations of backup models. If you have ever saved changes in a file and then realized your changes were incorrect and mourned the loss of the correct file, you may already be doing this! When developing a model, each time you begin to

TABLE 10.2
Comparison of Cell Address and Range Name Formulas

=10^E34*E36
=10^Log_vibrio_per gram*Size_of_meal

	D	E	F	G	H	I	J
27		1.350015					
28	n	4.625453	log counts/gram			rapid cool	
29		10	hours	Heat	Freeze	mitigation	
30		1.0742511		treatment	treatment		
31		5.699704		four.five log	two log	4.349689	
32		1.6	days	reduction	reduction		
33							
34		5.598846	log counts/gram	1.098846			
35							
36		145.33966	grams				
37							
38							
39		57707270	log counts	1824.864098			
40		256309		6			
41		5.4087639	log counts	0.77815125	[illegible]	[illegible]	
42		0.0012592		2.74382E-17	4.735E-08	2.30006E-06	
43		-2.8999184		-16.56164511	-7.3246562	-5.63826049	
44		0		0	0	0	

Charles Yoe:
Note that this value is the pathogen load shown in E34 reduced by a 4.5 log reduction due to cooking.

FIGURE 10.15 Use comments to document and explain the model.

revise it, save the revision as a new version, e.g., MyModel01, MyModel02, etc. Then when your work in progress veers off in a bad direction you can simply return to an earlier generation of your model.

Be careful with external links. Minimize the number of links you have to other workbook files. Limit yourself to links with external databases if you must link at all. Once you link to another workbook, it may be linked to yet another workbook, and you may be entering a web of interconnections that you do not understand. While speaking of data, make sure your data are in a legitimate number format. Data can be formatted as text when imported carelessly. Text will not perform numerically!

Do not use cell formatting (see Figure 10.16) to round numbers. They are not truly rounded, they are simply displayed without trailing decimals. Figure 10.17 shows a hypothetical binomial distribution. Values for n and p are given in column L. The exact syntax of the column L entry is given in column M. Thus, if you wanted to change the number 17.931 to the rounded-off integer, 18, use the ROUND function of your software. Cell formatting will display an 18 in the cell, but it will treat the cell value as 17.931. To illustrate this problem, imagine you want to use a number like 17.931 as an argument for the sample size, n, in a binomial distribution as shown in Figure 10.17. The binomial requires n to be an integer. In the top example, cell formatting is used for faux rounding. Notice it returns an error message (#VALUE in cell

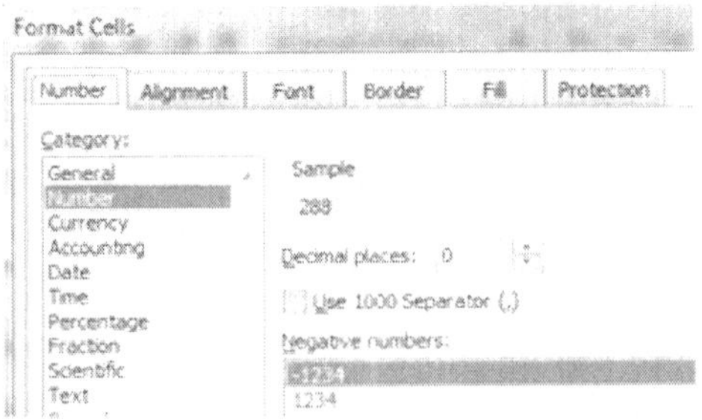

FIGURE 10.16 Truncating a value using cell format to mask decimals.

	K	L	M
15		**Cell Format Formula**	
16	Parameter	Value	Cell Formula in L
17	n	18	17.931
18	p	0.2	
19	Binomial	#NAME?	=riskbinomial(L16,L17)
20		**Rounding the Value**	
21	n	18	=ROUND(17.931,0)
22	p	0.2	
23	Binomial	4	=RiskBinomial(L21,L22)

FIGURE 10.17 Comparing the effect of cell format and rounding in a binomial distribution.

L18) when entered into the binomial distribution. The second example uses the round function of Excel and enables the binomial function to work as expected.

Use the function wizard to minimize syntax errors. It can be found by clicking the $\mathbf{f}_x$ on the standard Excel toolbar. Figure 10.18 shows the function wizard listing all the @RISK probability distributions. Beneath the list is the Excel template that can be used to enter @RISK functions, as well as any of the Excel functions. Using the wizard will help minimize typos and inadvertent errors in formulas.

Be careful with parentheses in long formulas. It is common practice to use parentheses to establish the order of operations. It is easy to make mistakes with the placement of parentheses.

When you are building a model, it can be useful to "break your model" at times. To do this, you enter intentional error values into a cell. An example would be to enter a probability outside the acceptable range or to attempt to divide by zero. Once in awhile error messages can be comforting. Your model should fail when it is supposed to. In a related fashion, it is useful to use simple values like 0.1, 1, 10, and other multiples of 10 to test your calculations. They need not be realistic values when you are building your model. Numbers like this make it easy to spot-check the accuracy of calculations.

10.5.2.2.8 Test and Audit the Model

Expect your spreadsheet model to contain errors. If you do, you'll make a more serious effort to find and fix those errors. Few things are worse for a modeler than to discover an error well after you've begun to use your model for decision analysis.

This step was called verification earlier in the chapter. Its purpose is to verify and ensure the model is free from logical errors. If your spreadsheet model is a simulation model, your goal is to ensure that every iteration of the simulation is both accurately calculated and possible. This is when the care taken in designing your model for use and communication will first pay off.

Testing and verifying your model can be laborious and monotonous. It is all the more difficult to catch errors when your focus wanders. One of the best ways to test your model is to have experts review the model. This is a luxury few risk management activities can afford. It is also difficult to find these experts. They usually need software, modeling, and subject-matter expertise. This is hard to find in a single

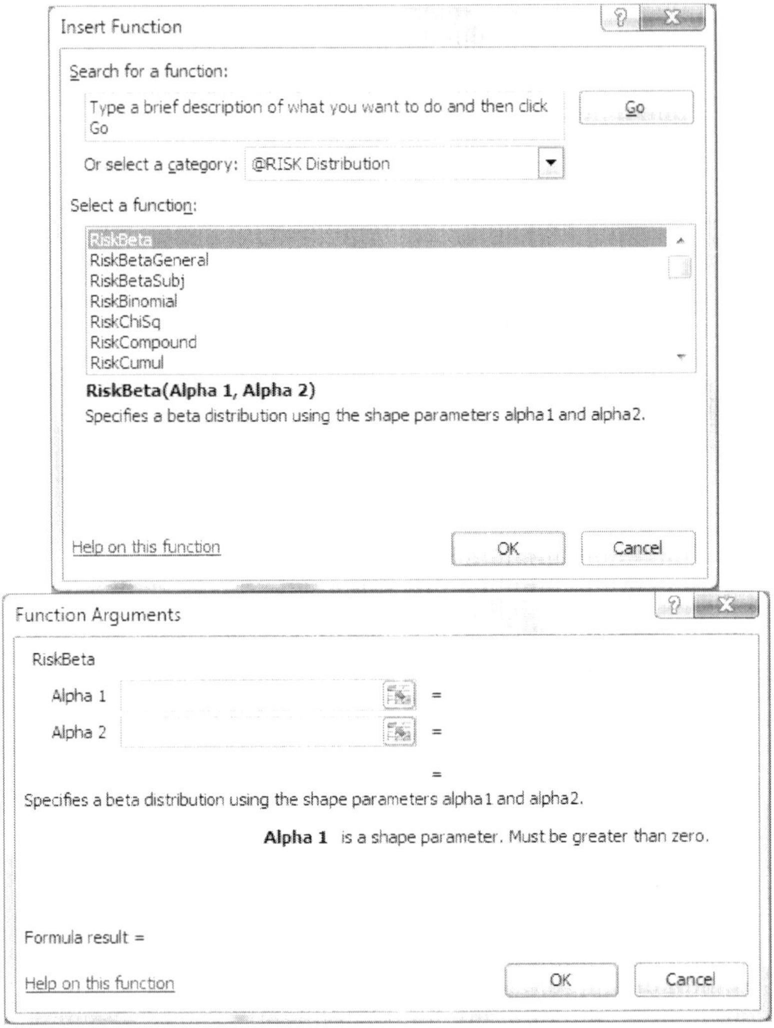

FIGURE 10.18 Excel's function wizard helps minimize typographical and syntax errors.

person. So, assuming you will have to verify your model, how can you best do that? Approach your model with a skeptical attitude. Powell and Baker (2007), once again, offer the single best description of a plan for testing your spreadsheet model. In the following discussion, I have supplemented their suggestions with those of the Fuqua School of Business at Duke University (2006), Mather (1999), and a few of my own.

10.5.2.2.9 Debugging

Debugging is a methodical process for finding and correcting errors in your model's logic and data. There is a rich literature on debugging, but the truth is the method of that process is highly individualized and often idiosyncratic. There is no simple or foolproof way to debug a model. The simplest advice is to build your model in small modules and debug each module carefully before you link them together. The most basic form of debugging is to methodically check and verify every line of code in your model, or in the case of a spreadsheet model, the contents of every cell.

As you link your models, check carefully to ensure that interactions between the modules do not create new bugs. Debugging, as used here, is usually an ongoing process rather than a discrete event in the life of a model. You are always debugging your model right up to the time that it is verified.

10.5.2.2.10 Make Sure Numerical Results Are Reasonable

As your model takes shape and numerical results (either intermediate calculations or outputs) begin to appear, or when it is initially completed and you begin verification, check the reasonableness of your numbers. Prepare rough estimates of what the values should be before you complete the calculation. Probabilities must lie between zero and one. We usually have or can estimate a range of plausible outcomes for a calculation. If that reasonable range is exceeded, verify the formula.

Check your calculations with a calculator. You may have a formula that functions exactly the way you programmed it, but if you programmed it incorrectly you may never catch the errors. Calculators can catch errors in formulas.

Sound logic in your model should provide logical results even for unrealistic sets of values. One way to test this logic is to use the zero test, where zero is regarded as an extreme value. Set one or more variables or groups of variables to zero and make sure outputs are as they should be if this value is zero. In our *Vibrio* model, zero *Vibrio* at the time of harvest should produce zero *Vibrio* at the time of consumption given the structure of the model, which does not permit cross-contamination. Likewise, zero *Vibrio* should never result in a positive probability of illness. Thus, the zero test often provides a quick check of the model logic at this extreme value. When zero is not a reasonable extreme value, set the variables to known or reasonable extremes and make sure the intermediate calculations and outputs are as they should be.

Logic tests can be used to debug your model. Increase and then decrease the values of each input value one at a time to ensure that the changes to the outputs are in the logical direction. For example, as water temperature or air temperature increase in the *Vibrio* model, the number of bacteria should increase as well. They should not only increase but be a reasonable magnitude as well.

Excel has some useful auditing capabilities. Two of them are the Trace Precedents and Trace Dependents commands. They can be used to graphically display or trace the relationships between precedent and dependent cells and formulas with tracer arrows. This enables you to ensure that the proper relationships are in place. An example of Trace Precedents is shown for the *Vibrio* model in Figure 10.19. It shows all the cells that influence the probability of illness and the route of their influence. You need not understand the nature of those complex relationships from the figure, just appreciate the potential utility of the feature. Trace Dependents, as you might imagine, shows all the values that depend on a given cell. With a little practice, these tools can be very helpful in the verification process.

10.5.2.2.11 Check the Formulas

Perhaps this section should have been titled, "No Duh!" Formulas can be checked visually. This is the tedious and monotonous part of the verification. It is also the dangerous

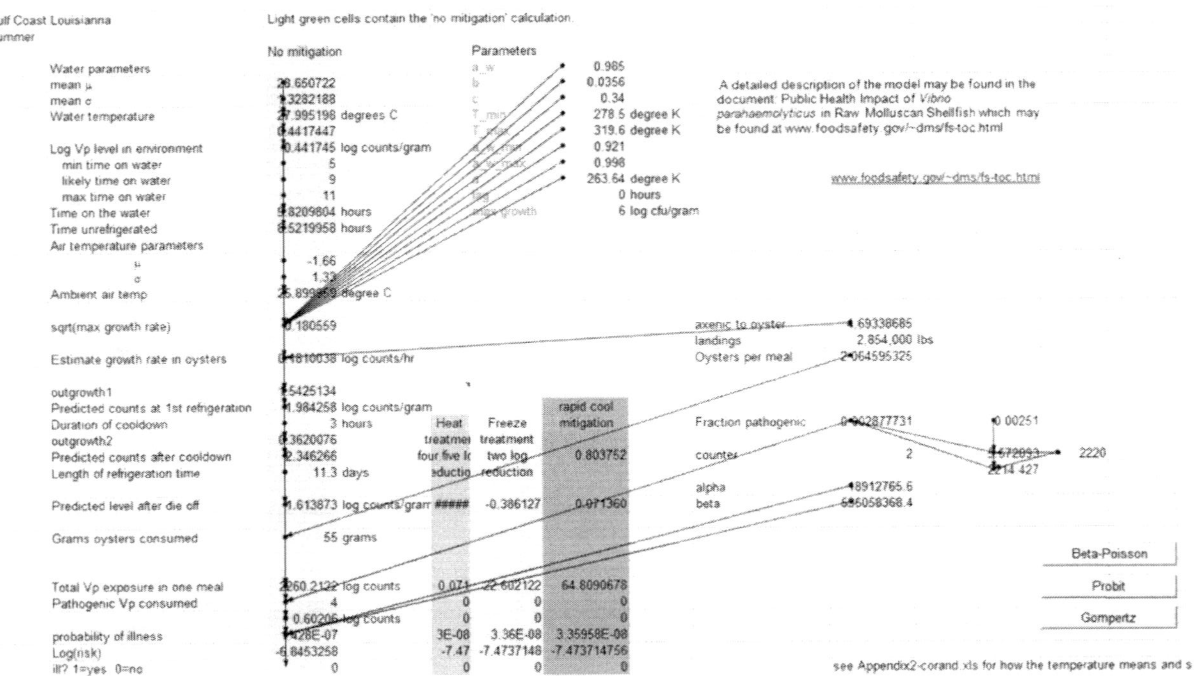

FIGURE 10.19 Trace precedents shows all the variables required to calculate the value in a single cell.

part, because if you built the model you may not be able to see your errors. Nonetheless, visual inspection of each cell formula is a minimal expectation for verification.

Excel has several nice features to aid this task. Putting your cursor in a model cell and pressing F2 or double-clicking in that cell will reveal the cell formula, and it will use a color-coded system to highlight every cell in the model that is referenced in that formula.

To reveal all the cell formulas at once, press Control ~. This is especially useful when cells in adjacent rows or columns have similar formulas. In that case, a break in the formula pattern may indicate an error.

Excel also offers an error-checking feature. If errors are found, each potential error is flagged by the appearance of a small triangle in the upper left corner of the cell. Excel's error checking can find eight different errors. Excel 2007 has an error-tracing feature as well. This feature most often points to the source of the error.

The Simulation Settings of Palisade @RISK offer a Pause on Output Error option that is very useful. If a logical error is found in any cell, the simulation will stop on that iteration, enabling you to find the cell with the error message. Excel's error-checking capabilities or your own trace back can then reveal the logical source of the error.

Excel 2007 offers a helpful Evaluate Formula option in the Formula Auditing group of the Formulas tab. To use it, position the cursor in the cell of interest and select Evaluate Formula. Evaluate, Step In, and Step Out options provide an excellent audit trail for checking formulas across a chain of cells. Using name ranges is a decided advantage for getting maximum value from this feature. If these built-in auditing features are not sufficient, there is a wide array of spreadsheet auditing software available on the Internet.

10.5.2.2.12 Test the Model Performance

If you cannot wait to begin to use your model before verifying your model—an ill-advised approach—examine your results as you would intend to do when using your model to answer the risk manager's question(s). Simply using the model may reveal some logic flaws, and unusual results may help you find bugs in the calculations. If your model is logical and accurate, your results should be also. Doing some sensitivity analysis may help uncover problems in the model. Note that this is not considered an adequate method for verifying your model. It is, however, sometimes helpful for finding gross errors in a model.

If you build models, you will spend a lifetime developing and refining your craft skills. Like model building itself, these can tend to be rather idiosyncratic. The ideas offered here are but the tip of the craft skills iceberg. But, if you actually follow them, they will advance you up the spreadsheet modeler's learning curve more quickly than your own trial-and-error methods will.

10.5.2.2.13 Documentation

Verification and validation are important steps in model development. They are often overlooked in the more casual application of risk analysis principles and in the less formal risk assessments. When a model has been verified and/or validated, the methods and time frame for accomplishing these tasks should be documented and appended to the model.

10.6 Summary and Look Forward

Not many risk assessors are trained as model builders. This is generally a skill that is picked up in practice by people with a quantitative orientation. As a result, the model builder is usually not the subject-matter expert. Model building is, consequently, often a group effort.

A 13-step modeling process is offered in this chapter to help those struggling with the notion of what it means to build a risk assessment model. Model building is iterative. It begins with a conceptual model, evolves to a specification model, and ends up as a computational model. Most model builders manage to muddle through those steps somehow, although it is rare to see the iterations as discrete and separate tasks. Verification, validation, and documentation are, in my experience, the most often overlooked tasks in model building.

The craft skills for model building are what experience teaches you. This chapter attempts to assist your development of craft skills by building on the eight heuristics offered by Powell and Baker (2007). Chapter 11 addresses one of the technical skills quantitative risk assessors must have: a basic understanding and command of the essentials of probability. That is followed by two more chapters on probability topics that will expand your technical skills while, hopefully, helping you to develop your craft skills. These are "Choosing Probability Distribution" (Chapter 12) and "Probability Elicitation" (Chapter 13).

REFERENCES

Evans, James R., and David L. Olson. 2002. *Introduction to simulation and risk analysis.* 2nd ed. Upper Saddle River, NJ: Prentice Hall.

Food and Drug Administration. 2005. Center for Food Safety and Applied Nutrition. Quantitative risk assessment on the public health impact of pathogenic *Vibrio parahaemolyticus* in raw oysters. http://www.fda.gov/Food/ScienceResearch/ResearchAreas/RiskAssessmentSafetyAssessment/ucm050421.htm.

Fuqua School of Business. 2006. Excel design & audit tips. Durham, NC: Duke University.

Hartman, J. M., Forsen, J. W., Wallace, M. S., and J. G. Neely. 2002. Tutorials in clinical research: Part IV: Recognizing and controlling bias. *Laryngoscope*, 112: 23–31.

Mather, D. 1999. A framework for building spreadsheet based decision models. *Journal of the Operational Research Society* 50:70–74.

Powell, Stephen G., and Kenneth R. Baker. 2007. *Management science: The art of modeling with spreadsheets.* 2nd ed. Hoboken, NJ: John Wiley and Sons.

Ragsdale, Cliff T. 2001. *Spreadsheet modeling and decision analysis: A practical introduction to management science.* Cincinnati, OH: South-Western College Publishing.

Sanchez, Paul J. 2007. Fundamentals of simulation modeling. In *Proceedings of the 2007 winter simulation conference*, ed. S. G. Henderson, B. Biller, M. H. Hsieh, J. Shortle, J. D. Tew, and R. R. Barton. Monterey, CA: Operations Research Department Naval Postgraduate School.

11

Probability Review

11.1 Introduction

Risk analysis is decision making under uncertainty. Resolving and addressing that uncertainty is a primary reason for risk assessment. Probability is the language we use to express our uncertainty. Someone has to understand probability if risk assessment is going to address the knowledge uncertainty and natural variability in our decision problems. Quantitative risk assessment uses the language of probability, and it is essential to know the structure of any language so we do not misuse it. Learning that basic structure is the purpose of this chapter.

Risk has been described as the product of a consequence and its probability. Assessors need to understand probabilities to assess them honestly. Managers need to understand probability to manage risks effectively. Risk communicators have to understand probability to explain it to others. Not everyone needs to know how to do good probabilistic risk assessment, but everyone does have to be conversant and knowledgeable about some very basic facts about probability.

One problem with probability is that it does not lend itself well to intuition. If there is a 10% chance of rain and we get caught in a downpour, we are more inclined to think the forecast was wrong than we are to think the rain was an event that had only a 10% chance.

To further illustrate the nonintuitive nature of probability, consider the so-called Monty Hall problem, which I have used in risk assessment training for years. A letter writer to Parade magazine (Whitaker 1990) posed the following question. "Suppose you're on a game show, and you're given the choice of three doors: Behind one door is a car; behind the others, goats. You pick a door, say No. 1, and the host, who knows what's behind the doors, opens another door, say No. 3, which has a goat. He then says to you, 'Do you want to pick door No. 2?' Is it to your advantage to switch your choice?"

The answer is yes, you win twice as often if you switch. This just does not make sense to most people. The usual logic goes something like this. When presented with the original choice of one door from among three, we have a 1/3 chance of winning the car. Once there are only two doors left, most people tend to believe each one now has a ½ chance of hiding the car, and so one perceives that each door has a ½ chance of winning. This is not so.

There was a 1/3 chance the car was behind the contestant's door and a 2/3 chance it was behind one of the other two. When the host opens one of these other two doors he is providing the contestant with new information, the car is not behind door 3. The observant contestant now knows that 2/3 chance all resides in door 2. Given a choice between a 1/3 chance of winning and a 2/3 chance of winning, who would not prefer the better odds? So you should always switch in such a game.

Here is how it works. Monty Hall, the game show host, always knows where the car is, and this is critical. Let the car be behind door 1. If you correctly chose the door with the car (door 1), he can open either of the other doors to reveal a goat. If you decide to switch when you pick door 1 you will give up the car and end up with a goat. Switching is a losing strategy in this case. The score is switch 0 and stay 1.

Suppose, now, you choose door 2, which is hiding a goat. Monty, knowing the car is behind door 1, opens the third door to show you a goat. This time switching doors wins for you. The score is now switch 1 and stay 1. Now imagine you choose door 3. Monty now opens door 2 to show you the goat. Once again, switching doors means you trade a goat for a car and win. The score is now switch 2 and stay 1. No matter which door the car is behind, the logic here always leads to a score of switch winning twice and stay winning once. If you switch, you will win 2/3 of the time; if you stay with your original choice you win 1/3 of the time. You are not guaranteed a win if you switch, but you will win twice as often as you lose in the long run, and that is not intuitively obvious to many people.

This chapter provides a review of probability concepts that you'll need to understand to do basic quantitative risk assessment. It is also a good review for managers and communicators as well. The chapter does not attempt a rigorous theoretical development or treatment of probability. What it does do is offer a survey review of the kinds of concepts anyone working in risk analysis is likely to need.

11.2 Two Schools of Thought

Probability is the chance that something will or will not happen. There are two schools of thought on the nature of probability. They go by many names, but we'll call them the *frequentist* and *subjectivist* schools. The frequentist approach to probability is based loosely on the notion that true probability values are "out there" and that we can discover them through data. Specifically, we can calculate the long-run expected frequency of occurrence. The probability of an event A, P(A), is equal to n/N, where n is the number of times event A occurs in N opportunities. It is the frequency with which A occurs out of the number of times it could occur. So the annual probability of a hurricane hitting your community is estimated by the number of years a hurricane strikes out of the number of years observed. The probability of an accident per vehicle mile is the number of accidents divided by the number of vehicle miles. The frequentist view of probability works quite well with repeatable events.

There is also a subjective or degree-of-belief view of probability. This is based loosely on the notion that probability is an intrinsic phenomenon, i.e., it is a device we humans use to explain and deal with natural variability and knowledge uncertainty. Probability is not out there; it is in us. Probability is a measure of the plausibility of an event given our incomplete knowledge.

Some events have a uniqueness about them that denies the existence of frequency. There are many things that have never happened before or things that may only happen once, so we cannot estimate their probabilities with frequencies of occurrence. A subjectivist view of probability is especially useful for these unique events. Bayesians favor this view of probability.

What is the probability I will obtain a head before I toss a fair coin? It is .5; this is a frequency. We can verify this probability by observing a great many flips and counting the proportion of heads. Once I flip the coin, what is the probability it is a head? The true probability is now either 0 if it is a tail or 1 if it is a head. The true probability is uncertain, and now a degree-of-belief view of probability applies. With this simple example we have natural variation in the large, and this is amenable to the relative frequency view. We also have uncertainty in the small (a single completed toss). Thus, there is room for both schools of thought in my own approach to risk assessment. You will have to make your own choice!

Two people can have different degrees of belief about the probability of the same uncertain event and both can be right. If I believe there is an 80% chance of rain today and you believe there is a 50% chance of rain, we are both right if it rains. In fact we are both right if it does not rain. If reading that hurts your head it may be because you are beginning to understand. Probability is not easily intuited.

The mathematics of probability are pretty well settled. The two schools of thought pretty much agree on these mathematical matters. It is the philosophy of probability that causes the problems. Discussions of this philosophy can bring out an intensity of emotions unrivaled by any other topic in mathematics, an intensity often reserved for politics and religion. If you want to be a good risk manager, never seat a frequentist next to a subjectivist at a dinner party.

11.3 Probability Essentials

Probability is measured as a number between zero and one. Zero means there is no chance the event will occur, i.e., it is impossible. Numbers close to zero are, therefore describing events that are close to impossible. One means the event has happened or is sure to happen. Numbers close to one describe events that are almost certain to occur.

What then is the most uncertain number of all? Is it your chance of winning the Powerball grand prize, which is 1 in 195,249,054 (5×10^{-9})? Surely not, as that probability is pretty definitive; winning that prize is as close to impossible as you are likely to get! Small probabilities mean unlikely events; they do not convey great uncertainty, however. The most uncertain probability of all is .5. If the probability of rain is 50%, it is as likely to rain as to not rain. Once the probability shades a little one way, .500001, or the other, .499999, we can see it is slightly more like to rain than not rain or to not rain than rain, and the uncertainty slowly begins to resolve as probabilities move toward zero or one.

FREQUENTIST AND BAYESIAN MEANS

A frequentist believes a population mean is real but unknown and often unknowable. It can only be estimated from the data using confidence intervals. A Bayesian believes the population mean is an abstraction. Only the data are real.

TABLE 11.1

Sample Space of Outcomes for Tosses of One Red and One White Die

Red	White	Red	White	Red	White	Red	White	Red	White	Red	White
1	1	2	1	3	1	4	1	5	1	6	1
1	2	2	2	3	2	4	2	5	2	6	2
1	3	2	3	3	3	4	3	5	3	6	3
1	4	2	4	3	4	4	4	5	4	6	4
1	5	2	5	3	5	4	5	5	5	6	5
1	6	2	6	3	6	4	6	5	6	6	6

The "sample space" is one probability theory concept it would be unwise to overlook. Let's start with something simple like tossing two dice, one red and one white. There are exactly 36 possible outcomes, no more and no less. Table 11.1 enumerates the possibilities. The sample space is the set of all the possible outcomes. The sum of the probabilities of all the possible outcomes must equal one. This is equivalent to saying that one and only one of these dice rolls must happen if we roll the dice. A sample space must be mutually exclusive (only one of them can happen at a time) and collectively exhaustive (there is no result that is not included in the sample space).

There are 36 possibilities. Each outcome in the space has an equal chance, $1/n$, where n is the number of possible outcomes. In this case the probability of any one outcome is 1/36. We can define an event as rolling a six, and using a frequentist approach we can see five different ways of doing this, each with a probability of 1/36. So the probability of this event is 5/36.

Now let's use an event tree to identify a sample space. Figure 11.1 shows an example for tossing three coins in sequence. The figure shows how the sample space (circled) is derived through the various pathways. All possible outcomes are shown; there are no other possibilities. The probability of each triplet outcome is equally likely, $1/n$. This time, n is 8, and the probability of three heads (HHH) is 12.5% or 1/8. We can again count up different outcomes to define events such as getting two heads. There are three ways to do that, each with a probability of 12.5%, so the probability of the event two heads and one tail is 37.5% or 3/8.

Event trees are handy risk models for a wide variety of problems. The endpoints of such a model define the sample space for the risk problem. The endpoints need not be equally likely in a risk problem, but the sum of the probabilities of all endpoints must sum to one, just as they do here.

Note that Figure 11.1 shows a sample space constructed of numerous pathways comprising nodes and branches. When we arrive at a node, something has to happen. That is, we move forward on one branch or another. Thus, all branches must be mutually exclusive and collectively exhaustive. The probabilities of all the branches coming out of a single node must also sum to one, because something has to happen at each node, and the branches define all the possibilities. So an event tree, in this sense, has sample spaces within sample spaces. The endpoints of each sample space must have probabilities that sum to one.

When using tree models to represent risks, the same simple probability rules hold. Identifying the mutually exclusive and collectively exhaustive set of endpoints that will define the sample space is not often a simple exercise in counting as it is with

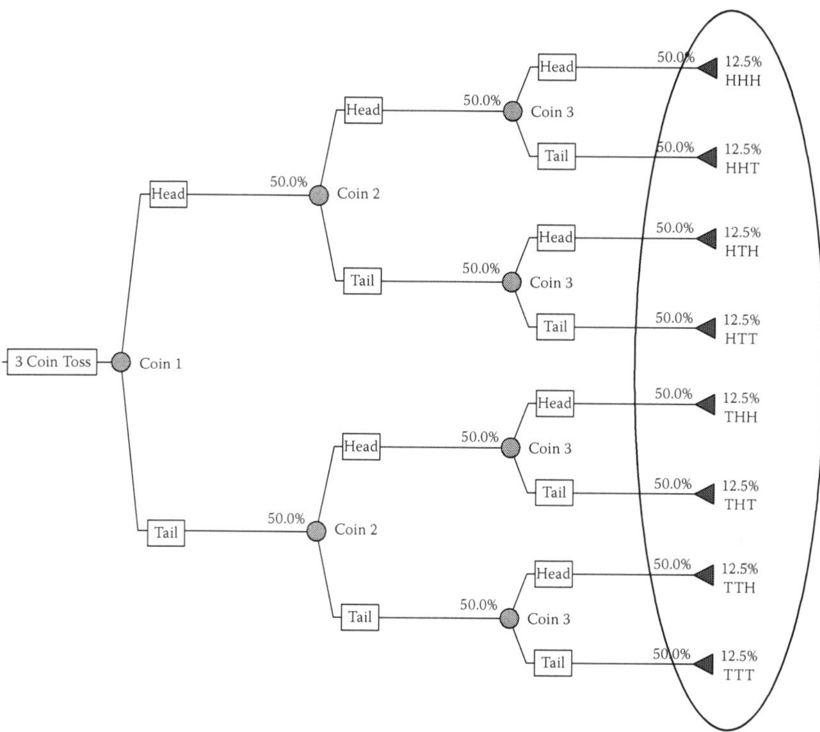

FIGURE 11.1 Event-tree sample space for a three coin toss experiment.

dice and coins. Nonetheless, it is important to understand that such models are just defining a sample space of outcomes of interest to risk managers. These models must obey the mathematical laws that order probability. It is not okay to simply build a model and start filling in numbers. The numbers must behave properly and follow the laws of probability.

Figure 11.2 shows a simplified and hypothetical earthquake risk for a structure. There are four possible endpoints for this simple risk model. The probabilities of these four endpoints must sum to one.

Each node in an event tree identifies a new event, and each has its own unique probability, as seen in Figure 11.2. The first event is that the soil does (30%) or does not (70%) liquefy as the result of an earthquake. If it does, the second event is that the structure does (80%) or does not (20%) crack. If the result of the first event is no liquefied soil, then the probability of cracking (60%) and not cracking (40%) change because of different precedent conditions. Unlike the coin example, where each event is independent of every other event and probabilities are constant, the world of risk is full of dependencies. Preceding events and conditions usually influence the probability and nature of antecedent events. This leads to a sample space where the outcomes are not all equally likely.

The most likely outcome in the sample space is structure cracking with no liquefiable soil. The least likely outcome is no cracking when the soil does liquefy. There are principles and facts that underlie the probability values in a risk model like the event tree of Figure 11.2.

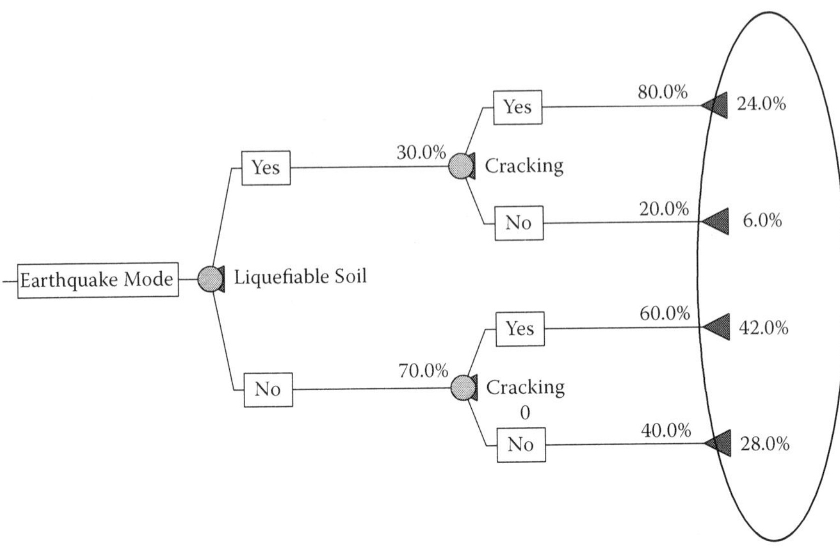

FIGURE 11.2 Event-tree sample space for structure cracking after an earthquake.

Probabilities can be expressed as a decimal, percentage, fraction, or odds. Using the model in Figure 11.2, the probability the soil will liquefy and the structure will crack can be expressed as follows:

- Decimal = .24
- Percentage = 24%
- Fraction = 24/100 = 6/25
- Odds = 6:19 (x:y based on $x/[x + y]$)

In the United States, the most common way of expressing probabilities may be odds. Games of chance and gambling are at the root of much of our knowledge about probability, and gamblers understand odds. In risk analysis the other three forms are more commonly used.

Decimals are sometimes preferred when speaking about the probability of a single event, e.g., the probability of a single egg containing *Salmonella* Enteritidis is about 5×10^{-5} (FSIS 1998). Percentages are sometimes preferred to convey information about a population, e.g., 0.005% of all eggs contain *Salmonella* Enteritidis. Fractions are another popular alternative for conveying risk information to the public, e.g., about 1 in 20,000 eggs contain *Salmonella* Enteritidis.

11.4 How Do We Get Probabilities?

How do we manage to identify these numbers between zero and one? Where do they come from? There are three basic ways to estimate probabilities, although these three ways really just reflect the two schools of thought on probability mentioned earlier.

RANK AND ODDS OF POKER HANDS

Royal flush	1:649,739
Straight flush	1:64,973
4 of a kind	1:4,164
Full house	1:693
Flush	1:508
Straight	1:254
3 of a kind	1:46
2 pair	1:20
1 pair	1:1.37
High card	1:1

Classical or analytical probabilities are mathematical calculations that are used by both schools. I like to describe analytical probabilities as the kinds of probabilities that "smart people" can do with pencil and paper or a calculator. When you open a new deck of cards and find the odds of being dealt different poker hands (see text box) before the draw, these probabilities have been calculated analytically.

Analytical probabilities rely on identifying sample spaces and taking subsets of them. Combinatorics like the factorial rule of counting, permutations ($n!/[n-r]!$), combinations ($n!/(r![n-r]!$), and other counting techniques are used to estimate probabilities of specific events, like a full house. Unfortunately, risk assessors do not too often get the opportunity to work with analytical probabilities, although random processes like the binomial, Poisson, and hypergeometric processes do provide opportunities to calculate these analytical probabilities at times. So, although analytical probabilities occupy a great deal of our higher education exposure to probabilities, they do not come up as often as we would like when doing risk assessment.

A second source of probabilities is empirical or frequentist probabilities. These are based on observation. How many times did the event of interest happen out of the number of times it could have happened? Think of a traffic light near your home. What is the probability you will catch it red the next time you encounter it? All we need to do is keep a little pad of paper on the seat of the car. Make a strike mark under "red" each time it is red and under "not red" when it is not. After a hundred or so observations, we can calculate the relative frequency of a red light. This is now our approximation of the true probability of catching the light red. As we record more observations over the years, our estimate gets better. A relative frequency is nothing but an estimate of the true probability for a frequentist. It is but data to a Bayesian.

Empirical probabilities are especially useful when the process of interest is repeated many times under the same circumstances. Empirical probabilities are good for estimating the reliability of electrical components, stream flows, causes of death, probabilities of cancer from animal toxicity studies, and the like. This makes relative frequencies or frequencies one of the most common sources of probabilistic information for risk assessment.

The third source of probability estimates is subjective probabilities. Subjective probability is based on evidence and the experience of the estimator. It relies heavily on expert opinion. It is most useful when we deal with the uncertainty that surrounds events that will occur once or that have not yet occurred.

Subjective probability estimates are especially useful for filling in data gaps and supplementing data with experience and judgment. Risk assessors must deal with different kinds of events as well as different levels of data and information about these events. Subjective probabilities are most useful for those unique events for which there are no relative frequency data and for which analytical calculations are not possible.

Suppose we want to estimate the probability that the channel bottom is 30% or more rock, or that there will be structural damage to a building if an earthquake less than 6.2 on the Richter scale occurs? We might want to estimate the probability of a fatal accident on a dangerous curve if the curve is redesigned and eased. We might need to estimate the probability of illness from a low-dose exposure to pathogenic bacteria, and so on. These lend themselves well to subjective probability estimates.

11.5 Working with Probabilities

If it was as simple as the previous material might suggest, anyone could work with probabilities. It is not that simple. There are rules and theories that govern our use of probabilities. Estimating probabilities of real situations requires us to think about complex events and to apply these rules carefully. Most of us do not naturally assess probabilities well. Hence, it is critical to good quantitative risk assessment that you have people who can work effectively with probabilities. Some of the fundamental axioms and rules of probabilities are described below.

11.5.1 Axioms

Probability density functions and cumulative probability distribution functions (these ideas are discussed at length in Chapter 12) and their properties are all essentially developed from three fundamental axioms of probability. An event, E_i, is anything for which you want to know the probability. S is the sample space that includes all events of interest. The probability of an event, $P(E_i)$, is defined such that:

1. $0 \le P(E_i) \le 1$
2. $P(S) = 1$
3. If A and B are mutually exclusive events, then P(A or B) = P(A) + P(B)

The first axiom means the probability of an event is a nonnegative real number between 0 and 1. The second axiom says the probability that some event in the sample space will occur is 1, and there are no events that can occur that are outside the sample space. The third axiom says the probability of two mutually exclusive events occurring together is the sum of their individual event probabilities. P(A or B) is the probability that A or B occurs.

TABLE 11.2

Levee Ownership and Maintenance Condition

	Inadequate Maintenance	Adequate Maintenance	Total
Private	80	20	100
Locally constructed	50	50	100
Federal construction	10	90	100
Total	140	160	300

11.5.2 Propositions and Rules

In addition to these axioms, there are some consequences of these axioms that prove to be essential for quantitative risk assessment.

11.5.2.1 Marginal Probability

A marginal probability is the probability of a single event, P(A) (read, the probability of A), where A stands for any event whose probability we want to know. To illustrate this and the following concepts, we'll use the data in Table 11.2. The hypothetical data shows ownership of the levees in a region in the rows and the maintenance condition of those levees in the columns.

Two examples of marginal probabilities are shown as follows:

$$P(\text{Private}) = 100/300 = .333$$

$$P(\text{Adequate}) = 160/300 = .533$$

11.5.2.2 Complementarity

The rule of complementarity ensures that the probability of an event and its complement in the sample space sum to one. The probability that event A does not happen (~A means not A) is:

$$P(\sim A) = 1 - P(A) \quad (11.1)$$

Two examples follow:

$$P(\sim\text{Private}) = 1 - .333 = .667$$

$$P(\sim\text{Adequate}) = 1 - .533 = .467$$

11.5.2.3 Addition Rules

There are many times when assessors will be interested in more than one event occurring. For two different events, A and B, the probability that A or B or both occur is defined by the addition rule. The form of the rule depends on whether A and B are mutually exclusive or not. When events are mutually exclusive they cannot occur simultaneously. When they are not they may both occur.

The addition rule for mutually exclusive events is:

$$P(A \text{ or } B) = P(A) + P(B) \tag{11.2}$$

An example is:

$$P(\text{Private and Local}) = 100/300 + 100/300 = 200/300 = .667$$

The rule works the same for N mutually exclusive events.

$$P(A_1 \text{ or } A_2 \text{ or} \ldots A_N) = P(A_1) + P(A_2) + \ldots + P(A_N) \tag{11.3}$$

The addition rule for nonmutually exclusive events is:

$$P(A \text{ or } B) = P(A) + P(B) - P(A \text{ and } B) \tag{11.4}$$

Because the events can occur at the same time we must avoid double counting outcomes. For example, consider the probability a levee is both private and inadequately maintained. There are 100 private levees and 140 inadequately maintained levees. There are eighty levees that are both, so they are counted twice and we must subtract out these joint events.

An example is:

$$P(\text{Private and Inadequate}) = 100/300 + 140/300 - 80/300 = 160/300 = .533$$

The rule works for *N* nonmutually exclusive events, but all multiple counts of the same elements must be accounted for.

11.5.2.4 Multiplication Rules

Multiplication rules apply when we are interested in the probability that two things occur together. The proper formula depends on whether the events A and B are independent or dependent events.

Two events, A and B, are **independent** if the fact that A occurs does not affect the probability of B occurring. When we have independent events, the probability of both occurring, i.e., the multiplication rule, is:

$$P(A \text{ and } B) = P(A)P(B) \tag{11.5}$$

The probabilities in Table 11.2 are dependent, so we must vary the example for a moment, and then we will return to the table. In Figure 11.3 let a head for coin 1 be event A and a head for coin 2 be event B. They are independent events. The result of the first coin toss has no effect on the probability of the outcome of the second coin toss. Note that the probability of a head is the same regardless of what the result of the first coin toss was. In this example of independent events:

$$P(A \text{ and } B) = P(\text{coin 1 is a head and coin 2 is a head}) = (.5)(.5) = .25 \text{ or } 25\%.$$

Two events, A and B, are dependent if the outcome or occurrence of A affects the outcome or occurrence of B so that the probability of B is changed. The earthquake event tree of Figure 11.2 is an example of dependent events. We saw that liquefiable soil made structure cracking more likely. If A and B are dependent events, then the probability of both occurring is:

$$P(A \text{ and } B) = P(A)P(B \text{ after } A) \tag{11.6}$$

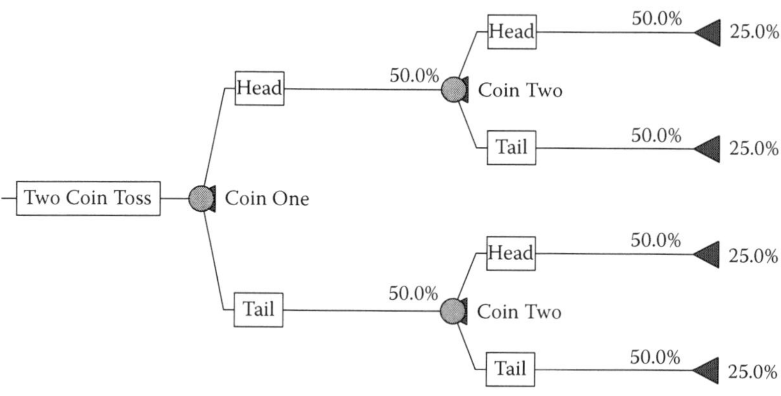

FIGURE 11.3 Probability tree for a two coin toss experiment.

P(B after A) can also be written as P(B|A) (read, the probability of B given that A has happened), and then the multiplication rule for dependent events is rewritten as:

$$P(A \text{ and } B) = P(A)P(B|A) \tag{11.7}$$

We can now return to our example using Table 11.2. However, let's display it as the probability tree in Figure 11.4. Note that the probability of inadequate maintenance varies depending on ownership. If ownership and maintenance were independent, the probability of inadequate maintenance would be the same for each kind of ownership and it would equal the marginal probability P(Inadequate) = .467.

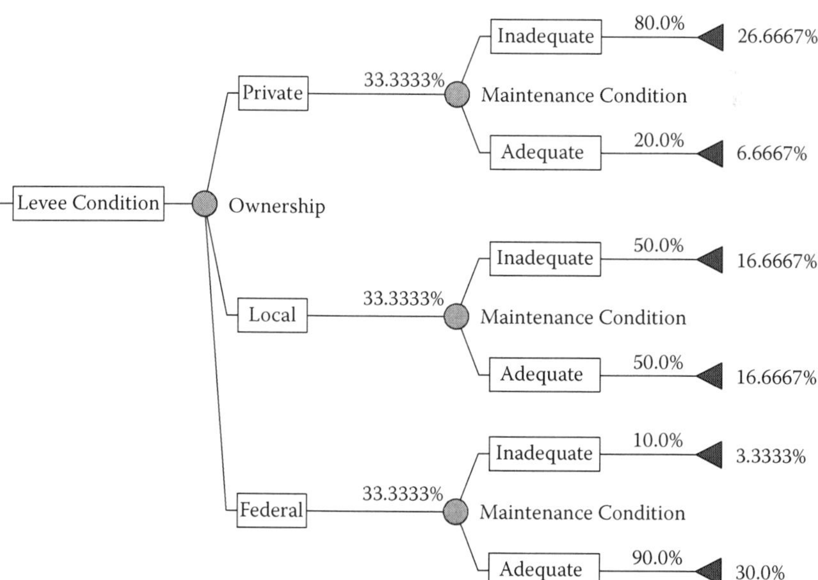

FIGURE 11.4 Probability tree for levee ownership and maintenance condition.

The P(private and inadequate) can be read directly from the tree: it is .267. If we look at the table, it is even easier to find. There are 80 levees that are both private and inadequate out of a total of 300 levees, 80/300 = .267.

Unfortunately, most probability problems do not come with a neatly worked out table or probability tree. That is when the formula becomes important. The term P(B|A) in equation (11.7) is called a conditional probability. It is the subject of the next section.

11.5.2.5 Conditional Probability

Information changes probabilities. We see that from the previous discussion. The P(Inadequate) = .467, but when provided with information about ownership, the probability that a levee is inadequate changes. To reflect this fact we use conditional probability.

The probability that event A happens, given the condition that event B happens, is written P(A|B). In our examples above we could say:

$$P(\text{Inadequate}|\text{Private}) = .8$$

$$P(\text{Inadequate}|\text{Local}) = .5$$

$$P(\text{Inadequate}|\text{Federal}) = .1$$

Let's revisit the multiplication rule for dependent events using the conditions of inadequate maintenance and private ownership.

$$\begin{aligned} P(\text{Inadequate and Private}) &= P(\text{Inadequate})P(\text{Private}|\text{Inadequate}) \\ &= (140/300)(80/140) = 80/300 = .267 \end{aligned}$$

P(Private|Inadequate) is a conditional probability, and it is a switch of the earlier conditional probability P(Inadequate|Private) shown previously. Using Table 11.2, we can see the condition that the levee is inadequately maintained in the second column. This tells us the levee of interest is one of the 140 levees in that column. We are no longer dealing with all 300 levees. Thus, the conditional information changed the probability of a private levee from 100/300 to 80/140. Information changes probabilities.

Conditional probabilities can also be defined with a formula. Starting from the multiplication rule for dependent events we have:

$$P(A \text{ and } B) = P(A)P(B|A)$$

Rearranging we obtain

$$P(B \mid A) = \frac{P(A \text{ and } B)}{P(A)} \tag{11.8}$$

Thus, substituting into equation (11.8), we get:

$$\begin{aligned} P(\text{Private}|\text{Inadequate}) &= P(\text{Inadequate and Private})/P(\text{Inadequate}) \\ &= (80/300)/(140/300) = 80/140 = .571 \end{aligned}$$

11.5.2.6 Bayes' Theorem

Building on the notion that information can change probabilities, we introduce Bayes' theorem, which is useful for updating probabilities on the basis of newly obtained information. Often we begin with an initial or prior probability that an event will occur. Then, as uncertainty is reduced or new information comes in, we revise the probability to what we call the posterior probability. This revision can be done using Bayes' theorem.

Bayes' theorem is

$$P(A \mid B) = \frac{P(B \mid A)P(A)}{P(B)} = \frac{P(A \text{ and } B)}{P(B)} \tag{11.9}$$

To illustrate the theorem, let's vary the example and consider the hypothetical event that a randomly selected crate of imported produce has some form of pathogenic *E. coli* to be .001. Thus,

P(Ec) = .001 and P(~Ec) = .999

Suppose a diagnostic test can accurately detect *E. coli* 99% of the time. Further assume that 5% of crates that do not have *E. coli* will test positive. That is,

P(+|Ec) = .99 and P(+|~Ec) = .05

Now suppose the test is administered to a randomly selected crate of imported produce, which may or may not have pathogenic *E. coli* on it, and the test is positive. What is the probability that the crate has pathogenic *E. coli*? Using the concepts we've developed, we are looking for the probability, P(Ec|+).

This is sometimes called the Bayesian flip because it is the opposite of the known probability P(+|Ec). Note also that we know P(Ec), which is the prior probability that is being updated with new information, i.e., that the crate tested positive. Substituting into Bayes' theorem in equation (11.9) we are calculating

$$P(Ec \mid +) = \frac{P(+ \mid Ec)P(Ec)}{P(+)} = \frac{P(Ec \text{ and} +)}{P(+)} \tag{11.10}$$

Substituting the available values we get

$$P(Ec \mid +) = \frac{(.99)(.001)}{P(+)}$$

So far we do not know P(+). There are two ways a crate can test positive: if it has *E. coli* and tests positive or if it has no *E. coli* and tests positive. These two possibilities are defined as:

$$\text{P(+) = P(Ec and +) + P(\sim Ec and +)} \tag{11.11}$$

Recall from equation (11.7) that P(A and B) = P(A)P(B|A), enabling us to rewrite this as

$$\text{P(+) = P(Ec)P(+|Ec) + P(\sim Ec)P(+|\sim Ec)} \tag{11.12}$$

Substituting, we have:

$$P(Ec \mid +) = \frac{(.99)(.001)}{P(Ec)P(+ \mid Ec) + P(\sim Ec)P(+ \mid \sim Ec)}$$

$$P(Ec \mid +) = \frac{(.99)(.001)}{(.001)(.99) + (.05)(.999)} = .019$$

Although the prior P(Ec) = .001, we now have an updated probability, conditioned on the knowledge that this crate has tested positive, and we see the probability that it is actually contaminated is only .019. This is the posterior of Ec given the positive test result. This is essentially telling us that if all crates were tested, only 1.9% of those that tested positive would actually be contaminated. That means that 98.1% of all positive testing crates would actually be free of pathogenic Ec. This somewhat surprising result is because so few crates are actually contaminated. Most crates are free of the organism, and these yield false positives at a low rate but in rather large numbers.

To see this, suppose we import 1,000,000 crates. At a rate of .001, we have 1,000 that are contaminated, and our test picks up 99% of them. So we have 990 true positives. But 999,000 are free of contamination and 5% of them show up positive: That is 49,950 false positives. There is a total of 50,940 positive tests, but only 990 or 1.9% of them are actually contaminated. This is a powerful argument against 100% inspection. We would be destroying a lot of good product.

There is an entire body of statistics based on Bayes' theorem. It is sometimes controversial if the prior probability is based on subjective considerations. This example began with a probability imbued with some credibility. When there is little to no data to support a prior probability, some people become uncomfortable with this approach.

11.6 Why You Need to Know This

You can't build valid models unless you know and follow the axioms and propositions of probability. It is essential to understand and obey the rules of probability when building models. It is not acceptable to simply apply probabilities in an uninformed fashion. It is not possible to do credible probabilistic risk assessment without a careful knowledge of and adherence to the rules of probability.

Consider the levee ownership model once more, this time as seen in Figure 11.5. Every rule discussed here is employed to build this simple model. Complementarity must be observed on the branches from a node and in the model endpoints. Probabilities of endpoints are determined via multiplication. Probabilities of dependent events rely on conditional probabilities. Probabilities throughout the model, but especially for endpoints, can be added to obtain other probabilities of interest. And this is a simple model. A risk assessor cannot know too much about probability.

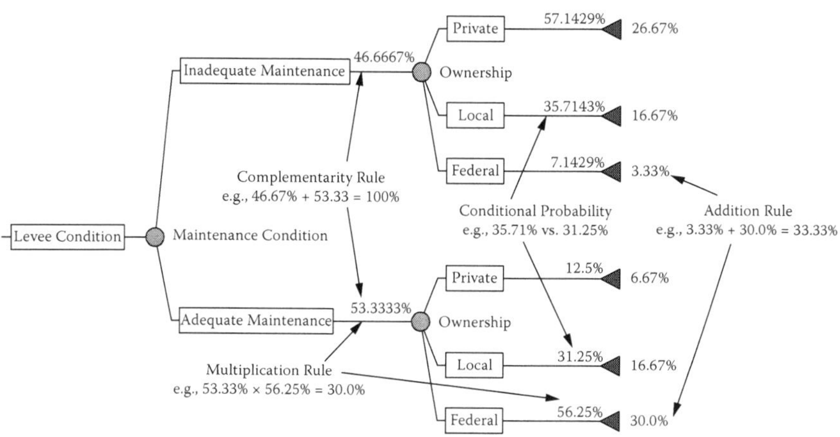

FIGURE 11.5 Levee-condition event tree showing application of rules of probability calculation.

11.7 Summary and Look Forward

Probability is the language of uncertainty. It is used to characterize our knowledge uncertainty and to describe the natural variability in the universe. Someone on the risk assessment team has to understand the basic laws, axioms, propositions, and rules of probability. A brief review of these essentials is presented here.

Chapter 12 is devoted to helping the self-taught or trained-on-the-job risk assessor to accomplish one of most challenging tasks in quantitative risk assessment: choosing the right probability distribution. This is a subject that is not treated very comprehensively in the literature. A nine-step process is offered and illustrated in Chapter 12. Chapter 13 follows with a discussion of subjective probability elicitation.

REFERENCES

FSIS. 1998. Department of Agriculture. Salmonella Enteritidis Risk Assessment Team. Salmonella *Enteritidis risk assessment: Shell eggs and egg products.* Washington, DC: Food Safety Inspection Service.

Whitaker, Craig F. 1990. Letter in "Ask Marilyn" column. *Parade Magazine.*

12

Choosing a Probability Distribution

12.1 Introduction

Quantitative risk assessment requires probability estimation, either explicitly or implicitly. If you're going to be involved in risk analysis, you've got to be comfortable with probabilistic thinking. To do probabilistic risk assessment well, someone on the team has to be proficient in choosing the appropriate probability distributions to describe data, model natural variability, and represent knowledge uncertainty about model inputs for the problems you are addressing. One of the most confounding tasks, for new and experienced risk assessors alike, is learning how to choose the right probability distribution for a risk assessment.

Choosing a probability distribution is sometimes called choosing a probability model; this is not to be confused with building a probabilistic model. This chapter presents a strategy to help you to choose probability distributions when you have some information about the quantity you are modeling. Information may be recorded in the form of data; it could be general or specific knowledge; or it might be experiential wisdom or professional judgment. The strategy described in this chapter is primarily for situations where you have some data, but it should help you in any situation where you must choose a probability distribution.

The chapter begins by considering different graphs used to display probability information. This is followed by a nine-step strategy for selecting a probability distribution. It includes a simple method for developing an empirical distribution as well as steps to take when an empirical distribution will not do. Two example applications of the strategy are presented before the chapter ends by presenting and briefly describing several distributions that have proven useful to risk assessors.

12.2 Graphical Review

Graphics are a principle means of conveying probabilistic risk information. Both risk assessors and managers need to be comfortable reading and interpreting probability graphs. So we begin by reviewing the most common probability graphs, i.e., the probability density function, cumulative distribution function, survival function, probability mass function, and the cumulative distribution function for a discrete random variable. These graphs gives us more options for thinking about probability distributions and communicating the meaning of our risk assessment results, the topic of Chapter 17.

12.2.1 Probability Density Function

The probability density function (PDF) for a continuous random variable is represented by a curve such that the area under the curve between two numbers is the probability that the random variable will take a value between those two numbers. Consider Figure 12.1. The PDF shown represents the distribution of the population of battery lives measured in hours in a population that is normally distributed with a mean life of 70 hours and a standard deviation of 4 hours.

Note that lifetimes of 62.16 hours and 77.84 hours are identified by the delimiters and that 95% of all batteries have lifetimes between these two values. Because there is an infinite number of values for this continuous variable, the probability of any one specific value cannot be read directly from this graph. The probability of obtaining any one of an infinite number of possible values approaches zero. It is important, therefore, to understand that the numbers on the vertical scale of PDF curves have no useful probability meaning. The vertical numerical scales are simply for scaling the figure and are not probabilities, despite the vertical axis label.

The total area under a density function equals 1. This is because the curve covers all the possible values, and one of them must occur. The sum of the probabilities of all possible values (i.e., the sample space) sums to 1, and this figure represents all those possibilities.

It is also possible to estimate probabilities of open-ended intervals. For example, there is a 2.5% probability a battery will last 62.16 hours or less and a 2.5% probability it will last 77.84 hours or longer. The shapes of a PDF can vary widely, depending on the mathematical form of the function that underlies it and the function's parameter values. The depiction of the distribution of a sample from a population is more appropriately called a histogram or relative frequency distribution.

Informally, you can think of a PDF as a picture of how a population is distributed. This is sometimes a helpful way to explain a distribution to those with little quantitative background. The distribution shows that the values of a variable fall along a specific

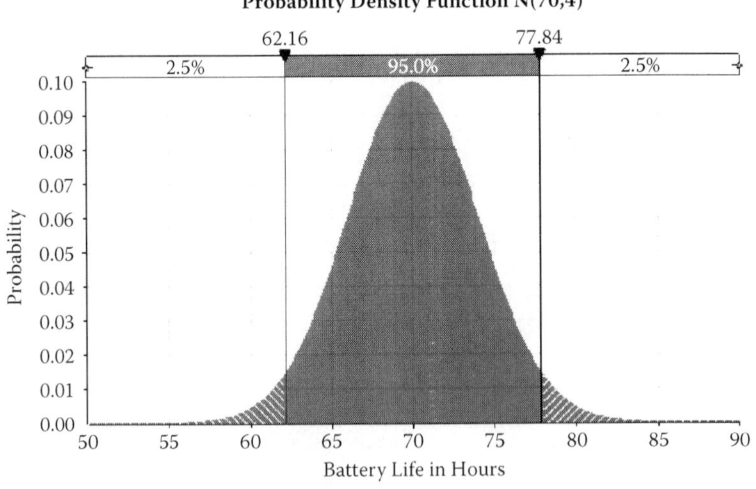

FIGURE 12.1 Probability density function of battery life in hours.

segment of the number line. The height of the distribution (the fat parts) shows which values come up most often. The low or skinny parts of the graph indicate that these values occur less often. Figure 12.1 suggests that your battery will most likely last somewhere close to 70 hours. It also suggests it would be very unusual for your battery not to last 55 hours, and it would be just as unusual for it to last more than 85 hours.

12.2.2 Cumulative Distribution Function

The continuous data in the PDF can also be displayed as the cumulative distribution function (CDF), seen in Figure 12.2. It is mathematically equivalent to the PDF; the PDF is the integral of the CDF and the CDF is the derivative of the PDF. The horizontal axis remains unchanged. Unlike the PDF, the vertical axis of the CDF has a meaningful scale. The vertical axis now represents a cumulative probability from 0 to 1. It may be easier for some to think of the vertical axis in terms of percentiles.

There are two ways to read this distribution. Beginning from the horizontal axis, choose a value of interest, say 62.16 hours (bottom horizontal arrow), and note its corresponding probability on the vertical axis. This value was conveniently chosen because its probability is also shown at the top of the graph, .025. The population interpretation of this value means that 2.5% of all batteries last 62.16 hours or less. The individual element interpretation is that the probability a random battery will last 62.16 hours or less is 2.5%. A third interpretation is that 62.16 hours is the 2.5 percentile value for the population.

The second way to read the curve is to choose a probability value on the vertical axis, find its intersection on the curve (top horizontal arrow), and read the corresponding variable value from the horizontal axis. In this case the 97.5 percentile, i.e., the cumulative probability of .975, value is 77.84 hours. Many CDFs have a more or less exaggerated S shape to them. Because the vertical axis has a meaningful scale, some people find this curve easier to work with to obtain probability

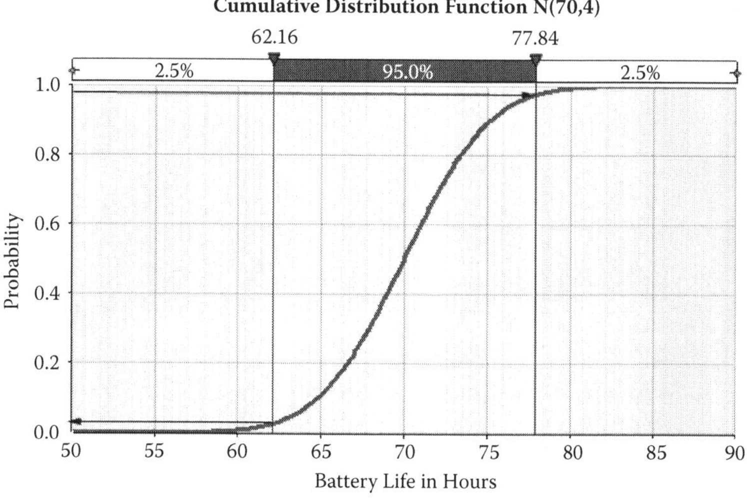

FIGURE 12.2 Cumulative distribution function of battery life in hours.

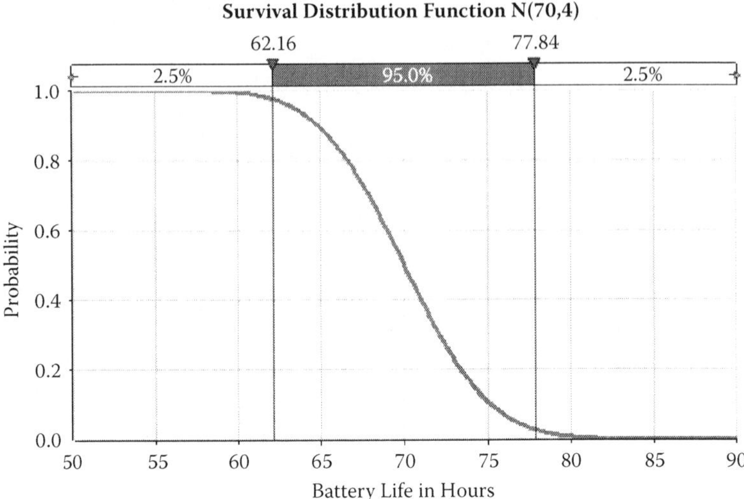

FIGURE 12.3 Survival function of battery life in hours.

estimates. It does not, however, reveal much about the shape of the distribution to the untrained eye.

12.2.3 Survival Function

The CDF curve in Figure 12.2 is sometimes called an ascending cumulative distribution. The same data can be plotted in reverse order as a descending cumulative distribution function, which is also called the survival function. An example is presented in Figure 12.3. The interpretation of the graph of continuous data changes in a nuanced way now. Look at the delimited value of 62.16 hours. The curve now says 97.5% of all batteries survive at least 62.16 hours. By contrast, only 2.5% of all batteries survive up to 77.84 hours. You can see batteries "dying off" as battery life is extended. Thus, this curve reveals the likelihood of surviving to a certain age.

The interpretation of the vertical axis has changed. In the CDF it is a cumulative probability, whereas the survival function is 1 – CDF. Choosing the value 62.16 hours, the CDF of Figure 12.2 says 2.5% last 62.16 hours or less. With the survival curve, the vertical axis is no longer a cumulative probability; it is now a survival rate. At a value of 62.16 hours, we now see that 97.5% of all batteries survive at least this long. The survival function is often preferred for issues that involve failures.

12.2.4 Probability Mass Function

A discrete random variable does not have a probability density function; it has a probability mass function (PMF). There are two major differences between a PDF and a PMF, the first being that a discrete random variable has a finite number of variable values. As a result, the second difference is that the vertical dimension of a PMF is a meaningful measure of probability. It shows the probability a specific value will occur.

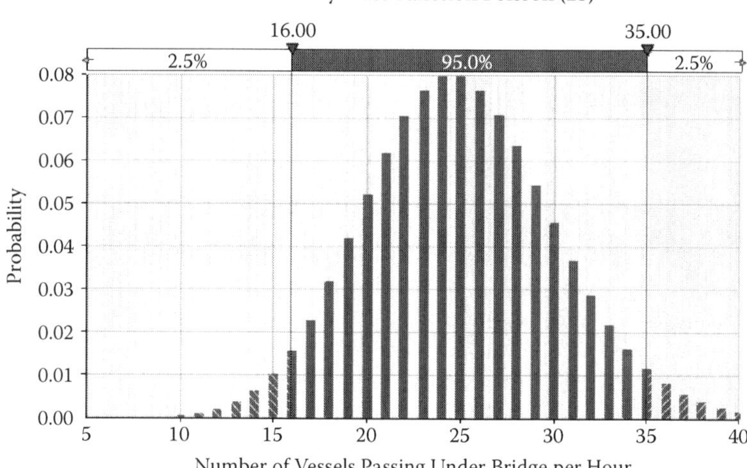

FIGURE 12.4 Probability mass function of number of vessels passing under bridge per hour.

The example in Figure 12.4 shows a Poisson distribution with a mean (lambda) of 25 watergoing vessels that pass beneath a bridge in an hour. The probability that exactly 15 vessels will pass beneath the bridge can be read directly from the vertical axis; it has a probability of .01. A population interpretation means 1% of all hours have exactly 15 vessels passing. Alternatively, a randomly chosen hour has a 1% chance of having exactly 15 vessels pass under the bridge. It is also possible to calculate ranges of values as with the PDF. We can say 2.5% of all hours have fewer than 16 vessels. This is the sum of the individual probabilities of all counts of 15 or less. A sample from the population of a discrete random variable when graphed in this fashion is also called a histogram or relative frequency distribution.

12.2.5 Cumulative Distribution Function for a Discrete Random Variable

Cumulative ascending (shown in Figure 12.5) and descending (not shown) distribution functions can also be generated for discrete random variables. They are read and interpreted as described previously for the continuous random variable. The shape of the step function is typical for a discrete random variable. As the number of discrete values rises, the steps get smaller and the curve will smooth out and begin to look as smooth as a CDF for a continuous variable.

Armed with this review information, let's now turn our attention to how to choose a probability distribution to represent a quantity in a probabilistic risk assessment. A distribution can be displayed in any of the ways described previously. The shape is most revealing in the PDF and PMF view. The data, especially probabilities, are easier to read in the CDF and survivor function views.

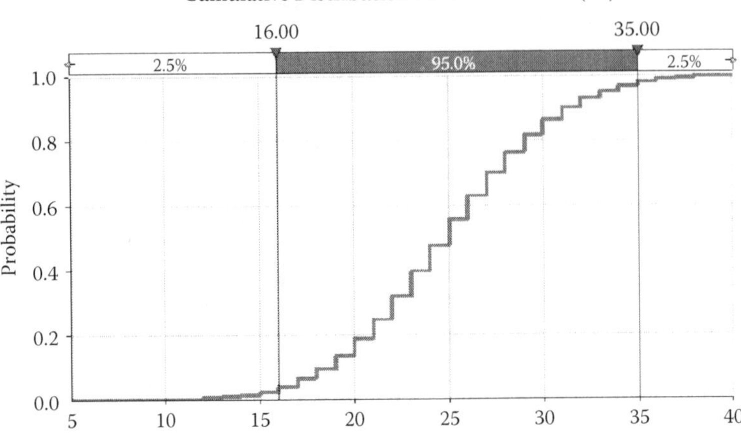

FIGURE 12.5 Discrete cumulative distribution function of number of vessels passing under bridge per hour.

12.3 Strategy for Selecting a Probability Distribution

When I began to do risk assessment I was not sure how to choose the proper probability distribution to represent the quantities in my models. I searched high and low for that definitive procedure that would guarantee I had not made a mistake. Guess what? It does not exist, and you will not find it here, either. I have managed, however, to develop a checklist for selecting a distribution. It was compiled over a number of years from a variety of sources, including my own imagination. It offers a sound basic strategy for choosing a probability distribution. Whether you use all, some, or none of this checklist for your own work it is essential that you document your rationale for selecting the probability distributions you do use.

Let's begin with the situation where you have some data to help you represent the natural variability or knowledge uncertainty in a quantity. Now you want to define a probability distribution to use in your risk assessment model. First, you should think critically about your quantity's uncertainty along the lines presented in Chapter 2. If you have some data, you are most likely modeling natural variability in your variable, although you could be representing your beliefs about the remaining knowledge uncertainty about your value, or possibly some combination of the two. With this as our starting premise, here are a series of steps to follow in order to select a probability distribution to use in your model. Each step is discussed in more detail in the pages that follow.

1. Use your data if possible.
2. Understand your variable and data.
    a. What is the source of your data?
    b. Is your variable continuous or discrete?
    c. Is your variable bounded or unbounded?

   d. Do your data have a parametric or nonparametric distribution?
      i. Do you know the parameters of your data (first or second order)?
   e. Are your data univariate or multivariate?
3. Look at your data—plot it.
4. Use theory.
5. Calculate some statistics.
6. Use previous experience.
7. Try distribution fitting.
8. Use expert opinion.
9. Do a sensitivity analysis.

12.3.1 Use Your Data

Many quantitative risk assessment models use simulation. Data can be used directly in a simulation or they can be used to define a probability distribution. Let us consider both these situations.

Some simulations use raw data values directly without any distribution. For example, if the data represent the time required to travel a certain distance or route in a model, then one of the data values is used each time a travel time is needed in the simulation model. In some cases, the data may be used in a very precise order, perhaps the order in which it was recorded. In other instances the data may be used in a different or random order, but no values will be used that are not members of the data set. This is sometimes called a trace-driven simulation (Law and Kelton 1991). This is the most literal meaning of "use your data."

An alternative to a trace-driven simulation is to use your data to define an empirical distribution. If you have data that are reasonably extensive and representative of the population of interest, consider just using your data to define the distribution. You do not have to choose one of the standard probability distributions if you use your data to construct an empirical distribution. In such a case, each time you need a value for your simulation, a travel time value would be sampled from the empirical distribution you have defined. Some of these values may be identical to the values in your data set, and others will be interpolations or extrapolations of the actual values you have collected.

One common problem encountered in constructing an empirical distribution, even with extensive data sets, is bounding the data. It's not unusual to be unsure about the absolute maximum and minimum values for a population. Consider the problem of estimating the minimum and maximum possible weights for an adult (18 years or older) male. It is not always easy to bound your data, but it is always important to do so when you use an empirical distribution.

Any data set can be converted into an empirical distribution using the cumulative distribution. This is a valuable skill to have. Consider the data in Table 12.1. The data set is kept small for the convenience of the example. Let it measure consumption of a chemical food additive in mg/kg of body weight daily for a lifetime (bw/d/l). For the purposes of the sample calculation, let's assume they are representative of the population of interest to the risk assessor who wants to eventually estimate the following relationship:

$$\text{Consumption}_i \times \text{Body weight}_i = \text{Total intake}_i \tag{12.1}$$

TABLE 12.1

Hypothetical Consumption Data for a Food Additive Measured in mg/kg bw/d/l

Sample Data					
33.3	44.2	33.4	0.9	16.2	11.7
22.2	40.2	16.5	24.5	24.9	6.0
25.8	34.7	3.6	5.0	22.7	23.2

Note: bw/d/l = body weight per day for a lifetime.

Imagine this as a simple spreadsheet model, where there is uncertainty about both consumption and body weight. We want to create an empirical distribution for consumption that can be inserted into our risk assessment model by creating a cumulative distribution function. The first step is to decide whether your consumption data are representative of the population. The simplest way to think about this question is to consider whether the shape of a plot of your data would reasonably mimic the shape of the entire population distribution. Although 18 observations would never likely be considered representative of the population, let us suspend disbelief for the sake of keeping the data set small enough to demonstrate the technique compactly and assume that it is representative. It is important to consider whether your data are extensive enough as well. Is it possible to observe values for your variable outside the range of your data? In other words, do your data likely represent the true minimum and maximum values for your variable? If they do not, you should enter a minimum and maximum bound, because extreme events are often of great interest in a simulation.

Suppose the example data do not contain the true minimum and maximum values. Then we must provide them. Let's estimate that zero consumption is a logical minimum and 60 mg/kg/bw/d is a reasonable maximum value. Using Excel or any similar tool, we add the minimum and maximum values to the data set, then sort and number the data starting at zero, as seen in Table 12.2. This index number (i) is then used to calculate the cumulative distribution probabilities using the formula "i/maximum index number."

This procedure works for a data set of any size. Note that the third column shows the cumulative probability percentile value for the corresponding data value. Thus, $x = 16.5$ is the 36.8 percentile value. In other words, 36.8% of all values in the empirical data set are 16.5 or less.

All that remains is to graph the curve as seen in Figure 12.6.* As an empirical distribution, it tends to lack the S shape seen previously. Real-world data do not always follow the neat mathematical form of a well-defined function. Reality is splendidly messy.

The PDF can be easily generated from the CDF, as shown in Figure 12.7. Note that this distribution does not resemble any of the familiar distribution shapes. If this is how the world behaves and how the real data look, then this is the distribution to use. Do not feel compelled to fit your data to a preexisting probability model. If you have good, extensive data that represent the population reasonably, then just use it as an empirical distribution. If you use an empirical distribution, there is no reason to proceed to the subsequent steps.

* This curve was created by entering the data in Table 12.2 into the cumulative distribution function of Palisade Corporation's @RISK software.

TABLE 12.2
Emprical Distribution Derived from Consumption Data

Index	Data Value	Cumulative Probability $F(x) = i/19$
0	0	0
1	0.9	0.053
2	3.6	0.105
3	5.0	0.158
4	6.0	0.211
5	11.7	0.263
6	16.2	0.316
7	16.5	0.368
8	22.2	0.421
9	22.7	0.474
10	23.2	0.526
11	24.5	0.579
12	24.9	0.632
13	25.8	0.684
14	33.3	0.737
15	33.4	0.789
16	34.7	0.842
17	40.2	0.895
18	44.2	0.947
19	60.0	1.000

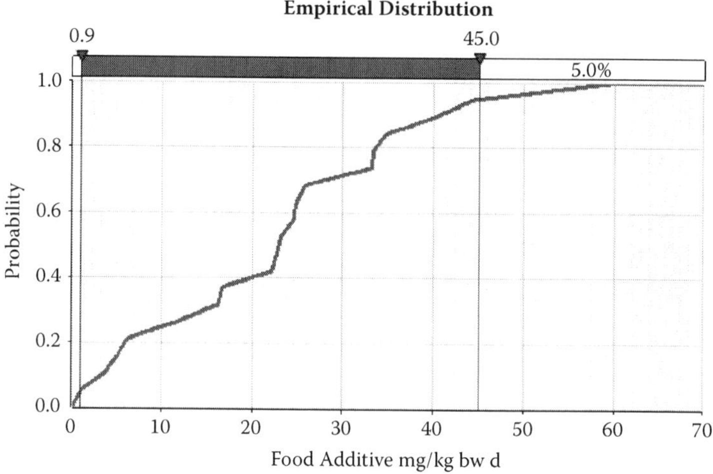

FIGURE 12.6 Empirical CDF of food-additive data.

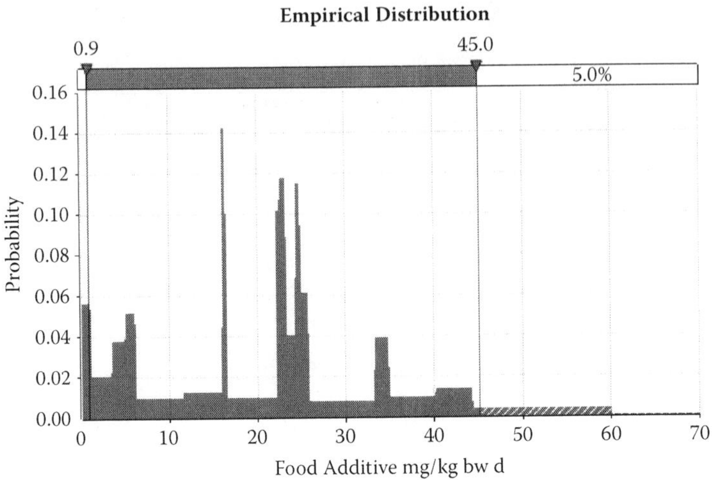

FIGURE 12.7 Empirical PDF of food-additive data.

12.3.2 Understand Your Data

Once you have decided it is not a good idea to use an empirical distribution, you will need to choose a probability distribution from among the bewildering list of candidate distributions. Figure 12.8 shows the probability distributions available with the popular Monte Carlo simulation software @RISK. A quick check of Wikipedia's "list of probability distributions" adds to this array of choices. Choosing a probability distribution from among this array that best suits your needs begins by understanding the data you do have.

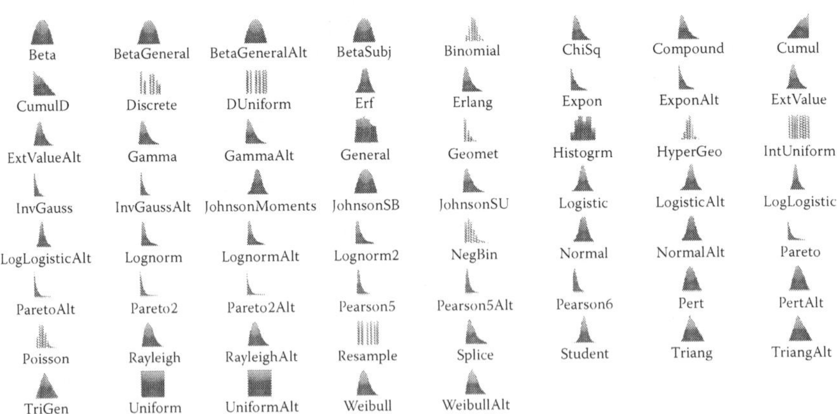

FIGURE 12.8 Palette of distributions available from Palisade's @RISK software.

12.3.2.1 What Is the Source of Your Data?

Where do your data come from? How were they collected? Under what conditions and when were they gathered? It should go without saying that data-gathering efforts should be designed to collect enough information to adequately define the tails of the distribution whenever it is feasible to do so. There is any number of data-quality issues you may concern yourself with, but the source of your data is far more basic. Its primary purpose is to ascertain if you can and should use your data to help you define a probability model.

For example, are your data quantitative or qualitative? Do your data come from experiments, observation, surveys, computer databases, literature searches, simulations, or are they test-case data? The last three sources of data are questionable and will affect your strategy. The literature rarely presents entire data sets. If your data come from a literature search, it is quite likely you are dealing with something more accurately called information data. Data that are the result of a simulation must be used with care; this would rarely be a high-quality source of data. Test-case data are not real data. They are fine for developing prototype models, but they should not be used in a risk assessment. As long as your data have credible information content, they can be used to help you choose a probability distribution. If that content is missing because of your data's source, do not use them to define a distribution.

If your data were collected selectively for any reason, they may not represent the full distribution of values in the population of interest to you. When this is the case, it is important to identify and assess the differences between your survey or sample data and the population you wish to represent in your model.

12.3.2.2 Is Your Variable Discrete or Continuous?

Discrete random variables tend to be things that are counted, while continuous random variables tend to be things that are measured like distance, time, area, weight, qualities, and statistics. Think of a portion of the number line that is relevant for your variable. If there are values on that portion of the line that your variable cannot take, it is discrete. If every value is possible, you have an infinite number of possible values and a continuous random variable. See Figure 12.9.

Discrete random variables should be represented by discrete probability distributions. Examples of discrete random variables include such things as barges in a tow,

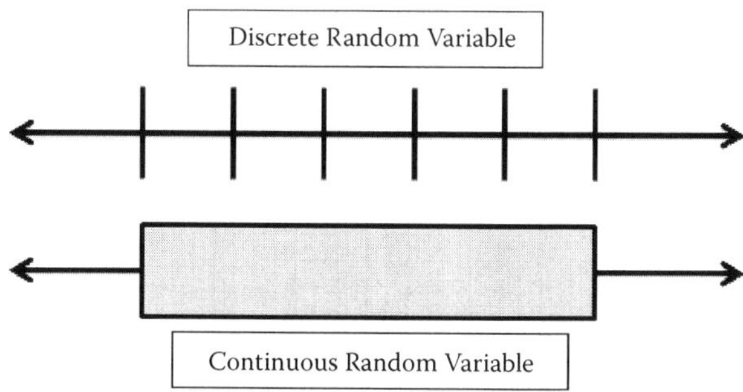

FIGURE 12.9 Discrete and continuous random variables' coverage of the number line.

houses in the floodplain, people at a meeting, results of a diagnostic test (positive or negative), casualties per year, relocations and acquisitions, quarantine centers, animals per quarantine center, sperm per sample, illnesses per year, monthly sales, and the like.

Continuous random variables need to be represented by continuous probability distributions. Examples of continuous random variables include such things as the average number of barges per tow, weight of an adult striped bass, sensitivity or specificity of a diagnostic test, transit time, expected annual damages, duration of a storm, area of shoreline eroded, sediment loads, average number of animals per quarantine center, weight of a sperm sample, amount of a food additive present in a sample, prevalence of a pathogen or disease, a serving size, and the like.

This seemingly obvious step is frequently overlooked, but it is an important consideration in the choice of a probability distribution. Examples of some probability distributions were shown in Figure 12.8. The discrete distributions are represented by vertical lines.

It is worth noting that although a random variable is clearly of one type or the other, a modeler may choose at times to treat it as if it was effectively its opposite type. For example, a dollar value is actually a discrete random variable. On the number line between $5 and $10, there are many values that cannot occur because dollars are only denominated in hundredths of a dollar. When dealing with large sums of dollar values, however, the number of possible values can grow very large and can be reasonably well approximated by a continuous distribution. As long as values obtained from a continuous distribution are converted to discrete values, via rounding for example, no harm is done by using a continuous distribution to represent a discrete variable. In fact, some discrete distributions (e.g., the binomial) approach a continuous distribution (e.g., the normal) as the sample size gets larger. Conversely, some continuous values like time may be sufficiently measured for some model purposes in whole hours or some other discrete time element. In such cases, it may be sufficient to represent a continuous variable with a discrete distribution.

12.3.2.3 Is Your Variable Bounded or Unbounded?

Boundedness concerns the range of values your quantity can assume. What values are possible? A variable with a minimum value is left-bounded. If it has a maximum value, it is right-bounded. Variables with a minimum and a maximum are bounded, and variables with neither are unbounded. Stated somewhat differently, the values of a bounded variable are confined to lie between two determined values. Unbounded variable values theoretically extend from minus infinity to plus infinity. A partially bounded variable is constrained at one end, usually the left.

It is more important at the outset to ascertain whether your variable has bounds than it is to worry about what they are. Most physical and human phenomena are bounded. Unbounded phenomena tend to be confined to relevant conceptual contexts such as an infinite number of tosses of a coin, annual stream flows into eternity, or servings of ground beef into infinity. When dealing with finite periods of time, most variables are bounded in some way. Choose a distribution with bounds that match your variable. Zero is a common minimum bound. Physical quantities, for example, do not ordinarily assume negative values.

If your variable is bounded, carefully consider and identify what those bounds are. Some bounds are easy. We know that the sensitivity of a test, the proportion of a population, or the probability of an event will be confined to the interval from zero to one. Other bounds are more difficult. We know that the travel time between two points has a nonzero minimum and some practical maximum, but we are unlikely to know exactly what they are. These are values worth sweating over. I wish I had more definitive guidance to offer, but I do not although bounds can be estimated via an expert elicitation (see Chapter 13). Establishing a physical or plausible range of values for a variable is as much art as science. Generally, the range in a model's output depends more on the bounds of the model's inputs than it does on the actual shape of the input distributions (EPA 1997). Estimating the bounds for a variable is not a trivial matter.

12.3.2.4 Parametric and Nonparametric Distributions

Parameters are numerical characteristics of populations. The most well-known parameters are measures of central tendency (mean, median, and mode) and measures of dispersion (range, standard deviation, and variance). In fact, any numerical characteristic at all—the minimum, the fifth largest, or the 37th percentile—could be a parameter. Population parameters are constants. The notion of a parameter is sometimes extended to any constant that appears in a model. Thus, in an equation that converts feet to inches, e.g., feet = 12*inches, 12 is a parameter. The language can be confusing because the word *parameter* is often used without its contextual meaning being entirely clear to the reader.

When it comes to describing probability distributions, it is convenient to break them into parametric and nonparametric groupings (Vose 2008). A parametric distribution is defined by a mathematical model. A nonparametric distribution is defined by its shape.

The normal distribution, for example, follows a specific mathematical model that describes much of what happens in the natural universe. Many phenomena of interest in risk assessment involve measurement of the duration of time, x, between some initial point and the occurrence of the phenomenon of interest. Examples include how long between failures, between entering and leaving a store, between floods, between switch malfunctions, and so on. Such activities are described by the mathematics of the exponential function. Mathematical model-based, i.e., parametric, distributions require greater knowledge of the underlying math and assumptions of the function if they are to be used properly. The shape of the suspected distribution is driven by physical or biological properties or other mechanisms. You may well be dealing with a parametric distribution.

Other times, there is no underlying mechanism or mathematical process guiding the distribution, and we simply use a shape, like a triangular distribution, a uniform distribution, or a histogram. The shapes of the data define the distributions; thus the empirical distribution described previously would be a nonparametric distribution. Nonparametric distributions are useful when the researcher knows nothing about the population parameters of the variable of interest. Their principle advantage may be that they require fewer and less stringent assumptions than their parametric counterparts. They are also intuitively easy to understand and are flexible.

A NOTE ON PARAMETERS

Words are used in varying contexts. You may find that you have chosen a nonparametric distribution like a triangular distribution only to be surprised to see its parameters defined as the minimum, most likely (mode), and maximum values. All distributions are defined by constants/parameters. Only parametric distributions are driven by an underlying mathematical process.

When the parameters of the chosen distribution are known, it is sometimes called a first-order distribution. When the parameters are subject to knowledge uncertainty and are themselves represented by a distribution, this is called a second-order distribution.

A normal distribution with a mean of 100 and a standard deviation of 10 is abbreviated N(100,10) and is a first-order distribution. If the mean and standard deviation are uncertain and represented by uniform distributions, it would be abbreviated as, for example, N(U[80,105],U[8,11]), and it is a second-order distribution.

Table 12.3 summarizes the type of variable, its boundedness, and its parametric nature for several common distributions. If you can identify these three attributes of your variable, you'll have a pretty good idea of the relevant distribution choices from the table. The shape column indicates whether the distribution assumes one basic shape or if the values used to define it can result in different shapes. Note that five nonparametric distributions can be used to define an empirical distribution.

In general, you would choose a parametric distribution if any of the following hold:

1. You have theory that supports your choice.
2. The distribution has proven accurate for modeling your specific variable despite a lack of specific theory to support its choice.
3. The distribution matches the observed data well.
4. You need a distribution with a tail extending beyond the observed minimum or maximum.

Likewise, choose a nonparametric distribution if:

1. Theory is lacking.
2. There is no commonly used model.
3. Data are severely limited.
4. Knowledge is limited to general beliefs and some evidence.

Parameters are used to define most probability distributions. The normal distribution, for example, is defined by two parameters: the mean (μ) and the standard deviation (σ). Each distribution, other than empirical ones, has its own set of defining parameters. These are entered as constants except in the case of second-order distributions, where the parameter itself is uncertain (see text box).

It is useful to recognize that although these parameters have different definitions and names that are not always revealing, they tend to play one of three basic functions in defining a distribution. They either identify the location, the scale (i.e., the range or extent), or the shape of the distribution.

The location parameter identifies the central location of the variable on the x-axis. It tells you where on the number line your data tend to fall. For the normal distribution, the mean is the location parameter. Figure 12.10 shows three identical distributions, each with a different location parameter (mean). A change in the location parameter shifts the distribution left or right without changing it. A shift parameter is sometimes called a second location parameter.

The scale parameter controls the spread of the data on the x-axis. A change in this parameter compresses or expands the distribution without altering its basic form. Think of it as telling you how much of the x-axis your data cover. The standard deviation is the scale parameter for the normal distribution. All three distributions in Figure 12.11 center over the same location parameter (mean = 0), but the majority of the data are distributed over different portions of the number line because they have different scale parameters.

The shape parameter, if there is one, governs the shape or basic form of the distribution. Think of it as indicating which values on the number line are most likely to occur. The normal always has the familiar symmetrical bell shape. The normal distribution does not have a shape parameter, nor does the exponential or several other distributions. Their shapes are governed by the mathematical model (function) that defines the distribution. A change in the shape parameter alters the form of the distribution, for example, its skewness. Some distributions, like the beta, have more than one shape parameter.

All of the distributions in Table 12.3 that are identified as shape shifters in the second-to-last column will have a parameter that serves as a shape parameter. The Weibull distribution shown in Figure 12.12 demonstrates the effect of a shape parameter

The first parameter identified in the legend is alpha; the second one is beta. These are not nominally very meaningful to anyone unfamiliar with the Weibull distribution, and in fact they have somewhat involved functional definitions. However, note that changing the alpha parameter, holding beta constant, changes the shape of the function appreciably. It also alters the scale of the distribution as well. Parameters may have roles more complex than fixing simple location, scale, and shape. This is governed by the mathematics of the different functions.

12.3.2.5 Univariate or Multivariate Distributions

Can you consider the distribution of your variable or your uncertainty about the value of an unknown parameter by themselves, or must you consider the values of other variables? A univariate distribution describes a single variable or parameter that is not probabilistically linked to any other variable or parameter in your model. Multivariate distributions are needed to address interrelationships with other variables. Many of the most important questions in science and technology concern how the values of one quantity can be used to control, explain, or predict another. Engineers are interested in how environmental factors affect tensile strength. Biologists are interested

TABLE 12.3

Summary of the Characteristics of Random Variables and Their Selected Probability Distributions

Distribution	Type	No Bounds	Bounded: Left and Right	Bounded: Left Bound Only	Category	Shape	Empirical Distribution
Beta	Continuous	No	Yes	No	Nonparametric	Shape shifter	No
Binomial	Discrete	No	Yes	No	Parametric	Some flexibility	No
Chi-square	Continuous	No	No	Yes	Parametric	Basic shape	No
Cumulative ascending	Continuous	No	Yes	No	Nonparametric	Shape shifter	Yes
Cumulative descending	Continuous	No	Yes	No	Nonparametric	Shape shifter	Yes
Discrete	Discrete	No	Yes	No	Nonparametric	Shape shifter	Yes
Discrete uniform	Discrete	No	Yes	No	Nonparametric	Basic shape	No
Erlang	Continuous	No	No	Yes	Parametric	Basic shape	No
Error	Continuous	Yes	No	No	Parametric	Some flexibility	No
Exponential	Continuous	No	No	Yes	Parametric	Basic shape	No
Extreme value	Continuous	No	No	Yes	Parametric	Basic shape	No
Gamma	Continuous	No	No	Yes	Parametric	Shape shifter	No
General	Continuous	No	Yes	No	Nonparametric	Shape shifter	Yes
Geometric	Discrete	No	No	Yes	Parametric	Some flexibility	No
Histogram	Continuous	No	Yes	No	Nonparametric	Shape shifter	Yes
Hypergeometric	Discrete	No	Yes	No	Parametric	Some flexibility	No
Integer uniform	Discrete	No	Yes	No	Parametric	Basic shape	No
Inverse Gaussian	Continuous	No	No	Yes	Parametric	Basic shape	No
Logarithmic	Discrete	No	No	Yes	Parametric	Some flexibility	No
Logistic	Continuous	Yes	No	No	Parametric	Basic shape	No
Lognormal	Continuous	No	No	Yes	Parametric	Basic shape	No

Lognormal 2	Continuous	No	No	Yes	Parametric	Basic shape	No
Negative binomial	Discrete	No	No	Yes	Parametric	Some flexibility	No
Normal	Continuous	Yes	No	No	Parametric	Basic shape	No
Pareto	Continuous	No	No	Yes	Parametric	Basic shape	No
Pareto 2	Continuous	No	No	Yes	Parametric	Basic shape	No
Pearson V	Continuous	No	No	Yes	Parametric	Some flexibility	No
Pearson VI	Continuous	No	No	Yes	Parametric	Some flexibility	No
PERT	Continuous	No	Yes	No	Nonparametric	Shape shifter	No
Poisson	Discrete	No	No	Yes	Parametric	Basic shape	No
Rayleigh	Continuous	No	No	Yes	Parametric	Basic shape	No
Student	Continuous	Yes	No	No	Parametric	Basic shape	No
Triangle (various)	Continuous	No	Yes	No	Nonparametric	Some flexibility	No
Uniform	Continuous	No	Yes	No	Nonparametric	Basic shape	No
Weibull	Continuous	No	No	Yes	Parametric	Shape shifter	No

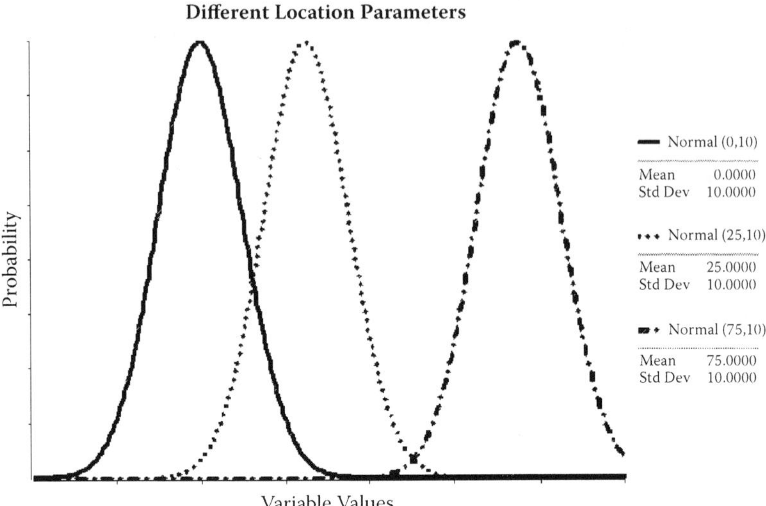

FIGURE 12.10 Plot of three PDFs with different location parameters.

in how organisms react to changes in their environment. Sometimes it is not possible to properly consider variables in isolation. Take the time to identify and consider the presence or absence of significant correlations or interdependencies between input variables. If such relationships are not properly addressed by multivariate distributions, be sure to address them in the structure of the model.

Mechanistic models express relationships between variables in terms of exact mathematical formulas (force = mass × acceleration) or in terms of the logic of causation (a causes b). Stochastic models express such relationships in terms of statistical

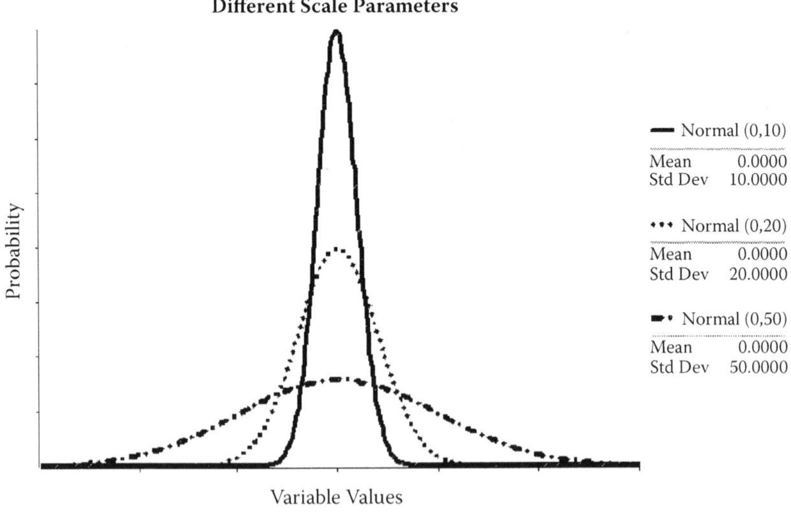

FIGURE 12.11 Plot of three PDFs with different scale parameters.

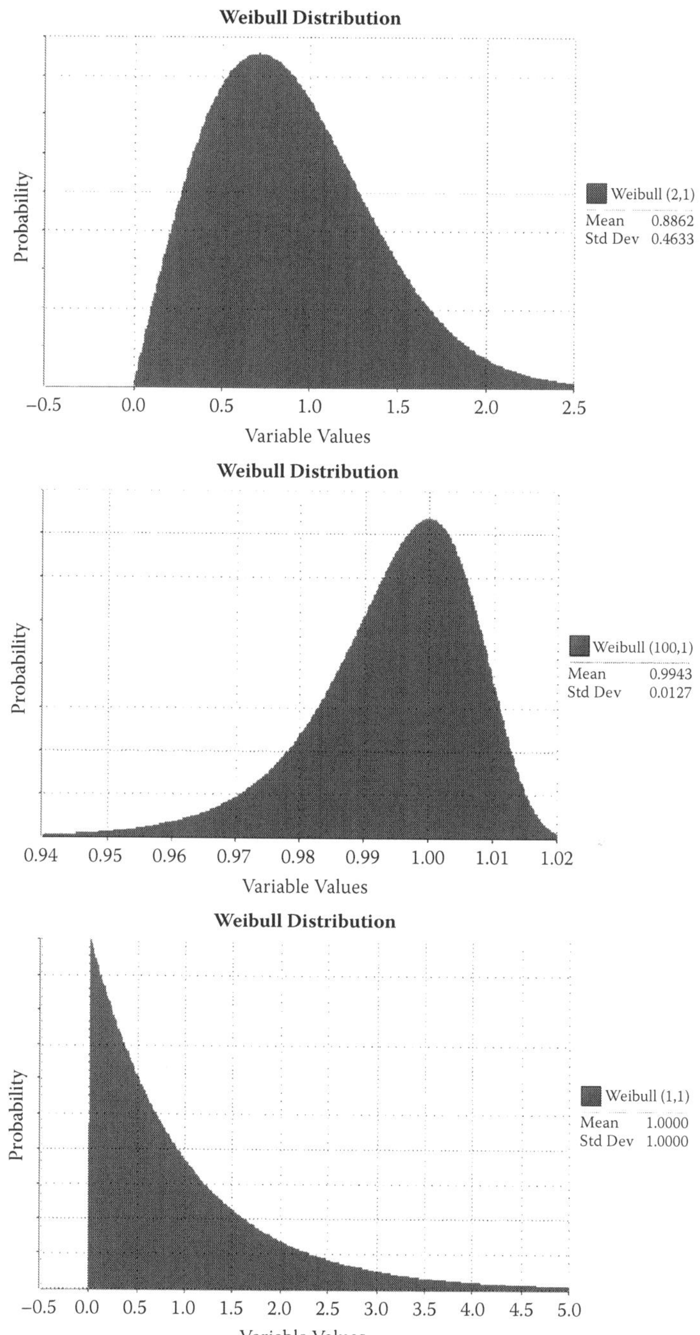

FIGURE 12.12 Plot of three shape-shifting Weibull PDFs.

tendencies of joint occurrence (e.g., high values of x tend to be associated with low values of y). Expressed in the more formal language of probability, the probabilistic framework for these stochastic situations involves the notion of composite events. These events give rise to two or more random variables.* The resulting probability model is a joint or multivariate distribution (Derman, Gleser, and Olkin 1973).

The multivariate distributions are few and include the multivariate normal distribution, Dirichlet distribution, multinomial distribution, multivariate hypergeometric, and the negative multinomial distribution. They tend to be more difficult for most people to work with, so there is a bias toward handling such interrelationships mechanistically whenever possible. Thus, many modelers would use mass × acceleration rather than a multivariate distribution for force.

If you are interested in the intersection of two or more events, or the outcome of two or more random variables at the same time, you may build your model with the appropriate dependencies reflected in the model structure. Alternatively, you may use a multivariate distribution.

12.3.3 Plot Your Data

Lest the point be lost, as you address the previously discussed data issues, you are converging on a smaller set of candidate distributions for your variable and your data. As you better understand your data from having considered the identified characteristics, it is time to take a look at the data. You may already have some notion about how your data ought to be shaped before you do so. Are your data known or thought to be symmetric or skewed? If skewed, in which direction? What else do you know or suspect about the shape of the distribution? It is useful to consider these things before you actually plot your data. You will be more likely to scrutinize surprises more carefully if you do.

Always plot your data to see what they look like. Data for the time between eruptions at the Old Faithful geyser in Yellowstone National Park (Howell 1998) are plotted in Figure 12.13. The mean time between eruptions for a sample of 222 data points was 71 minutes, with a standard deviation of 12.8 minutes. With these two parameter estimates and an assumption of a normal distribution, one could end up pretty embarrassed, as the data plot shows.

The actual data are bimodal and closer to two separate distributions of eruption times. One is centered around 55 minutes and the other is centered around 80 minutes. A normal distribution N(71,12.8) is superimposed for dramatic effect. Always make sure to plot your data and look at it. Also, remember that the number of bins used in a histogram plot can influence or even mislead your judgment with too much jaggedness or too much smoothness. It is wise to vary the number of bins before reaching any conclusions about the shape of your data. A wide variety of techniques are available to help you understand the shape of your data. The histogram is likely to be the most useful, but consider stem and leaf plots, dot plots, box and whisker

* The simplest examples might be tossing two six-sided die. Each die has its own simple probability distribution, but in the game of craps we would be inclined to treat the situation as a single joint probability experiment. In this bivariate case we are, in effect, finding a volume under two curves in three-dimensional space, as opposed to finding the area under a single curve in the univariate case.

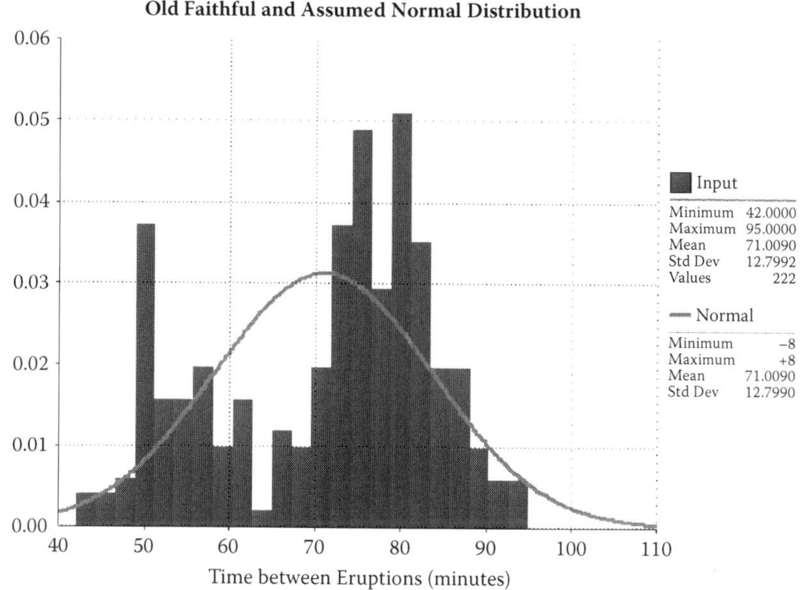

FIGURE 12.13 Time between eruptions for Old Faithful geyser in Yellowstone Park compared to a normal distribution of time.

plots, scatter plots, and other graphical techniques when the histogram is not revealing enough. Several of these plots are demonstrated in Chapter 17.

What do your data look like? Are they single-peaked? Symmetrical? Skewed in one direction or another? Look for distinctive shapes for your data and use that information to eliminate potential distributions. Clearly the Old Faithful data will not be described by any single-peaked or symmetric distributions. In fact, there are no familiar shapes that match the Old Faithful data. Comparing the shape of these data to the distributions in Figure 12.8 would steer us toward a cumulative or general distribution, which can be used to define an empirical distribution.

Using the information about your variable to narrow the candidate list of distributions, shape can then be used to further refine the candidate distributions. Be aware of the fact that some distributions are shape shifters, and a simple graphic like Figure 12.8 will not indicate all the possible shapes a shape-shifting distribution can assume. The Gamma, Weibull, and beta distributions are sometimes useful and flexible forms because of their shape-shifting properties.

12.3.4 Theory-Based Choice

This is usually going to be the most compelling reason for choosing one distribution over another when it is available to you. On rare occasions there may be very formal theory about a variable. For example, the results of the central limit theorem posit conditions under which a sampling distribution is a normal distribution. This type of formal theory is a very powerful argument for a distribution choice, but it is rarely available in risk assessment.

More likely, and less formally, you may have theoretical knowledge of the variable. Each discipline has developed its own body of theory and knowledge of the behavior of different phenomena. Predictive microbiologists will have theoretical knowledge of the growth and attenuation of bacterial populations. Wildlife biologists and biostatisticians will have well-developed theories about population responses to habitat changes. Economists have knowledge of financial and others systems; engineers know the behaviors of materials and physical systems under a wide variety of conditions. And so it goes for most risk-relevant disciplines. This specialized knowledge can lead to working theories about the distributions of variables and uncertain parameters. These working theories lack formal proofs, but they are supported by the knowledge and theories of the disciplines from which they arise. Many parametric distribution choices could be based on these working theories.

In addition to formally proven theories and working theories, let's introduce the notion of informal/ad hoc theory and rules of thumb. It has been observed, for instance, that many biological phenomena tend to have normal distributions and many physical phenomena have lognormal distributions. Likewise, variables that are sums of other random variables often tend to have normal distributions. Variables that are products of other random variables tend to be lognormal. These sorts of observations and rules of thumb may be useful aids for thinking about the best choice of a distribution. Is the process that generates your data an additive one, as was suggested for risk consequences in Chapter 9? Is it a multiplicative one, as suggested for some risk probabilities? Such insights may aid your distribution choice.

Informal theories tend to be study-specific. They are often ad hoc notions that may not even be suitable for experimental design and testing. They are also often expedient reasons for handling variables that are not among the critical uncertainties. For example, we might develop a simple ad hoc theory about the number of caffeine drinks consumed daily by members of the population. Common sense suggests that large numbers of drinks are not as likely as smaller numbers of drinks, so we expect a skewed distribution. Casual observation may also suggest that relatively few people drink just one caffeine drink per day, so we are not expecting the variable to fall from an initial peak, but to peak asymmetrically. Ad hoc theories like this might be more properly considered logic chains. Whatever they are called, they should be carefully documented when invoked as an additional piece of evidence for the choice of a distribution.

Take care to explicitly explore and develop the best reasonable theory for the shape of your data and the nature of your distribution. It may be the single most compelling piece of evidence for your choice.

12.3.5 Calculate Statistics

Summary statistics may sometimes provide clues about the probability distribution from which your data have come. A normal distribution has a low coefficient of variation and a mode, mean, and median that are identical. When these values are close to one another, it is an indication of single-peaked symmetry. An exponential distribution is positively skewed with an identical mean and standard deviation; thus it has a coefficient of variation equal to 1. Positive skewness measures indicate data skewed to the right; negative measures indicate data skewed left. So when the sample statistics from your data come close to meeting these conditions, that can be helpful information.

If only you could read a long, handy list of such rules of thumb, the task of choosing a distribution would be easier. Alas, that list has not yet been compiled, largely because it does not exist. Nonetheless, examining measures of central tendency and dispersion may help the assessor get a better feel for the data that are available and the distribution that best describes them.

Outliers can be problematic at times. Extreme observations could drastically influence the choice of a probability model. An outlier is defined as an observation that "appears" to be inconsistent with other observations in the data set. It is usually left to the analysts to define what is unusual. An outlier has a low probability that it originates from the same statistical distribution as the other observations in the data set. On the other hand, outliers can provide useful information about the process that produced your data.

There is no prescriptive method for addressing outliers. What is the data point telling you? What about your worldview is inconsistent with this outlying result? Should you reconsider your perspective? What possible explanations have you not yet considered? An outlier can be created by a shift in the location (mean) or in the scale (variability) of the process. It can also be a gross recording or measurement error. If an observation is an error, remove it.

Sometimes choosing a different distribution can eliminate an outlier. Values that seem to be outliers may be tail values of a skewed distribution. Try choosing a distribution with and without the outlier(s) to see if it makes a significant difference in your choice of distributions or the estimated parameters for the distribution. There are more-sophisticated statistical methods for dealing with outliers. See Iglewicz and Hoaglin (1993) for examples.

Ultimately, the explanation for your outlier(s) must be correct, not merely plausible. If you can't explain it and must keep it, use the conventional practices described in this chapter and live with the skewed consequences. Another alternative to consider is to choose distributions that are less sensitive to such extreme observations like the Gumbel or Weibull.

12.3.6 Previous Experience

Have you dealt with this same situation successfully before? What distribution did other risk assessments use? What does the literature reveal? It is always a good idea to see how others have handled similar situations in choosing a distribution. As the number of risk assessments grows, there is a rich body of experience growing, and this can be a source of useful insight in the choice of a probability distribution. The Beta-Poisson is a common choice for a dose-response curve largely because it has been successfully used in a number of risk assessments,

12.3.7 Distribution Fitting

Distribution fitting relies on goodness-of-fit testing (GOF) to provide statistical evidence to test hypotheses about the nature of the parent distribution for your data. Once you have worked your way through the preceding steps, you should have some

TABLE 12.4

Interarrival Times in Seconds of Passengers at an Airline Counter During Peak Demand

Interarrival Times of Passengers (s)									
10.4	40.5	95.6	0.0	27.0	11.9	7.3	15.5	31.8	37.9
156.8	168.8	6.7	3.0	14.0	157.9	7.8	41.0	237.6	57.1
58.0	40.8	86.7	17.2	157.9	2.1	83.3	40.5	61.4	11.3
88.3	28.4	87.0	135.0	1.4	104.0	337.3	103.4	24.7	116.8
16.6	121.0	258.8	28.2	88.4	8.2	44.6	179.4	26.1	25.5

ideas about the specific distributions from which your data may have come. At this point, and only at this point, it is time to use GOF to test the null hypothesis (H_0) that your data come from a specific kind of distribution.

Three tests are widely used for providing such evidence. They are the chi-square test, Kolmogorov-Smirnov test, and the Andersen-Darling test. All work in essentially the same way by comparing the distribution of your actual data to the theoretical distribution from which you think they may have come. Differences between these two distributions are examined to see if they are statistically significant; think of that as meaning the differences you see are too large to be the result of simple chance differences. It is important to note that a GOF test is at best another piece of evidence to consider in your strategy for choosing distribution and never a determining factor. The available software is so easy to use that many people are tempted to make the mistake of running their data through a fitting routine and letting the software identify the best distribution fit. This is the worst possible way to identify a distribution, and it should never be done in that manner.

Suppose the 50 data points shown in Table 12.4 are the times in seconds between arrivals of passengers at an airline ticket counter during peak demand. Further imagine that we have followed the above strategy and we are leaning toward an exponential distribution, but it could also be lognormal or loglogistic.

The top of Figure 12.14 shows the plot of the actual data (histogram bars) and an exponential distribution with a mean of 70.25, the mean of the data set. You may be the judge of how well the two curves match for the moment. Be advised that the number of bins used in your graph can have a substantial visual impact on the apparent quality of the fit, as seen in the middle of Figure 12.14. For this reason, it is often best to display the data in CDF format, the bottom graph in the figure, where the differences between the input data and the hypothetical distribution are not masked by the number of bins.

The chi-square test is the most common GOF test. It is valid for both discrete and continuous distributions. Like the other tests, it tests the H_0 that your sample data come from a specific distribution versus the alternative hypothesis, H_1, that they do not. It is a nonparametric, one-sided test. A good chi-square test usually requires about 50 observations or more.

The Kolmogorov-Smirnov test (K-S) is for continuous sample data and is more suitable for small samples than the chi-square test. It works by sorting data in ascending order and finding the greatest difference between the theoretical value for each ranked observation and the actual corresponding data point. In other words, it plots two CDFs, as shown in the bottom graph of Figure 12.14, and looks for the largest horizontal difference between the two. The test essentially examines the likelihood of

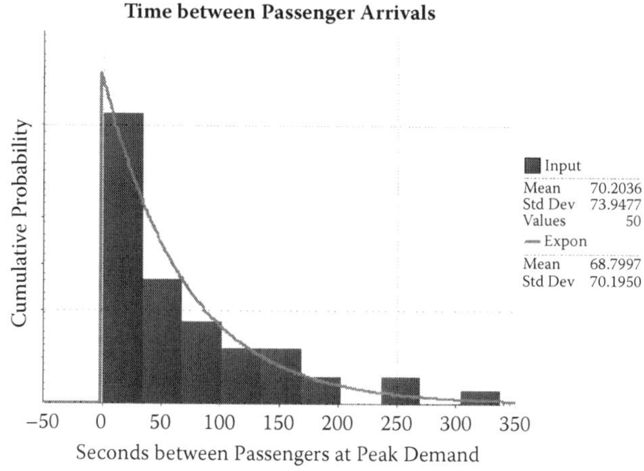

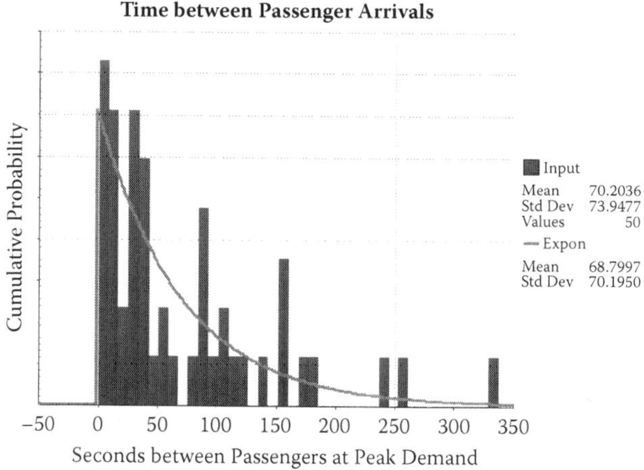

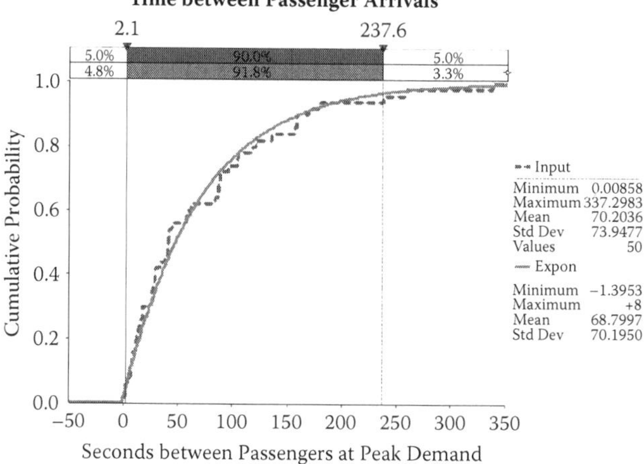

FIGURE 12.14 Three plots of chi-square goodness-of-fit test for passenger arrival data.

TABLE 12.5

Results of Three Different Goodness-of-Fit Tests for Three Parent Distributions

	Exponential	Loglogistic	Lognormal
Chi-square statistic	3.12	5.04	7.6
P value	.874	.655	.369
K-S statistic	.0977	.0897	.1354
P value	>.15	NA	NA
A-D statistic	.5433	NA	NA
P value	>.25	NA	NA

finding a difference that large by chance in a sample of that size if the data are truly from the hypothesized distribution.

The α = .05 critical value for our sample of n = 50 for the chi-square statistic is 14.1. Thus, a value greater than this would lead us to reject the hypothesis that our data come from a specific distribution. Selected GOF statistics are shown in Table 12.5.

The chi-square test results show our data could be from any one of these three distributions. From the table we see that the sample data have a chi-square statistic of 3.12 for the hypothesis that the data come from an exponential distribution. This is less than the critical value of 14.1, so we accept the H_0. The p value can be interpreted to mean the probability of getting a chi-square statistic of 3.12 or greater (i.e., seeing the differences we see in the two curves) by chance is .874, and that is not at all unusual. Typically we use a p value of .05 or less as our subjective definition of what would be too unusual to consider the result of chance. Note that the GOF test proves nothing with regard to our choice of distribution. The sample data could have come from any one of these three probability distributions. This is why it is so important to build a case for your choice based on the preceding steps.

In general, a small test statistic and large p are "desirable" for accepting H_0. The K-S and A-D tests also show that our data could be from an exponential distribution. We would use the results of this GOF testing to add one more piece of evidence to the argument that our data come from an exponential distribution.

The K-S test provides a better fit for mean values than for tail values. The A-D test weights differences between the theoretical and empirical distributions at their tails greater than at their midranges. It is the preferred test when a better fit at the extreme tails of a distribution are desired.

12.3.8 Seek Expert Opinion

Hiring a qualified consultant is always an option for supplementing the risk assessment team's knowledge of probability distributions. This step is not to be confused with expert elicitation of subjective probability information. That is the subject of the next chapter.

12.3.9 Sensitivity Analysis

If you have reached this point in the strategy without arriving at a clear choice of a distribution, then use a sensitivity analysis. Run your model with each candidate

TABLE 12.6

A Sample of Daily High Water Temperatures in July (°C)

Summer Water Temperture in Degrees Celsius										
28.1	28.3	29.0	28.8	29.1	28.8	28.9	28.7	28.5	28.8	28.3
28.6	29.1	28.9	28.7	29.3	28.9	29.4	29.4	28.8	29.5	28.5
29.0	28.6	28.7	29.1	29.1	28.6	28.6	28.8	29.2	29.2	28.4
29.5	28.9	28.9	28.9	28.7	28.3	28.7	28.9	28.4	28.7	29.2
29.2	29.7	29.1	29.2	28.5	28.9	28.6	28.7	28.5	28.5	29.1
28.8	28.2	28.6	28.9	29.5	28.9	29.0	29.1	28.8	29.2	28.7
29.3	28.7	28.8	29.1	29.4	29.0	29.2	28.8	28.5	29.0	29.3
29.1	29.0	28.6	28.6	29.5	29.5	28.8	29.6	29.0	29.5	28.7
28.9	28.2	29.2	29.0	28.7	28.9	28.6	28.5	29.6	29.6	28.3
28.7	29.0	29.0	29.3	28.5	28.9	28.4	28.7	28.9	28.9	29.0

distribution to see if the choice of distribution matters to the model outputs and thus the answers to the risk manager's questions. If the choice does not matter, use the most conventional distribution. If the choice does matter and there is no reasonable option for further discerning the choice, then simply document the differences in results and highlight this result as a significant uncertainty.

12.4 Example 1

Your final choice of a distribution for a specific quantity should be systematic, thoughtful, and documented. The strategy presented here enables you to choose in this way. Consider the situation where the assessor has the 110 data points for daily high water temperatures in July shown in Table 12.6 and is seeking a distribution to model the variability in high water temperatures for that month.

Let us suppose the assessor does not consider these data extensive enough to use as the basis for an empirical distribution. If she had thought otherwise, the method for developing an empirical distribution with the CDF for these data would be identical to that demonstrated earlier in the chapter.

12.4.1 Understand Your Variable and Data

What is the source of your data? These data were recorded via buoys on the waterway as part of routine data collection. They are a random sample of real data. I begin suspecting they may be normally distributed simply because the world often is.

Is your variable continuous or discrete? The data are continuous.

Is your variable bounded or unbounded? There are practical minimum and maximum temperatures for these coastal waters during the summer. Those bounds are, however, indefinite. Fuzzy bounds like that mean I could treat my quantity as unbounded over a limited range of the number line. This is a finesse point that is not evident from earlier discussion in this chapter. It basically means a bounded distribution is more logical, but I do not yet have to eliminate the unbounded normal distribution I have begun to suspect.

Do your data have a parametric or nonparametric distribution? Although appearing early in the process, this question often cannot be answered until later in the process. Support for the arguments here will be seen shortly. I lean toward a parametric distribution because I have some theory that supports my choice, that distribution matches the observed data well, and I need a distribution with a tail extending beyond the observed minimum or maximum.

Do you know the parameters of your data (first or second order)? Parameter estimates can be obtained from these data, so it will be a first-order distribution.

Are your data univariate or multivariate? This quantity is univariate.

At this point, I am looking for a continuous bounded parametric distribution. I am also willing to consider an unbounded distribution like the normal.

12.4.2 Look at Your Data

The data are shown in the histogram in Figure 12.15. The bins are 0.2 degrees wide. The data appear to be somewhat left-skewed. It is unclear if we have two peaks or just a random spike at 28.4–28.6 degrees. Plotting the data with more bins might help to resolve this issue. Based on the graph alone, it is difficult to call these data single-peaked and symmetrical.

The box plot in Figure 12.16 confirms a slight skew to the left in the sample data and in the interquartile range. The skew does not look extreme in either view.

12.4.3 Use Theory

My theory is quite informal for this data set. Quite simply, the "system" that produces a daily maximum temperature is complex enough and random enough that we believe deviations about the mean are likely to be symmetrical. A below-average temperature is as likely as an above-average temperature.

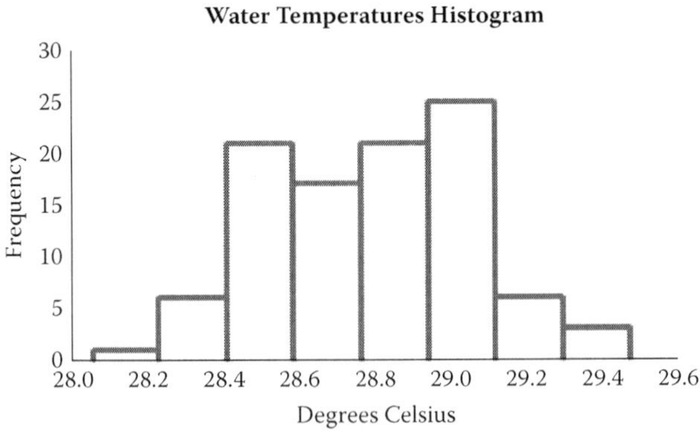

FIGURE 12.15 Histogram of maximum high daily water temperatures in degrees Celsius.

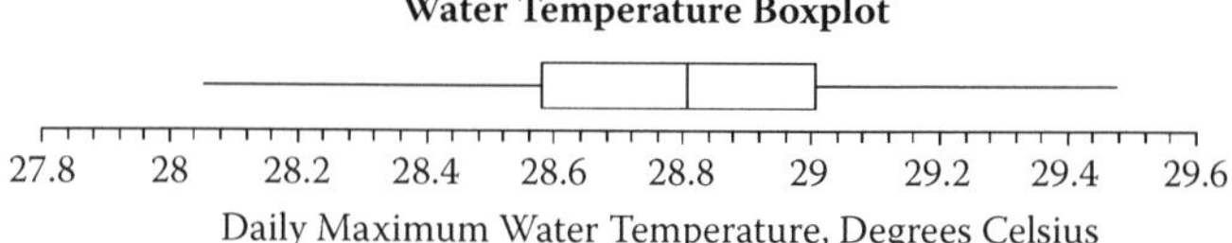

FIGURE 12.16 Box plot of maximum high daily water temperatures in degrees Celsius.

In addition, I think temperature is the cumulative effect of many environmental variables like cloud cover, ambient air temperature, currents, wind speed and direction, and so on. I believe these factors are likely to affect water temperature in a fashion more additive than amplifying, as a multiplicative relationship would be. This ad hoc theory or logic chain continues to nudge me toward a normal distribution.

12.4.4 Calculate Some Statistics

Selected descriptive statistics are shown in Table 12.7. The mean and median are approximately equal, and the coefficient of variation (0 to 100+ scale) is small. These indicate a potentially single-peaked and symmetric distribution like the normal. There are no outliers in this data set.

12.4.5 Use Previous experience

A U.S. FDA (2005) risk assessment of pathogenic *Vibrio parahaemolyticus* in raw oysters used a normal distribution to model water temperatures in the coastal waters of the United States. This is another piece of evidence for choosing a normal distribution.

12.4.6 Distribution Fitting

At this point I have eliminated all discrete and nonparametric distributions. I have also eliminated anything with a right skew. I have shaky visual evidence of a normal distribution, but I also have my informal theory pulling me back in that direction.

TABLE 12.7

Descriptive Statistics for Daily High Water Temperature in July (°C)

Count	110
Mean	28.8
Median	28.8
Sample standard deviation	0.3
Minimum	28.1
Maximum	29.5
Range	1.4
Standard error of the mean	0.03
Confidence interval 95% lower	28.7
Confidence interval 95% upper	28.9
Coefficient of variation	1%

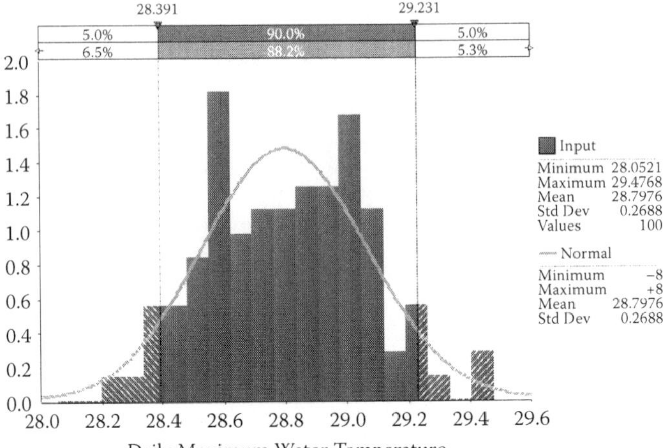

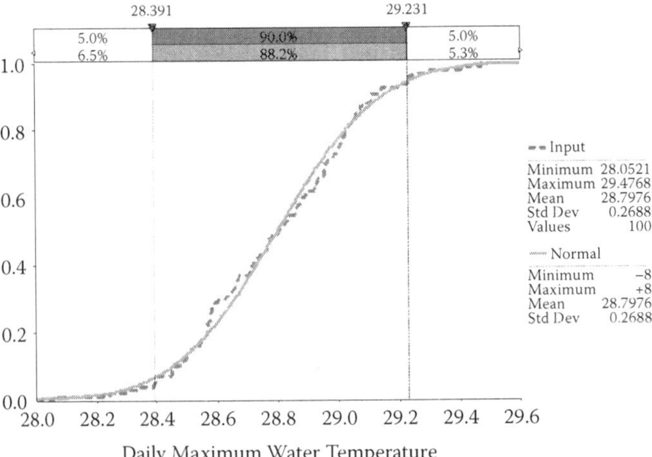

FIGURE 12.17 Two plots of the GOF test for the water temperature data coming from a normal distribution.

Some statistics and the prior experience of another risk assessment make me lean toward a normal distribution.

A GOF test was run using several candidate distributions, including the normal, extreme value, gamma, inverse Gaussian, logistic, Pareto, and Weibull. The top of Figure 12.17 shows a theoretical normal distribution with the same mean and standard deviation as the data superimposed on the input data.

There are a few places where we seem to have too many observations and a few where there are too few. This is to be expected from a random sample. What we want to know is if the differences we are seeing are statistically significant differences, i.e., differences so big it is unlikely these data come from a normal distribution. Note

TABLE 12.8

Chi-Square Test Results for GOF Test for July Water Temperature Data and Five Selected Distributions

	Normal	Weibull	Logistic	ExtValue	Pareto
Chi-Square Test					
Chi-square statistic	7.14	8.02	11.98	17.26	121.89
P value	.7122	.6269	.2864	.0688	0
Crit. value @ .750	6.7372	6.7372	6.7372	6.7372	6.7372
Crit. value @ .500	9.3418	9.3418	9.3418	9.3418	9.3418
Crit. value @ .250	12.5489	12.5489	12.5489	12.5489	12.5489
Crit. value @ .150	14.5339	14.5339	14.5339	14.5339	14.5339
Crit. value @ .100	15.9872	15.9872	15.9872	15.9872	15.9872
Crit. value @ .050	18.307	18.307	18.307	18.307	18.307
Crit. value @ .025	20.4832	20.4832	20.4832	20.4832	20.4832
Crit. value @ .010	23.2093	23.2093	23.2093	23.2093	23.2093
Crit. value @ .005	25.1882	25.1882	25.1882	25.1882	25.1882
Crit. value @ .001	29.5883	29.5883	29.5883	29.5883	29.5883

the probabilities associated with the delimiters at 28.391 and 29.231 degrees: They give an indication of the fit in the tails of the distribution. My sample data have fewer observations below and above these values than the theoretical distribution would have. That also is not unusual in a sample the size of ours (n = 110). Always bear in mind that the number of bins in your graphic can have a powerful influence on the appearance of a fit. The CDF view, seen in the bottom of Figure 12.17, provides a different perspective on the quality of the fit.

A statistical summary of the chi-square test is provided in Table 12.8. The test statistic for the null hypothesis that my data come from a normal distribution is 7.14. Using a critical value of .05, I see that my test statistic is less than the critical value of 18.3, and so we cannot reject the null hypothesis. This is another piece of evidence to suggest my data could be from a normal distribution.

Note that the test statistics also suggest my data could come from a Weibull, logistic, or extreme value distribution as well. Only the Pareto distribution has a test statistic greater than the critical value of 18.3. This limits the utility of GOF testing as a primary piece of evidence for distribution choice. It is rarely, if ever, definitive evidence.

12.4.7 Expert Opinion

An expert is usually only consulted when we have been unable to identify a reasonable candidate distribution. I am satisfied with the body of evidence building toward a normal distribution. An expert opinion is not necessary.

12.4.8 Sensitivity Analysis

Absent a set of viable alternative distributions, sensitivity analysis will not be necessary.

12.4.9 Final Choice

The normal distribution is continuous, parametric, consistent with my theory, successfully used in the past, and statistically consistent with my data. Therefore I will use it. The differences and apparent skew in the plot of the data are considered mere random effects. The normal distribution describes the variability in water temperature. Because I am confident in the parameter estimates for this parametric distribution, there is no uncertainty reflected in the parameters of this distribution. I have a first-order distribution.

The normal distribution is an unbounded distribution. Its minimum and maximum values are minus and plus infinity, respectively. To ensure that I will not be causing any serious logic problems by using an unbounded distribution to model a bounded phenomenon, I must consider values that can result from a Monte Carlo process (the subject of Chapter 14).

The choice is to use a normal distribution with a mean of 28.8°C and a standard deviation of 0.3°C. If such a distribution can produce values that exceed the minimum and maximum bounds on the daily high water temperature in July, it cannot be used. Values within ±5 standard deviations of the mean include virtually 100% of all possible values. This temperature range extends from about 27.3°C to 30.3°C. Values within that range are not outside the minimum and maximum bounds for a daily high water temperature in July. The mean ±3 standard deviations accounts for all but about one-quarter of 1% of all possible temperature values. This smaller range (27.9°C–29.7°C) neatly matches the range of observed values. Thus, I am satisfied that using a normal distribution will not cause any boundedness problems in the model, and I am confident in the choice of a normal distribution.

12.5 Example 2

We will not have data every time we need a distribution. Knowledge uncertainty and data gaps are common hurdles in risk assessment. Let's consider a quantity that varies, about which we have knowledge uncertainty. Imagine we are working on a risk assessment where we need to know how much time watermen spend on the water harvesting oysters on any given day. This is important in determining the potential outgrowth of pathogenic bacteria that may be on the oyster flesh when it is harvested.

Assume that the only information we have is what we gained from a couple of informal conversations with watermen. They have suggested they will not spend less than 5 hours on the water or more than 11, with 9 being the most common amount of time they spend harvesting oysters. Clearly we lack the data for an empirical distribution, so we will choose a probability distribution from the universe of available distributions.

12.5.1 Understand Your Variable And Data

What is the source of your data? It is anecdotal information from watermen.

Is your variable continuous or discrete? It is continuous.

Is your variable bounded or unbounded? It is bounded.

Do your data have a parametric or nonparametric distribution? This is a nonparametric distribution. Our uncertainty is too great to consider anything more sophisticated.

Do you know the parameters of your data (first or second order)? Because we will use a nonparametric distribution, we will be working with a first-order nonparametric distribution.

Are your data univariate or multivariate? This is a univariate quantity.

12.5.2 Look at Your Data

There are no data to plot.

12.5.3 Use Theory

There is no formal theory about watermen's behavior. There is a great deal of uncertainty here, and we will use the information provided. The only crude theory I can offer is that I expect relatively few watermen to spend the minimum time on the water. In relative terms, I expect fewer values from the low end and more from the high end of the distribution. The significance of this theory will become more clear in the sensitivity discussion. I base it on the fact that if watermen decide to go out onto the water, they are going to work for as long as possible barring bad weather, equipment failure, or personal emergencies. I expect these limitations to be relatively rare.

12.5.4 Calculate Some Statistics

We have been able to anecdotally estimate a minimum, most likely (mode), and maximum value for time on the water. There is no other information currently available. Watermen have told us that if they go out onto the water, they will never be there for less than 5 hours. If conditions are not favorable, they will not go at all, and if they do go, travel time, setup, and a minimal harvesting would be no less than 5 hours. Visibility and sometimes regulations constrain the time they can leave or spend on the water. In addition, they must return in time to get their product to market the same day; this caps the maximum time at 11 hours. Most of the time a waterman spends 9 hours on the water and often several more hours at the dock or in other aspects of the business. Thus we have a minimum of 5, a most likely value of 9, and a maximum of 11 hours spent on the waterway.

12.5.5 Use Previous Experience

The previous experience of watermen was the basis for the input data, but let us imagine there is no previous specification of such a distribution that we can find.

12.5.6 Distribution Fitting

There are no data for distribution fitting.

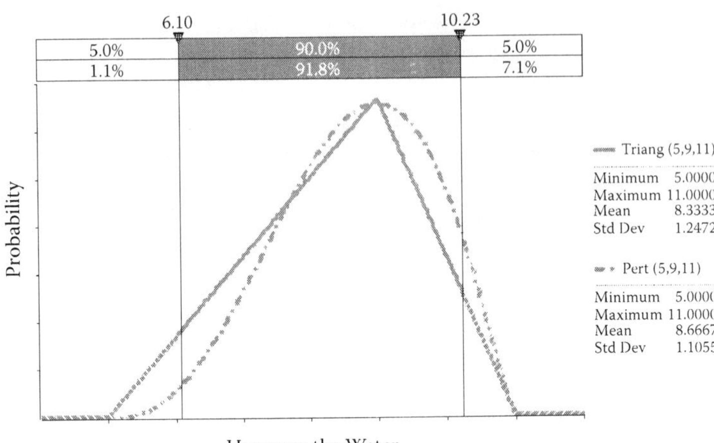

FIGURE 12.18 Comparison of triangle and pert PDFs for time on water data.

12.5.7 Expert Opinion

The options for choosing a distribution are restricted to those nonparametric distributions that require little data. An expert is not required to make this choice. The candidate distributions include the uniform, triangle, pert, and beta. Because I have three bits of information, I will reject the uniform and beta, which require only two. My information is sparse and I do not want to throw any of it away.

12.5.8 Sensitivity Analysis

Sensitivity analysis can always be conducted by inserting different distributions into the model. Sometimes it is possible to do some sensitivity analysis outside the model. Figure 12.18 compares the two candidate distributions. Both use the same three parameters (5,9,11). The triangle shows that smaller values are more likely than in the pert and larger values are less likely than in the pert.

12.5.9 Final Choice

Based on the available information and my crude theory, the pert distribution will be used to express my beliefs about how much time watermen spend on the water in a given day. This distribution is representing natural variability in the time spent as well as knowledge uncertainty about that variability. The amount of time spent on the water is clearly a source of uncertainty that needs to be communicated to the risk manager. With additional research we could reduce the uncertainty and better define the variability.

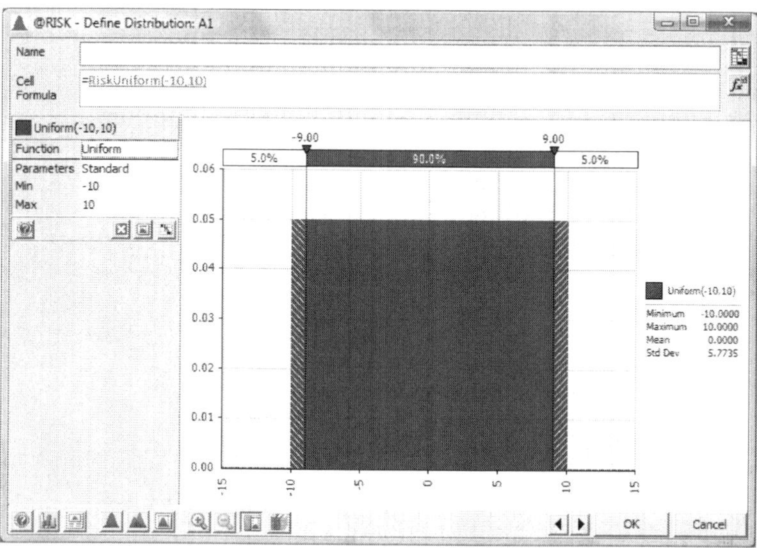

FIGURE 12.19 Uniform distribution.

12.6 A Dozen Useful Probability Distributions for Risk Assessors

Probabilistic risk assessment requires us to specify the distributions of random variable model inputs. The strategy above describes a systematic approach to identifying an input distribution. This final section comments on some of the more common and useful distributions used for risk assessment. Four distributions are offered for those instances when there are relatively little data or information available for the variable of interest. Then, four discrete and four continuous distributions that are likely to be most useful to risk assessors are presented.

12.6.1 Four Useful Distributions for Sparse Data

The uniform distribution (Figure 12.19) is used when data are sparse or absent. Note that the absence of data need not mean we lack information. The uniform describes a situation where a quantity is believed to vary between a minimum value and a maximum value (the distribution parameters) and little else is known about it. All values between the minimum and maximum occur with equal likelihood. In this sense it is a maximum ignorance distribution.

The triangular distribution (Figure 12.20) is also used in the absence of data and entails a little more knowledge about the quantity than the uniform does. It describes a situation where you know the minimum, maximum, and most likely values (the parameters) to occur. The most likely value is the mode, not the mean. The mode value is elevated to form the peak of the triangle.

The beta (Figure 12.21) is a family of distributions that is also useful in the absence of data. It's a very flexible distribution commonly used to describe variability or uncertainty over a fixed (bounded) range. It is convenient for modeling uncertainty

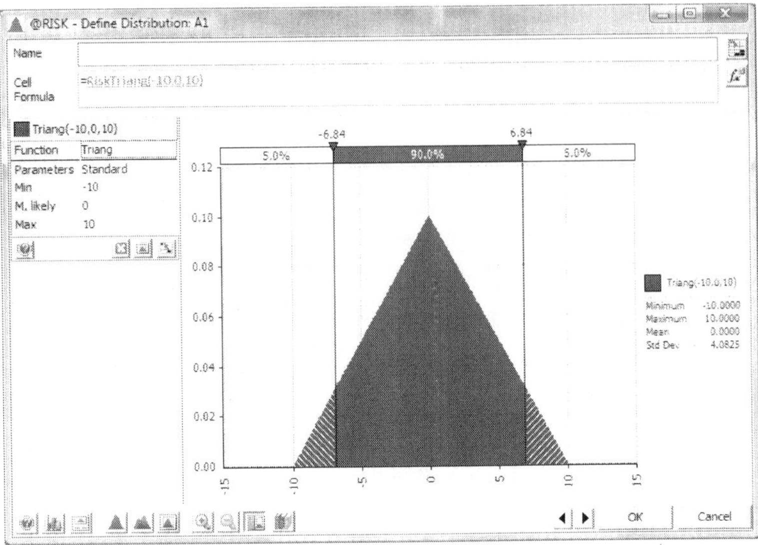

FIGURE 12.20 Triangular distribution.

about the probability of the occurrence of an event, as it is naturally defined over the 0 to 1 range. It is also useful for representing values expressed as fractions or percentages, such as incidence, sensitivity and specificity of tests, population proportions, and the like. Its parameters, α_1 and α_2, are not intuitively obvious, but they function as shape and scale parameters.

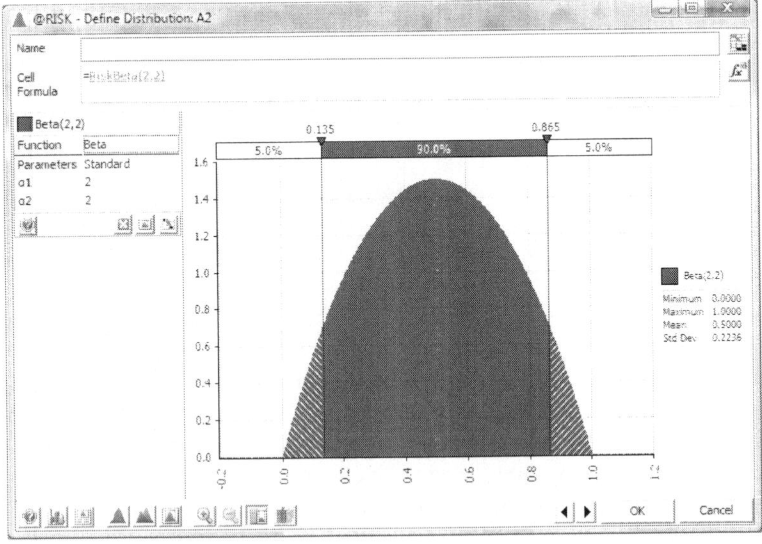

FIGURE 12.21 Beta distribution.

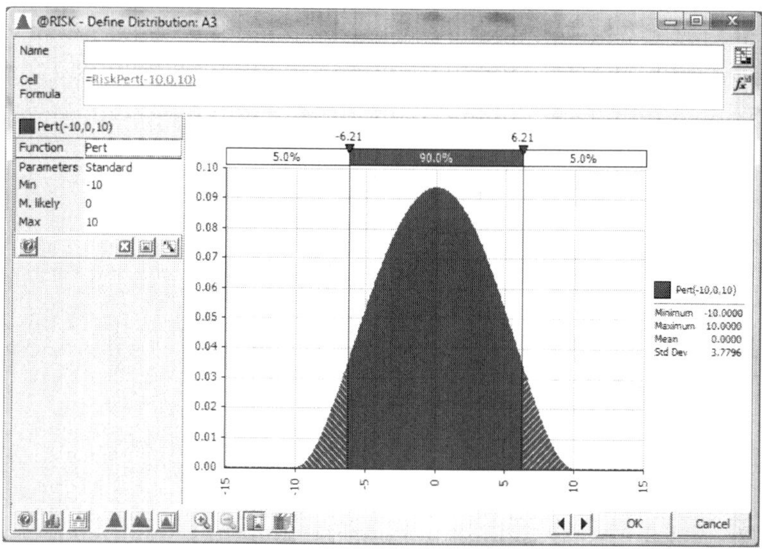

FIGURE 12.22 Pert distribution.

The pert or betaPERT distribution (Figure 12.22) is based on the same information as the triangle distribution. It describes a situation where you know the minimum, maximum, and most likely values to occur. While the triangle "connects" these three points with straight lines, the pert uses the changing slope of a curve. As a result, the pert and triangle tend to differ in the percentage of values near the minimums and maximums.

These four nonparametric distributions will meet a great deal of the risk assessor's needs for representing knowledge uncertainty in a model.

12.6.2 Four Useful Discrete Distributions

The yes-no distribution (Figure 12.23) describes observations that can have only one of two values, such as yes or no, success or failure, true or false, gets ill does not get ill. It is a special case of the binomial distribution where the sample size, n, equals 1. It is often called a Bernoulli distribution. Its parameters are the probability of a yes and the probability of a no.

The discrete distribution (Figure 12.24), not to be confused with the distribution of a discrete random variable, is made up of a limited number of values or alternative outcomes (A, B, C in the figure). Each of these values/alternative outcomes, which need not be sequential, has a probability of occurring, and that probability can vary. The discrete uniform distribution is a special case of the discrete and is the discrete equivalent of the continuous uniform distribution. All integer values in the discrete uniform distribution are equally likely to occur. Its parameters include a number of x-values and the probability of each x-value.

The binomial distribution (Figure 12.25) describes the number of times a particular event with a fixed probability occurs in a fixed number of trials. Examples include the

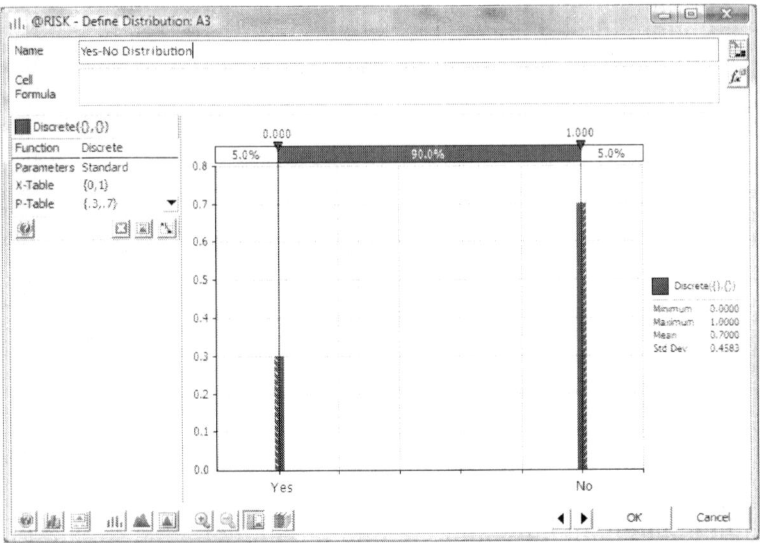

FIGURE 12.23 Yes-no distribution.

number of defective items in a shipment of 50 items for a given defect rate, the number of contaminated eggs in a gross for a given prevalence, the number of floods with an exceedance frequency of 0.02 or smaller in a 100-year period, and so on. The binomial is one of the most important distributions for risk assessors to learn. Its parameters include a sample size, n, and a probability of occurrence for the event of interest, p.

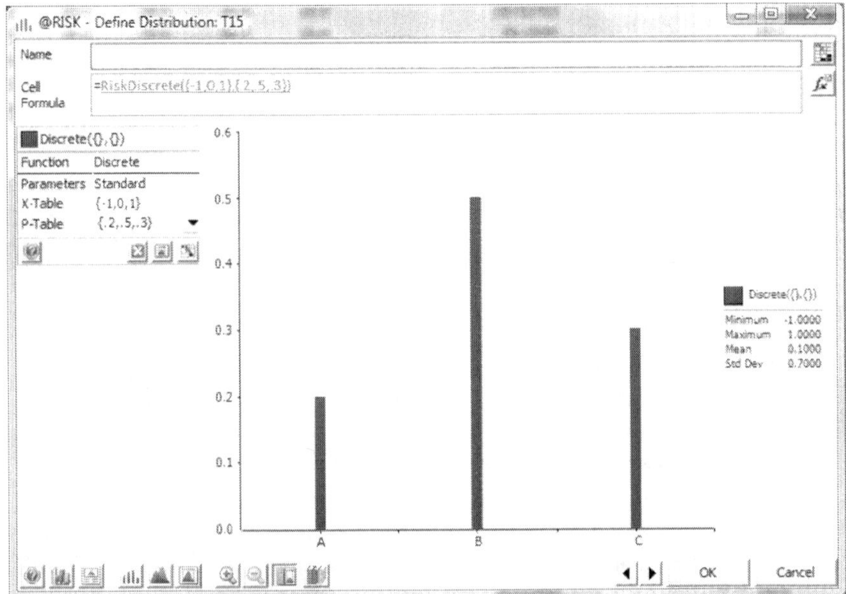

FIGURE 12.24 Discrete distribution.

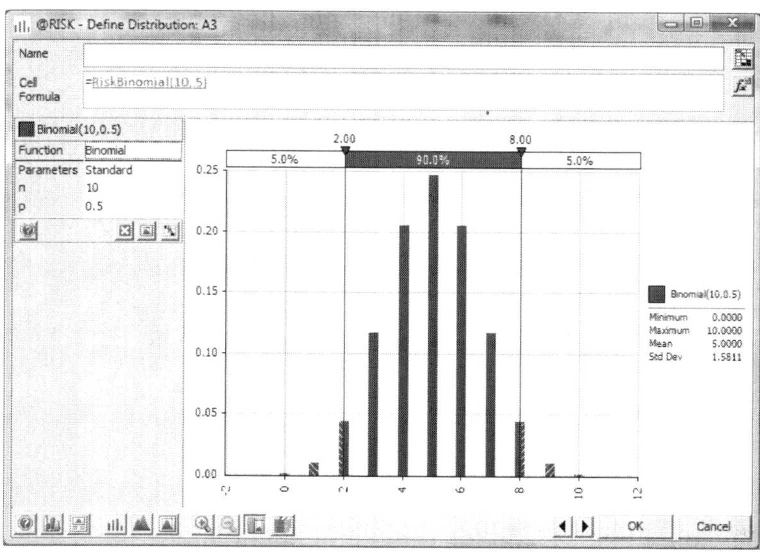

FIGURE 12.25 Binomial distribution.

The Poisson (Figure 12.26) is useful for describing a number of events per unit of measure, i.e., a given interval of time or space. Specific examples include such things as vessels or cars per day, houses per hectare, annual shipments, outbreaks/month, flaws per square meter, errors per page, spores per 50-g sample, and so on. The mean (lambda) is the only parameter for this distribution.

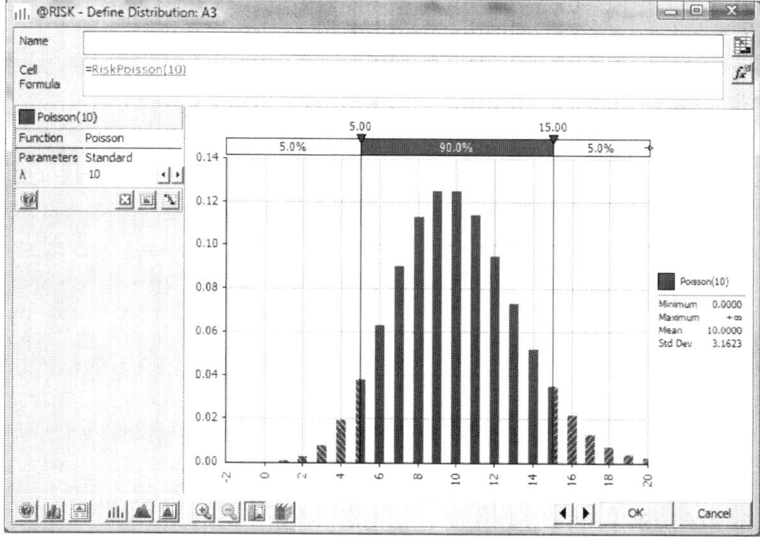

FIGURE 12.26 Poisson distribution.

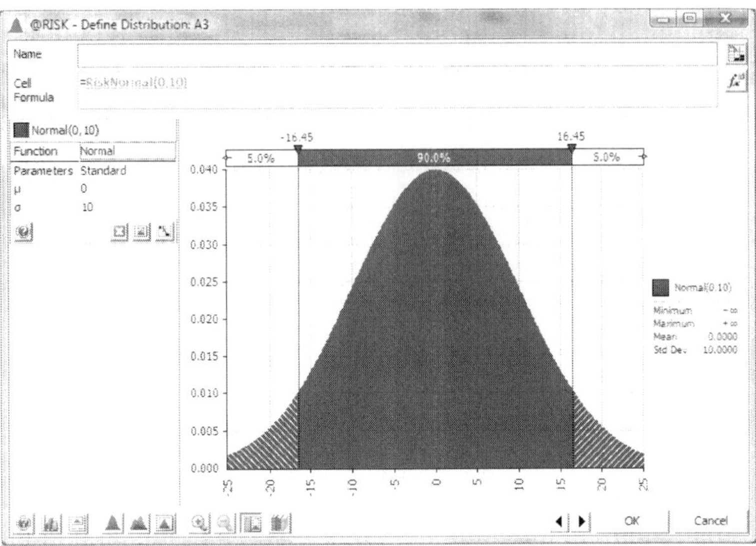

FIGURE 12.27 Normal distribution.

12.6.3 Four Useful Continuous Distributions

The normal distribution (Figure 12.27) may be both the most important and the most commonly used distribution. It is important because it describes many natural phenomena, especially biological variables such as population heights, weights, IQs, brain size, and the like. It is also useful for pure random processes and it often describes errors of various types quite well. Quantities that are the sum of a large number of other quantities are also normally distributed by virtue of the central limit theorem. Its parameters are the population mean and standard deviation.

The lognormal distribution (Figure 12.28) describes values that are positively skewed and nonnegative. Charitable giving may be well described by a lognormal distribution. There are relatively few very small contributions which rise to a large number of moderate contributions and then fall off to a declining number of larger contributions. The lognormal also describes many natural physical quantities like particle sizes and chemical concentrations, as well as financial quantities. Quantities that are the product of a large number of other quantities are lognormally distributed by virtue of the central limit theorem. Its parameters are also the mean and standard deviation of the population.

The exponential distribution (Figure 12.29) describes events that occur at random points in time or space. The exponential measures the duration of time (or space) between an initial point and the occurrence of some event of interest. For example, time between failures of a piece of equipment, time between arrivals of vessels or people, time between disease outbreaks, the duration of a phone call or other event, and so on are all well represented by an exponential distribution. The mean is the only parameter of this distribution.

Weibull distributions (Figure 12.30) are a family of distributions used to describe nonnegative quantities that are not well served by other distributions.

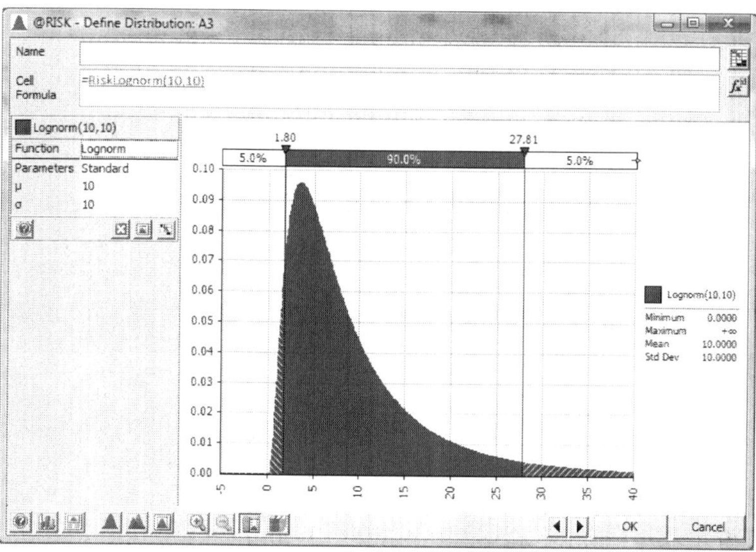

FIGURE 12.28 Lognormal distribution.

It is often used for length of life and endurance data. It is commonly used to describe failure time in reliability studies as well as the breaking strengths of materials in reliability and quality control tests. Weibull distributions are also used to represent various physical quantities, such as wind speed. It has two non-intuitive parameters, α and β.

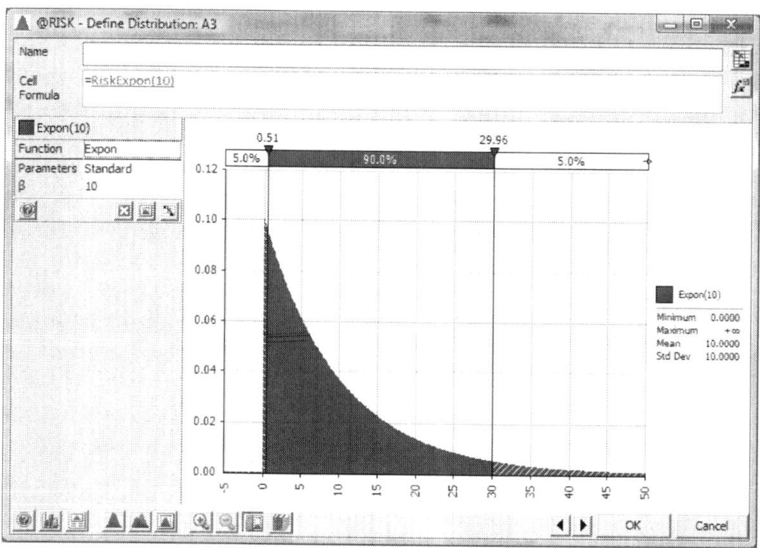

FIGURE 12.29 Exponential distribution.

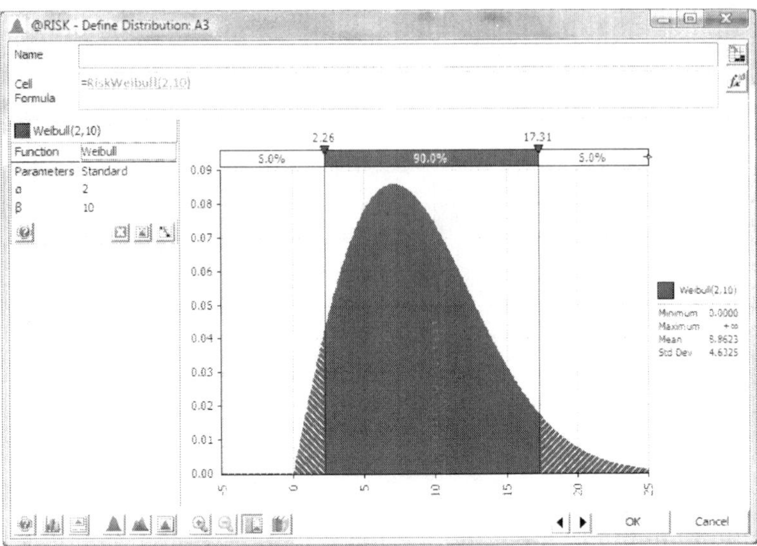

FIGURE 12.30 Weibull distribution.

12.7 Summary and Look Forward

If you have good data, consider using an empirical distribution in your risk probabilistic assessment models. If your data are not extensive enough to do that, then devise and follow a careful procedure for choosing the probability distribution to represent the natural variability and knowledge uncertainty in your model inputs. A nine-step process presented in this chapter is offered for anyone without a better process. Most risk assessors can usually get by on a small handful of distributions. Four nonparametric distributions are suggested here for use along with four each of discrete and continuous distributions.

The process described in the chapter relies on the risk assessors having some data. Voids and gaps in data are all too common in risk assessment, and we need methods for representing our uncertainty in these situations as well. The next chapter describes the use of subjective probability elicitations to tease useful knowledge out of an expert. This is a rapidly growing and frequently abused practice in risk assessment. We humans are proven to be rather poor at estimating probabilities, especially small ones. It is important for risk assessors to understand the heuristics we rely on when forced to consider things we are uncertain about so that their effects can be countered and accounted for when we elicit information from experts. You'll learn how to do that in the next chapter.

REFERENCES

Derman, Cyrus, Leon J. Gleser, and Ingram Olkin. 1973. *A guide to probability theory and application*. New York: Holt, Rinehart and Winston.

Howell, David C. 1998. Old Faithful at Yellowstone: A bimodal distribution. http://www.uvm.edu/~dhowell/StatPages/More_Stuff/OldFaithful.html.

Iglewicz, B., and D. C. Hoaglin. 1993. *How to detect and handle outliers*. Milwaukee, WI: American Society for Quality Control.

Law, Averill M., and W. David Kelton. 1991. *Simulation and modeling analysis*. New York: McGraw-Hill.

U.S. Environmental Protection Agency. 1997. Risk Assessment Forum. *Guiding principles for Monte Carlo analysis*. Washington, DC: U.S. Environmental Protection Agency.

U.S. Food and Drug Administration. 2005. Center for Food Safety and Applied Nutrition. Quantitative risk assessment on the public health impact of pathogenic *Vibrio parahaemolyticus* in raw oysters. http://www.fda.gov/Food/ScienceResearch/ResearchAreas/RiskAssessmentSafetyAssessment/ucm050421.htm.

Vose, David. 2008. *Risk analysis: A quantitative guide*. 3rd ed. West Sussex, U.K.: John Wiley & Sons.

13

Probability Elicitation

13.1 Introduction

How do you address probability in a risk assessment when you have no data to estimate an event's probability? You can elicit probability information from experts. Subjective probability estimates are an important part of risk assessment, and formally eliciting them from experts is the subject of this chapter.

We saw in Chapter 11 that the frequentist view of probability is not always a useful one. We cannot use this approach to estimate the probability of another major terrorist attack in the next three years. Nor is it useful for estimating sea level rise of two feet or more along Florida's coast by 2075. There are no data for these situations, but a subjective or degree-of-belief approach to probability is useful for situations that have never happened before, may only happen once, or, interestingly, for situations that have already happened. What is the probability that the New York Yankees played the St. Louis Cardinals in the 1927 World Series? If you do not know, you would express this probability informally as your personal belief that this happened. Any number other than zero would be wrong, however; the Yankees played the Pittsburgh Pirates.

We all have gotten quite used to making informal probabilistic judgments about the uncertainty in our own lives, and we do this with surprising regularity. When you don't take an umbrella, you believe it will not rain hard enough to need one. When you cross the street, you believe you will not get hit. When you handicap your favorite team's chances to win their next game or for the incumbent party to retain the White House, you are addressing uncertainty through the subjective assessment of probability. Over time you learn where and when your instincts are more or less good, and you make decisions accordingly.

It is an altogether different situation, however, when the uncertainty we are assessing can have significant implications for others. It is one thing to say I do not think I need an umbrella and another to say it ought to be safe to launch the space shuttle. In these situations, it is important to be more systematic and more careful about how we estimate these probabilities and other uncertain quantities. Bear in mind that our uncertainty is not always expressed in terms of a probability. There are many instances when a critical value is unknown and we use expert elicitation techniques, like those that will be described in this chapter, to estimate them. How many spores are required on a hectare of land to enable a fungus to become established in an agricultural crop? What will a kilowatt-hour (kW×h) of energy cost two years from today? How much oil will be recovered in the next 25 years if the United States allows drilling offshore of the Atlantic coast? Although we will focus on probabilistic elicitation, thresholds and other values can be elicited in the same way.

INFORMAL ELICITATIONS

Expert elicitation can be divided into informal and formal elicitations. Informal elicitations, which are not the focus of this chapter, include self-assessment (you decide the values), brainstorming, causal elicitation without structured efforts to control biases, and discussion among peers and colleagues (EPA 1997). These informal techniques are perhaps the most common means of supplementing the distribution-choice techniques of Chapter 12. The formal elicitations of this chapter are generally reserved for the most critical variables in high-visibility and controversial analyses.

This chapter begins by considering what the words used to describe probabilities mean. Then it returns to the topic of subjective probability to consider it in relation to an expert elicitation. A checklist is offered to help you decide when to do an elicitation. We then consider some of the literature's more interesting findings on how people, including experts, make judgments under uncertainty. It turns out that our brains like to rely on some common heuristics. These heuristics present very common pitfalls for estimating subjective probabilities, so elicitations should be designed to avoid these pitfalls. Describing such an elicitation protocol is the subject that follows the heuristics. The chapter concludes by considering the calibration of experts and combining the judgments of multiple experts.

13.2 Probability Words

In our normal lives away from risk analysis, it is common practice for most of us to address uncertainty in our natural language. Things are probable or unlikely, and within the family circle these words function adequately. It makes sense, then, that so many people would approach probability using the qualitative judgments of our natural language. You will find attempts to do so in this book and throughout the risk analysis literature. Many risk assessors have tried to describe probability using natural language terms rather than the more precise statistical notions of probability that are counterintuitive and inaccessible to so many people. To say that an event is probable, likely, or has a good chance of occurring may give the speaker a comfortable feel for the situation, but its meaning is not likely to be clear to anyone. In addition to the speaker's inherent difficulties in dealing accurately with uncertainty, there is the issue of semantics, when the same subjective word or phrase can have different meanings to different people.

Kent (1964), writing for the intelligence community, described three situations that arise in that field. The first was a statement of indisputable fact: "There is an airfield with a 10,000-foot runway." The second was a judgment, a knowable thing not known precisely: "It is almost certainly a military airfield." The third was another estimate, this one made without any direct or indirect evidence: "It would be logical for the bad guys to extend this field sooner or later." The intelligence community was struggling with how to convey their uncertainty about such situations, finding narrative discussion unavoidable. The challenge was to convey what is certain knowledge or reasoned

TABLE 13.1

Five Gradations of Uncertainty Proposed by the Intelligence Community

100%		Certainty
The general area of possibility		
93%	Give or take about 6%	Almost certain
75%	Give or take about 12%	Probable
50%	Give or take about 10%	Chances about even
30%	Give or take about 10%	Probably not
7%	Give or take about 5%	Almost certainly not
0%		Impossibility

Source: Kent (1964).

judgment. They sought to standardize the language of "estimate probability" so that a chosen word or phrase would accurately describe the degree of certainty that would be understood similarly by writer and reader.

They considered using odds to convey meaning. For example, characterizing a statement as more likely than not, more not than likely, and a coin toss. In the intelligence world of closed covenants secretly arrived at, even these statements eluded analysts. They began with words and then tried to assign numbers to them, as seen in Table 13.1.

In what Kent colorfully described as a debate between the poets of risk and the mathematicians of risk, variants for the chosen words grew. *Possible* was also conceivable, could, may, might, and perhaps. *Almost certain* became virtually certain, all but certain, highly probable, highly likely, and odds overwhelming. *Probable* was likely, "we believe," and "we estimate." A *50-50 chance* was chances about even, chances a little better than even, improbable, and unlikely. *Probably not* was "we believe that...not," "we estimate that...not," "we doubt," and doubtful. *Almost certainly not* was also virtually impossible, almost impossible, some slight chance, and highly doubtful. Expanding the formalized language to include all these and other variants was impossible. Many of these phrases represented heresies to the mathematicians of risk. In essence, banning the use of some phrases and enforcing the common understanding of words that belong to the natural language proved impractical.

Since Kent, numerous authors have done studies in an attempt to add precision to the words of probability. These include Lichtenstein and Newman (1967), Moore (1983, 34–35), Boehm (1989, 133), Hamm (1991), British Standards Institute (2000) and Conrow (2003). They have all tried to quantify the numerical range of probability covered by the words of probability. Finding a lack of coherence among these studies that made comparison somewhat problematic, Hillson (2005) e-mailed a survey to members of the Risk Doctor network that asked them to quantify 15 terms, which were presented in alphabetical order. Table 13.2 shows the average of all minimum and maximum scores for each phrase.

Hillson finds good agreement for low-probability terms and less consistent agreement for large-probability terms. Interestingly, *definite*, which usually means 100%, and *impossible*, which means 0%, were interpreted differently. Having worked in areas where probabilities of 1% or less are common, I am struck by how contextual a continuum of phrases must of necessity be. In applications where a 10^{-6} risk is a

TABLE 13.2

Hillson's Results for Values of Probability Phrases

Descriptive Phrase	Range
Definite	76.6%–83.5%
Almost certain	73.6%–84.1%
Highly probable	64.2%–78.3%
A good chance	54.3%–74.3%
Likely	49.9%–68.4%
Quite likely	51.8%–66.4%
Probable	47.5%–66.7%
Better than even	47.1%–65.6%
Possible	28.8%–57.5%
Improbable	10.6%–25.3%
Highly unlikely	9.8%–23.3%
Unlikely	6.6%–20.4%
Seldom	6.2%–17.1%
Impossible	5.5%–11.4%
Rare	3.9%–12.2%

Source: Hillson (2005).

reasonable threshold or where floods with a recurrence interval of 0.002 are considered, it would seem that new vocabulary is needed.

What this quantification research suggests is that even well-defined, commonly used words are interpreted by individuals in unreliable and unpredictable ways. This clearly suggests that there is no consensus about the quantitative and true meaning of these words. Consequently, using natural language words to identify probabilities is not feasible unless they are carefully and precisely defined. A challenge to bear in mind for all would be users of the risk matrix of Chapter 9.

If word categories of probabilities are to be used, they are of dubious value unless they are defined quantitatively. This begs the question, "If numerical definitions are required to define these categories, why give them names?" The reason is that it is normal to express ourselves in words. The definition of relevant probability ranges is a matter of judgment to be determined by the risk assessors and risk managers. The greater difficulty is often obtaining a numerical estimate of a subjective probability, the subject of the remainder of this chapter.

13.3 Subjective Probability

The point of elicitation is to capture an expert's knowledge about some uncertain event or quantity. An event either occurs or not. An uncertain quantity is a random variable, and we are often interested in the probability that it takes a certain value or range of values. Uncertainty about an event is described by specifying a probability. The uncertainty about a random variable X is described by specifying a probability distribution for that variable.

We have previously said that natural variability (i.e., aleatory uncertainty) is due to randomness. Knowledge uncertainty (i.e., epistemic uncertainty) is due to imperfect information about something that is not in itself random. The frequentist view of probability is applicable only to random events and variability. Knowledge uncertainty is more often associated with things that do not repeat, such as the value of a parameter in a risk assessment. To estimate the probabilities of nonrepeating events, subjective probability is used. Subjective probability estimates have also been used to estimate natural variability in the absence of better data.

These probabilities are not preformed numbers lying hidden in our brain cells waiting to be revealed to the world. They are values that must be carefully constructed when needed. According to the subjectivist view, the probability of an event (including an event such as $P(X) > x$) is a measure of a person's degree of belief that it will occur. Thus, probability is not a property of the event but a property of the expert's judgment (Morgan and Henrion 1990).

Experts may vary in their judgments, and so my subjective probability of an event may be different from your subjective probability of the same event. Thus, there are no "correct" subjective probabilities. Subjective probability estimates of knowledge uncertainties will always depend on the knowledge and experience a person has.

Suppose we are interested in repairing Tainter gates on the dams in our country.* A Tainter gate is a cylindrical gate supported by trunnions to control the flow of water over a dam. So the variable of interest is the tensile strength of the trunnions in all the dams in the country.

To differentiate natural variability and knowledge uncertainty, let's distinguish the tensile strength of one trunnion from the average tensile strength of all trunnions. That is, we want to distinguish an individual observation from the sample mean. The tensile strength of the one trunnion is uncertain because of natural variability. That natural variability is completely described by the distribution of tensile strengths in the population of trunnions in the country if we know that distribution.

Now suppose that distribution is not known. This introduces another source of uncertainty to the natural variability. The tensile strength of trunnions from other countries may be known, and the expert may have some knowledge of these data. There still remains knowledge uncertainty about the tensile strength of trunnions in this country. Specifically, the mean tensile strength is unknown. Let us suppose this mean is the value the expert is being asked to estimate.

Is the expert's uncertainty about mean tensile strength natural variability or knowledge uncertainty? We could bump this problem up a notch and imagine a new population, this one of all national means. At this all-nations level we have introduced yet a new notion of uncertainty about the means due to the natural variability among the means of different nations. If your head hurts, reread slowly, this is an important idea.

We have just argued that uncertainty about the mean could be due to knowledge uncertainty, or if we reframe the thought experiment it could be due to natural variability. In the current example focusing on the U.S., uncertainty about the mean is predominantly knowledge uncertainty because this country is not a randomly chosen

* The Tainter gate example is an adaptation of the O'Hagan et al. (2006) example of tree yields that begins on page 12 of their book.

one. Our uncertainty is driven by our lack of knowledge about the tensile strength of trunnions in this country.

The frequentist view of probability is appropriate to use to describe the distribution of tensile strengths, but it cannot be used to describe the mean tensile strength for trunnions in this country. The only way to make probability statements related to this mean value is through subjective probability. However, if the expert is asked about the tensile strength of any one trunnion when the mean strength is unknown, we have both natural variability and knowledge uncertainty. If the mean strength is known and we're asked the strength of a specific trunnion, we are dealing only with natural variability.

The mean tensile strength of the population is a parameter and a constant. Probability distributions are specified by their parameters, as we saw in the previous chapter. These parameter values are often what we ask experts to estimate in a probability elicitation. The uncertainty about them is always knowledge uncertainty because each population is unique and its parameters are constant; there is no natural variability for a constant. Subjective probability, sometimes called personal probability, is needed to address knowledge uncertainty.

Subjective probabilities are often used in risk assessments. To be useful they must be coherent, i.e., they must obey the laws and theories of probability such as those laid out in Chapter 11. Lindley, Tversky, and Brown (1979) posited that the expert subject has a set of coherent probabilities that can be distorted by the elicitation process, so that process must be carefully constructed. Savage (1954) and DeGroot (1970) have argued that a person who wants to make rational decisions when faced with knowledge uncertainty must act as if they have a coherent set of probabilities. Some, like Gigerenzer and Todd (1999), are more forgiving about coherence of the individual and are willing to live with reasonable adaptive inferences about our real-world environments.

If subjective probabilities are not based on formal and coherent constructions of the theory and may in fact be noncoherent, then what status do these probabilities have? Winkler (1967) said there is no built-in prior distribution there for the taking, and different elicitation techniques may produce different distributions. He goes on to suggest that these different distributions are equally valid, even though research has shown that some forms of questioning can lead to invalid responses.

This all means that the elicitation technique matters because experts' statements of probability are likely made in response to the question asked rather than based on preanalyzed and preformed coherent beliefs. The purpose of elicitation is to represent the expert's knowledge and beliefs accurately in the form of a good probability distribution (O'Hagan et al. 2006).

13.4 When to Do an Elicitation

Not every uncertain quantity for which you lack data will be estimated in a formal elicitation process. In a normal risk assessment, most uncertain values will be described using uniform, triangular, pert, or other nonparametric distributions for which the assessors estimate the defining values. Such uncertainties occur routinely and they are treated routinely.

As part of the evaluation of an ecosystem restoration project downstream of Tenkiller Dam in Oklahoma, our team developed habitat evaluation procedure (HEP) models (U.S. Fish and Wildlife Service 1980) for catfish, bass, and trout. The model consisted of hundreds of variables about which there was uncertainty, either natural variability, knowledge uncertainty, or both. The model output was only one of the decision criteria for this project. Most variables in the HEP models were described by minimum, most likely, and maximum estimates provided by two local resource agency employees. There was no formal elicitation process. They were more than competent to answer questions like, "What is the minimum water temperature in this stretch of the river during the summer months?" In addition, the HEP models were not terribly sensitive to minor changes in the input variables, and the budget for the analysis was about $14,000 in a multimillion dollar project.

You do not need to complete a formal elicitation process for every uncertain variable in your model for which you lack sufficient data to choose a distribution in the manner described in the previous chapter. So when do you need to do an elicitation? That depends on your decision context and the goals of the risk management activity. The more the list below tends to describe your decision problem, the better served you'll be by a formal elicitation process.

- Your problem is complex and highly visible.
- Trust in your analytical work is an issue.
- Your risk assessment will provide the basis for public decision making.
- Your risk assessment will provide the basis for a formal rule or regulation affecting industry, nongovernmental organizations (NGOs), or others.
- Your organization has stewardship responsibility for some public value, e.g., public health, public safety, water resources, homeland security, and the like.
- There are values in conflict for your issue.
- Your issue is a controversial one.
- There are many stakeholders with diverse views.
- Your issue is subject to intense media or other scrutiny.
- One or more stakeholders are likely to challenge your decision in a court or administrative setting.
- Human life, health, or safety may be affected by your decision.
- There is a clearly critical uncertain variable or two in your risk assessment model.

The ecosystem restoration project mentioned above met none of these criteria and could be handled through the available sparse data and the expert judgments of those responsible for the analysis. As the importance of the decision grows, as indicated by the preceding list, so does the need for a more formal and credible process for estimating probabilities and uncertain values in our models.

13.5 Making Judgments under Uncertainty

To develop useful elicitations procedures and to obtain useful probability estimates, it helps to understand how people form judgments about uncertain values. Research suggests that experts and laypersons alike rely on the use of some common heuristics in forming judgments about uncertain quantities (Tversky and Kahneman 1974). Knowing about these shortcuts and their pitfalls is the first and most important step in avoiding them. Elicitation processes that avoid reversion to these heuristics or at least minimize their impact are better than those that do not.

Our brains are wired in such a way that it is easier to think dramatically than it is to think quantitatively, and especially probabilistically. Probability is not intuitive, as we have seen. It is not easy to think in probabilistic terms. Instead, we tend to use some rather crude cognitive techniques to help us along. The most common of these heuristics are discussed below.

13.5.1 Availability

Estimates of uncertain quantities are driven by the ease with which previous similar events or instances can be recalled. The more easily examples come to mind, the more probable we think the event is. Similarly, the easier it is to recall instances of a class of events, the larger we estimate that class to be. When the expert has extensive personal experience, this heuristic can be helpful. For the rest of us, this heuristic is used instead of the more complex processes we might use to estimate probabilities.

There are certain biases that can show up when the availability heuristic is used. When low-frequency events receive extensive media coverage—terrorist attacks, levee failure in New Orleans, plane crashes, oil-drilling disasters—it is common to overestimate their probability of occurrence because it is easy to call examples of these things to mind. On the other hand, the probabilities of more common events that do not get the same attention—seasonal flu, automobile accidents, choking on food—are often underestimated.

Recent events or events that have affected the expert personally may be emotionally paper-clipped memories that are easier to recall. This can cause a temporary increase in the estimates of the likelihood of these events. If a family member is diagnosed with leukemia, it is more likely that a person will overestimate the probability of this disease. The flipside is also true. If there have been no recent occurrences (of floods, for example) or it is a problem that has not touched the expert personally (e.g., fire in the home), probabilities may be underestimated.

Co-occurrences of events can also distort estimates through an availability bias. I grew up at a time when many family members and neighbors associated space launches and weather events. This connection of events was more illusion than cause and effect. After awhile, the hurdle for unusual weather was dropped and the window of opportunity for an occurrence was extended, all of which had the effect of a self-fulfilling prophecy. Eventually people saw more co-occurrences than actually happened. A heavy rain, an extreme temperature within weeks of a launch, or a heavy snowfall in the same year could be associated with the space launch. These events are more strongly associated in the mind than in fact. Beliefs about relationships can affect judgments for all kinds of uncertainties. If availability misleads the expert to misinterpret the strength of a relationship, it can affect probability estimates (Gilovich, Griffin, and Kahneman 2002, Sedlmeier and Betsch 2002). Linked events in a scenario, for example, are easier to imagine than the individual events considered in isolation.

The availability bias often operates indirectly on judgments when an expert is asked about an event that has never happened before and prior events are unavailable. If the expert uses classes of similar events that are available to reason about this particular event, the availability bias can still affect the estimate. Experts are prone to extrapolate from one situation to another. Good elicitation protocols should neutralize the availability heuristic.

13.5.2 Representativeness

We tend to expect relationships that hold true in the large to be reflected in the small. Tversky and Kahneman (1983) define representativeness as "an assessment of the degree of correspondence between a sample and a population, an instance and a category, an act and an actor, between an outcome and a model." The way this works is that we tend to compare information we have about a category of events or things with the stereotypical member of the category. The more similar the two, the higher we estimate the probability of membership in the category.

People can be inclined to assess the representativeness of an event of some more familiar class instead of the probability of the event. This is especially relevant to the estimation of conditional probabilities. Common representativeness biases are summarized in the following sections based primarily on the work of O'Hagan et al. (2006).

13.5.2.1 Conjunction Fallacy

Tversky and Kahneman (1983) offered this oft-repeated example of the conjunction fallacy.

> Linda is 31 years old, single, outspoken, and very bright. She majored in philosophy. As a student, she was deeply concerned with issues of discrimination and social justice, and also participated in antinuclear demonstrations.
>
> Which is more probable?
>
> Linda is a bank teller.
> Linda is a bank teller and is active in the feminist movement.

Most people (usually 80% or more) select the second answer, and this is impossible. How is it possible that Linda is more likely to be a teller (A) and active (B) than that she is just a teller (A). The second answer is a joint probability, the conjunction of two conditions (A and B). There is no way there can be more people who are tellers and activists than there are people who are just tellers. Surely some tellers are not activists. So it must be more likely that she is a bank teller.

What is at work here is that the description of Linda is unrepresentative of what most of us think a bank teller to be. This tends to decrease our belief that she is a teller. However, adding the active feminist detail provides a description of Linda that is more representative of an activist. We tend to rely more on the representativeness than we do on the simple facts of probability to gauge the likelihood of the right answer.

13.5.2.2 Base-Rate Neglect

A base rate is a relative frequency. People will often ignore information about the base rate and estimate probabilities based solely on the information about the event or instance before them. If you have ever been to the doctor and been instructed to have a laboratory test for a rare and serious disease, the chances are good that your emotional response illustrated base-rate neglect. The simple fact is, absent other specific information, it is highly unlikely you have the rare disease. When the doctor asks you to have the test, you ignore this fact and focus on the fact that she has asked you to get this test. We are likely to overestimate the likelihood that we have the disease.

Tversky and Kahneman (1982) offered the following example. Suppose the percentage of homosexuals who have the disease is three times higher than the percentage of heterosexuals who have the disease. Pat is diagnosed with the disease. This is all that you know about Pat; you do not even know if Pat is male or female, much less Pat's sexual orientation. What is the probability that Pat is homosexual?

The disease is more representative of homosexuals than heterosexuals, so the representativeness bias would cause most people to overestimate the probability Pat is homosexual. The actual answer depends on the real numbers, so let's use a population of 100, 10% of whom are homosexual (10). Of them 3 have the disease (30%). There are 90 heterosexuals with a 10% rate of disease (9). Three of 12 people (3 + 9) who have the disease are homosexual, so 25% of all diseased people are homosexual, and the probability that Pat is homosexual is 25%. Base-rate neglect pushes the probability estimate closer to 75% for many people (75% is three times 25%, reflecting the higher prevalence for homosexuals).

13.5.2.3 Law of Small Numbers

People often expect the properties of a population to be manifest in a small sample. We are inclined to think sample data are representative of the population and do not understand how the variability in a small sample can be very unrepresentative of the population's true parameters.

People who have lived a long time on the coast tend to think they have seen just about every kind of storm. So when they are warned to evacuate, many think they can ride it out like they have the vast array of other storms they've seen in their lifetime. Although 70 years on the coast may be a long time in human terms, it is a trivially small sample in geologic time, and many people have lost their lives to the law of small numbers by overestimating their experience and underestimating the strength of storms.

Take another example exaggerated for effect. The United States is a country of over 300 million people. Imagine we take a random sample of ten people and use it to profile the age, gender and race attributes of the United States. If five of our people are, by chance, African American, two are Asian, and three are white, we would have a very biased view of the population. We would overestimate the

TABLE 13.3

Contingency Table for Disease Condition and Test Result

	Test+	Test–	Total
Disease+	800	200	1,000
Disease–	9,900	89,100	99,000
Total	10,700	89,300	100,000

probability of some subpopulations and underestimate others like the Hispanic and white populations.

Things that by chance show up in fewer than representative numbers in our small samples get underestimated. Likewise, things that show up more in the sample than the population are overestimated. The tendency to expect samples to be representative of populations can distort our view of reality.

13.5.2.4 Confusion of the Inverse

People tend to confuse conditional probabilities with their inverse. They tend, for example, to think the P(T+|D+) is the same as P(D+|T+). Even doctors have been known to fall prey to this error. A doctor may see a positive test as being representative of having the disease, and she may see having the disease as being representative of a positive test. This could simply be semantic confusion or it could be failure to apply Bayes' theorem. In either event it can lead people to misstate probability estimates.

Consider Table 13.3 for an example. The probability that a person is disease-positive is 1%. The test accurately predicts the disease when present (sensitivity) 80% of the time. It correctly predicts the absence of disease 90% of the time. Imagine a patient is given the test for the disease and it comes back positive. What is the probability that the person has the disease? In a study by Plous (1993), doctors said it was about 75%. In fact, the value we seek is P(D+|T+) and that is 800/10700 or 7.5%! The inverse, P(T+|D+) = 800/1000 = 80%. Clearly, the doctors had inverted the conditional probability, another form of the representativeness bias that can seriously distort probability estimates.

*13.5.2.5 Confounding Variables**

When trying to predict one value based on other values known to be related, we rarely give sufficient consideration to confounding variables. Most people recognize a relationship between height and weight. There is a tendency to make simple translations from one variable to the next. For example, if a person weighs 10% more than average, there is a tendency to expect them to be 10% taller than average. This translation from weight to height simply does not bear up in reality. There are many other variables that can affect or confound the relationship between height and weight, rendering it far from the representative translation our minds tend to prefer. When we estimate uncertainties based on imperfect relationships, we can over- or underestimate them.

* The title for this effect is the author's invention and not that used by O'Hagan et al. (2006).

There are, as you see, several forms of representativeness bias. They can be subtle and often difficult to counteract in a probability elicitation.

13.5.3 Anchor and Adjust

When asked to estimate an uncertain quantity, we tend to come up with an initial estimate that is often our estimate of a most likely value like a mean, median, or mode. Call this an anchor. We then tend to make adjustments up or down to this value based on new circumstances. The major problem with this bias is that we rarely make sufficient adjustments for extreme values.

This heuristic cuts down on mental processing time. It avoids the need to reprocess information and recalculate values. The problem is that adjustments are anchor-centric, and we tend to stay too close to the anchor. The anchors we come up with are often arrived at rather spontaneously, often with little analytical thought. Unless our anchors are challenged, we often do not understand the role they play in subsequent judgments.

Don't use the Internet to answer this question; use your expertise. What is the driving distance from Sacramento, California, to Little Rock, Arkansas? I do not expect you to be able to guess such a quantity precisely, so give me the minimum mileage it could be and the maximum mileage it could be so that you are 90% sure you have estimated correctly. In other words, estimate a realistic interval for the distance, not it is somewhere between zero and a million miles!

If you are like most people, when the original question was asked, a number popped into your mind. You somehow came up with a most likely or best-guess number. That is your anchor. When asked for a minimum and a maximum, most people tend to subtract a number from the anchor and then add the same number to the anchor to get their interval. Did you? Did you make enough of an adjustment to capture the true distance?* Even experts rarely do so.

Assessments of probabilities as well as other quantities are subject to the anchor-and-adjust bias. Many elicitation questions come with anchors embedded in them. For example, we may ask the towboat captain the probability it takes more than 2 hours to get from point A on the waterway to point B. Two hours has now been established as an anchor. Alternatively, the question could be asked as, "State the time at which there is a 60% chance you could travel from point A to point B on the waterway." The anchor is now 60%. These values could affect the response to all subsequent questions.

13.5.4 Motivational Bias

In addition to the cognitive biases identified here, there is the possibility of motivational bias. People may want to influence a decision; they may perceive that they will be evaluated based on the outcome; they may suppress their uncertainty to appear more knowledgeable; or they may have a strong position on a question (Morgan and Henrion 1990). Some people may just be cockeyed optimists while others are perpetual pessimists. Any of these outlooks can bias an estimate.

* Google Maps reports the distance as 1,960 miles by car.

People, including experts, sometimes find an incentive to offer information that does not reflect their expert knowledge. People who have figured out what kinds of answers are most likely to secure more desirable outcomes may be motivated to provide those answers regardless of their true beliefs. In some situations, the rewards and punishments associated with over- or underestimating quantities vary significantly. Underestimating a human health risk has a greater risk to the expert than overestimating that risk, for example. Criminal lawyers will rarely tell you things will be fine. They are inclined to emphasize the seriousness of the defendant's situation. It makes higher fees easier to swallow. Estimates of uncertain quantities may be affected by implicit incentives as well as explicit ones.

13.6 Responding to Heuristics

Although the research on heuristics that can distort our probability estimates is fascinating, our concern as risk assessors is: "Can we ask elicitation questions in such a way as to remove these biases? What can we do with this knowledge of these heuristics to get better probability estimates?" Koehler (1996) found that several of the problems associated with representativeness can be improved by eliciting frequencies instead of single-event probabilities. For example, instead of asking the probability for "Linda is a teller and an activist," we might ask, "Of 100 individuals like Linda, how many would be bank tellers and how many would be active in the feminist movement and bank tellers?" (Fiedler 1988).

Frequency formats help reduce representativeness bias as compared to probability of single-event question formats. Unfortunately, they also increase availability problems. The anchor-and-adjust heuristic has important implications for the order and nature of questions posed in an elicitation. It can be reduced by avoiding anchors, i.e., exploring the upper and lower limits the quantity can take before examining central tendencies.

Clemen and Reilly (2001) offer suggestions for improving estimates based on knowledge of these heuristics. First, awareness of the heuristics can help people learn to become better assessors of probabilities. Knowing how not to think does not guarantee one will not think in these ways, but it does give the conscientious assessor a fighting chance. If assessors can recognize the effects as they occur, they may be able to avoid or countervail them.

Second, the technique used to assess probabilities may reduce the effects of the heuristics. Requiring people to think about probabilities in a more structured way, e.g., using lotteries to estimate probabilities, is one option.

Third, decomposing a problem into several component events may help reduce the effects of overconfidence. It means conducting more but simpler elicitations.

O'Hagan et al.'s review of the literature finds that many of the biases mentioned here are ephemeral and can be eliminated entirely in some instances. The literature also suggests some people are more prone to these biases than others. More intelligent people are more likely to be able to reason through the biases and avoid them. The book by O'Hagan et al. (2006) is an excellent resource for accessing that literature. The bottom line, however, remains the same. A careful elicitation protocol is the best way to address these problems.

13.7 The Elicitation Protocol

People do not have well-formed subjective probabilities in their heads. Asking people to make up numbers out of the blue is not a good idea. There is a great deal of literature that says people are not especially good at estimating probabilities, especially small ones. We tend to be amazingly overconfident in our estimates of uncertain things, and we rely on the heuristics discussed above.

Using elicitation techniques that avoid reliance on heuristics and that force people to think about the numbers and calibrate themselves can be useful. Our goal is to improve the elicitation process and its results. If there are systematic ways we tend to err, perhaps we can avoid them with well-structured formal protocols for assessing uncertainty and eliciting probabilities.

There are any number of elicitation protocols described in the literature. Three are briefly summarized here. The first is a seven-step method proposed by Clemen and Reilly (2001). The steps are:

1. Background identification of variables requiring expert assessment; review of scientific literature
2. Identification and recruitment of experts
3. Motivating experts—establish rapport with experts and generate enthusiasm for the process
4. Structuring and decomposition—explore the experts' understanding of causal and statistical relationships among the relevant variables
5. Probability assessment training—explain the principles of the assessment, the biases, and ways to counter them; provide opportunities for experts to practice making probability assessments
6. Probability elicitation and verification—experts make required assessments under the guidance of a facilitator; assessments checked for consistency and coherence
7. Aggregation of experts' probability distributions—if multiple experts were used, it may be necessary to aggregate their assessments

ROLES IN ELICITATIONS

Decision makers will use the results of the elicitation (the client).
Substantive experts have the knowledge about the uncertain quantities.
Normative experts provide probabilistic training, assess the process, validate results, and provide feedback.
Facilitators are experts in the elicitation process and manage the elicitation.

***Source*: O'Hagan et al. (2006).**

The second is the Stanford/SRI Assessment Protocol (Morgan and Henrion 1990), which consists of five phases:

1. Motivating—establish rapport with experts, explain need and process, explore motivational bias
2. Structuring—develop unambiguous definition of quantity to be assessed in a form meaningful to experts
3. Conditioning—get experts to think fundamentally about their judgments while avoiding cognitive biases
4. Encoding—elicit and encode the expert's probabilistic judgments
5. Verifying—test judgments to see if they reflect expert's beliefs

The third is a five-step model for elicitation developed by O'Hagan et al. (2006).

1. Background and preparation—the client identifies variables to be assessed
2. Identify and recruit experts—the choice may be obvious or it may require some effort; six criteria for experts:
   a. Tangible evidence of expertise
   b. Reputation
   c. Availability and willingness to participate
   d. Understanding of the general problem area
   e. Impartiality
   f. Lack of economic or personal stake in the potential findings
3. Motivating and training the experts—assure the experts that uncertainty is natural; training should have three parts:
   a. Probability and probability distributions
   b. Introduction to most common judgment heuristics and biases as well as ways to overcome them
   c. Practice elicitations with true answers unlikely to be known by the experts
4. Structuring and decomposition—spend time exploring dependencies and functional relationships that meet agreement by experts; precisely define quantities to be elicited
5. The elicitation—an iterative process with three parts
   a. Elicit specific summaries of expert's distribution
   b. Fit a probability distribution to those summaries
   c. Assess adequacy: if adequate, stop; if not, repeat the process with experts making adjustments

Elicitations are best in a face-to-face setting that allows interactions between the expert and the facilitator. They may be one-on-one or group elicitations. All protocols have several stages, but the elicitation always comes at or near the end of the protocol. Its success depends on preparation and foundations established in the preceding steps. Each of these protocols is described in greater detail in the reference documents.

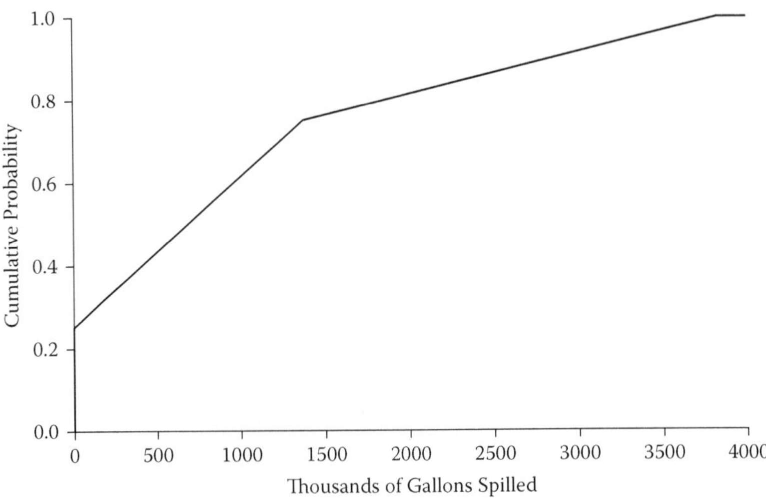

FIGURE 13.1 CDF of the size of an oil spill in the Delaware River.

13.7.1 Eliciting Probabilities

The elicitation step is where the information is developed. The expert's statements of probability are judgments made in response to the facilitator's questions. They are not pre-analyzed and preformed beliefs of the expert. Thus, the matter of how one asks the questions is of some importance.

The elicitation is likely to be seeking one of three kinds of information: a point estimate of some sort, a probability distribution, or the parameters of a specific probability distribution. To better frame the discussion, assume we are interested in estimating the size of a crude oil spill in a waterway that serves several refineries. A point estimate might be used to estimate the probability one or more oil spills will occur this year. A probability distribution that represents the expert's beliefs about the total number of gallons spilled in an event might look like the example in Figure 13.1.

This is a free-form estimate based on an expert's estimation of the quartile values shown Table 13.4. The curve was constructed by linearly interpolating between the points provided. Note that to define such a distribution, the facilitator can elicit values for x_i, in this case gallons of oil spilled, associated with a given probability (p_i), or the facilitator could provide the x value and ask for associated p values.

TABLE 13.4

Oil Spill CDF Elicited from Experts

Cumulative Probability	Thousands of Gallons Spilled
0	0
.25	5
.5	668
.75	1353
1	3850

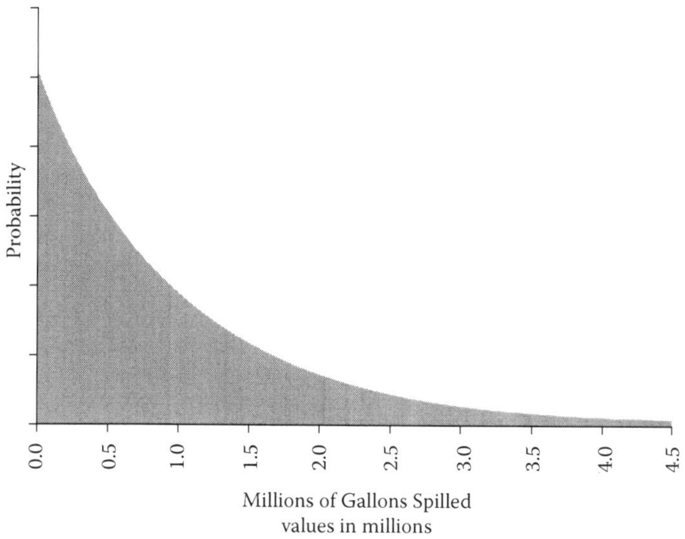

FIGURE 13.2 Exponential distribution describing the size of an oil spill in the Delaware River.

The third basic elicitation would be to estimate the uncertain parameters of a distribution. This is somewhat similar to the point-estimate elicitation in concept. For this example, imagine it has been determined the size of an oil spill is best represented by an exponential distribution. This assumption imposes a structure on the oil spill problem. The experts are needed to estimate the parameters of the distribution. In this instance, the exponential has only one parameter, the mean. Assuming our experts have estimated this value to be 988,000 gallons, the exponential distribution would be the one shown in Figure 13.2.

We'll focus on issues associated with the second kind of elicitation. That will subsume many of the issues associated with the other elicitations, because it is the more involved process. The size of an oil spill is a continuous random variable. Assuming we are going to elicit the expert's CDF, there are an infinite number of points that could be elicited. The strategy is to elicit information about a relatively few number of individual (p,x) pairs that enable us to reproduce the expert's beliefs in the form of a probability distribution.

The information sought is called a summary. The idea is that a few well-chosen summaries will enable us to identify the expert's beliefs with a high degree of precision. Then, using these summaries, a curve is fitted to them. This can be done by interpolation as shown previously or by fitting a general form function to the points.

The summaries we obtain can, as noted, be based on probabilities or on quantity values. Summaries based on probabilities may be the most common form of elicitation. A basic strategy is to seek specific quantiles that can be used to construct a CDF, beginning with the median value, $p = .5$. Note that this kind of questioning invites the anchor heuristic if one is not careful to neutralize it. This divides all possible values in half. The expert is then asked to find the midpoint of each remaining range. This method produces the median, then the quartiles, then the 12.5, 37.5, 62.5, and 87.5

percentiles, etc. It becomes difficult for the expert to go much beyond these values. Sample questions adapted from O'Hagan et al. (2006) for the oil spill example follow.

> Can you determine an oil spill size such that any spill is as likely to be larger than this number as smaller than this number?
>
> Suppose we have learned the oil spill is smaller than your median. Can you now determine a new oil spill size (lower quartile) such that it is equally likely that this oil spill is larger than or smaller than this value?
>
> Suppose we have learned the oil spill is larger than your median. Can you now determine a new oil spill size (upper quartile) such that it is equally likely that this oil spill is larger than or smaller than this value?

For contrast, an alternative formulation of the second question might be, "Give an oil spill size such that you think an actual oil spill has a 25% chance of being less than it." There can be many formulations of your specific questions. None of them are transparently simple to comprehend.

In such a fashion a distribution can be elicited. However, there are a great many nuances and lessons learned that are documented in a large and growing literature. The O'Hagan et al. text, referenced so often in this chapter, is essential reading for anyone who intends to work in this area. The Ayyub (2001) text is also an essential reference for those interested in more details on the topics introduced in this chapter.

Fixed-value methods work a little differently than the quantile method. If we are deriving a CDF we might ask, "What is the probability that an oil spill will be 5,000 gallons or less?" Successively larger spill sizes can be introduced to flesh out the shape of the CDF. My experience, and that of others, suggests that an assessor can fall into a biased rhythm of answering if the facilitator's questions have a discernible pattern to them. The pattern could be as rigid as asking about oil spills in 5,000-gallon intervals or as subtle as asking questions in a monotonically increasing (or decreasing) way. Varying the pattern of your questions can encourage a more analytical response.

For example, one might ask the probability that a spill is 5,000 gallons or less, followed by 150,000 gallons or less, then 70,000 gallons or less, followed by 1,000,000 gallons or less, and so on. The idea is to vary the pattern of the questions to avoid simple interpolation or other biased answering schemes and to require more careful thinking about the probabilities. One could vary the question by asking the probability that a spill is 70,000 gallons or more, instead of less. Such questions can help identify problems with coherence in the expert's assessment of the uncertain quantity. Bounding the CDF, i.e., establishing minimum and maximum values, is usually going to be a sensitive aspect of the elicitation. Given the anchor-and-adjust heuristic, it is especially important to probe the limits carefully (see text box).

It is also possible to ask questions that approximate a PDF for an unknown variable. In this instance we might ask, "What is the probability that if an oil spill occurs it will be between 10 (or whatever the agreed upon minimum size of an oil spill is) and 5,000 gallons? Greater than 5,000 gallons but less than 25,000 gallons?" and so on. The notion for the PDF elicitation is to seek the probability of a quantity being within a fixed interval.

BOUNDING A DISTRIBUTION

When eliciting an expert's distribution, it is often necessary to establish a minimum and maximum value for the quantity of interest. Neither data nor experience are likely to reveal absolute extreme values when logical limits do not exist (0 is a natural limit for many quantities, and the [0,1] interval applies to a number of problems).

When experts must estimate these values, it is important to find nonthreatening ways to challenge their estimates. In the oil spill example, one might ask, "If you were gone for a year and returned to learn there had been a spill of X gallons (some value noticeably larger than the expert's maximum estimate), how could you explain that?" If the expert offers a plausible explanation, offer him the opportunity to amend his estimate.

Probe the limits carefully.

An alternative approach to elicitation is to ask the expert to draw points for the CDF. This is generally not a terribly realistic way to approach the problem, but it has been used in the past.

Estimates of probabilities can sometimes be better analyzed by decomposing the problem. An oil spill problem can be decomposed by focusing on the vessel's size or the cause of the spill. This can increase the number of elicitations required, or it can entail elicitation of multivariate distributions, which are far more complex. If a problem is to be decomposed, the decomposition can be prescribed by the investigators or by the experts. The latter is more feasible when a single expert is being elicited.

Think about whether probability values will be sought as probabilities, percentages, or odds to improve the quality of the estimate. Choose a method that suits the expert assessor's cognitive styles, skills, and preferences (Morgan and Henrion 1990). I have worked with experts who did not finish high school (towboat operators) and Ph.D.s (epidemiologists and engineers). No one-size probability elicitation fits all needs. It is best to establish the preferred or most promising question style in the foundational work that precedes the elicitation, including the practice elicitations.

13.8 Calibration

Morgan and Henrion (1990) reported, "The most unequivocal results of experimental studies of probability encoding has been that assessors are poorly calibrated." If we go to all the trouble of a formal subjective probability elicitation, it would be nice if we could have some confidence in its results. The tricky part is if we say there is a .9 chance an oil spill will be less than 1 million gallons and a 2-million gallon spill occurs, does that mean we were wrong? The answer, of course, is no. Only statements of absolute certainty, i.e., $p = 0$ or $p = 1$, can be proven right or wrong by subsequent observations.

That means the strength or weakness of our elicitation more or less rises and falls with our process and our experts. Experts who are overconfident overestimate the actual frequency of events in their subjective probabilities. They might estimate a spill size that occurs with a .4 probability as having a .7 probability. Underconfident experts do the opposite. They might estimate the probability of that spill as .3. Both the overconfident and underconfident exhibit this tendency for the entire curve.

Overextremity is another common pattern of estimation by experts. This is neither over-nor underconfident; rather it is a mix of both. These folks tend to overestimate the likelihood of low-probability events and underestimate the likelihood of high-probability events.

If it is important enough, your experts can be calibrated so you and they can learn how well they tend to recognize and evaluate their uncertainty. One common way to calibrate an expert is to ask a series of questions with factual answers that are unlikely to be known by the expert.

There are three general styles of calibration. On relies on asking a large number of a variety of questions as well as the expert's confidence in the correctness of the answers. The questions might include such things as how many gallons in an acre-foot of water, what is the capital of Zambia, and what is the currency used in Thailand? Experts are asked to express the likelihood that their answer is correct. With enough such questions, one would expect a well-calibrated expert to get about 60% of the questions she said she was 60% confident in to be correct, and so on.

Binary calibrations are a second common technique. This consists of a series of true/false questions such as the state of Idaho borders Iowa, the Baltimore Colts won the 1973 Super Bowl, blue light has a shorter wavelength than green light, and so on. The expert answers each question and then identifies the confidence that the answer is correct as 50%, 60%, 70%, 80%, 90%, or 100%. You total all of their percentages and convert it to a decimal to identify the expected number of correct answers. Suppose there are 100 questions and the sum of all confidence ratings is 6240%. As a decimal that is 62.4, so that is the expected percentage of correct answers. An expert with fewer than 62 correct answers is overconfident. One with more than 62 correct answers is underconfident. The expert with close to 62 correct answers is well calibrated.

A third technique requires experts to estimate quantitative questions with 90% confidence intervals. They might be asked the distance from Kiev to Paris, the height of Mount Denali, the mean weight of an adult male boxer dog, and so on. A well-calibrated expert will get 90% of the answers within his interval estimate. Those with fewer than 90% correct are overconfident, and those with more than 90% are underconfident. Other more sophisticated techniques and scoring rules, like the Brier Score (Brier 1950), have also been used.

The calibration process, prior to an elicitation, is one of the most effective ways to make an expert aware of the overconfidence bias. Once aware, many educated experts are able to neutralize that bias.

13.9 Multiple Experts

Multiple experts will frequently be used when an elicitation is needed. Assuming the task is to elicit a probability distribution, the group can be used to create a single distribution (behavioral aggregation), or each individual expert may elicit her own distribution. In this latter case it may be desirable or necessary to eventually combine them all into a single distribution (mathematical aggregation).

One common model for elicitation is to have individual experts develop their own distributions and then to share these results, anonymously, with the group of experts. In a plenary session, they consider their results in discussion and then allow the experts to modify their elicitations based on what they learn from the discussion that ensues. This method still produces as many distributions as there are experts.

There are a number of techniques available for aggregating these distributions. Bayesian methods (Clemen and Winkler 1999), opinion pooling that assigns weights to the distributions (including equal weights), and Cooke's method (Cooke 1991) are all in common usage. O'Hagan et al. (2006) suggest that the average of the distributions provides a simple, general, and robust method for aggregating expert knowledge. Averages may be computed across probabilities (average all x values for each selected probability) or across x value (average all probabilities for a given x value). The former is most common. I have also had success using each of the individual distributions as a sort of sensitivity analysis, to establish whether or not the differences among the experts make much of a difference in model outputs and decisions. This method has the advantage of providing what amounts to high and low estimates of the outputs of interest, which may help bound the uncertainty about these values.

Group elicitation seems to offer substantial and relatively untapped potential for modelers. O'Hagan et al. argue that it offers better synthesis and analysis of knowledge through group interaction. The facilitator's role grows ever more important when group dynamics play such a significant role. The process should encourage sharing knowledge as opposed to opinions. It should also recognize expertise and differences in expertise as well as make effective use of feedback from the group. At the same time, the process must avoid domination by individuals or shared knowledge. Groups can have an even stronger tendency toward overconfidence, especially if group-think sets in. The challenge of avoiding group reliance on heuristics remains with a group process.

13.10 Summary and Look Forward

There will be times when you simply do not have all the data you wished you had for your risk assessment. Some of those times you will not have the luxury of being able to commission and wait for research to fill those voids. Subjective probabilities can be useful in filling in data gaps and voids.

Spoken language has proven to be notoriously imprecise when trying to express our beliefs about things that are uncertain. Hence, we have come to rely on numerical estimates of probabilistic information and probabilistic estimates of numerical information. The most consistent result in all the related literature is that we are notoriously bad at estimating probabilistic information. There are several well-known heuristics we have come to rely on that short-circuit any analysis of these uncertain data. Consequently, it is important to devise and use elicitation techniques that can limit the damage done by these heuristics.

Not every subjective probability estimate warrants a formal elicitation. Nonetheless, assessors are well advised to follow good elicitation techniques whenever they estimate subjective probabilities.

Now that we have a pretty good idea how probability distributions are chosen and elicited, we turn in the next chapter to the Monte Carlo process to see how this technique is used. Probability distributions are used more often in simulations than analytical calculations in risk assessment, and the Monte Carlo process is an extremely powerful and useful technique for propagating the uncertainty in our model inputs through a model. Chapter 14 takes a peek behind the curtain of the Monte Carlo process to see how it works so you can better understand how your risk assessment models work.

REFERENCES

Ayyub, Bilal M. 2001. *Elicitation of expert opinions for uncertainty and risks*. Boca Raton, FL: CRC Press.

Boehm, B. W. 1989. *Software risk management*. Piscataway, NJ: IEEE Computer Society Press.

Brier, Glenn W. 1950. Verification of forecasts expressed in terms of probability. *Monthly Weather Review* 78:1–3. http://docs.lib.noaa.gov/rescue/mwr/078/mwr-078-01-0001.pdf.

British Standards Institute. 2000. *British Standard BS6079-2:2000 Project Management—Part 2: Vocabulary*. London, U.K.: British Standards Institute.

Clemen, Robert T., and Terence Reilly. 2001. *Making hard decisions with DecisionTools*. Pacific Grove, CA: Duxbury Thomson Learning.

Clemen, Robert T., and Robert L. Winkler. 1999. Combining probability distributions from experts in risk analysis. *Risk Analysis* 19 (2): 187–203.

Conrow, E. H. 2003. *Effective risk management: Some keys to success*. 2nd ed. Reston, VA: American Institute of Aeronautics and Astronautics.

Cooke, Roger M. 1991. *Experts in uncertainty: Opinion and subjective probability in science*. Oxford, U.K.: Oxford University Press.

DeGroot, Morris H. 1970. *Optimal statistical decisions*. Hoboken, NJ: Wiley Interscience, 2004.

Fiedler, Klaus. 1988. The dependence of the conjunction fallacy on subtle linguistic factors. *Psychology Research* 50 (2): 123–129.

Gigerenzer, Gerd, Peter M. Todd, and ABC Research Group. 1999. *Simple heuristics that make us smart*. New York: Oxford University Press.

Gilovich, Thomas, Dale Griffin, and Daniel Kahneman, eds. 2002. *Heuristics and biases: The psychology of intuitive judgment*. Cambridge, U.K.: Cambridge University. http://assets.cambridge.org/97805217/92608/sample/9780521792608ws.pdf.

Hamm, R. M. 1991. Selection of verbal probabilities: A solution for some problems of verbal probability expression. *Organisational Behaviour & Human Decision Processes* 48 (2): 193–223.

Hillson, David A. 2005. *Describing probability: The limitations of natural language*. Originally published as a part of PMI Global Congress 2005 EMEA Proceedings, Edinburgh, U.K. http://www.risk-doctor.com/pdf-files/emeamay05.pdf.

Kent, Sherman. 1964. Words of estimative probability. *Studies in Intelligence* 8 (4): 49–65. https://www.cia.gov/library/center-for-the-study-of-intelligence/csi-publications/books-and-monographs/sherman-kent-and-the-board-of-national-estimates-collected-essays/6words.html.

Koehler, J. J. 1996. The base rate fallacy reconsidered: Normative, descriptive and methodological challenges. *Behavioral and Brain Sciences* **19 (1)**: 1–53.

Lichtenstein, S., and J. R. Newman. 1967. Empirical scaling of common verbal phrases associated with numerical probabilities. *Psychonomic Science* 9 (10): 563–564.

Lindley, D. V., A. Tversky, and R.V. Brown. 1979. On the reconciliation of probability assessments. *Journal of the Royal Statistics Society* 142 (2):146–180.

Moore, P. G. 1983. *The business of risk*. Cambridge, U.K.: Cambridge University Press.

Morgan, M. Granger, and Max Henrion. 1990. *Uncertainty: A guide to dealing with uncertainty in 874 quantitative risk and policy analysis*. Cambridge, U.K.: Cambridge University.

O'Hagan, Anthony, Caitlin E. Buck, Alireza Daneshkhah, and J. Richard Eiser. 2006. *Uncertain judgements: Eliciting experts' probabilities*. West Sussex, U.K.: John Wiley & Sons.

Plous, Scott. 1993. *The psychology of judgment and decision making*. New York: McGraw-Hill.

Savage, L. J. 1954. *The foundations of statistics*. New York: Wiley.

Sedlmeier, Peter, and Tilmann Betsch. 2002. *Etc: Frequency processing and cognition*. Oxford, U.K.: Oxford University Press.

Tversky, Amos, and Daniel Kahneman. 1974. Judgment under uncertainty: Heuristics and biases. *Science* 185 (4157): 1124–1131.

———. 1982. Evidential impact of base rates. In *Judgment under uncertainty: Heuristics and biases*, ed. Daniel Kahneman, Paul Slovic, and Amos Tversky, 153–160. Cambridge, U.K.: Cambridge University Press.

———. 1983. Extensional versus intuitive reasoning: The conjunction fallacy in probability judgment. *Psychological Review* 90 (4): 293–315.

U.S. Environmental Protection Agency. 1997. Risk Assessment Forum. *Guiding principles for Monte Carlo analysis*. Washington, DC: Environmental Protection Agency.

U.S. Fish and Wildlife Service. 1980. *Ecological services manual: Habitat as a basis for environmental assessment*. Washington, DC: Fish and Wildlife Service. http://www.fws.gov/policy/ESM101.pdf.

Winkler, Robert L. 1967. The assessment of prior distributions in Bayesian analysis. *Journal of the American Statistical Association* 62 (319): 776–800.

14

Monte Carlo Process

14.1 Introduction

We have considered the art of modeling and a method for choosing a probability distribution. What we need now is a way to propagate the information in our input distributions to characterize the range of potential results for our model outputs. Interval methods and fuzzy methods are quantitative techniques that can be used, but they are not probabilistic. Methods that are quantitative and probabilistic can be divided into analytical and numerical methods.

Analytical methods are used when we are solving explicit equations. There are a good number of risk problems that can be solved to yield exact solutions. If the probability of an electrical component failing is 1% and we have 1,000 components, we can calculate the expected value (.01 × 1000). We can also use the binomial distribution to calculate the probability of any number of failures from 0 to 1000. Unfortunately, few risk problems are simple enough to solve for specific solutions, in part because of the ubiquitous nature of uncertainty. Many risk problems are simply too difficult or impossible to solve through analytical methods.

Numerical methods provide a versatile alternative solution method with broad applicability and flexibility for characterizing the effects of natural variability and knowledge uncertainty. The Monte Carlo process is the most common numerical method for propagating distributions through a model, and it is the subject of this chapter. It is essentially a sampling process. Random input values are sampled from input distributions and are used to calculate output values. A Monte Carlo simulation produces a sampled distribution of output values.

14.2 Background

In the last part of the nineteenth century, several people used simulation to approximate the value of pi. This was done by randomly dropping needles on a board with parallel lines. In the early twentieth century, British statisticians advanced some unsophisticated Monte Carlo work. W. S. Gosset, a/k/a Student, used experimental sampling for several applications, including his now famous *t*-distribution (Pllana 2010).

The theoretical basis for the Monte Carlo method was understood in the 19th century, long before the ability to perform the numerical calculations was practical. Modern Monte Carlo methods can be traced to work on the atomic bomb during World War II. Kochanski (2005) describes Manhattan Project scientists trying to calculate the probability that a neutron from one splitting uranium atom would

ANALYTICAL METHODS AND NORMAL DISTRIBUTIONS

Normal distributions lend themselves handily to exact analytical solutions. The sum of normal distributions is also a normal distribution. The mean of the sum of normally distributed inputs is the sum of the means of each input distribution. The variance of that sum is the sum of the variance of the inputs. Such simple relationships, unfortunately, do not hold for nonnormal distributions of inputs. This limits our ability to use analytical methods for many risk problems.

cause another atom to split. The equations had to mirror the complicated geometry of the actual bomb, and the answer had to be right. If the first test failed, it would take months to acquire enough uranium for another attempt.

The scientists solved the problem starting in about 1944 by using many mechanical calculators and the Monte Carlo process to follow the trajectories of individual neutrons. At each step, they calculated the probabilities that a neutron either was absorbed, escaped from the bomb, or started another fission reaction. Using random numbers and appropriate probabilities, they observed simulated neutrons stopping their fission reactions or starting new fission chains. The great insight here was that simulated trajectories would have identical statistical properties to the real neutron trajectories, and reliable answers could be had for important questions about the bomb. The code name for the method, Monte Carlo, was taken from the city in Monaco, where roulette wheels routinely generate random numbers.

14.3 A Two-Step Process

The Monte Carlo process begins by replacing parameters with distributions that represent natural variability or knowledge uncertainty in the model inputs. The Monte Carlo process is, in essence, a sampling method. Given a probability distribution, the Monte Carlo process randomly selects a value from that distribution.

In a model with many probability distributions, a random value is selected from each distribution in the model. The programmed calculations are then executed and outputs are generated. These outputs can be saved. A new set of random values is selected from the input distributions, and the model's calculations are executed again until a sufficient number of output values have been obtained.

The Monte Carlo process consists of two steps. The first step is to generate a simple random number. The second step is to transform it into a useful value using a specific probability distribution.

14.3.1 Random Number Generation

To generate a random value from a probability distribution, the Monte Carlo process begins with a simple random number between 0 and 1. Imagine a complex spreadsheet model with hundreds of probability distribution inputs in it. Now imagine that we might want to run 10^5 or more iterations of the model. Each input for each iteration

SOME TERMINOLOGY

Iteration: One recalculation of the model during a simulation. Uncertain variables are sampled once during each iteration according to their probability distributions. The sampled values are used to complete the model's calculations.

Simulation: A collection of iterations, i.e., a technique for calculating a model output value many times with different input values. The purpose of a simulation is to get a complete range of all possible scenarios and their resulting outputs.

requires its own unique simple random number. The Monte Carlo process begins with a method to generate this very long list of random numbers. A 100,000-iteration simulation with 200 probability distributions in its inputs would need 20 million random numbers, for example.

There has been much discussion about the ability of various methods to generate such a set of truly random numbers. What is needed is a series of random numbers that are statistically independent, that are not autocorrelated, and that show no cycles or periodicity in the numbers generated. The numbers so generated are often called pseudo-random numbers because the same sequence of numbers can be replicated by using the same seed and the same calculation method. A good pseudo-random number generator should be able to produce many millions or billions of numbers before the cycle repeats. Random and pseudo-random are used synonymously in this chapter when referring to number generators.

There are many algorithms available for generating simple random numbers. The mid-square method, attributed to John von Neumann, who worked on the Manhattan Project, is one of the first methods used. It is no longer in use because of problems that will soon become obvious. Nonetheless, it is handy to use for an example because it is easy to follow. Just understand that much more sophisticated and efficient algorithms are used now.

Let us suppose we have a risk assessment model and one of the inputs is a random number uniformly distributed between the values 10 and 50. We want to see how the Monte Carlo process generates a number in this range.

The mid-square method needs a seed value to begin to generate the sequence of random numbers. This seed can be any number at all. Let us use a seed = 2123. The method proceeds simply by taking the square of this seed:

$$2123^2 = 4507129$$

Because we started with a four-digit number, we take the middle four digits of this value. If there is an uneven number of digits, we simply need to decide beforehand how to identify the middle four digits. In this instance, if an odd number of digits occurs we will have one more digit in front than behind our number. Accordingly, we obtain 0712 as the middle of the square. Because we began with four digits, we divide this number by 10^4 and obtain a random number between 0 and 1, namely, 0.0712.

Of course, this value does not meet the needs of our input distribution, so we will examine how this value is transformed into a value that is useful to us in the next section. What we need now is our next random number. The mid-square value becomes the seed for the subsequent value.

$$712^2 = 506944; r_2 = 0.0694$$

$$694^2 = 481636; r_3 = 0.8163$$

And so on it goes.

The problem with this technique is that some seeds generate very short sequences of numbers before the pattern repeats itself. Other sequences end abruptly. Imagine we chose a seed of 1000. Follow the steps above and see how many random numbers it generates. This is one reason that this method is no longer used.

Now that we have a sequence of numbers 0.0712, 0.0694, 0.8163, and so on, we need a method to transform them into numbers we can use in our models.

The details of the random number generator are usually handled by the software you use. Palisade's @RISK software offers a number of random number generators, as seen in Figure 14.1, but few users of the software will have any reason to use any but the default Mersenne Twister. This generator was developed in 1997 to correct flaws found in older algorithms. It provides fast generation of very high-quality pseudo-random numbers.

Random number generators require a seed to initiate them. This seed is often chosen at random from the computer's internal clock. However, it is sometimes useful to be able to duplicate the sequence of random numbers so that the same values are

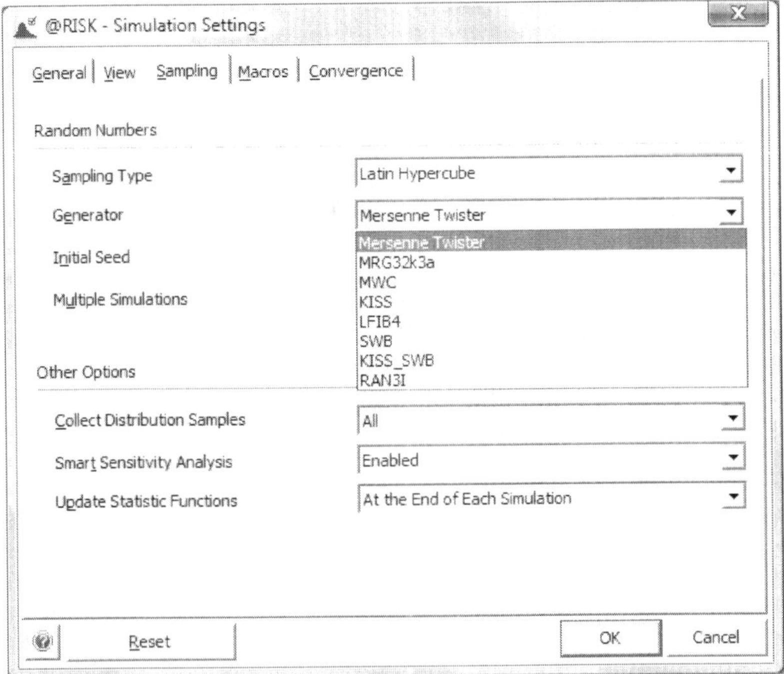

FIGURE 14.1 Random number generators available with Palisade's @RISK software.

generated for model inputs in subsequent model runs. For example, it is common practice to run the same model several times to investigate the efficacy of different risk management options. When this is done, we'd like to ensure that the only differences in the model runs are those that can be attributed directly to the risk management option. When this is desired, a fixed seed should be used. The @RISK software provides this option, as described in Appendix B.

14.3.2 Transformation

Imagine that spreadsheet model with hundreds of inputs once more. Once a pseudo-random number generator has produced a sequence of numbers long enough to meet the needs of the simulation, a different random number from your list of random numbers is assigned to each cell for each iteration. A random sample of one value is drawn from each input distribution based on the simple random number assigned to the cell for that iteration. That sampled value is entered into the risk assessment model to calculate one estimate of the model output(s). In most cases the simple random number, like the value 0.0712 above, will need to be transformed by the Monte Carlo software to create a number that is actually useful to the assessment model. In this case we want a number uniformly distributed between 10 and 50, and 0.0712 does not meet this criterion.

The Monte Carlo process can use several methods for transforming a simple random number to a random value from the probability distribution of each model input. The easiest method is to use the cumulative distribution function (CDF) method. The random number 0.0712 is the 7.12 percentile of the uniform distribution, so the software would calculate the 7.12 percentile for the input distribution and use it.

The function of random variables methods is another way to generate a random value. Assume a uniform distribution, U(a,b), where the minimum a = 10 and the maximum b = 50. To obtain a value, x, from this distribution we use the function:

$$x_i = a + (b - a)r_i$$

where r; is a simple random number. In this case, $x = 10 + 40r_i$.

Substituting the r_i assigned to that particular cell, we would see the following values sampled from U(10,50).

$$x = 10 + (50 - 10).0712 = 12.9$$

$$x = 10 + (50 - 10).0694 = 12.8$$

$$x = 10 + (50 - 10).8163 = 42.7, \text{etc.}$$

Similar functional transformations are done for other distributions. The math simply gets more difficult for more complex distributions.

If you are working with a spreadsheet risk assessment model with several probability distribution inputs and hit the recalculate key, you will see the numerical values in these input cells changing. This two-step process is being executed for each cell with a distribution in it. Each time you recalculate the values, a new and unique simple random number is assigned to each cell and then transformed into a value that is consistent with the probability distribution in that cell. Each random number is used in only one transformation in one cell. This is why you need a good random number generator.

14.4 How Many Iterations?

How many times do you have to run your model? How many iterations do you need? You have two options for running a simulation. You can predetermine the number of iterations you want, or you can predetermine a level of precision for selected outputs, like the mean or standard deviation of the distribution for an output. Iterations continue until the simulation results achieve convergence with the desired precision.

Most new users of the Monte Carlo process are likely to use a predetermined sample size, i.e., a number of iterations. But how large should the sample size be? The answer depends on the nature of the desired output information and the desired numerical precision for the model output.

As a rule of thumb, based on experience and absent more rigorous proof, I have used the following order-of-magnitude guidelines.

- Means often stabilize quickly—10^2
- Estimating outcome probabilities—10^3
- Defining tails of output distribution—10^4
- If extreme events are important it may take many, many more—10^5

If you are only interested in expected values in a relatively simple linear model, these often converge quickly with only a few hundred iterations. If you want to be able to have a reasonably good definition of the CDF for the output, so that you can estimate a range of outcome probabilities, you need several thousand iterations.

When the shape and thickness of the tails of your distribution are important, i.e., when you care about the less frequent events, you may need tens of thousands of iterations. If extreme events are your primary interest, it could take in excess of 100,000 iterations.

Given the power and speed of computers and the available software, it is no longer unusual to run large simulations. However, some models are large and complex, and it may not be possible to run a large number of iterations. In these latter cases, it may be more desirable to establish a desired level of precision in your outputs.

During the first several iterations, output statistics, like the mean or standard deviation, can jump about and change dramatically. As the number of iterations increases, these values tend to stabilize. After awhile, an unusually high or low value has a diminishing impact on the sample statistics. Once these statistics have stabilized, the time spent running additional iterations may yield little information of added value, especially when you are most interested in those very statistics.

To use the convergence option with @RISK you can set the convergence tolerance. This allows you to specify how close you want to come to the distribution's true parameter value. Figure 14.2 shows a 3% convergence tolerance for the mean and standard deviation. This means that once the output distribution yields means and standard deviations within 3% of their true value, it can stop. If you are more interested in tail values, you would choose a percentile value in those tails as the statistic to monitor. The confidence level specifies that you want your estimate of the mean of each output simulated within the 3% tolerance to be accurate 95% of the time. As the figure shows, you can monitor all outputs or you can choose to designate specific ones.

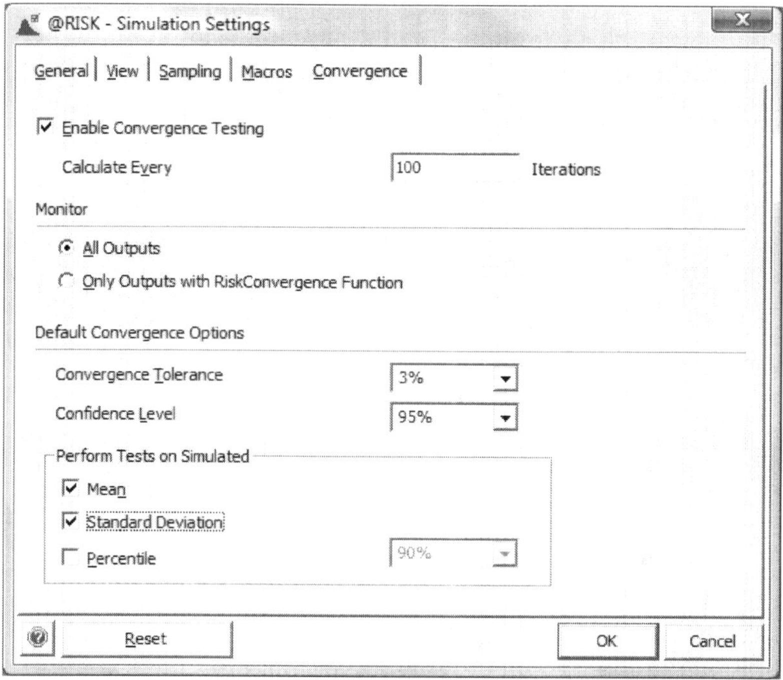

FIGURE 14.2 Using convergence tolerance to control the number of iterations.

There is an important caveat to consider. It may not make sense to run a large number of iterations to obtain a highly precise numerical estimate of a model output if your assessment is based on sparse data. If you have data quality limitations for critical model inputs or if your model depends on a large number of assumptions, you should not spend too much time and effort trying to achieve a high degree of precision for the 99.99 percentile of a model output, for example. If your model input distributions are highly uncertain, more iterations are not going to improve the precision of your output estimates.

14.5 Sampling Method

Most Monte Carlo software chooses your number of iterations in one of two ways: Monte Carlo sampling or Latin hypercube sampling. Latin hypercube sampling (LHS) uses stratified sampling without replacement (Iman, Davenport, and Zeigler 1980). For simplicity, suppose we want to sample five values from a normal distribution with a mean of 100 and a standard deviation of 10. With LHS, the probability distribution is split into n intervals of equal probability, where n is the number of desired samples. In this example, the CDF is split into five equiprobable intervals, as shown on the vertical axis of Figure 14.3. The projection of these intervals on the horizontal axis will vary in size as shown.

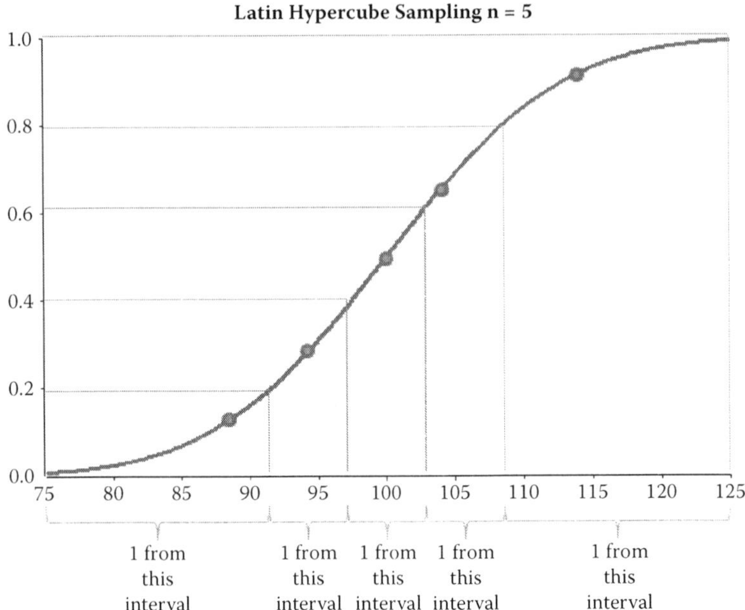

FIGURE 14.3 Latin hypercube sampling for $n = 5$.

During the simulation, each of the n intervals is sampled once. Within each equal probability interval the value chosen can be based on the median value (median Latin hypercube sampling) or it can be chosen at random (Latin hypercube sampling). LHS has the advantage of generating a sample that more precisely reflects the shape of a sampled distribution than simple random (Monte Carlo) sampling does. The advantage of LHS is that the mean of a set of simulation results approaches the "true" value more quickly.

Monte Carlo sampling does not stratify the distribution into n intervals of equal probability. Figure 14.4 shows a hypothetical result that by chance samples all five points between the values 85 and 100. When n is large there is no practical difference between the two sampling methods. In general, LHS is considered more efficient in the sense that it can reveal more information about the distribution it samples from for any sample size.

14.6 An Illustration

Figure 14.5 presents a simple picture of a Monte Carlo simulation model. It shows two random variables being multiplied together. One is the number of coffee drinkers in the United States. The other is the average number of cups consumed by coffee drinkers (not per capita) on a daily basis. Both numbers are uncertain and are represented by triangular distributions.

A value from each is randomly chosen via the two-step process described above and then the model's calculations are executed. In the example, the number of drinkers

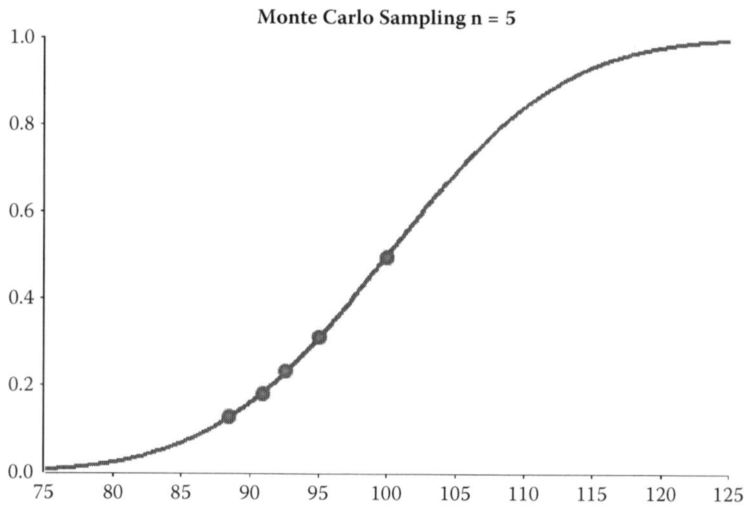

FIGURE 14.4 Monte Carlo sampling for $n = 5$.

sampled is 105 million and the average number of cups of coffee is 3.1 cups/day. The two values are multiplied together to yield 325.5 million total cups. This is a single estimate of the output value and it constitutes one iteration. This process was repeated 5,000 times and the results are shown in the output distribution. Monte Carlo simulations are a preferred method for probabilistic risk assessment.

It is not always or even often necessary to specify a distribution for all or most variables in a risk assessment model. Doing so may be useful for exploring the full range of natural variability and knowledge uncertainty, but it is not always cost-effective or necessary to do so. It is often advisable to restrict the use of probability distributions to significant variables and parameters (EPA 1997).

14.7 Summary and Look Forward

The Monte Carlo process samples individual values from a probability distribution so that they can be used to characterize the range of potential outputs in uncertain situations. It is a two-step process that requires the generation of a simple random number and a means to convert that number to a useful value from the chosen distribution. It is a calculation-intensive numerical process that has become immensely popular and easy to use with advances in personal computers and spreadsheet software.

The Monte Carlo process can be added to a wide variety of model structures. When probabilistic methods like that are added to scenario-structuring tools, like event trades, for instance, they create a powerful bundle of tools called probabilistic scenario analysis (PSA). PSAs are the subject of the next chapter. It begins by considering several different kinds of scenarios and different ways of analyzing and then comparing them. It then presents an introduction to tree models and presents an example of a PSA using an event-tree structure.

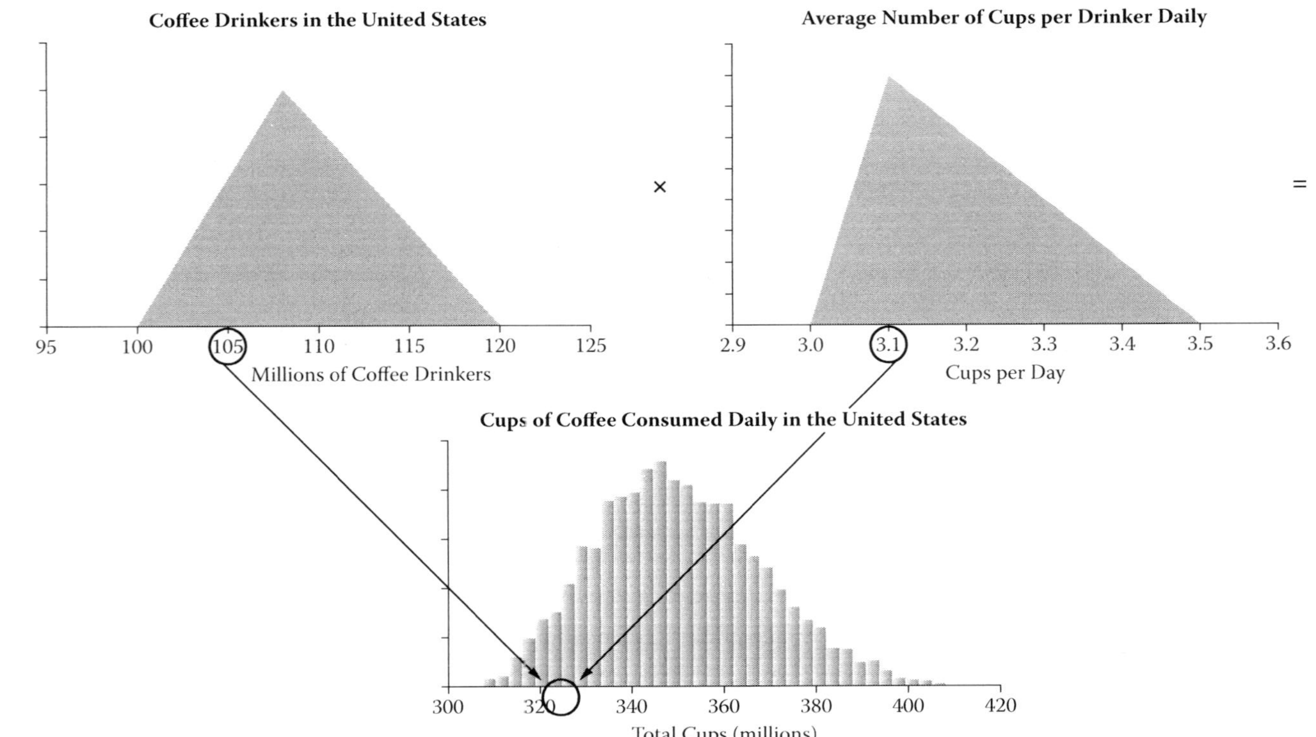

FIGURE 14.5 A simple illustration of the Monte Carlo process is a two-variable model..

REFERENCES

EPA. 1997. Risk Assessment Forum. *Guiding principles for Monte Carlo analysis.* Washington, DC: U.S. Environmental Protection Agency.

Iman, R. L., J. M. Davenport, and D. K. Zeigler. 1980. *Latin hypercube sampling (A program user's guide).* Technical Report SAND79-1473. Albuquerque, NM: Sandia Laboratories.

Kochanski, Greg. 2005. Monte Carlo simulation. http://kochanski.org/gpk/teaching/0401 Oxford/MonteCarlo.pdf.

Pllana, Sabri. 2010. History of the Monte Carlo method. http://stud2.tuwien.ac.at/~e9527412 /history.html.

15

Probabilistic Scenario Analysis

15.1 Introduction

Probabilistic scenario analysis (PSA) refers to a bundle of tools and techniques that make use of scenarios, probabilities, and probabilistic analysis. Most quantitative risk assessments are likely to be some form of probabilistic scenario analysis. Deterministic quantitative risk assessments would be the exception.

Many decisions are hard because they are complex. They involve inherent uncertainty and have conflicting objectives. The various stakeholders with an interest in the decision are likely to hold many different perspectives. Scenarios can address these aspects of a decision problem.

A scenario is literally an outline or synopsis of a play. Scenarios can be used to describe the present or the past. They are most often used to describe possible futures.

Risk scenarios are readily used to provide the answers to our four informal questions:

1. What can go wrong?
2. How can it happen?
3. What are the consequences?
4. How likely is it?

Think of scenarios as the stories we tell about risks. They are a series of events that could happen. In risk assessment, a scenario is defined by a set of assumptions about model input values and how those variables are related to one another. Thus, scenarios could be different model formulations that tell substantially different stories. Or they could simply involve different values for uncertain inputs in a model with an unvarying structure.

We begin the chapter by considering different types of scenarios and proceed to considering three types of scenario analysis. Scenarios are often developed so that they can be compared to help risk managers choose the most desired future. Three common methods of comparing scenarios are reviewed.

Tree models are one of the easier and more useful techniques for constructing orderly scenarios. This chapter focuses on event trees as one of the most useful and accessible tools for structuring risk scenarios. Once the basics of event trees are presented, the probabilistic methods of preceding chapters are added to them in an example that demonstrates a very powerful and useful set of risk assessment tools.

15.2 Common Scenarios

If a scenario is an outline for a play, it is convenient, then, to think in terms of well-established plot lines to begin to consider the different types of scenarios that are used most commonly. Three types of scenarios come up over and over again in risk assessment. They are as-planned, failure, and improvement scenarios.

The first story we might want to tell about our systems is the one where everything works exactly as planned. This as-planned scenario describes the system we are interested in operating free of any failures. It is a surprise-free scenario. In the example of an engineering system, all the loads on the structure are as anticipated, and every feature functions as it was designed to function. In a food safety system, every feature of the system functions as planned and consumers experience no exposure to a hazard.

An as-planned scenario is not necessarily a risk-free scenario, as many systems are designed with some degree of risk built into them. Airplane design trades off wing strength and body weight, for example. Levees are designed to be overtopped in a specific location. The principle characteristic of an as-planned scenario is to show every risk management feature of the system functioning just as it was designed to do.

To illustrate the idea, let's introduce the example we'll use later in the chapter. The system illustrated in Figure 15.1 is designed to prevent imported contaminated carcasses from entering the food supply. It begins with the unknown condition of the carcass's contamination. All the as-planned outcomes are shown in the figure.

The second plotline for a class of scenarios is the failure scenario. The model in Figure 15.1 shows one failure scenario. Failure scenarios tell the story of how various elements of the system might interact under certain conditions that result in undesirable outcomes. Thus, failure scenarios intentionally challenge the notion that a system will function as planned. Any aspect of an as-planned scenario may be challenged. Failure scenarios may illustrate one or multiple modes of system failure. These failures are usually sources of risk.

The best known failure scenario may be the worst-case scenario. The worst-case scenario is usually constructed using the analyst's judgment to choose that set of circumstances or, once the model is specified, the set of input values that will yield the worst possible outcome from a model. For an engineering design, this could be the

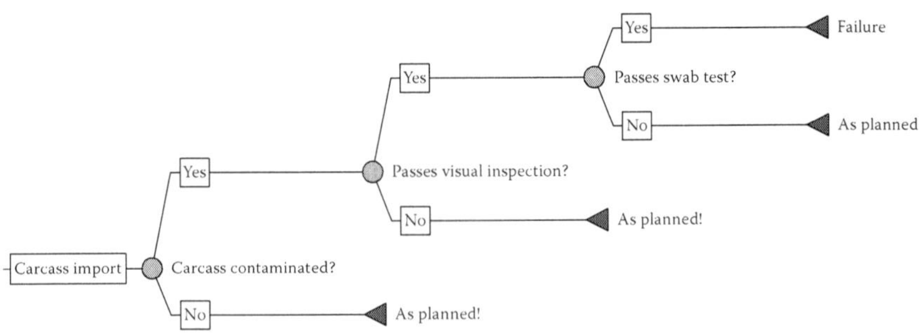

FIGURE 15.1 As-planned scenarios for carcass import model.

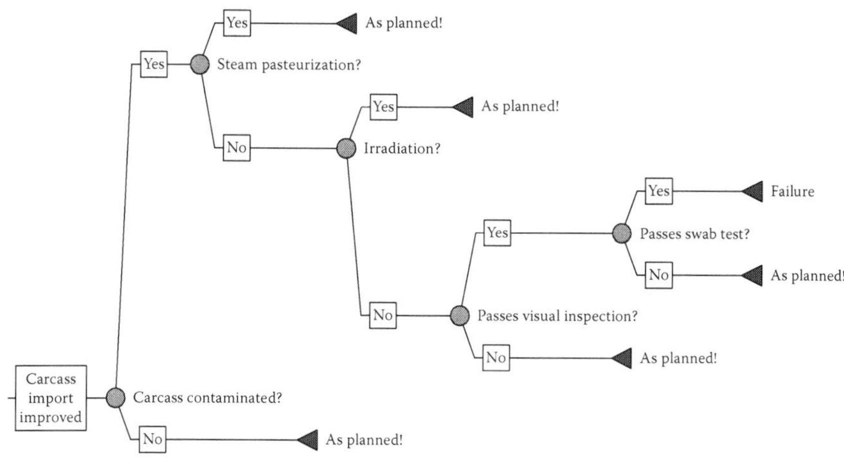

FIGURE 15.2 Improvement scenarios in carcass import model.

highest possible loads during a category 5 hurricane that coincides with a magnitude 8 earthquake. The basic appeal of the worst-case scenario is that if it occurs and the outcomes are still acceptable, then we think we have little to worry about. The worst-case scenario deliberately exaggerates the risk in a given situation. If the system continues to function at an acceptable level during the worst-case scenario, decision makers have often assumed that there is no need to further assess the risk. This is often flawed thinking, so we'll return to the worst-case scenario again in the next chapter.

The third type of scenario plotline is the improvement scenario. Risk analysis often results in new risk management options (RMOs) that have a direct effect on risks. Improvement scenarios often describe how the RMOs affect the failure scenario. They are used to evaluate risk management options. The "best" improvement scenario often points to the RMO that will be implemented. Let's reconsider our simple model with some improvement scenarios as seen in Figure 15.2.

Kill steps have been added to the process in the form of steam pasteurization and irradiation. Often a separate improved scenario is developed for each RMO rather than combining them as was done here. In some cases the structure of the scenarios will change; in other cases only the input values will differ. For example, the original model we began with could be used to represent an improved scenario if live animals were vaccinated, thereby reducing the number of contaminated carcasses, or if a better visual inspection method or swab test was introduced. In any of these cases the structure of the model would remain as shown in Figure 15.1. The input values (not shown in the figure) along the branches would change to reflect the improvements.

In addition to the broad categories of scenario plotlines, there are some other commonly used scenarios. The best known of these is the existing condition. Others include the worst case, best case, most likely, do nothing, future without additional risk management, locally preferred, industry-preferred scenarios, and so on. These scenarios tend to be deterministic rather than probabilistic, which brings us to our next topic.

15.3 Types of Scenario Analysis

There are three kinds of scenario analysis: monolithic, deterministic, and probabilistic. The first requires the invention of language. I doubt you will find monolithic scenario analysis anywhere in the literature. We have always used scenarios for decision making. Before the rise of risk analysis, it was common practice to develop and present a scenario as if it were a fait accompli, when in fact it was simply one of many possible story lines about a problem or opportunity. So I will use the notion of a single unchallenged scenario, or a monolithic scenario, to represent decision making where uncertainty is not explicitly recognized. Monolithic scenarios are often used by the political system to enact legislation. National health care reform in the United States was enacted in 2010 on the basis of a more-or-less monolithic scenario analysis. The outcome of the program was presented without consideration of alternative outcomes. Monolithic scenario analysis uses only one scenario for decision making.

When we move from a single scenario to a few selected scenarios we are engaging in deterministic scenario analysis (DSA). DSA defines and examines a limited number of specific scenarios. This can be a useful way to organize and simplify an avalanche of data into a small number of possible future states of the system being modeled. The scenarios so identified are usually chosen for specific reasons. They may be exploratory, such as with the worst-case, most likely, and best-case scenarios; or they could be chosen for strategic or tactical reasons. For example, we might look at the effects of three different flu vaccines on morbidity rates.

There are some serious limitations to deterministic scenario analysis. First, only a limited number of scenarios can be considered. Second, the likelihoods of these scenarios cannot be estimated with much confidence. Third, this approach is inadequate for describing the full range of potential outcomes.

Probabilistic scenario analysis, on the other hand, overcomes these limitations by combining probabilistic methods, for example the Monte Carlo process, with a scenario generation method like event tree models to produce a PSA. A PSA may produce measures of probability, measures of consequence, or measures of risk, which combine the probability and consequence measures into a single value.

15.4 Scenario Comparisons

The effectiveness of an RMO is often judged on the basis of changes in decision criteria observed through scenario comparisons. Take the example in Table 15.1. If no additional risk management is undertaken (most likely future), there will be 5,000 illnesses but there will be no costs associated with reducing them, as no effort will be made to do so.

If risk management option A is implemented, illnesses will be reduced to 2,500 at a cost of $1 million. If we compare these two scenarios we get the changes noted in the table. Illnesses are cut in half at an additional cost of $1 million. Option B eliminates all the illnesses but at a much greater cost. Risk managers would be expected to choose the best option based on differences revealed by scenario comparisons.

If scenario comparisons are to be useful for decision makers, they must identify features in scenarios that make a difference, i.e., show things that are important and

TABLE 15.1
Simple Scenario Comparisons

	Illnesses	Cost
Most likely future	5,000	$0
Future under RMO A	2,500	$1,000,000
Change due to RMO A	−2,500	−$1,000,000
Future under RMO B	0	1,000,000,000
Change due to RMO B	−5,000	−$1,000,000,000

that matter to decision makers. That means that risk assessors, with input and feedback from risk managers, need to carefully identify decision criteria. These are the measurable criteria upon which decisions will be based. In best practice, these metrics will reflect some or all of the risk management objectives. Recall that when these objectives are met, problems are solved and opportunities are realized. Scenario comparisons must begin by comparing things that matter, and that will usually include some comparison of risk estimates.

The complexity of some problems combined with lack of data and other uncertainties make risk assessment a rigorous and often difficult undertaking in many situations. Assessors are often happy if they can even provide an estimate of the existing risk. Projecting risks over time is not yet common practice in a lot of applications. Nonetheless, the best comparisons of scenarios would consider whether a risk was static, growing worse, or self-attenuating over time.

There are three basic comparison methods, first introduced in Chapter 3. We'll revisit them here in a bit more detail. For simplicity, the discussion will focus on a single decision metric. In actual practice, a scenario comparison would likely involve multiple decision criteria, each of which would be compared.

Current practice often relies on a static estimate of the existing risk. This estimate often becomes the baseline scenario against which all comparisons are made. The most common scenario comparison is a before-and-after comparison, as illustrated in Figure 15.3. This takes an estimate of the risk before any additional risk management measures are implemented (the baseline scenario) and compares it to the risk estimate

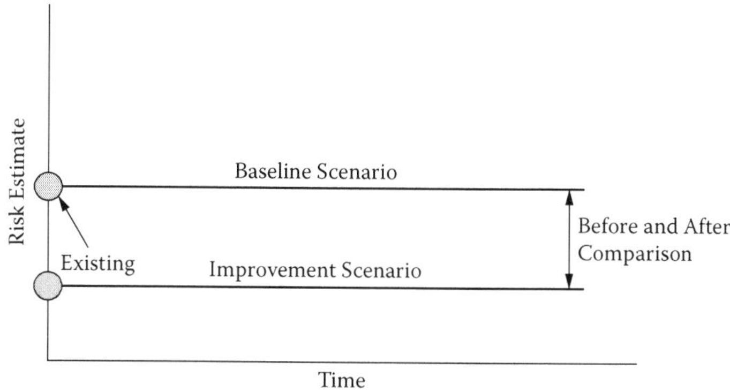

FIGURE 15.3 Conceptual before-and-after scenarios comparison.

that would occur under the improved conditions with the risk management option in place and functioning (the improvement scenario). The difference between these two risk estimates is calculated in a before-and-after comparison. That value is most often a static estimate of the difference in the risk. It usually ignores any trends in the risk itself as well as any phase-in of the risk management option's effectiveness over time.

The use of this comparison method is widespread. Its greatest weakness is that it does not take into account changes in the risk over time. In dynamic systems, risks may increase or decrease with environmental or other factors. RMO effectiveness may also vary depending on the phase-in period for the RMO and associated compliance rates. In the case of mechanical or other physical risk management measures, their effectiveness may change over time with wear and tear.

To account for these kinds of changes, a without-and-with comparison is preferred (Figure 15.4). This example shows that if no additional action is taken to manage the risk, the risk estimate increases steadily over time. This is called the "without condition" and it represents the most likely future condition without any additional risk management options implemented. For the convenience of understanding the without-and-with comparison, let's assume that this future can be represented by a single path. The concepts also hold for uncertain futures, but the explanation just grows more complex without adding much understanding; so let's just follow one path.

Figure 15.4 shows a risk that is growing worse, although some risks may actually be self-attenuating and could have a negative slope into the future. Likewise, the path of the risk estimate with RMO A in place, the "with condition," is stylistic as well. It shows a future in which it takes some time to realize maximum risk reductions. This could be because measures that comprise the RMO are phased in over time, or it could reflect the fact that those required to implement the RMO comply at different rates.

Under the without-and-with condition scenarios comparison, a proper analysis would have to estimate the changes in risk (and any other dynamic decision criteria) over time. The original baseline estimate is still shown (the unlabeled line) to provide a reference point. If, for the convenience of this argument, we consider the baseline as the before condition and the lowest position of the with condition as the after condition, it is easy to see that the previously presented static view of the RMO's performance provides a different picture than the without-and-with condition comparison.

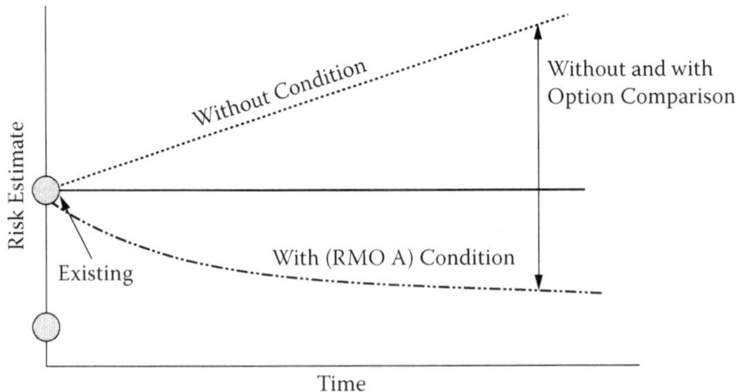

FIGURE 15.4 Conceptual without-and-with condition scenarios comparison.

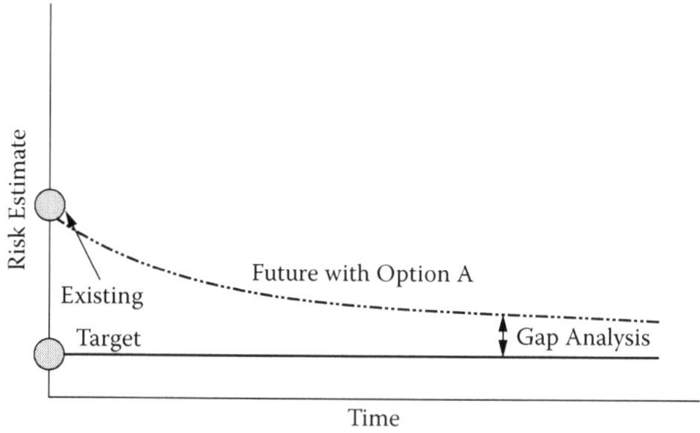

FIGURE 15.5 Conceptual gap-analysis scenarios comparison.

In general, the without and with condition comparison is more accurate and preferred. In practice it is often too difficult to do well, and a before-and-after estimate is used in its place. However, it is important to note that some risks, by their nature, are relatively static. For example, many public health issues affect a relatively fixed percentage of the population, so these risks will only change as the population changes. Most food safety risk assessments use a before-and-after comparison. If population growth is low, then treating these problems as static problems does no great disservice in trade for the computational simplicity of assuming a static problem. These are assumptions that should be made explicitly clear and documented in any risk assessment.

A third kind of comparison is predicated on some higher authority establishing a tolerable level of risk or a risk reduction target. As seen in Figure 15.5, once a target is established, risk managers try to hit the target. When the target is ambitious, some RMOs may fall short of the target, establishing a gap between the desired level of performance of the RMO and its actual performance. In other instances, the target may be exceeded. Gap analysis is a comparison technique that focuses on the distance between the desired target and the actual performance. Frequently, when an option falls short of the target, additional measures will be considered to mitigate this shortfall. Water quality targets are often set administratively or by legislation. Targets are not uncommon for environmental issues, and they are also common organizational performance measures. Gap analysis is frequently used in conjunction with mitigation banking for environmental issues.

Risk managers would presumably adopt one of these comparison methods for a given risk management activity and apply it across all the risk management options it is considering. Table 15.2 is replicated from Chapter 3. It shows the effects associated with three different RMOs using a without-and-with condition scenario comparison. Each column was prepared using the same without condition. It is the with condition (or the after condition or the gap in the other comparison methods) that varies for each and every RMO.

There are five decision criteria in the table. This means that a without-and-with analysis was done for each of them. For any one column, the improvement scenario was the same, and analysts focused on different decision metrics to show how they would most likely be affected under that scenario. When the scenario comparisons

TABLE 15.2

Without-and-With Condition Scenario Analysis for Alternative Risk Management Options

	RMO 1	RMO 2	RMO 3
Illnesses reduced	−30,000	−40,000	−10,000
Illnesses remaining	20,000	10,000	40,000
Costs	+$150 million	+$500 million	+$100 million
Benefits	Decrease	Decrease	No change
Jobs	−2,000	0	−500

TABLE 15.3

Actual Scenario Comparisons Taken from FDA's Risk Assessement of *Vibrio* in Oysters

	Predicted Mean Number of Annual Illnesses			
Region	**Baseline**	**Immediate Refrigeration**	**(2-log reduction) Freezing**	**(4.5-log reduction) Heat Treatment**
Gulf Coast (Louisiana)	2050	202	22	<1
Gulf Coast (non-Louisiana)	546	80	6	<1
Mid-Atlantic	15	2	<1	<1
Northeast Atlantic	19	3	<1	<1
Pacific Northwest (dredged)	4	<1	<1	<1
Pacific Northwest (intertidal)	192	106	2	<1
Total	2,826	391	30	<1

Source: Food and Drug Administration (2005).

are based on a probabilistic scenario analysis, the individual numbers in the example would be replaced by distributions.

The FDA's risk assessment of *Vibrio* in oysters summarizes a number of scenario comparisons. Table 15.3, taken from that risk assessment, provides a good example of a scenario comparison. The columns show risk management scenarios. They include a without condition (the baseline) risk estimate with three risk management options: immediate refrigeration, freezing, and heat treatment. These comprise the with conditions. The rows show the results from geographic scenarios. The reader is left to calculate the impacts of the various scenarios by comparing the improvement scenarios to the existing scenario.

15.5 Tools for Constructing Scenarios

Scenarios can be activity-driven, process-driven, or event-driven. These sorts of models are often built in spreadsheets or more sophisticated programming environments. Working with spreadsheet models is discussed in Chapter 10. Two useful models amenable to the spreadsheet environment are influence diagrams and tree models. Each is described in the following sections. Trees are especially useful for visually splitting and separating the uncertainty in a problem into enough levels for intuition to function most effectively.

15.5.1 Influence Diagrams

An influence diagram provides a simple visual representation of a decision problem and its uncertainty. Influence diagrams are a simple tool for identifying and displaying the essential elements of a decision problem, i.e., the decisions to be made, the payoffs to be realized, and the uncertainties that influence these elements and the relationships among them. As a tool, influence diagrams help the assessor find the structure of problems and identify subsequent analytical tasks, including data collection. They are often useful for conceptualizing the problem before modeling its details. As a decision aid, they pose concise problem statements.

Influence diagrams consist of nodes and arcs that connect the nodes. Nodes typically include decisions drawn as squares, uncertainty or chance shown as circles, and payoff nodes as diamonds. The relationships among the different nodes are shown by arcs between nodes. An arc points from a predecessor node to a successor node. Unlike a flow chart, however, influence diagrams are not necessarily sequential.

Palisade PrecisionTree uses arcs to identify three different kinds of influence between two nodes. When the value of one node is influenced by a predecessor node, there is a value influence. When one node must occur before (or after) another, there is a timing influence. When the structure of an outcome is affected by the outcome of a predecessor node, there is a structure influence. The direction of an arc is important.

A simple influence diagram is shown in Figure 15.6. It represents the problem of estimating the net value to society of using a pesticide on agricultural products. The basic decision is whether to use a pesticide or not. If it is used, yields may increase but so could exposure to the carcinogenic pesticide. If the decision is not to use the pesticide, exposure is decreased but yield might be also.

The diagram is compact and it efficiently portrays major elements of the decision problem. It is also easy to understand and therefore is accessible to nonexperts.

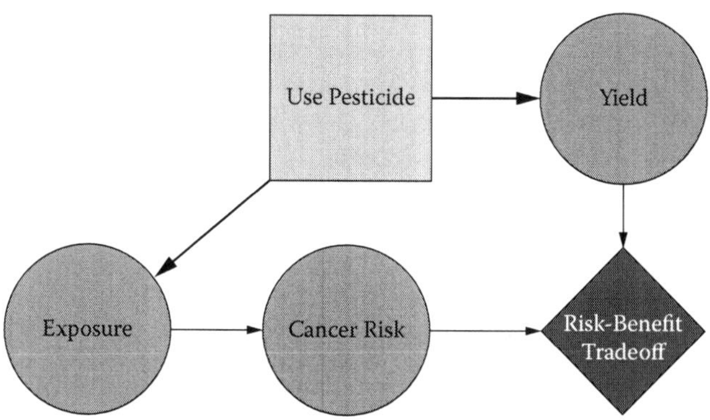

FIGURE 15.6 Influence diagram for a pesticide risk.

GUIDELINES FOR DESIGNING INFLUENCE DIAGRAMS

1. It should have only one payoff node.
2. It should not contain any cycles. A cycle is a "loop" of arcs in which there is no clear endpoint. To recognize a cycle, trace back from the payoff node. If you come across the same node more than once in the same path, your diagram contains a cycle.
3. It should avoid barren nodes. Barren nodes are chance or decision nodes that do not have successors, and thus do not influence the outcome of the model.

Source*: Palisade Corp. (2009), *PrecisionTree User's Guide.

This makes problems easier to explain to others. Although, in fairness, more complex influence diagrams can be somewhat bewildering to those unfamiliar with the problem.

The diagrams are also useful for explaining what influences what. Pesticide use, for example, does not directly affect the cancer risk unless there is exposure to the pesticide. Thicker arcs indicate a structural influence; the thinner ones show a value influence. A drawback to influence diagrams is that while they are excellent for showing the structure of a problem, they fail to reveal many details that are sometimes quite important. For example, we do not know the nature of the dose-response relationship from this diagram. An influence diagram is often an excellent tool for beginning to structure a decision problem. They are especially useful when constructing the conceptual model for a risk assessment.

15.5.2 Tree Models

Tree models are used to explore how systems respond to challenges. They are handy ways to catalogue as-planned, failure, and improvement scenarios. Tree models comprise a series of nodes, branches, and endpoints used to map an issue to aid our understanding and solution of a problem. There are four relatively common kinds of tree models:

1. Event trees
2. Probability trees
3. Fault trees
4. Decision trees

Event trees use forward logic. There is a single initiating event, (a flood occurs or a load is exceeded, and so on) that can lead to many possible outcomes (endpoints). The nodes in an event tree are all uncertain or chance events; there are no decisions to be made. There are no events that are controlled by the decision makers. The model shows the sequence of chance events that can lead to desirable or undesirable outcomes. The outcomes of an event tree can include the probabilities of arriving at the various endpoints as well as consequences and other impacts of interest. These may include things like profits or dollar damages, numbers of deaths, miles driven, or any other metric germane to the decision problem.

Probability trees are a specific kind of event tree. The only outcome of a probability tree is the estimated probability of arriving at an endpoint. No other impacts are quantified.

Fault trees differ from the others in that they use backward logic, as shown in Figure 15.7. This figure shows the different pathways by which a hamburger could have become contaminated with a pathogen like *E. coli* O157 H7. Fault trees begin with an endpoint, outcome, or result and work backwards to find the most likely cause of the fault. Murder investigations, structure failures, airplane crashes, food-borne illness outbreaks, and the like are examples of outcomes that set off investigations back toward the cause. Any of these might use a fault tree. As with event trees, all nodes are uncertain or chance events. Although different impacts can be tracked in a fault tree, they commonly focus on identifying the most likely path from an endpoint back to its originating event; thus probabilities are usually the only metrics in a fault tree. For more on fault trees see the U.S. Nuclear Regulatory Commission's *Fault Tree Handbook* (1981).

In a single-stage decision tree, all decisions (there can be more than one) are made at the beginning of the model. Then all uncertainties are resolved in the remainder of the model. Examples can include making insurance purchase decisions or operating a water control structure to regulate water levels. Multistage decision trees are characterized by opportunities to make decisions as uncertainties are resolved. Their logical structure follows some sort of decision, chance, decision, chance pattern. Managing a response to an evolving natural disaster like a flood, drought, wildfire, or earthquake would involve numerous decision opportunities, as would managing a stock portfolio. Decision trees differ from the others in that there are nodes that can be controlled by the decision maker, so there are decision nodes and chance nodes. Decision trees use forward logic. They usually include impacts in addition to probability measures.

While fault trees have their own symbol sets, event trees are composed of nodes and branches. Typically the nodes are:

- Circles to represent uncertain, chance, or probability events
- Squares to represent decision opportunities
- Triangles to represent endpoints

Nodes also represent points in logical time, which may differ from chronological time. Nodes that appear before (i.e., to the left of) other nodes are predecessor (or parent) nodes. Those that appear after a node are successor (or children) nodes. Choose any node in a model, and every successor to that node is assumed to have already occurred. Decisions that precede it have been made, and the chance or uncertain events before it have been resolved. For nodes to be considered predecessors or successors, they must be directly connected by one or more branches.

"Tree time" is a logical time and may not be a chronological time. For example, if we wanted to develop a model that shows how construction material and neighborhood quality affect the likelihood that a home will be broken into and entered (B&E), which of these attributes should come first in the model: construction material or neighborhood quality?

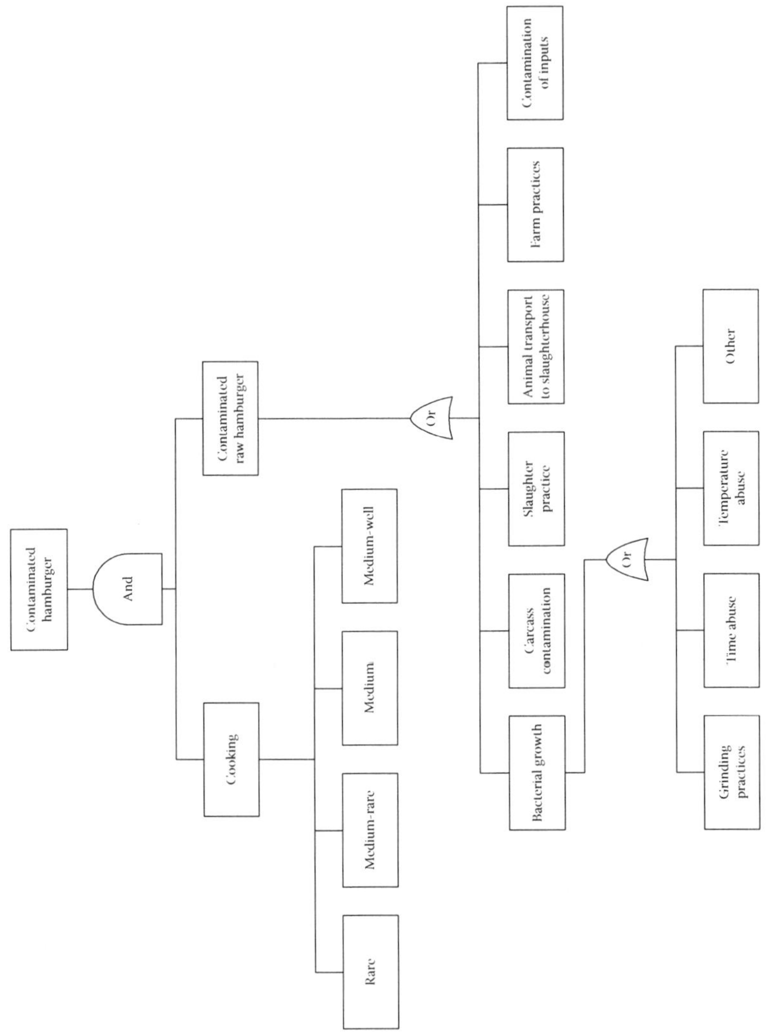

FIGURE 15.7 Fault tree analysis for contaminated hamburger.

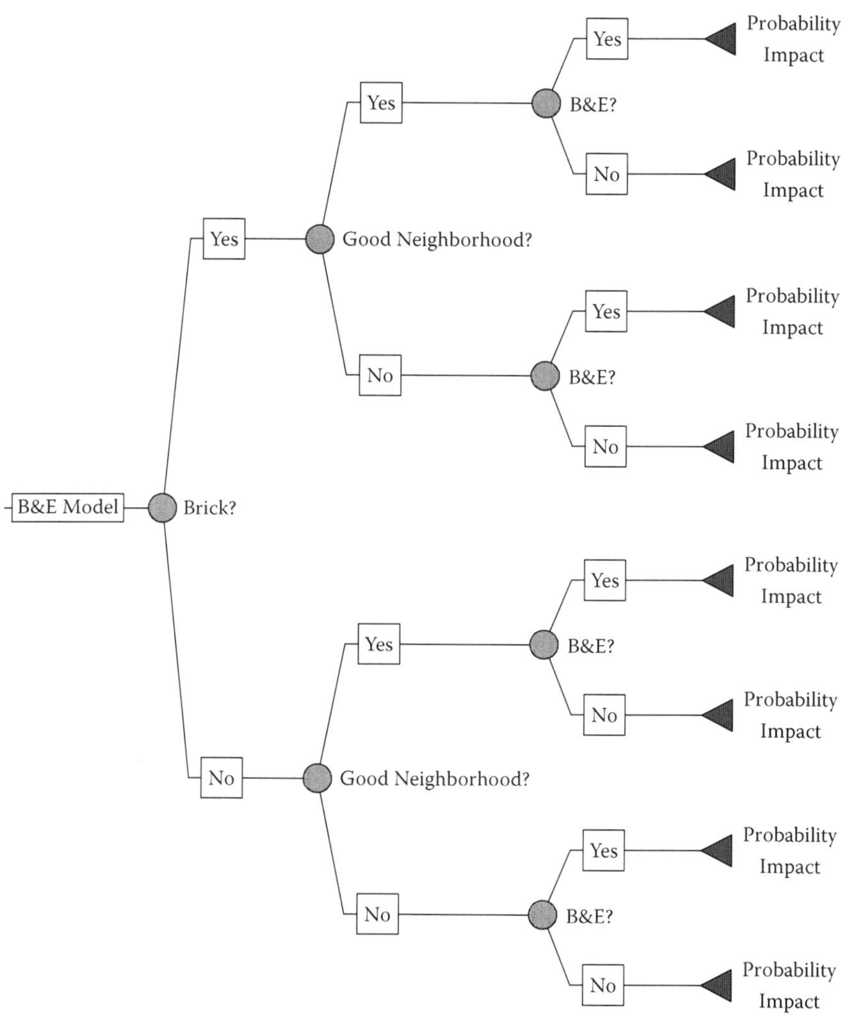

FIGURE 15.8 An example of logical time in an event tree model.

Figure 15.8 shows the model with construction material first. There is no reason other than the modeler's choice for putting that before the neighborhood quality. Thus, not all tree models need to have a literal chronological sequence. However, once a model is built, the sequence of elements imposes a logical time on the model's structure. Consequently, branches leading into a node have already occurred and represent knowledge that the analyst now has. Branches leading out of or following a node have not occurred yet. In this example, that means construction material is resolved and has become known before the quality of the neighborhood is resolved.

The location of the node represents a point in the logical sequence and in the logical time of the model. A decision node represents the point in time when the decision maker makes a decision. A chance node shows the point in time when the result of an uncertain event becomes known. An endpoint marks that point in time when the process has ended, the problem is resolved, or the eventual fate beyond this point is no longer of concern to decision makers.

Branches from a chance node show the possible outcomes of uncertain events. You have no control over these. Branches from decision nodes are the possible decisions that can be made and you can control these. Branches have values associated with them. Probabilities are listed on top of the branch. They must obey all the laws of probability theory. Quantitative estimates of other impacts are listed on the bottom of the branch. Use Bernoulli pairs for branches when possible, e.g., branches labeled yes/no, failure/success, and so on. This suggested style, however, is not always possible or desirable.

Nodes, branches, and endpoints must be mutually exclusive and collectively exhaustive. Pay particular attention to endpoints. They define your sample space; make sure they identify and include all possible outcomes of interest to your decision problem.

As noted previously, probability values must obey the laws of probability. Other quantitative impacts must obey the laws of logic. Pay attention to the units of measurement you will use. It is easy for novice model builders to lose the continuity of units in a model. One may begin a model using bushels of oysters as an impact metric, while the endpoints are expected to be the number of people who get sick. Oysters cannot suddenly change into people in a tree model. Be consistent in your units throughout the model. Multiple impacts can be tracked through the model. A tree model may constitute the entire risk assessment model or it may be but one module in a larger model.

Models end when you have answered the risk manager's questions or you no longer care what happens next. A tree model need not trace the entire history of every object or entity that enters the model. For example, suppose we are interested in estimating the number of vessels that run aground in a channel because their draft exceeds the channel's controlling depth. This tree may begin by asking if the vessel depth exceeds the controlling depth. If a vessel does not, you may not be interested in what happens to it even though some of those vessels may run aground for other reasons. The relevant point for the model may be that our risk management option, deepening the channel, may have no effect on groundings that are due to operator error, mechanical failure, bad weather, and causes other than channel depth.

15.6 Adding Probability to the Scenarios

Up until now we have emphasized scenarios and tools for their construction. It is time to consider why and how probability enters the picture. The simple answer is that we use probabilities because there is natural variability in natural and anthropogenic systems, and there is knowledge uncertainty virtually everywhere we turn. Uncertainty gives birth to many possible scenarios. We cannot describe them all in a DSA, and some of them may be important to the decision process.

Probability is the language of natural variability and knowledge uncertainty. If we want our scenario models to deal effectively with these two phenomena they must be probabilistic. Probability can be added to a scenario model in a variety of ways; one of those we have examined is the Monte Carlo process. Once a scenario structure has been built, we can readily replace point estimates with probability distributions that represent the natural variability and knowledge uncertainty inherent in these values. Different scenarios yield different outcomes. The probabilities of particular outcomes or values greater or less than specific outcomes may be important. The example of the next section will demonstrate how this all comes together.

15.7 An Example

Consider a hypothetical example familiar from earlier in the chapter. Suppose you import carcasses of meat for human consumption from another nation. Those carcasses may or may not be contaminated by a particular pathogen that is a public health concern in your country. The herds from which the carcasses come may or may not be vaccinated against the pathogen. The carcasses themselves may or may not be pathogen-free. When the carcasses arrive in your country they are first subjected to organoleptic testing (look, smell, feel) for pathogen presence. Carcasses that pass that test are next subjected to a swab test that is considered more sensitive.

Carcasses that are not contaminated are of no concern to us. Even if we incorrectly think it is contaminated, we do not care. A carcass that fails the first test is removed from the human food supply chain. Likewise, a carcass that fails the swab test is removed. We have been asked to answer the following questions:

- What is the probability that an imported carcass contaminated with the pathogen will enter our food supply?
- How many infected carcasses could enter the food supply annually?

Our task is to design a model that enables us to answer these questions.

There are some data available to us. Over the last ten years we have imported an average of 2,000 carcasses with a standard deviation of 200 annually. The prevalence of the pathogen in the herds from which our carcasses are culled is no less than .01 and no more than .03. Most of the studies show it has been about .015. Our inspection service claims that it is somewhere between 80% and 87% effective in detecting this pathogen through visual inspection. The literature suggests the swab test is somewhere between 93% and 98% effective when properly administered.

We saw the basic model, reproduced in Figure 15.9, earlier in the chapter. This describes and structures a specific scenario. The first node asks if the carcass is contaminated or not. This is an example of logical time that often confuses new modelers. Bear in mind that a model is an abstraction from reality. This model simply shows that some carcasses are contaminated when they enter the country. We do not know how many or which ones they are, but that does not prevent us from thinking about the problem in this way.

Structuring the model in this way enables us to track the contaminated carcasses in the model. An alternative formulation of the model could have begun with the visual inspection and then the swab test, waiting to reveal the state of the carcass after all the examinations had been done. This would have revealed false positive results for the inspection and the swab test. Had that been part of the decision problem, i.e., one of the questions we were trying to answer, it would have been the preferred formulation.

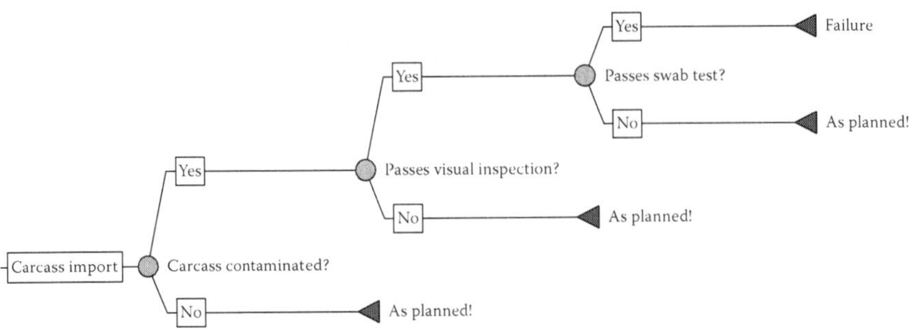

FIGURE 15.9 Carcass import event tree model.

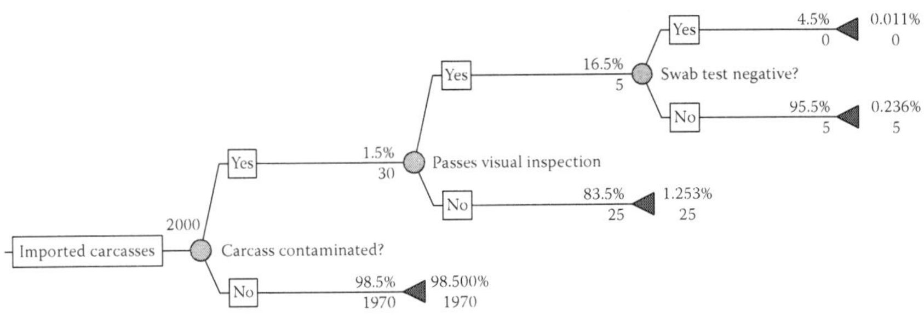

FIGURE 15.10 Prototype model without natural variability and knowledge uncertainty.

Let's begin by using a single best estimate for each value, as shown in Figure 15.10, and ignore natural variability and knowledge uncertainty in this first formulation of the model. In other words, let's build our prototype model. The values entered are the expected values calculated from the data listed above. The number of contaminated carcasses in this model (30) is the product of the prevalence (1.5%) and the number of imports (2000). Of these, 16.5% (5) pass the visual inspection. Continuing with a model free of natural variability, 4.5% of these 5 are rounded to zero, and no contaminated carcasses enter the food supply. As this deterministic model is structured here, there will never be a contaminated carcass entering the food supply. The probability of this happening is low, 0.011%, and 0.011% of 2000 is zero.

The deterministic model is not realistic because it does not address the natural variability in the world. Even with a given prevalence of the pathogen and a fixed number of carcasses, we would not always expect exactly 30 contaminated carcasses. If you observed 31 or 29 contaminated carcasses would you be surprised? I would suspect not, as it is but a chance occurrence. Introducing variability enables us to better model the possible outcomes.

Using the simple binomial distribution in Figure 15.11 to model the variability with $n = 2000$ and $p = .015$, we can see the range of possible results and their likelihoods. For 90% of the time we'd expect to see between 21 and 39 contaminated carcasses in a year. With incredibly good luck we might observe fewer than 10, and with incredibly bad luck we might have more than 50.

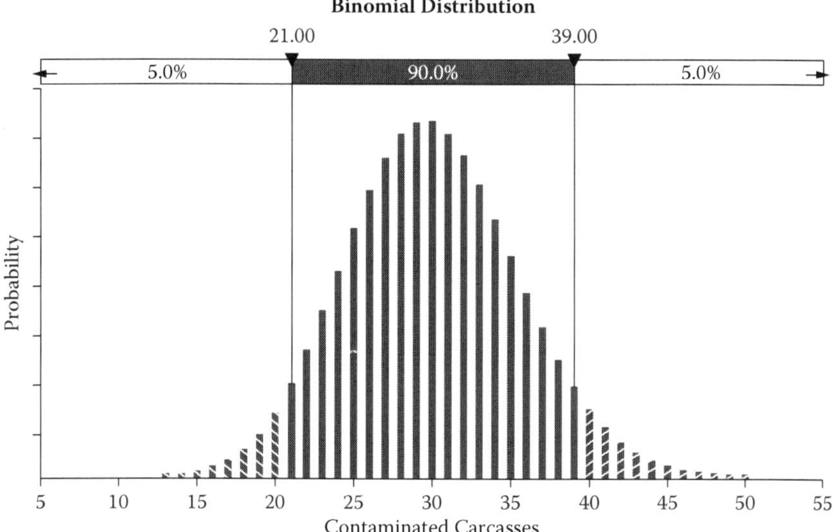

FIGURE 15.11 Binomial distribution with n = 2000 and p = .015.

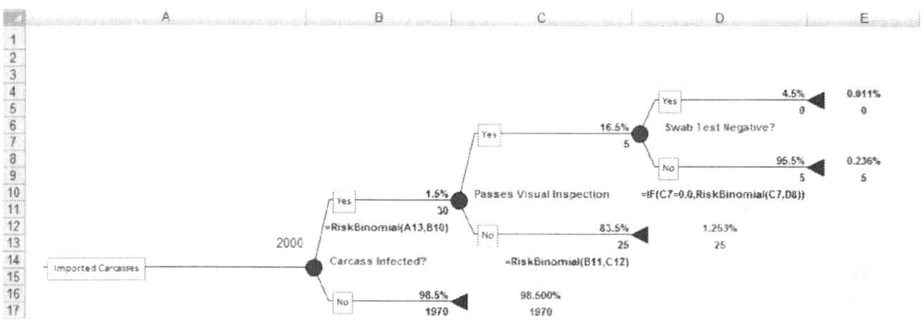

FIGURE 15.12 Carcass import model specified with some natural variability in it.

Likewise, there is variability in the number of contaminated carcasses that are not detected in inspection or by the swab test, so we replace the products in the prototype model with binomial distributions for these activities as well. Figure 15.12 shows the model with natural variability incorporated. The syntax* for the binomial distributions is shown beneath the cell where the binomial distribution appears. This is one example of how probability is added to a scenario tool to account for some of the natural variability in the world.

Using the same expected values and allowing for natural variability, the model result changes from the constant zero seen in Figure 15.10 to the 5,000-iteration simulation result shown in Figure 15.13. The most common result is far and away no carcasses entering the food supply. The maximum number of carcasses observed in

* The syntax used in the figures is that of Palisade Corporation's @RISK software.

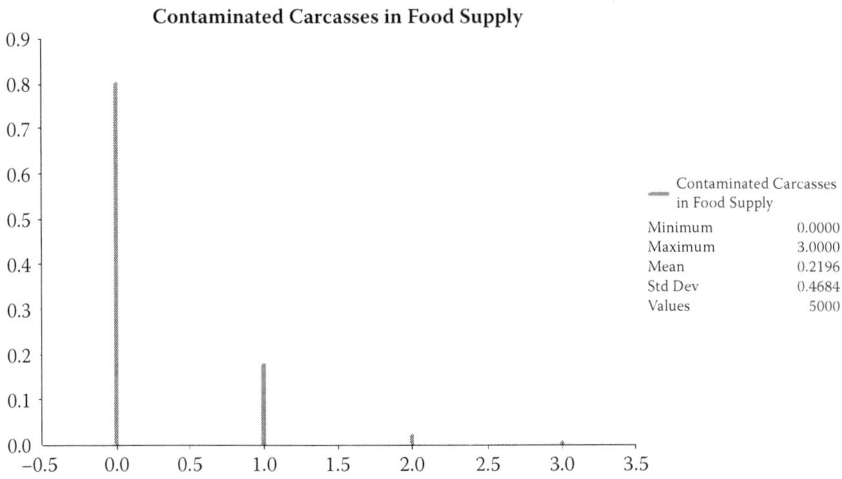

FIGURE 15.13 Output of import model specified with partial variability.

5,000 iterations was three carcasses. Once we account for natural variability in the world, the expected value rises from zero to 0.22 carcasses annually.

The rightmost binomial distribution of Figure 15.12 is preceded by a logical IF statement. It effectively says that if there are no carcasses that passed the visual inspection, then just put a zero here as well; do not use the binomial distribution. If a zero is entered for the sample size, *n*, in the binomial, you will receive an error message. The IF statement prevents this error.

These results were obtained accounting only for natural variability in the model. We have not yet let the number of carcasses imported vary nor have we addressed the uncertainty in the other values. The next step is to incorporate all variability and to honestly portray our uncertainty about the model inputs using probability distributions.

Figure 15.14 shows the syntax that describes the distributions (in the adjacent cell) used to represent the variability and uncertainty in this model. This is how probability is layered over a scenario model to produce a probabilistic scenario analysis.

The newly introduced distributions, seen in Figure 15.15, show the variability in the number of animals imported (Normal(2000, 20)). Note that the normal distribution is a continuous variable while carcasses are discrete. The value sampled from the normal distribution is converted to a discrete value using the =ROUND function of Excel. The other three newly added distributions reflect the available data and the relatively broad range of uncertainty we have about these three continuous values in the model.

All that remains at this point is to run the model and answer the question posed at the start of this section, which was: What is the probability an imported carcass contaminated with the pathogen will enter our food supply?

The probability that a contaminated carcass will enter the food supply is not the same as the frequency with which that will happen, because there is variability in the world. This value is found just to the right and above the topmost endpoint of the model seen in Figure 15.14. (Rounding prevents you from reproducing the shown values precisely from the figure.) Each iteration of the model produces a new estimate

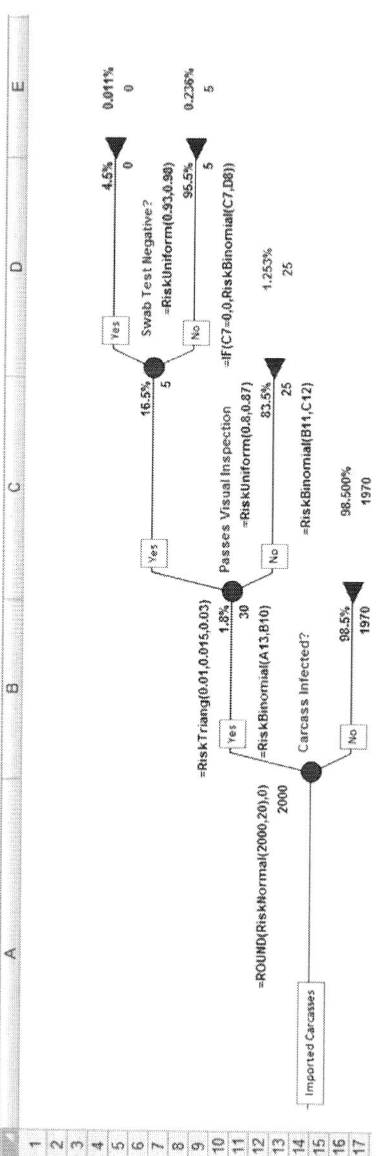

FIGURE 15.14 Final carcass import model accounting for natural variability and knowledge uncertainty.

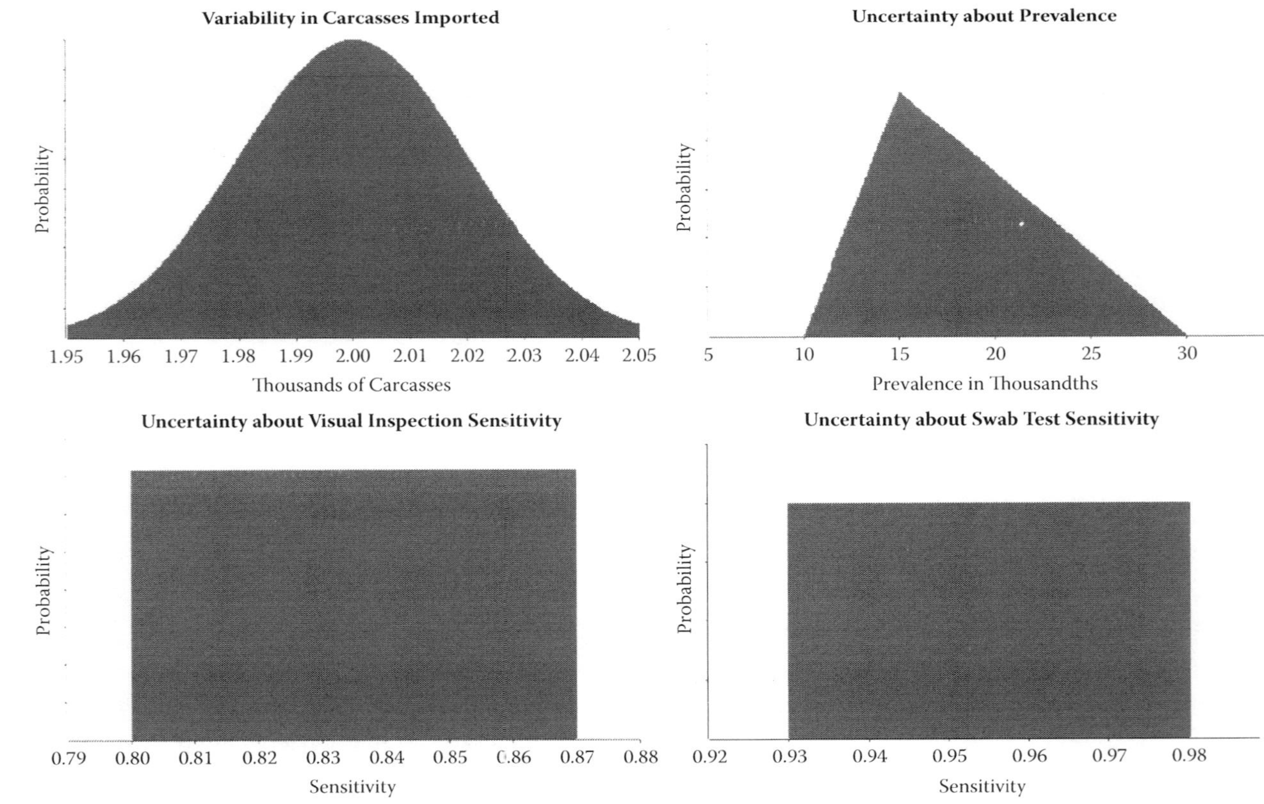

FIGURE 15.15 Probability models for one variable and three uncertain model inputs.

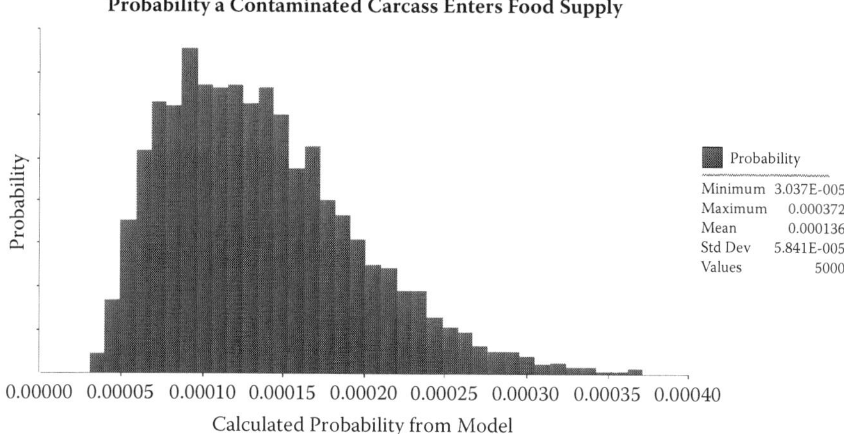

FIGURE 15.16 Estimate of the probability that a contaminated carcass enters the food supply.

for this output. The results of the simulation are shown in Figure 15.16. The minimum observed value was .00003, and the maximum was about .0004. The expected value is .0001, essentially unchanged from the first crude formulation of the model, shown in Figure 15.10. This makes sense, however, because we used expected values in the prototype. In the long run, about 1-in-10,000 carcasses will be contaminated and slip into the food supply. What that might look like in actual fact is addressed in the next question and answer.

How many infected carcasses could enter the food supply annually? The minimum number of contaminated carcasses to enter the food supply is, as before, zero, which happened 76.5% of the time. The maximum number to enter the food supply in this simulation was four. The expected value is 0.27 carcasses per year. This figure differs from the previous result (.22) from the model that incorporated only part of the natural variability. When we finally expressed our uncertainty, the likelihood of one or more contaminated carcasses entering the food supply was greater. The model output value for the second question is presented in Figure 15.17.

Event trees are visually instructive and easy to work with for a wide variety of risk issues. Carefully layering probability onto the event tree models will sometimes require the analyst to modify the software's built-in function and capability. Examples of that were seen in this simple model, where rounding functions and a logic statement were used. As a bundle of tools and techniques, probabilistic scenario analysis currently remains the methodology of choice for most probabilistic risk assessment.

15.8 Summary and Look Forward

Scenario analysis is an essential risk assessment tool. Constructing as-planned, failure, and improvement scenarios is an essential part of both qualitative and quantitative risk assessment. Risk assessors make extensive use of deterministic and probabilistic scenario analysis. Once in a while a monolithic scenario is used for decision making, but hopefully not in risk analysis. DSA and PSA rely on scenario comparisons

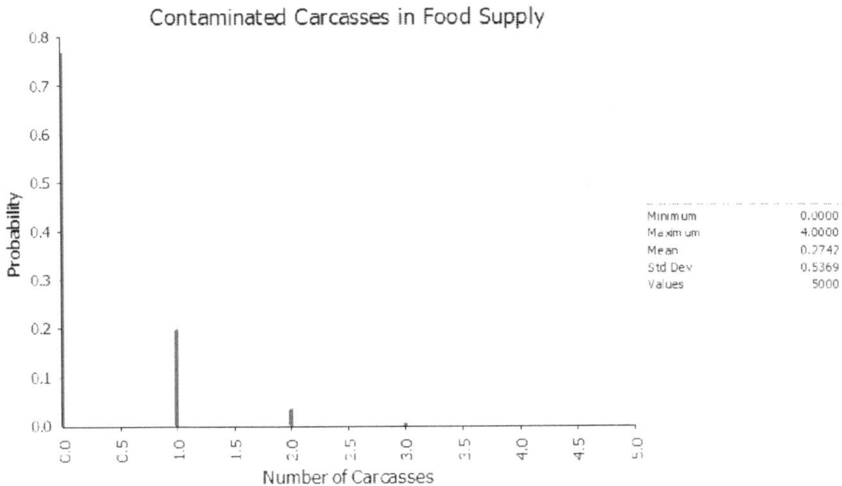

FIGURE 15.17 Estimate of the annual number of contaminated carcasses entering the food supply.

to place risks in a useful context for decision makers. Without-and-with condition comparisons are preferred for exploring the efficacy of risk management measures. When probabilistic methods are added to the available suite of scenario structuring tools and techniques, this produces a very powerful suite of tools called probabilistic scenario analysis. Most probabilistic risk assessments are some form of a PSA.

Chapter 16 covers sensitivity analysis. Sensitivity analysis is the study of how the variation in a risk assessment output can be apportioned to different sources of variation. This is the bare minimum expectation for addressing uncertainty in an explicit fashion in any kind of risk assessment, qualitative or quantitative. Numerous techniques are presented for both qualitative and quantitative sensitivity analysis.

REFERENCES

Food and Drug Administration. 2005. Center for Food Safety and Applied Nutrition. *Quantitative risk assessment on the public health impact of pathogenic* Vibrio parahaemolyticus *in raw oysters*, chap. 5. http://www.fda.gov/Food/ScienceResearch/ResearchAreas/RiskAssessmentSafetyAssessment/ucm185190.htm.

Palisade Corp. 2009. *User's guide PrecisionTree decision analysis add-in for Microsoft® Excel Version 5.5*. Ithaca, NY: Palisade Corp.

U.S. Nuclear Regulatory Commission. 1981. *Fault Tree Handbook*. Washington, DC: Nuclear Regulatory Commission.

16

Sensitivity Analysis

16.1 Introduction

Expect that knowledge uncertainty will never be reduced to zero. Know that natural variability will never disappear. These two simple facts mean that the risk characterization from your risk assessment is, at best, an informed estimate. At worst, it may be well-intentioned speculation. What assessors and managers both need to know is that the outputs of a risk assessment are conditional answers based on the data and data gaps, on the assumptions and estimation tools, and on the techniques and methodologies used to arrive at the answers. They also need some idea of where the assessment answers lie on the continuum between the best estimate and well-intentioned speculation.

A simplistic schematic of a decision situation is shown in Figure 16.1. A risk assessment has several inputs that include knowledge, data, policy, and information in many forms. Data gaps and other forms of uncertainty are ubiquitous characteristics of these inputs.

Assumptions made to help address uncertain inputs find their way into the model and affect model outputs, e.g., risk characterizations and the like, which, in turn, influence the risk management decision. Any methodologies used to address the uncertainty and variability in inputs will result in a similar chain of events. Likewise, assumptions made about the model's structure will influence outputs and decisions.

Sensitivity analysis is the study of how the variation* in a risk assessment output can be apportioned, qualitatively or quantitatively, to different sources of variation. Complex risk assessments may have dozens of input and output variables that are linked by a system of equations and calculations. Risk assessors need to consider how sensitive a model's output, a risk characterization, or other important assessment outputs are to changes or estimation errors that might occur in model inputs, model parameters, assumptions, scenarios, and the functional forms of models. This information must then be effectively conveyed to risk managers so they can explicitly consider its significance for their decision making.

Some risk assessment outputs and the decisions that rely on them may be sensitive to changes in assumptions and input values. However, it is not always immediately obvious which assumptions and uncertainties most affect outputs, conclusions, and decisions. The purpose of sensitivity analysis is to systematically find this out.

* Variation includes the effects of natural variability and knowledge uncertainty.

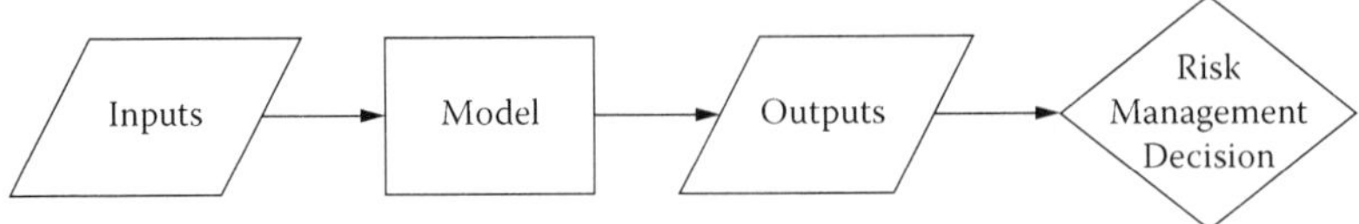

FIGURE 16.1 Decision model schematic.

A good sensitivity analysis increases the assessor's and manager's confidence in the risk assessment model and its predictions. It provides a better understanding of how model outputs respond to changes in the inputs. Because risk assessments can be qualitative or quantitative, sensitivity analysis can likewise be qualitative or quantitative. In a qualitative analysis, the assessor identifies the uncertainties affecting the assessment and forms a judgment of their relative importance. A quantitative sensitivity analysis quantifies the variation in model outputs that is caused by specific model inputs and the model structure.

Some sensitivity analysis should be an integral component of every risk assessment. This is the point in a risk management activity when the risk analysis team focuses intentionally on better understanding the things we don't know and their importance for decision making. The results of the sensitivity analysis will provide insight into the importance of different sources of uncertainty.

Sensitivity analysis has at times been called what-if analysis. It can be used to answer questions like those below, which build on questions first provided by Mokhtari and Frey (2005).

- How might the decision be changed by increases or decreases in selected input values?
- What is the range of values a parameter can assume in a function without changing the decision?
- By how much must a value change to lead to an alternative best decision?
- How sensitive is our output to forecast error or other changes in inputs?
- Which inputs contribute most to the variation in the output?
- Which inputs are most responsible for the best (or worst) outcomes of the output?
- What is the rank order of importance among the model inputs?
- Are there two or more inputs to which the output has similar sensitivity, or is it possible to clearly distinguish and discriminate among the inputs with respect to their importance?
- Might changes in our decisions/actions improve our outputs?
- Does the model respond appropriately to changes in assumptions and inputs?

This chapter is divided into sections that address qualitative and quantitative approaches to sensitivity analysis. Material that could have been included in both of the sections will be addressed in the quantitative sensitivity analysis discussion. The most common sensitivity analysis methods are relatively simple techniques. There are others that are quite complex. The quantitative sensitivity analysis discussion summarizes methods from across this continuum. Examples are provided for the simpler methods; references are provided for the more complex ones.

16.2 Qualitative Sensitivity Analysis

Risk analysis is a framework for making decisions under uncertainty. Because your uncertainty cannot be eliminated, it is critically important to explore the importance of the manner in which you addressed the uncertainty as well as the uncertainty that remains prior to making a decision. A qualitative sensitivity analysis characterizes the uncertainty and its potential significance to decision making in nonnumerical ways. This is done to aid risk managers who need to make decisions in the face of this uncertainty. Quantitative sensitivity analysis is almost always preferred when it is feasible. Qualitative sensitivity analysis, being more subjective, is generally less reliable. At a minimum, qualitative sensitivity analysis provides a greater degree of confidence in assessment outputs and management decisions based on them for having identified and considered critical data gaps and other sources of uncertainty.

A basic methodology for qualitative sensitivity analysis includes:

- Identifying specific sources of uncertainty
- Ascertaining the significant sources of uncertainty
- Qualitatively characterizing the significant uncertainty

A reasonable objective for qualitative sensitivity analysis is to identify the sources of uncertainty that exert the most influence on the risk assessment outputs.

16.2.1 Identifying Specific Sources of Uncertainty

Identifying specific sources of uncertainty often begins with an acknowledgment that uncertainty exists. This is not always as easy as it seems it should be (see sidebar). A significant number of organizations still want to know "the number," and there is no incentive for acknowledging uncertainty in such a culture.

THE CHANNEL BOTTOM

In an early proof-of-concept risk assessment for the U.S. Army Corps of Engineers, an experienced engineer was questioned about a point estimate of the percentage of rock in a channel bottom. This is an important determinant of the cost of channel deepening. Offered the opportunity to bound this estimate, he refused, insisting that he had better information from sample borings and more data than he had ever had in a long and successful career. He insisted there was no uncertainty about this value and was offended that we might think otherwise. His estimate turned out to be almost less than half of the actual rock content. Costs quickly doubled, and he is now a proponent of risk assessment.

Once we are ready to acknowledge that uncertainty exists, we must recognize it when it is present. Knowledge is a relative commodity, and when we are used to working with relatively little data, for example, a lot of data can sometimes mask the uncertainty that remains. The assessor must be able to separate what is known from what is unknown in a decision problem.

We need to be able to identify and point to an input, assumption, scenario, or model and say this is uncertain. When we have done that, it is helpful to say why it is uncertain, i.e., identify what is the cause of the uncertainty (see Chapter 2). The initial goal is to honestly identify the things we do not know. A useful output from this step is a list of inputs recognized as uncertain, by type of input. The next step is to figure out which of them matter most.

16.2.2 Ascertaining the Significant Sources of Uncertainty

A significant uncertainty is one that could affect decision making. Thus, significant uncertainties are those that can affect model outputs, risk characterizations, conclusions, answers to the risk manager's questions, or other important decision criteria.

The best place to begin to identify what the most important source of uncertainty may be by considering, "What do people say is important?" Use the relevant theory, read the professional literature, talk to an expert, reason it through for yourself, and listen to what people are saying. Respect the wisdom of crowds and experts!

If you have used a mental or other model, look at the structure of your model. It reflects the extent to which a physical system or phenomenon is understood. Is there any aspect or element of your model that has more influence on the model outputs than any other? It is almost invariably the case that some model elements have a disproportionate influence on model outputs. If so, these elements may be potential sources of important uncertainties.

Which inputs can you control or affect? It is generally wise to pay extra attention to those things you can control, or at least affect. Inputs that people say are important, that are influential in your model, and that you can affect may well be significant uncertainties. Take that list of uncertain inputs you prepared and put an asterisk next to each one that you suspect may be significant. Now it is time to characterize that uncertainty in ways that risk managers can understand and consider in decision making.

16.2.3 Qualitatively Characterizing the Uncertainty

The World Health Organization (2006) offers a useful scheme for qualitatively characterizing the uncertainty in your assessment. Although developed for use with

DATA INPUTS

Data inputs can be either a parameter (constant) that enters the model somewhere or a value or distribution for a variable. The variables may be decision variables, which we can exert some control over, chance variables, which we have no control over, or influence variables, which we cannot directly control but may influence with our decision variables. In qualitative risk assessment data often appear in nonnumerical forms.

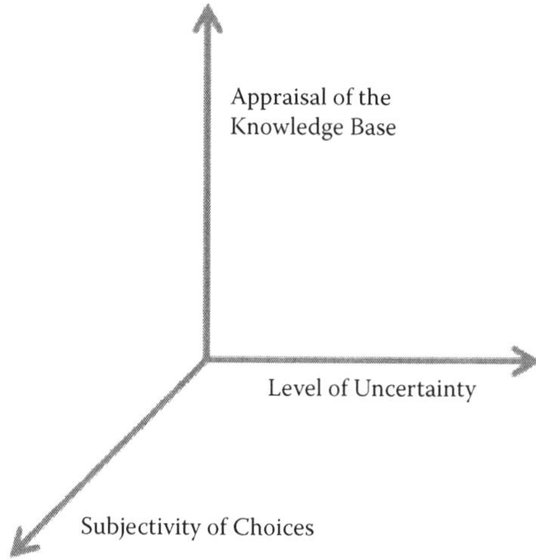

FIGURE 16.2 Three-dimensional space of qualitative risk assessment. (*Source*: World Health Organization 2006.)

chemical exposure assessments, it has general applicability. To qualitatively identify the most critical uncertainty, think of the process as locating each uncertain item in the three-dimensional space shown in Figure 16.2. Values near the origin are the least uncertain. A low, medium, or high score is determined by the assessor for each of the axes. The most significant uncertainties are those with the highest scores for all three dimensions. The axes may be differentially weighted or not. Following the WHO process, level of uncertainty > appraisal of the knowledge base > subjectivity of choices. Explicit weights need not be assigned because the characterization is a sequential one, as seen in the hypothetical example in Table 16.1.

The level of uncertainty dimension expresses the degree of severity of the uncertainty, from the assessor's perspective. A scale ranging from low to high has been suggested, as shown in Figure 16.3 (WHO 2006).

TABLE 16.1

Sample Summary of the Evaluation of Uncertainty in a Qualitative Risk Assessment

Sources of Uncertainty		Characteristics of Uncertainty		
		Level of Uncertainty	**Appraisal of Knowledge Base**	**Subjectivity of Choices**
Scenario		Medium	Low	Medium
Model	Conceptual	High	Medium	Medium
	Mathematical	NA	NA	NA
Inputs				
Input 1 (irreducible)		Low	Low	Low
Input 2 (reducible)		High	Medium	Low
Input 3 (reducible)		Medium	Medium	High

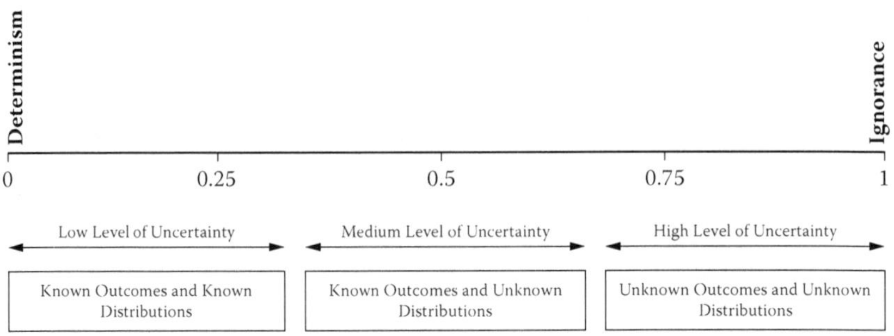

FIGURE 16.3 Rating the level of uncertainty. (*Source*: World Health Organization 2006.)

A low score means that a large change in the source of uncertainty would have a small effect on the results. Inputs scored as medium would have proportional effects on outputs. A high score implies that a small change in the source of the uncertainty would have a large effect on the results. As qualitative inputs are scored in this dimension, it is helpful to indicate whether the uncertainty is reducible or irreducible (see Table 16.1) so that risk managers better understand their options for addressing the level of uncertainty.

The appraisal of the knowledge base is based on accuracy, reliability, plausibility, scientific backing, and robustness, all seen on the vertical dimension axis of Figure 16.4 (WHO 2006). It is intended to rate the adequacy of the state of knowledge about the input. A low uncertainty in the knowledge-base rating suggests the qualities listed in the first column. The other columns illustrate the two other possible ratings (medium and large).

The final dimension along which to rate an uncertain input is the subjectivity of the choice the assessor had in making an assumption about this input. The aspects of the choice include the choice space, intersubjectivity among peers and among stakeholders, influence of situational limitations (e.g., money, tools, and time) on choices, sensitivity of choices to the analysts' interests, and the influence of choices on results.

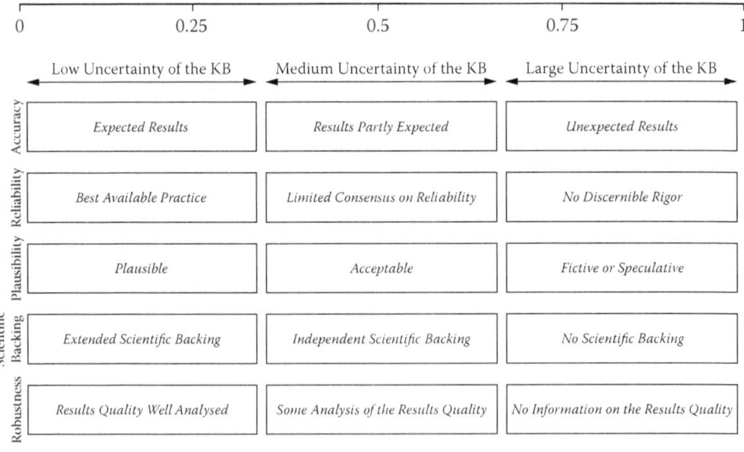

FIGURE 16.4 Rating the appraisal of the knowledge base. (*Source*: World Health Organization 2006.)

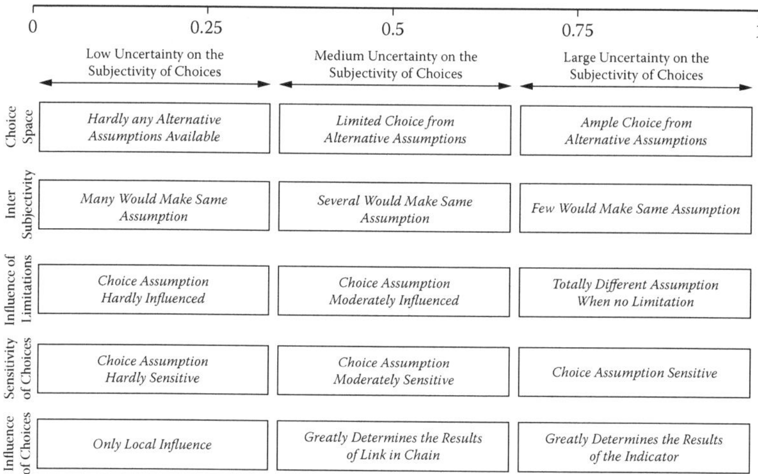

FIGURE 16.5 Rating the subjectivity of choice. (*Source*: World Health Organization 2006.)

Descriptions for each aspect can be read across the vertically arranged rows in Figure 16.5 (WHO 2006).

A hypothetical qualitative summary evaluation of the uncertainty in an assessment is shown in Table 16.1. Imagine that the preceding steps have identified the base scenario, the conceptual models, and three of its inputs as potentially significant sources of uncertainty. If we follow the WHO rationale and consider the three dimensions of uncertainty as sequential screening tools, we would find the model and input 2 of most interest because each scored high on the first screening criterion. Subsequent screening criteria are used to refine the sorting process. Both are rated medium on the knowledge-base criterion. Moving to the subjectivity-of-choices criterion identifies the conceptual model as the more significant uncertainty in this case. Between the two medium levels of uncertainty, input 3 is rated as more significant based on an appraisal of the knowledge base. The WHO method proceeds to rank sources of uncertainty by aspects; first, ranking by level of uncertainty, then appraisal of the knowledge base, and finally by subjectivity of choices.

Once the uncertainties have been identified in this or another manner, it remains for the assessor and the manager working together to subjectively evaluate the significance of these findings for the purposes of risk management. The risk manager should explicitly consider the results of such a sensitivity analysis. His decision could include additional research to minimize the reducible uncertainty. Progressing to a quantitative risk assessment may be an option for some situations. Systematically exploring the impact of alternative assumptions or ranges of values on the decision to be made is, hopefully, an obvious response to uncertainty in a qualitative assessment.

16.2.4 Vary the Key Assumptions

A sensitivity analysis structured as described in the previous discussion is not going to be done for every qualitative assessment. However, doing no sensitivity analysis is unacceptable. The absolute, bare minimum sensitivity analysis is to carefully and systematically explore and vary the assumptions that underlie the assessment. When we separate what we know from what we do not know during a risk assessment, it is necessary to find effective means for dealing with that which we do not know. Making assumptions about uncertain values is one of the most common and expedient ways of addressing uncertainty.

We make two basic kinds of assumptions. Those we know we are making, i.e., explicit assumptions, and those we do not know we are making, i.e., implicit assumptions. Explicit assumptions should be documented, written down, and preserved for the attention of assessors and managers and other interested parties. Implicit assumptions are often not recognized by the assessors or managers. Peer review by multidisciplinary reviewers can be an effective technique for detecting implicit assumptions. It often takes someone outside one's own discipline, specialty, or organization to recognize the assumptions we make as the common practice of our fields and organizations.

To the extent that we use assumptions to address uncertainty and to enable ourselves to move forward with our assessments, we should routinely test the sensitivity of our assessment outputs to those assumptions. The simplest way to do this is to first make a list of the key assumptions of your risk assessment. Next, explore what happens as you drop or challenge each assumption individually and replace it with alternative assumptions. Do your outputs change? Do the answers to the risk manager's questions change? Might any of these changes affect the risk management decision? If so, that information needs to be conveyed to the risk managers of a nuclear power plant site. For example, suppose we have assumed sea level rise of one foot or less in the next 100 years. If it is more than that, this site will be surrounded by water year round. This is information that needs to be carefully communicated to risk managers. We have assumed sea level rise will pose no hazard, but that is a remaining uncertainty.

It can sometimes be instructive to explore what happens as you drop/change your assumptions in combinations. Assumption dropping is a very effective kind of qualitative sensitivity analysis that can also be useful in quantitative assessments. When the assumptions do make a difference to your assessment outputs and answers, you have identified a critical uncertainty. Risk managers should respond accordingly either by reducing the uncertainty through additional data collection or research or by having the risk assessors address the assumptions appropriately in their assessment, perhaps with probabilistic risk assessment methods. The final obvious alternative for risk managers is to take the knowledge of that sensitivity into account when making a decision.

16.3 Quantitative Sensitivity Analysis

There are four classes of quantitative sensitivity analysis tools. These are scenario, mathematical, statistical, and graphical analysis. In the discussion that follows, example methods from each class are described. The more popular methods are illustrated by reference to a simple cost estimating example.* Two of the methods presented here rely on different examples that are better suited to the methods demonstrated.

* Cost-estimating examples are used in this chapter and the next one. Although they may not spring first to mind when thinking about risks, they have the advantage of being easily understood and accessible to readers from every discipline. The risk of a cost overrun or under-estimating the cost of an action can be significant. As such they provide a friendly context for learning the concepts.

16.3.1 Scenario Analysis

Analyzing the sensitivity of assessment results to the scenarios we use is called scenario analysis. The base-case scenario for most risk assessments is either an existing condition or the "without" additional risk management options (RMOs) condition. In the simplest scenarios, model inputs are entered as point estimates that represent the "best guess" as to the true but unknown value of each input. Rather than to alter these inputs one or two at a time, scenario analysis alters the entire scenario. If we think of scenarios as the stories we tell about risks, these alternative scenarios include different plot lines such as best case, worst case, common practice, the most likely case, a locally preferred scenario, a new policy case, and so on.

The alternative scenarios can vary markedly in their structure and details. When a different alternative is considered as a starting point for the assessment it can sometimes lead to a markedly different characterization of the risk. Existing, as-planned, failure and improvement scenarios are all subject to scenario sensitivity analysis. It is worth noting that scenario analysis does not require the construction of structurally different scenarios. The alternative scenarios may simply comprise alternative sets of input values within an unchanging scenario structure. The goal is to focus on the structural elements or inputs of a scenario that are most uncertain and to vary them to explore the sensitivity of risk characterizations and answers to the risk manager's questions to these plausible different scenarios.

One of the most commonly used alternative scenarios is the worst-case scenario. If we can identify and live with the worst-case scenario, a common (and sometimes wrong) presumption is that we can live with lesser scenarios as well. Consequently, assessors will sometimes try to envision the most extreme negative set of input values possible.

Engineers and public health officials, to name just two professions, are biased toward designing systems conservatively to try to minimize or eliminate the chance of adverse outcomes. The engineering profession in particular has regarded conservatism in design as the accumulated wisdom of centuries of experience that has taught that the conditions of the real world are not always predictable and it makes good sense to provide some margin of error for unforeseen events. This drive toward conservatism has led to the widespread propagation and use of worst-case scenarios.

There is no real formal definition of a worst-case scenario. It is simply that future in which everything that can reasonably go wrong does go wrong. If the worst-case scenario yields an acceptable result, decision makers in the past have often assumed there is no need to manage a risk. On the other hand, worst-case scenarios that result in unacceptable consequences often have lead decision makers to take precautions to preclude the worst-case scenario from occurring.

Despite its widespread usage, the worst-case scenario is not without its problems. First among these is that introducing this conservatism into an analysis focuses policy and possibly resources on what is often a deliberately unrealistic scenario. This is precisely what risk assessment is designed to prevent. Second, given any worst-case scenario, an even worse case can, paradoxically, still be defined. Third, the likelihood of a worst-case scenario may be so small as to lead to the waste of efforts, materials, and other resources in attempts to reduce it. Fourth, there is an almost hypnotic appeal to thinking that if we have covered the worst-case, we have covered everything. However, failure in the better-than-worst-case world is still possible and is often overlooked with a worst case orientation to risk management.

Nonetheless, worst-case scenarios are likely to remain useful and popular failure scenarios to investigate. In general, though, it is not wise to make policy based on worst-case scenario analysis.

Other scenarios can be evaluated for the benefit of risk assessors and managers alike. An optimistic or best case (if everything that can break our way does) scenario may be of interest to risk-taking risk managers, for example. Different stakeholders may proffer alternative views of the future that should be explicitly considered in scenario analysis. The risk assessors may develop a range of scenarios to reflect their concerns about the key uncertainties in a risk assessment. What the team assumes about climate change, sea level rise, geopolitical events, natural disasters, technological advancements and the like could have important ramifications on decision making. The most likely scenario is often the base case, but any number of scenarios can be developed, and deterministic scenario analysis is not an uncommon sensitivity analysis approach.

When assessors use scenario analysis, the idea is to compare the different situations to identify differences in important model outputs. Differences that make a difference for decision making are important, and assessors and risk managers must become aware of them in a sensitivity analysis. Risk managers then must decide how much to weigh the range of potential outcomes that could result from the remaining uncertainty in the decision-making process.

16.3.2 Mathematical Methods for Sensitivity Analysis

Mathematical methods rely on calculating outputs for a range of input values or for different combinations of input values. These methods differ from statistical methods, which rely on simulations in which inputs are represented by probability distributions. Although mathematical methods do not describe variance in outputs due to variance in inputs as statistical methods do, they are still useful in estimating the impact of a range of input values on model outputs. In addition, mathematical methods can help identify the most important inputs, and they are useful for verifying models. Sensitivity of model outputs to individual inputs or groups of inputs can be explored by various means (Frey and Patil 2002).

Graphical methods are not easily separated from mathematical or statistical techniques because they often rely on these techniques to produce the relationships to be graphed. Nonetheless, graphical methods and the visualization of data are becoming important techniques for representing sensitivity information visually and are treated separately in this chapter.

Several sensitivity analysis methods presented in the pages that follow use the cost estimate example seen in Table 16.2. These are the costs of dredging a navigation channel and using the material to create wetlands by placing it behind geotube barriers adjacent to the shoreline.

Note that row and column locations are provided. This may be helpful for interpreting some of the tables and figures that follow. The first column (A) describes the basic work item, column (B) provides the quantity in the units of column (C). The price (D) is per unit. The amount (E) is the product of the quantity and unit price. The subtotal sums the total amounts that precede it in the table. Of this total, 8% and 6% are calculated to cover, respectively, advanced engineering and design (AED) as well

TABLE 16.2

Cost Estimate for Channel Dredging

	A	B	C	D	E
8	**Description**	**Quantity**	**Unit**[a]	**Price**	**Amount**
9					
10	Lands and damages	0	LS	$...	$...
11	Relocations				
12	Lower 20 pipeline, 653+00	425	LF	$730.00	$310,250
13	Remove 8" pipeline, 678+00	1,000	LF	$50.00	$50,000
14	*Total—Relocations*				*$360,250*
15	Fish and Wildlife Facilities (Mitigation)				
16	Oyster reef creation	0	ACR	$...	$...
17	*Total—Fish and Wildlife Facilities (Mitigation)*				*$...*
18	Navigation, Ports and Harbors				
19	Mobe and demobe	1	LS	$500,000.00	$500,000
20	Pipeline dredging, Reach 1	576,107.00	CY	$2.78	$1,601,577
21	Pipeline dredging, Reach 2	1,022,769.00	CY	$2.60	$2,659,199
22	Pipeline dredging, Reach 3A	1,182,813.00	CY	$3.16	$3,737,689
23	Pipeline dredging, Reach 3B	736,713.00	CY	$2.76	$2,033,328
24	Scour pad, Reach 1	17,550	SY	25.69	$450,860
25	Geotubes, 30′, Reach 1	1,400	LF	$188.52	$263,928
26	Geotubes, 45′, Reach 1	4,912	LF	$222.18	$1,091,348
27	Scour pad, Reach 3	38,750	SY	$25.69	$995,488
28	Geotubes, 45′, Reach 3	13,940	LF	$222.18	$3,097,189
29	*Total — Navigation, Ports and Harbors*				*$16,430,606*
30	*Subtotal*				*$16,790,856*
31					
32	Engineering and Design	8%			$1,343,268
33					
34	Construction Management	6%			$1,007,451
35					
36	*TOTAL PROJECT COST*				*$19,141,576*

[a] LF = linear feet; LS = lump sum; CY = cubic yards; SY = square yards; ACR = acres.

as construction management (CM). They are added to the subtotal to obtain the total project cost, $19.1 million, the only output of interest for our example.

The numerical values shown in the table are the best point estimates of the cost estimators. For the mathematical methods, the model is treated as a deterministic one. Once we progress to the use of statistical sensitivity measures, these point estimates are replaced by probability distributions and a Monte Carlo process is then used to simulate results. Assume that there is knowledge uncertainty and natural variability "sprinkled" among the various inputs required to estimate the total project cost.

16.3.2.1 Nominal Range Sensitivity

Nominal range sensitivity is a mathematical method used for deterministic, rather than probabilistic, models. It is usually used to identify the most important input(s) (Cullen and Frey 1999), and this can be useful for setting research and data collection priorities.

Nominal range sensitivity analysis is also known as one-at-a-time analysis (OAATA). It works by evaluating the effect of changes in an individual input on an output variable. This local sensitivity is conceptually equivalent to a partial derivative. Although simple, this technique has at least two key shortcomings: (1) it does not account for simultaneous variation of multiple model inputs; and (2) it does not account for any nonlinearities in the model that create interactions among the inputs. Nonetheless, for simple linear models, OAATA can be instructive.

The selected input may be changed incrementally or allowed to vary across its entire range of plausible values while holding all other input values constant, usually at their nominal, mean, representative, or base-case values. This is sometimes done to identify threshold values that result in significant changes in model outputs. It can also be used to identify which uncertain variables may have the greatest impact on outputs of interest in the risk assessment. Alternatively, a plausible range of variation, say ±20%, may be applied to one or more inputs. Morgan and Henrion (1990) call the change in the model output due to a unit change in the input the sensitivity or swing weight of the model for the chosen input variable. This sensitivity analysis can be repeated for as many input variables as desired.

Bearing in mind that the sensitivity can be described in terms of a change in output for a unit or any given percentage change in the input variable value, the direction

LIMITATIONS OF ONE-AT-A-TIME ANALYSIS

Do not automatically equate the magnitude of an uncertain variable with its influence. Consider two random variables expressed by the following uniform distributions: $A = U(10^7,10^8)$, $B = U(2,6)$. Some might be tempted to assume A is more influential because of its sheer magnitude.

It is essential to know the structure of your model, and nonlinearities can change everything you think you know about sensitive variables. For example, consider a new variable, C, that is a function of A and B. If $C = A + B$; A dominates. If $C = A^B$; B dominates.

Dependence and branching in a model can also create flaws with the logic of OAATA. Consider this example:

If $X < 50$ then
$Y = Z + 1$
Else
$Y = Z^{100}$

What value will you set X equal to when you investigate the sensitivity of Y and Z?

TABLE 16.3

One-At-A-Time Analysis Example for Dredging Cost Estimate

Description	Quantity	Unit	Unit Price	Amount	Change
Original Estimate					
Pipeline Dredging, Reach 1	576,107	CY	$2.78	$1,601,577	NA
Total Project Cost				$19,141,576	NA
20% Increase in Reach 1 Quantity					
Pipeline Dredging, Reach 1	691,328	CY	$2.78	$1,921,893	$320,316
Total Project Cost				$19,506,736	$365,160

or sign of the change is especially useful to note. Nominal range sensitivity analysis works best with linear models where rank orders can be easily established based on this measure of sensitivity. In nonlinear models, output sensitivities may depend on interactions with other inputs that may not always be obvious and therefore cannot be placed in a rank order.

This method is easy to use. It is most reliable: when assessors have a good idea of the plausible range of input values; in linear models when the effect of a one-unit change in an input does not depend on the starting value of the input; and when there is no significant interaction of the chosen input with other input variables. When interactions are possible among variables, this method is likely to produce an inadequate description of the range of possible input values and, therefore, output responses.

To illustrate this technique, consider the quantity of pipeline dredge material in reach 1 from the cost estimation case study in (cell B20) Table 16.2. What happens to total costs if that quantity increases 20%? Table 16.3 shows the effect of a 20% increase in the reach 1 dredging quantity on total project costs. Only the reach 1 dredging cost line is shown as it is the only change in the model inputs.

A 20% increase in this one input causes total project cost to rise by $365,000, a 1.9% increase in total project cost. Note that the simple nature of this example renders the one-unit change in the input (from 576,107 to 576,108 CY) a trivial exercise. Every cubic yard costs the same $2.78 in the model. Of course this must be increased to reflect allowances for AED and CM.

The input changes investigated may be percentages of a value, like the 20% increase used here, or specific values of interest can be chosen. For example, we might ask what happens to total project costs if the dredging quantity in reach 1 rises to 750,000 CY. This process can be repeated for as many individual inputs as desired. It is easy to see that two or more inputs could be varied simultaneously in a similar fashion. This is a simple method, but it is tedious.

Palisade Corporation's TopRank 5.5 (see Appendix B for details on the use of this software) can be used to conduct both a one-way what-if analysis and a multiway what-if analysis. This is done by allowing the assessor to vary a fixed-point estimate of an input by some plus or minus percentage. To demonstrate this technique, every variable input in the model (excluding the percentages for engineering design and construction management and the number of mobilization and demobilizations of equipment, cell B19)) is allowed to vary by ±20%.

Allowing each input to change one at a time, first by its minimum and then by its maximum value, we see which inputs have the greatest potential impact on total project cost in Table 16.4. Only the top ten inputs are shown. The table identifies the input by name and location in the model. This name is chosen automatically by the software. The first input is the quantity of pipeline dredging material in reach 3a. It is found in cell B22 of the model. When it assumes its minimum value of 946,250 CY, total project cost is $18,289,383, a decrease of 4.45%. When the maximum of 1,419,375 CY is substituted into the model, costs rise to $19,993,769, an increase of 4.45%. The table shows the top ten most influential inputs. The software output evaluates every input one-at-a-time, so that if you are interested in the effect of a specific input you can find it readily. A graphic display of this result is shown under the discussion of graphic sensitivity techniques.

It is often common practice to build models initially using plausible values or point estimates of expected values. In a complex model there could be dozens or even hundreds of inputs. Many models lack the transparent simplicity of our example. When that is the case, the risk assessor may not want to go to the time and trouble to bound each input or to specify a probability distribution to use for each model input if it has little or no effect on the model output(s) of interest. A one-way what-if analysis (equivalent to OAATA) will quickly identify those inputs that the assessor ought to focus attention on when considering the effects of uncertainty.

Presuming no more data than the values shown in Table 16.2, the OAATA summarized in Table 16.4 provides the assessor with a clear identification of the most significant uncertainties among the model inputs. Table 16.5 goes one step further by showing additional percentage changes in the model inputs. The ranking of inputs is identical to that seen in Table 16.4. Here you see that the software enables you to examine a change of ±20% and ±10% in the same analysis. These expanded results support spider plots, which are discussed in section 16.3.4.3.

Analyses like those seen in the tables can be prepared for a variety of ranges of change in the input values. They need not be ±20%. In a linear model this produces a reliable indication of those inputs to which the model output will be most sensitive. If your assessment model has multiple outputs, a separate OAATA must be performed for each output. These results are a valuable guide for handling uncertainty in subsequent iterations of the model. This kind of analysis helps the risk assessor understand which uncertainties have the greatest influence on output quantities of interest. With this information the assessor can prioritize data gaps and future research needs.

As we have cautioned, there may be times when it does not make sense to vary one input at a time. The desired changes in a model can always be made manually, but this is very limiting, especially when the assessor would like to explore the sensitivity of model outputs to the many different inputs. TopRank offers the capability to conduct multi-way what-if analysis. As with the one-way analysis, the assessor can choose a percentage by which to vary the inputs, identify which variables to consider in a multiway analysis, and how many to consider at a time. An examination of all possible combinations of two-at-a-time variable inputs is summarized for the cost estimate in Table 16.6. This shows the top five input combinations from the model.

The report shown provides the results of an analysis of the extremes, i.e., when both inputs assume their minimum values and when both inputs assume their maximum values. It should come as no surprise that the ranking of paired variables mirrors the results of the OAATA. Detailed outputs are provided by TopRank for the multi-way what-if analysis, so that additional pairs of inputs and additional combinations of values (%'s) can be explored.

TABLE 16.4

Top Ten Inputs from a One-Way What-If Analysis for the Total Project Cost Output

			Minimum			Maximum		
			Output			Output		
Rank	Input Name	Cell	Value ($)	Change (%)	Input Value	Value ($)	Change (%)	Input Value
1	Pipeline Dredging, Reach 3A/Quantity (B22)	B22	18,289,383	–4.45	946250.4	19,993,769	4.45	1419375.6
2	CY/Price (D22)	D22	18,289,383	–4.45	2.528	19,993,769	4.45	3.792
3	Geotubes, 45′, Reach 3/Quantity (B28)	B28	18,435,417	–3.69	11152	19,847,735	3.69	16728
4	LF/Price (D28)	D28	18,435,417	–3.69	177.744	19,847,735	3.69	266.616
5	Pipeline Dredging, Reach 2/Quantity (B21)	B21	18,535,279	–3.17	818215.2	19,747,874	3.17	1227322.8
6	CY/Price (D21)	D21	18,535,279	–3.17	2.08	19,747,874	3.17	3.12
7	Pipeline Dredging, Reach 3B/Quantity (B23)	B23	18,677,977	–2.42	589370.4	19,605,175	2.42	884055.6
8	CY/Price (D23)	D23	18,677,977	–2.42	2.208	19,605,175	2.42	3.312
9	Pipeline Dredging, Reach 1/Quantity (B20)	B20	18,776,416	–1.91	460885.6	19,506,736	1.91	691328.4
10	CY/Price (D20)	D20	18,776,416	–1.91	2.224	19,506,736	1.91	3.336

Note: Top ten inputs ranked by percent change.

TABLE 16.5

An Expanded One-Way What-If Analysis for the Total Project Cost Output

			Input Variation			Output Variation		
Input Name	**Cell**	**Step**	**Value**	**Change**	**Change (%)**	**Value ($)**	**Change ($)**	**Change (%)**
Pipeline Dredging, Reach 3A/Quantity (B22)	B22	1	946250.4	−236562.6	−20.00	18,289,383	(852,193)	−4.45
		2	1064531.7	−118281.3	−10.00	18,715,479	(426,097)	−2.23
		3	1182813	0	0.00	19,141,576	(0)	0.00
		4	1301094.3	118281.3	10.00	19,567,673	426,097	2.23
		5	1419375.6	236562.6	20.00	19,993,769	852,193	4.45
CY/Price (D22)	D22	1	2.528	−0.632	−20.00	18,289,383	(852,193)	−4.45
		2	2.844	−0.316	−10.00	18,715,479	(426,097)	−2.23
		3	3.16	0	0.00	19,141,576	(0)	0.00
		4	3.476	0.316	10.00	19,567,673	426,097	2.23
		5	3.792	0.632	20.00	19,993,769	852,193	4.45
Geotubes, 45; Reach 3/Quantity (B28)	B28	1	11152	−2788	−20.00	18,435,417	(706,159)	−3.69
		2	12546	−1394	−10.00	18,788,496	(353,080)	−1.84
		3	13940	0	0.00	19,141,576	(0)	0.00
		4	15334	1394	10.00	19,494,646	353,080	1.84
		5	16728	2788	20.00	19,847,735	706,159	3.69
LF/Price (D28)	D28	1	177.744	−44.436	−20.00	18,435,417	(706,159)	−3.69
		2	199.962	−22.218	−10.00	18,788,496	(353,080)	−1.84
		3	222.18	0	0.00	19,141,576	(0)	0.00
		4	244.398	22.218	10.00	19,494,646	353,080	1.84
		5	266.616	44.436	20.00	19,847,735	706,159	3.69
Pipeline Dredging, Reach 2/Quantity (B21)	B21	1	818215.2	−204553.8	−20.00	18,535,279	(606,297)	−3.17
		2	920492.1	−102276.9	−10.00	18,838,427	(303,149)	−1.58

		3	1022769	0	0.00	19,141,576	(0)	0.00
		4	1125045.9	102276.9	10.00	19,444,725	303,149	1.58
		5	1227322.8	204553.8	20.00	19,747,874	606,297	3.17
CY/Price (D21)	D21	1	2.08	−0.52	−20.00	18,535,279	(606,297)	−3.17
		2	2.34	−0.26	−10.00	18,838,427	(303,149)	−1.58
		3	2.6	0	0.00	19,141,576	(0)	0.00
		4	2.86	0.26	10.00	19,444,725	303,149	1.58
		5	3.12	0.52	20.00	19,747,874	606,927	3.17
Pipeline Dredging, Reach 3B/Quantity (B23)	B23	1	589370.4	−147342.6	−20.00	18,677,977	(463,599)	−2.42
		2	663041.7	−73671.3	−10.00	18,909,777	(231,799)	−1.21
		3	736713	0	0.00	19,141,576	(0)	0.00
		4	810384.3	73671.3	10.00	19,373,375	231,799	1.21
		5	884055.6	147342.6	20.00	19,605,175	463,599	2.42
CY/Price (D23)	D23	1	2.208	−0.552	−20.00	18,677,977	(463,599)	−2.42
		2	2.484	−0.276	−10.00	18,909,777	(231,799)	−1.21
		3	2.76	0	0.00	19,141,576	(0)	0.00
		4	3.036	0.276	10.00	19,373,375	231,799	1.21
		5	3.312	0.552	20.00	19,605,175	463,599	2.42
Pipeline Dredging, Reach 1/Quantity (B20)	B20	1	460885.6	−115221.4	−20.00	18,776,416	(365,160)	−1.91
		2	518496.3	−57610.7	−10.00	18,958,996	(182,580)	−0.95
		3	576107	0	0.00	19,141,576	(0)	0.00
		4	633717.7	57610.7	10.00	19,324,156	182,580	0.95
		5	691328.4	115221.4	20.00	19,506,736	365,160	1.91
CY/Price (D20)	D20	1	2.224	−0.556	−20.00	18,776,416	(365,160)	−1.91
		2	2.502	−0.278	−10.00	18,958,996	(182,580)	−0.95
		3	2.78	0	0.00	19,141,576	(0)	0.00
		4	3.058	0.278	10.00	19,324,156	182,580	0.95
		5	3.336	0.556	20.00	19,506,736	365,160	1.91

TABLE 16.6

Top Five Inputs from a Two-at-a-Time Multiway What-If Analysis for Total Project Cost

		Output Variation			
		Minimum		Maximum	
Rank	Multi-Way Name	Value ($)	Change (%)	Value ($)	Change (%)
1	Pipeline Dredging, Reach 3A/Quantity [B22] CY/Price [D22]	$17,607,628	−9.79%	$21,016,401	9.79%
2	Pipeline Dredging, Reach 3A/Quantity [B22] Geotubes, 45′, Reach 3/ Quantity [B28]	$17,583,224	−8.14%	$20,699,928	8.14%
3	Pipeline Dredging, Reach 3A/Quantity [B22] LF/Price [D28]	$17,583,224	−8.14%	$20,699,928	8.14%
4	CY/Price [D22] Geotubes, 45′, Reach 3/Quantity [B28]	$17,583,224	−8.14%	$20,699,928	8.14%
5	CY/Price [D22] LF/ Price [D28]	$17,583,224	−8.14%	$20,699,928	8.14%

Note: Top five inputs ranked by percent change.

16.3.2.2 Difference in Log-Odds Ratio (ΔLOR)

A specific application of the nominal range sensitivity methodology is the log-odds ratio method. This mathematical methodology can be used when the output of interest is a probability. That means our cost example does not lend itself to this method, so we will briefly switch to a different example. First, let's understand the method.

The odds ratio or odds of an event is simply {P(A)/[(1 − P(A)]} or, in words, it is the ratio of the probability that the event occurs to the probability that the event does not occur (Gordis 1996). The log of the odds ratio or logit simply takes the log of the odds ratio, i.e., logit = log {P(A)/[(1 − P(A)]}.

With this definition ΔLOR can be defined as follows:

$$\Delta LOR = \text{Log}\left(\frac{\dfrac{P(A)}{1 - P(A)}}{\dfrac{P(B)}{1 - P(B)}}\right) \tag{16.1}$$

which simplifies to

$$\Delta LOR = \log\left(\frac{P(A)}{1 - P(A)}\right) - \log\left(\frac{P(B)}{1 - P(B)}\right) \tag{16.2}$$

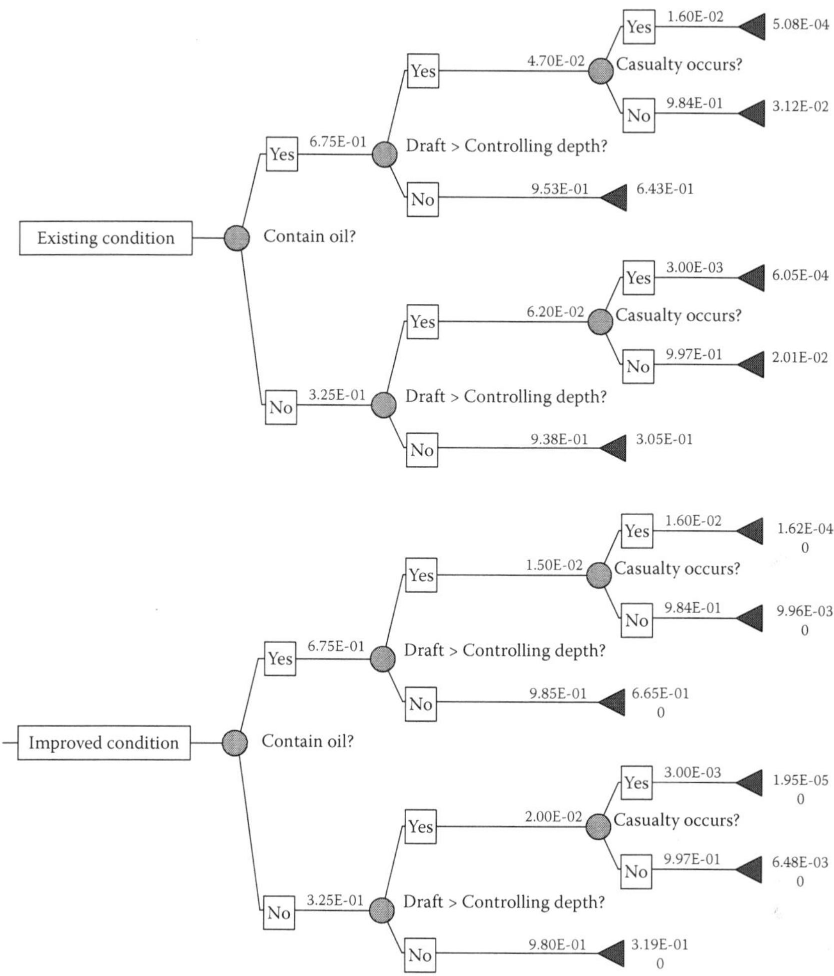

FIGURE 16.6 Example of event-tree models for the difference in log-odds ratio.

or

$$\Delta LOR = logit\ P(A) - logit\ P(B) \tag{16.3}$$

If we now define event B to be the original probability estimate and event A is the probability recalculated with changes in the sensitive input, then a positive ΔLOR means that a change in the selected input increases the probability of the event. A negative ΔLOR means that a change in the input decreases the probability of the event. The larger the ΔLOR value, the greater is the sensitivity of the probability output to changes in that input. ΔLOR is subject to many of the same limitations as the nominal range sensitivity when it comes to nonlinear models.

An example for a single variable is presented in the two event-tree models shown in Figure 16.6. This model represents a small boat harbor where shoaling has decreased the depth of the navigation channel. Vessels come into the harbor with oil and other

TABLE 16.7

Difference in Log-Odds Ratio Calculation

Condition	Probability		Ratio	Logit
	Casualty	No Casualty		
Existing	0.0006	0.9994	0.000568	–3.24537
Improved	0.0002	0.9998	0.000182	–3.74104

ΔLOR = *logit P(A)* – *logit P(B)* – -3.74 – (-3.25) – -0.5

commodities. Some vessels have drafts that exceed the controlling depth of the channel. This affects the probability of a marine casualty occurring. Casualties include groundings, allisions, and collisions.

The output of interest is the probability of a casualty. This is obtained by summing the relevant endpoint probabilities in each model. Note that there are two different scenarios here: an existing and an improved scenario. The improvement is considered to be maintenance dredging to increase the controlling dept of the channel. This alters several input values between the two scenarios, but it does not affect the possibility of a casualty.

Note that the model ends if the controlling depth of the channel is not a binding constraint. The assumption is not that these vessels have no casualties, but that the number and nature of the casualties will not be affected by dredging the channel. Note also that the only change captured in this model is the change in the probability that the controlling depths will be exceeded and the effect of this change on subsequent calculations in the models.

To calculate the ΔLOR = logit P(A) – logit P(B), the values shown in Table 16.7 are obtained from the model. The probabilities of a casualty are sums of two model endpoints. The ratio of P(Casualty)/P(No Casualty) is given in column 4. The log of these ratios is found in the last column. Subtracting the two produces the ΔLOR = –0.5. The negative value means that dredging the channel, thereby decreasing the probability of a vessel exceeding the controlling depth, reduces the probability of a casualty.

In this simple computational example there is no other input to change to produce another ΔLOR value to compare to this one. In a more complex problem or model, there might be multiple model inputs to change that would influence the output probability of interest. Finding the control variable that has the largest desirable effect on the probability output via the ΔLOR is a significant help to the risk manager. The largest absolute value of the ΔLOR would identify the most sensitive input by this method.

16.3.2.3 Break-Even Analysis

This mathematical method applies the familiar concept of breaking even to sensitivity analysis. The notion, borrowed from economics, is that at the break-even point we are indifferent to producing or not producing a good or service. The important idea here is to look for a break-even/cutoff/threshold value for a parameter or decision variable, i.e., a point where something interesting happens, such as, something goes negative, turns good, turns bad, equals zero, and the like.

Applied as a sensitivity method, break-even analysis requires one to find values of inputs that provide a model output for which the risk manager is indifferent between two (or more) risk management options. These could include choosing between accepting the risk or managing a risk, choosing option A or option B, and so on. Alternatively, a threshold might indicate a point at which the risk manager has a strong preference for one course of action over another.

The input value or combination of input values for which the risk manager is indifferent is called the switch-over or break-even value. Once the break-even input value(s) is determined, the risk manager must judge whether the most likely input values will lie above or below these break-even values. If an input's range of uncertainty includes the break-even point, that input is important for decision making. In other words, it will not be clear which is the best decision because the switch-over point may or may not be exceeded. In that case, additional research or efforts to reduce the uncertainty will be necessary for a more confident decision. Conversely, if the uncertainty about an input does not include the break-even point, a decision can be made more confidently.

Finding these break-even values is a unique endeavor for each risk assessment; there is no generalized technique applicable to all models. It can be difficult to find these values when the number of sensitive inputs increases. There is also no clear ranking to be obtained from this method. Frey and Patil (2002) provide an example of a patient choosing between a medication and an operation that produces an iso-risk line for the utility of the medication versus the probability of success for an operation. The application of this technique can be sophisticated.

Let's simplify here to aid understanding. Let's return to our dredging cost estimate. Suppose this project was to be constructed under a government program with a $20-million budget cap. A modified application of this break-even method would be to calculate what the dredging quantity in reach 1, for example, would have to be before the threshold (break-even point) is passed. If dredging in that reach exceeds 846,972 cubic yards,* costs will exceed $20 million. If our uncertainty about this value is, say, from 500,000 to 900,000 cubic yards, then this quantity alone is sufficient to exceed the budget constraint, and the project may not be able to proceed. Risk assessors may be motivated to estimate the probability that the dredging quantity will equal or exceed this amount. Risk managers would be well advised to do all they can to improve the estimate of the required dredging in this reach. It is not difficult to imagine that this technique can become tedious when multiple inputs are considered. It also does not lend itself readily to considering multiple inputs without the use of iso-risk lines.

16.3.2.4 Automatic Differentiation Technique

The automatic differentiation (AD) technique is a mathematical method that relies on the use of partial derivatives of outputs with respect to small changes in sensitive inputs to calculate local sensitivities. It is usually reserved for larger models. Because these models are not always well-behaved systems of equations, differentiation often relies on numerical techniques that can be time consuming and difficult to calculate.

* Costs are $858,424 below the cutoff. At $3.17 a CY ($2.78 plus 14% for AED and CM), the dredging quantity would have to rise by $858,424/$3.17 = 270,865 CY to exceed the budget cap. The current estimate of 576,107 CY would rise by this amount.

In addition, they may yield inaccurate results. What makes mathematical AD techniques different is that they rely on precompilers that analyze the code of the model and then compile algorithms for computing first or higher order derivatives in an efficient and accurate manner.

This is a software-intensive technique that is not going to be available to most risk assessors. For more information on this technique, see the work of Bischof et al. (1992, 1994, 1996). Applications also can be found in the work of Carmichael, Sandu, and Potra (1997); Issac and Kapania (1997); Ozaki, Kimura, and Berz (1995); and others.

16.3.3 Statistical Methods for Sensitivity Analysis

16.3.3.1 Regression Analysis

Regression analysis is one of the more common sensitivity methods used because simple linear regression sensitivities have been built into some of the commercially available risk assessment software packages. This statistical method can be a useful probabilistic sensitivity analysis technique.

There are many standard econometric textbooks that explain regression analysis techniques. In best practice, the assessor will specify a functional form for the cause-and-effect relationship between the output (dependent variable) and the relevant inputs (independent variables) based on sound theory. In a probabilistic risk assessment, a random sample of values for these variables will be obtained through some probabilistic analysis like a Monte Carlo simulation.

Using data from the simulation for inputs and outputs, a multiple regression model of the form

$$Y_i = \beta_o + \beta_1 X_{2,i} + \ldots + \beta_m X_{m,i} + \varepsilon_i \tag{16.4}$$

is estimated. The $X_{m,i}$ are the inputs, where m indicates the number of the individual input variable and i indicates the ith input data point. The Y_i indicates the ith output data point.

The beta values, β_m, are regression coefficients. Because they are estimated from a random sample, they are themselves random variables. When the beta coefficients are statistically significantly different from zero they provide a measure of the effect of the particular input variable X_m on Y, all other input variables being held constant. In fact, the β_m value shows the effect of a one-unit change in X_m on the output variable, when all the other input variables in the regression equation have been accounted for. This makes regression coefficients similar to the nominal range sensitivity technique in terms of interpretation, although it is a more sophisticated technique that can account for some nonlinearities (if they can be transformed to a linear form) in the model. It also can handle variation among multiple assessment inputs.

Input variables with beta regression coefficients that lack a statistically significant difference from zero relationship with the output variable are not sensitive. Those with significant regression coefficients indicate a sensitivity. The regression coefficients can be normalized over the [−1,1] interval to eliminate dimensional effects and to allow ranking based on the absolute value of the coefficient.

Some of the popular risk assessment software generates simple linear regressions as part of their sensitivity analysis package. A simple linear regression has only one independent (X) variable. The software will regress each input value against the same selected output using the ordinary least squares (OLS) estimation technique and rank them by normalized regression coefficients (β). If the output (dependent variable) is actually a function of several inputs (independent variables), this technique could violate one or more classical assumptions for OLS estimators (Stundenmund 2006) and provide misleading results. Consequently, one should be wary of these built-in regression results.

Multiple regression techniques allow evaluation of the sensitivity of individual model inputs, taking into account the influence of other inputs on the output. The cost estimate model does not provide a very interesting example for this technique because it is so linear, so let's switch examples again for the moment. The example used for this demonstration is based on an FDA (2001) risk assessment entitled "The Human Health Impact of Fluoroquinolone Resistant *Campylobacter* Attributed to the Consumption of Chicken." Fluoroquinolone (FQ) drugs have been administered to some domestic U.S. poultry at subtherapeutic doses. There was some evidence to suggest this was causing an increase in antibiotic resistance of *Campylobacter* to this drug. The risk hypothesis suggested that some consumers who were made ill by FQ-resistant *Campylobacter* in chickens would seek medical care and subsequently be prescribed FQ drugs, which would be ineffective against their illness. The FDA conducted a risk assessment to learn how many people may have been affected in this way.

Five input variables are used to estimate the number of people affected by this problem. They are the independent variables in the regression and are listed in Table 16.8. A sample of $n = 500$ values for each input and the output was obtained through a Monte Carlo simulation using the FDA model. A multiple regression was run with this sample using the number of people with campylobacteriosis seeking the care of a doctor and being administered FQ drugs for the dependent variable. The results in Table 16.8 show all independent variables, but the proportion of sick people seeking care from a doctor have effects that are statistically significantly different from zero. The beta coefficients have not been normalized over the [−1,+1] interval because all inputs have the same metric units, so the usual coefficients provide a ranking.

Table 16.8 shows that a 1% increase in the proportion of FQ-resistant *Campylobacter* infections from chicken has the potential to affect an average of 538 people, with all other inputs held constant. This makes it the most sensitive input of those investigated.

TABLE 16.8
Regression Results

Regression	Coefficient	p-Value
Constant	−25,523	<0.0001
Proportion of Campylobacter cases associated with chicken	171	<0.0001
Proportion of FQ resistant *Campylobacter* infections from chicken	538	<0.0001
Proportion sick seeking care	129	0.1164
Proportion treated with antibiotic	91	0.0029
Proportion receiving FQ treatment	189	<0.0001

Using simple (one independent variable) linear regression can produce less reliable results. In addition, the built-in software features may fail to select important intermediate calculations as inputs. In general, these software features identify inputs as those spreadsheet cells that have been assigned a probability distribution. We will return to this simple regression technique and the cost estimation example when we consider graphical sensitivity analysis techniques.

16.3.3.2 Correlation

Correlation is a statistical method used to determine whether two variables (an input and an output) tend to move together. If above-mean values of one variable are associated with above-mean values of the other, and vice versa, there is a positive correlation. A negative correlation results when above-mean values of one variable are associated with below-mean values of another variable.

The Pearson correlation coefficient is a numerical value between −1 and 1. The sign of the coefficient indicates whether the association is a positive one or a negative one. The size of the coefficient indicates the strength of the association. The top row of Figure 16.7 shows negative correlations with coefficients of −1, −.5, and −.25. The bottom row shows positive correlations of .8, .5, and .25. Note that as the absolute value approaches 1, the cloud of points approaches a straight line.

Data from a simulation that produces a sample of input and output values can be used to calculate correlation coefficients. This is a feature on some simulation software. Correlation is not causation, and correlation coefficients are best understood when accompanied by a scatter plot like those in Figure 16.7. In general, inputs that are highly correlated with the output are of potential interest in a sensitivity analysis.

For a correlation example, let's return to our cost estimate example. A 10,000-iteration simulation of costs produced the characterization of the possible costs of dredging the waterway and creating wetlands with the dredged material seen in Figure 16.8.

The simulation had 20 variable inputs and a single output. @RISK 5.5's built-in simulation sensitivity produced Spearman rank correlation coefficients for each input-output pair as shown in Table 16.9.

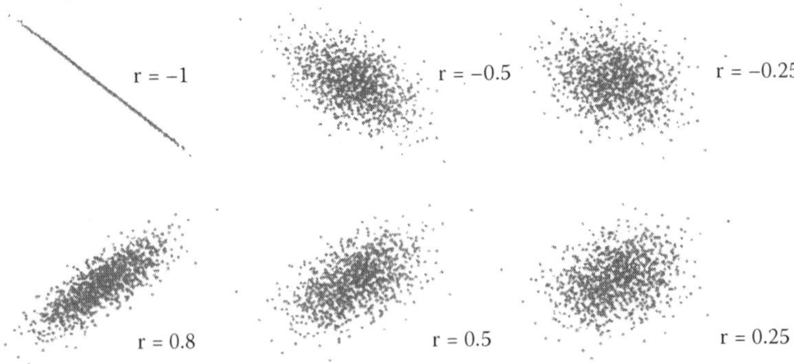

FIGURE 16.7 Selected scatter plots and Pearson correlation coefficient examples.

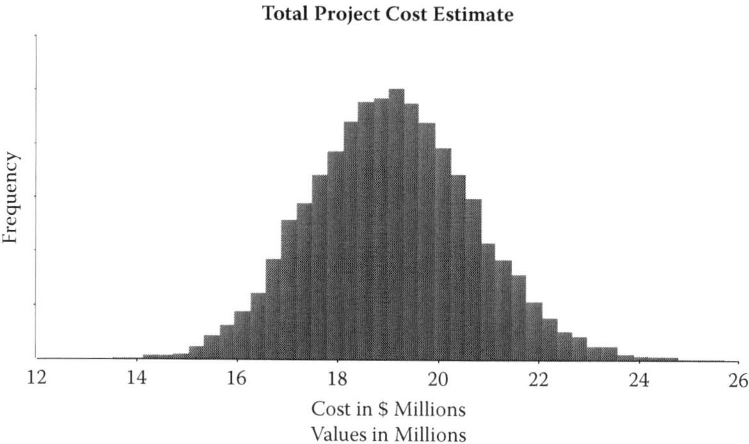

FIGURE 16.8 Distribution of dredging cost estimates.

This technique indicates that the total project cost is most highly correlated to the price of 45-foot diameter geotubes in reach 1, followed by the dredging quantity to be removed from reach 3A. Thus, any efforts to reduce uncertainty in these variables would produce a more reliable cost estimate. This sensitivity analysis can be equally useful at times in identifying variables that are of no special concern. Variables with low correlation coefficients can be treated as point estimates with no loss of fidelity. You need not develop probability distribution estimates for every uncertain input. The cruel irony is that you may not know this until after you have done so and conducted this kind of sensitivity analysis.

TABLE 16.9

Importance Analysis of Cost Estimate Inputs Using Spearman Rank Correlation Coefficient

Rank Four Output	Cell	Name	Spearman Correlation Coeff.
1	E26	Price of Reach 1 45′ Geotube	.529
2	C22	Quantity Pipeline Dredging, Reach 3A	.406
3	E24	Price of Reach 1 Scour Pad	.4
4	C21	Quantity Pipeline Dredging, Reach 2	.325
5	E22	Price per CY Reach 3A	.297
6	E21	Price per CY Reach 2	.208
7	C20	Quantity Pipeline Dredging, Reach 1	.164
8	E23	Price per CY Reach 3B	.145
9	E20	Price per CY Reach 1	.133
10	C23	Quantity Pipeline Dredging, Reach 3B	.091

16.3.3.3 Analysis of Variance

Analysis of variance (ANOVA) is another statistical method that is a model-independent probabilistic sensitivity analysis method. It is used to determine if there is a statistical association between an output and one or more inputs. Unlike regression analysis, ANOVA requires no assumption about the functional form of the relationships between inputs and the outputs.

Inputs are called "factors." Values of quantitative factors and categories of qualitative variables are called factor levels. The output is called a "response variable." Single-factor and multifactor ANOVA are options for the analyst. ANOVA is used to determine if values of the output vary in a statistically significant manner associated with variation in values for one or more inputs. If variation in the output is not statistically significant with respect to the input(s), then it is considered random. There are somewhat stringent assumptions required to validate the ANOVA process. If these assumptions are violated, corrective measures must be taken to address the problem. ANOVA is a technique preferred by certain disciplines. It is a topic that can be found in most standard statistics texts. Warner (2008) has an especially detailed and helpful treatment of the topic.

16.3.3.4 Response-Surface Method (RSM)

The response-surface method (RSM) is a statistical technique used to estimate the relationship between a response variable (output) and one or more explanatory inputs. It is a complex method. Think of it as the graph of a surface in n-space that identifies curvatures in this space by accounting for second-order effects that enable assessors to observe the effect on the output given selected effects in one or more inputs.

It is best to limit the number of inputs so as to limit the size of n-space. Therefore, RSM is best used after other sensitivity screening methods have identified the most important inputs. Frey and Patil (2002) suggest Monte Carlo simulation methods can be used to generate multiple values of each model input and the corresponding output, and then a least squares regression method is used to fit a standardized first- or second-order equation to the data obtained from the original model. If the classic assumptions of least squares regression are not satisfied, other techniques such as rank-based or nonparametric approaches should be used (Khuri and Cornell 1987, Vidmar and McKean 1996).

A response surface can be linear or nonlinear. Think of it as a "model of a model." Once generated, it is often easier to conduct sensitivity analysis of the response surface than it is of the original model. The sensitivity analysis of the response-surface analysis is often simpler and faster to execute than sensitivity analysis of the original model. This means that computationally intensive sensitivity analysis methods, such as Mutual Information Index (Finn 1993) or others, may be more readily applied to the response surface than to the original model. Applications of the RSM can be found in Gardiner and Gettinby (1998); Moskowitz (1997); Hopperstad et al. (1999); Williams, Varahramyan, and Maszara (1999); and others. This is not a technique that many risk assessors are likely to use due to its complexity.

16.3.3.5 Fourier-Amplitude Sensitivity Test

The Fourier amplitude sensitivity test (FAST) is a statistical method that can be used for both uncertainty and sensitivity analysis (Cukier et al. 1973, 1975, 1978). It is not for beginners. FAST is used to estimate the contribution of individual inputs to the variance of the output. It is independent of any assumptions about model structure. Assessors can study the effect of single or multiple inputs using FAST.

Frey and Patil (2002) describe the method as relying on a transformation function used to convert values of each model input to values along a search curve. The transformation specifies a frequency for each input, and using Fourier coefficients, the variance of the output is evaluated. The contribution of each input observation (x_i) to the total variance is also calculated based on the Fourier coefficients, fundamental frequency, and higher harmonics of the frequency, as explained by Cukier et al. (1975). The ratio of the contribution of each input to the output variance and the total variance of the output can be calculated and used to rank the inputs (Saltelli, Chan, and Scott 2000).

The model needs to be evaluated at enough points in the input parameter space that numerical integration can be used to determine the Fourier coefficients (Saltelli et al. 2000). For applications, see Lu and Mohanty (2001), Helton (2000), and Rodriguez-Camino and Avissar (1998).

16.3.3.6 Mutual Information Index

The mutual information index (MII) is a statistical sensitivity analysis method that produces a measure of the information about the output provided by a specific input. The MII is based on conditional probabilistic analysis. The magnitude of the MII for different inputs can be compared to determine which inputs provide useful information about the output. This is a computationally intensive method typically used for models with dichotomous outputs.

Frey and Patil (2002) describe MII as typically involving three steps: (1) generating an overall confidence measure of the output value; (2) obtaining a conditional confidence measure for a given value of an input; and (3) calculating sensitivity indices (Critchfield and Willard 1986a, 1986b). The cumulative distribution function (CDF) of the output is used to estimate the overall confidence in the output, where confidence is the probability of the dichotomous outcome of interest. The conditional confidence is estimated by holding one input constant at some value and varying all other inputs. The new resulting CDF of the output is a measure of the assessor's confidence in the output conditioned on the particular value of the input used to generate it.

The mutual information between two random variables is the amount of information about a variable that is provided by the other variable (Jelinek 1970). In other words it is a quantity that measures the mutual dependence of the two variables. A description of the calculation of the MII is found in Frey and Patil (2002).

Critchfield and Willard (1986a, 1986b) devised and demonstrated the application of the MII method using a decision-tree model. MII includes a more direct measure of the probabilistic relatedness of two random variables than correlation coefficients, and it can account for the joint effects of all inputs when evaluating sensitivities of an input. It is computationally complex and difficult to apply. It is not one of the more commonly applied sensitivity analysis methods.

16.3.4 Graphical Methods for Sensitivity Analysis

16.3.4.1 Scatter Plots

Scatter plots, shown earlier in Figure 16.7, can be used to visually assess the influence of individual inputs on an output. A Monte Carlo simulation, for example, can generate many input-output pairs, which when plotted can reveal potentially sensitive associations between these variables. Linear or nonlinear patterns, which may be observed, could depict potential dependencies between an input and an output.

There need to be enough points to show a pattern but not so many as to obscure the variability in the scatter. Pattern detection methods can be applied to help identify relationships between inputs and outputs (Kleijnen and Helton 1999, Shortencarier and Helton 1999). Applications are found in Sobell et al. (1982); Rossier, Wade, and Murphy (2001); Hagent (1992); Fujimoto (1998); Helton et al. (2000); and others.

When you have completed a probabilistic risk assessment, it can be helpful to plot the scatter plot for each input-output relationship as a first step in the sensitivity analysis of your statistical sample. It is a useful screening mechanism, but it can become quite tedious if the model has large numbers of inputs and outputs. When a pattern is evident, more rigorous sensitivity analysis is warranted.

For an example, consider the cost estimate again. Earlier, the Spearman rank correlation coefficient identified the price of 45-foot diameter geotube as a significant source of variation in the output. A plot of 500 data points from the 10,000-iteration simulation in Figure 16.9 shows a clear positive relationship between this input and

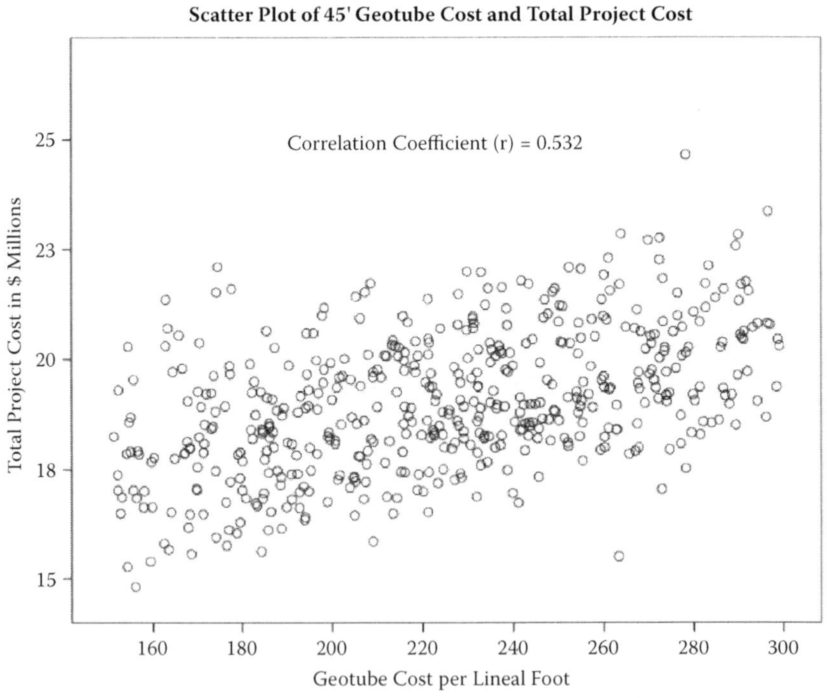

FIGURE 16.9 Scatterplot of total project cost and the price of a 45-foot-diameter geotube.

the output. Note that this correlation coefficient is a Pearson correlation and that it differs from the Spearman value often produced by commercial software packages.

16.3.4.2 Tornado Plots

Tornado graphs are a variation of a bar graph. They are used to show the relative sensitivity of an output to uncertain (or variable) inputs. They are normally centered over a vertical zero-point axis when sensitivity is measured by a correlation or normalized regression coefficient. Bars extending to the right indicate a positive relationship with the selected output; bars extending to the left of the zero axis indicate a negative relationship.

The length of the bar indicates the relative strength of the positive or negative relationship. It is often measured as a normalized regression coefficient, when the sensitivity is based on a normalized simple linear regression between an individual input and the selected output, or it may be measured by a correlation (Spearman or Pearson) coefficient.

The example shown in Figure 16.10 begins with the distribution of total project costs. There is about a $12-million range in the cost estimate outputs. To learn what contributes most to this variation, a regression analysis or correlation sensitivity method must first be completed and its results plotted, as shown in the bottom graph of Figure 16.10. This tornado plot is based on normalized beta coefficients from a sequence of simple linear regressions.

In this instance, the price of 45-foot geotubes is at the top of the list.* A tornado chart can be prepared for each model output of interest to risk managers. This statistical sensitivity analysis, which is based on data from a Monte Carlo process, yields different results than our initial OAATA did, as we should expect. These are all tools that require professional judgment. The price of 45-foot geotubes was not the most significant input during the OAATA analysis, although it was near the top. Had an assessor decided not to pursue the quantification of the uncertainty around that value, it would have been an error. We need to take into account not only the structure of the model, which OAATA did reasonably well, but also the magnitude of the uncertainty. As it turned out, the cost estimators were less certain about the price of the large geotubes than about most other inputs.

Let's return, briefly, to our OAATA analysis and consider another form of the tornado graph. Figure 16.11 shows a tornado graph prepared using a nominal range sensitivity (one-way what-if analysis) based on point estimates rather than a Monte Carlo process using probability distributions for the inputs.

In this particular graph, the inputs on the left were allowed to vary by ±20 percent. The impact of this variation on total project costs is shown on the horizontal axis of the graph. In this graph, the impact is measured as a percentage. It could have just as easily been measured in actual costs, with the deterministic cost estimate as the center of the graph. The length of the bar indicates the amount of change the input caused to the output measured as a percentage. The input with the largest effects (and longest bar) is shown at the top, and those with less impact are shown below. The

* If you are curious why these results are different from the OAATA results, recall that that is a mathematical method based on point estimates for each input. This is a statistical one that is based on the expression of each input value as a probability distribution. The actual uncertainty among the inputs is more variable than the ±20% used for each input in the OAATA.

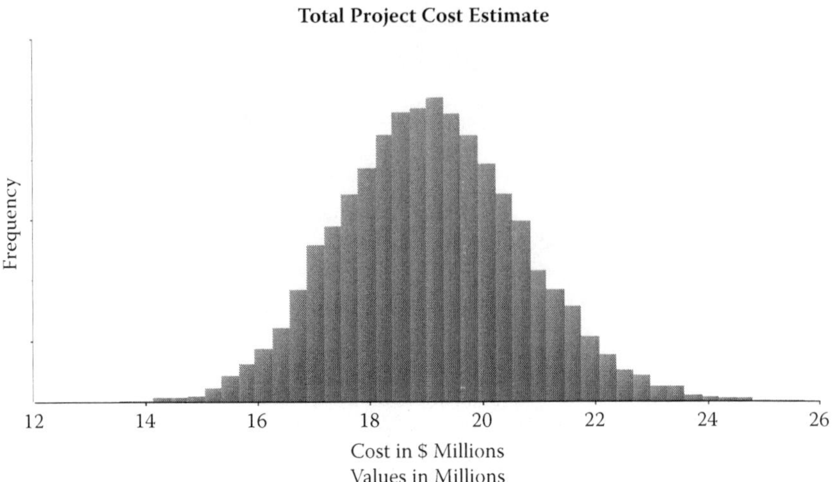

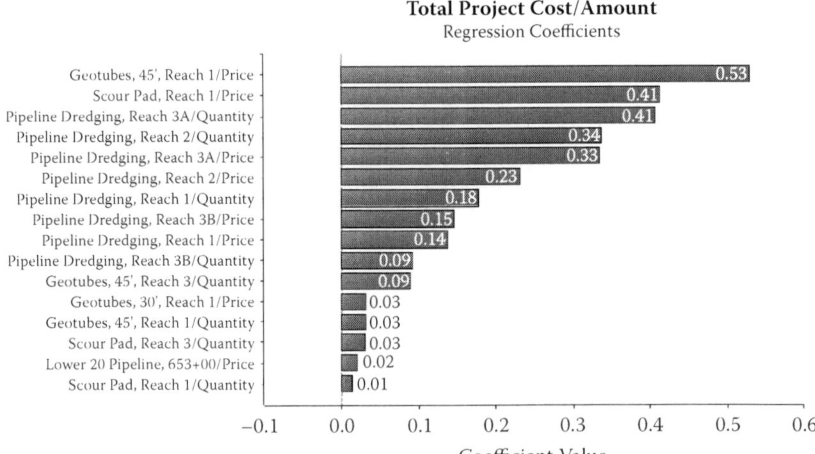

FIGURE 16.10 Total project cost distribution with a regression-based tornado chart.

diminishing influence of less significant inputs produces the tornado shape. Note that geotube prices are listed third on the tornado chart.

16.3.4.3 Spider Plot

Figure 16.12 shows a sensitivity graph. It shows the relationship between the quantity of dredged material in reach 3a and total project cost. The horizontal axis shows input changes of –20%, –10%, 0%, 10%, and 20%. The vertical axis shows the corresponding effect on total project costs. This is essentially the data from an analysis like that in Table 16.5.

A spider plot is a collection of multiple sensitivity graphs. The spider plot is an alternative way of visualizing effects of inputs on outputs that is useful for models

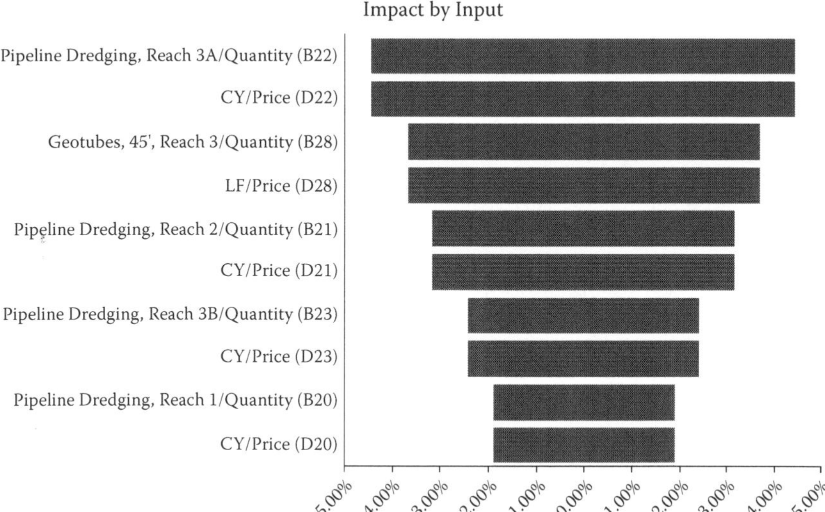

FIGURE 16.11 OAATA tornado graph for total project cost.

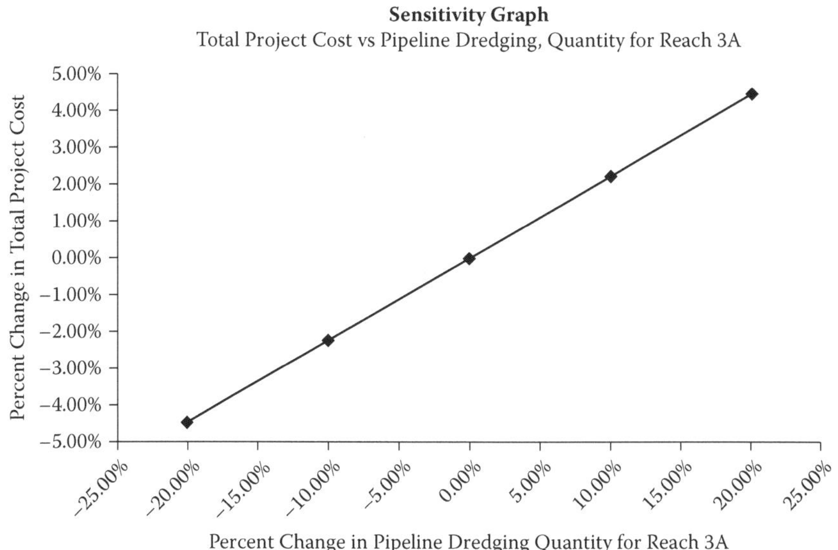

FIGURE 16.12 Sensitivity graph.

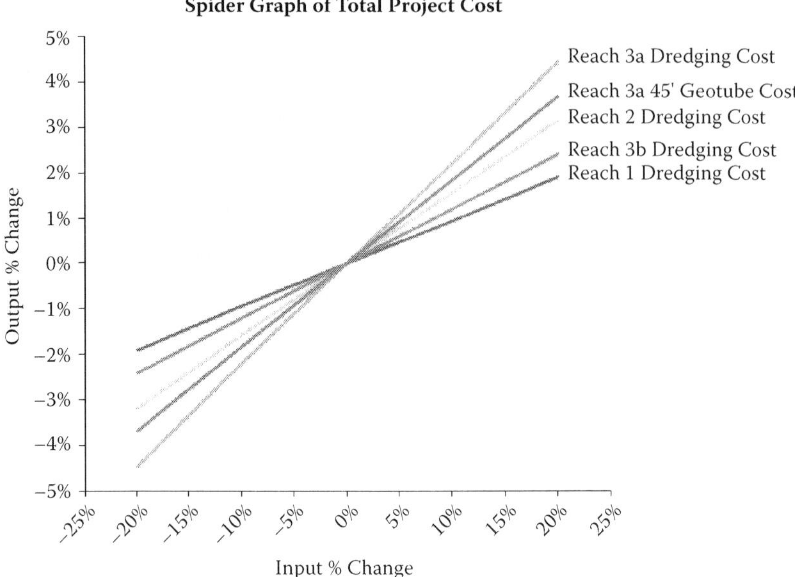

FIGURE 16.13 Spider plot for total project cost.

with fixed-point inputs. It relies on the nominal range sensitivity method to generate data for the plot. Variation in input effects often creates a spiderlike spread of effects about the no-change point (0%, 0%). An example is shown in Figure 16.13.

For each input, the percentage change in its value from the fixed-point estimate (base case) is plotted on the x-axis, and the percentage change in the output is plotted on the y-axis. Lines that are higher to the right of the no-change (0%) point indicate inputs with a greater impact on the output under investigation. Conversely, lines that lie highest to the left of the no-change point have the least impact on the output. This figure is showing that a 20% change in the cost of dredging in reach 3a causes over a 4% change in total project cost, making it the most sensitive input in the spider plot. This is the same result reported numerically earlier in the chapter.

16.4 The Point

The point of doing sensitivity analysis does not end with the identification of your most sensitive inputs. Finding out what your most significant uncertainties are is an important part of a sensitivity analysis. The real purposes of a sensitivity analysis, however, are to help develop a plan for addressing the uncertainty to inform decision makers about the significance of uncertainty for decision making.

Nominal range sensitivity, for example, can be used early in a risk assessment to help identify those variables for which the most effort should be made to describe the uncertainty. It is not always necessary, as mentioned earlier, to enter every uncertain input as a probability distribution. Some of these techniques can suggest which variables to concentrate on.

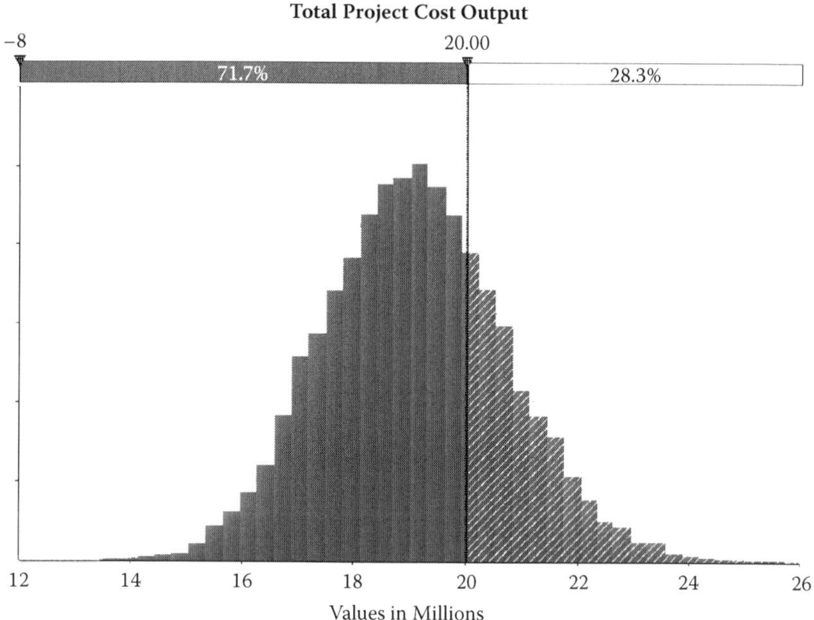

FIGURE 16.14 Distribution of total project cost estimates.

When risk managers are presented with output distributions like the one in Figure 16.14 for costs, it is helpful to be able to explain why there is so much variation in the output upon which they will base their decisions. Furthermore, if we can present risk managers with options for further reducing the most sensitive input uncertainties, the sensitivity analysis adds even more value to the risk assessment. Managers then have an opportunity to reduce the uncertainty or to make a decision based on the available information.

The best risk assessments will be subjected to sensitivity analysis that will discover the most important input variables. This helps everyone understand what contributes most to good and bad outcomes. Once these important variables have been identified, risk assessors should address the uncertainty systematically, by varying assumptions and examining their effects on outcomes, using one or more of the techniques here, or of course, by doing a probabilistic risk assessment. Good risk assessment fixes what can be fixed and addresses what can be addressed. If variation due to knowledge uncertainty can be reduced, risk managers may choose to do so. To the extent that the variation in output is attributable to natural variability, sensitivity analysis can help assessors describe that variability.

Risk managers need to understand the uncertainties that could influence their decisions. When a decision is sensitive to changes or uncertainties within the realm of possibility, then more precision and additional information may be required. In addition, they need to understand the potential quality of the risk management options they are considering. Thus, it is helpful, when conducting a sensitivity analysis, to identify sensitive inputs that are controllable, especially decision variables. When sensitivity analysis identifies those inputs with the greatest positive and negative effects on outputs as well as which of those we can influence, it adds value to risk assessment.

Consider a sensitivity analysis that suggests that the price of 45-foot geotubes is a significant source of output uncertainty. This could provide the impetus for an estimator to contact manufacturers of these tubes for a more reliable price quote. Sometimes sensitivity analysis can suggest new risk management options. For example, the uncertainty attending this cost estimate might be managed through an innovative futures contract arrangement. The responses to a good sensitivity analysis can influence both risk assessment and risk management.

16.5 Summary and Look Forward

Good risk assessment, whether qualitative or quantitative, must include some sensitivity analysis. Examining the effects of varying the assessor's assumptions should always be included in every sensitivity analysis. Qualitative sensitivity analysis has often been overlooked, and it should not be. A three-step process was suggested in this chapter. It includes identifying sources of uncertainty, identifying significant uncertainties and characterizing their effects on decision parameters.

Quantitative sensitivity analysis has a much larger toolbox available to it. Some of these tools can be quite sophisticated. My personal bias is to use the simplest quantitative technique you legitimately and usefully can. The purpose of sensitivity analysis is to support decision making that is better informed about the most significant uncertainties in an assessment. Not only are complex techniques more difficult to execute properly, they are often much more difficult for others to understand. There will no doubt be circumstances when the most sophisticated and robust techniques are warranted by the goals of the risk management activity, but when a good job can be done with simpler methods, use the simpler methods.

By now we have discussed a good many sophisticated analytical techniques capable of generating a great deal of decision-critical information. The next question is, "How do we best present this information to risk managers, and how do they use them to support decision making?" These are the topics covered in the next chapter.

REFERENCES

Bischof, C. H., A. Carle, G. Corliss, A. Griewank, and P. Hovland. 1992. ADIFOR generating derivative codes from FORTRAN programs. *Scientific Programming* 1 (1): 11–29.

Bischof, C. H., A. Carle, P. Khademi, and A. Mauer. 1994. The ADIFOR2.0 system for the automatic differentiation of FORTRAN77 programs. *IEEE Computational Science and Engineering* 3 (3): 18–32.

———. 1996. ADIFOR 2.0: Automatic differentiation of Fortran 77 programs. *IEEE Computational Science & Engineering* 3 (1): 18–32.

Carmichael, G. R., A. Sandu, and F.A. Potra. 1997. Sensitivity analysis for atmospheric chemistry models via automatic differentiation. *Atmospheric Environment* 31 (3): 475–489.

Critchfield, G. C., and K. E. Willard. 1986a. Probabilistic analysis of decision trees using Monte Carlo simulation. *Medical Decision Making* 6 (1): 85–92.

———. 1986b. Probabilistic analysis of decision trees using symbolic algebra. *Medical Decision Making* 6 (1): 93–100.

Cukier, R. I., C. M. Fortuin, K. E. Shuler, A. G. Petschek, and J. H. Schailby. 1973. Study of the sensitivity of the coupled reaction systems to uncertainties in rate coefficients: I. Theory. *Journal of Chemical Physics* 59 (8): 3873–3878.

Cukier, R. I., H. B. Levine, and K. E. Shuler. 1978. Nonlinear sensitivity analysis of multiparameter model systems. *Journal of Computational Physics* 26 (1): 1–42.

Cukier, R. I., J. H. Schailby, and K. E. Shuler. 1975. Study of the sensitivity of coupled reaction systems to uncertainties in rate coefficients: III. Analysis of approximations. *Journal of Chemical Physics* 6 (3): 1140–1149.

Cullen, A. C., and H. C. Frey. 1999. *Probabilistic techniques in exposure assessment*. New York: Plenum Press.

Finn, John T. 1993. Use of the average mutual information index in evaluating classification error and consistency. *International Journal of Geographical Information Science* 7 (4): 349–366.

Frey, H. C., and R. Patil. 2002. Identification and review of sensitivity analysis methods. *Risk Analysis* 22 (3): 553–577.

Fujimoto, K. 1998. Correlation between indoor radon concentration and dose rate in air from terrestrial gamma radiation in Japan. *Health Physics* 75 (3): 291–296.

Gardiner, W. P., and G. Gettinby. 1998. *Experimental design techniques in statistical practice: A practical software-based approach*. West Sussex, U.K.: Horwood Publishing.

Gordis, Leon. 1996. *Epidemiology*. Philadelphia: W. B. Saunders.

Hagent, N. T. 1992. Macroparasitic epizootic disease: A potential mechanism for the termination of sea-urchin outbreaks in northern Norway. *Marine Biology* 114 (3): 469–478.

Helton, J. C., J. E. Bean, K. Economy, J. W. Garner, Robert J. Mackinnon, Joel D. Miller, J. D. Schreiber, and Palmer Vaughn. 2000. Uncertainty and sensitivity analysis for two-phase flow in the vicinity of the repository in the 1996 performance assessment for the waste isolation pilot plant: Disturbed conditions. *Reliability Engineering and System Safety* 69 (13): 236–304.

Helton, J. C. et al. 2000. Uncertainty and sensitivity analysis for two-phase flow in the vicinity of the repository in the 1996 performance assessment for the waste isolation pilot plant: disturbed conditions. *Reliability Engineering and System Safety* 69: 236–304.

Hopperstad, O. S., B. J. Leira, S. Remseth, and E. Trømborg. 1999. Reliability-based analysis of a stretch-bending process for aluminum extrusions. *Computers and Structures* 71 (1): 63–75.

Issac, J. C., and R. K. Kapania. 1997. Aero-elastic sensitivity analysis of wings using automatic differentiation. *Journal of American Institute of Aeronautics and Astronautics* 35 (3): 519–525.

Jelinek, F. 1970. *Probabilistic information theory*. New York: McGraw-Hill.

Khuri, A. J., and J. A. Cornell. 1987. *Response surfaces*. New York: Marcel Dekker.

Kleijnen, J. P. C., and J. C. Helton. 1999. Statistical analysis of scatter plots to identify important factors in large-scale simulations. *Reliability Engineering & System Safety* 65:147–185.

Lu, Y. C., and S. Mohanty. 2001. Sensitivity analysis of a complex, proposed geologic waste disposal system using the Fourier amplitude sensitivity test method. *Reliability Engineering and System Safety* 72 (3): 275–291.

Mokhtari, A., and H. C. Frey. 2005. Recommended practice regarding selection of sensitivity analysis methods applied to microbial food safety process risk models. *Human and Ecological Risk Assessment* 11 (3): 591–605.

Morgan, M. Granger, and Max Henrion. 1990. *Uncertainty: A guide to dealing with uncertainty in quantitative risk and policy analysis.* Cambridge, U.K.: Cambridge University.

Moskowitz, H. R. 1997. A commercial application of RSM for ready-to-eat cereal. *Food Quality and Preference* 8 (3): 191–201.

Ozaki, I., F. Kimura, and M. Berz. 1995. Higher-order sensitivity analysis of finite-element method by automatic differentiation. *Computational Mechanics* 16 (4): 223–234.

Rodriguez-Camino, E., and R. Avissar. 1998. Comparison of three land-surface schemes with the Fourier amplitude sensitivity test (FAST). *Tellus Series A—Dynamic Meteorology and Oceanography* 50 (3): 313–332.

Rossier, P., D. T. Wade, and M. Murphy. 2001. An initial investigation of the reliability of the Rivermead extended ADL index in patients presenting with neurological impairment. *Journal of Rehabilitation Medicine* 33 (2): 61–70.

Saltelli, A., K. Chan, and E. M. Scott, eds. 2000. *Sensitivity analysis.* Sussex, U.K.: John Wiley and Sons.

Sanchez, Paul J. 2007. Fundamentals of simulation modeling. Presented at the 2007 Winter Simulation Conference, Washington DC, December 9–12.

Shortencarier, M. J., and J. C. Helton. 1999. *A FORTRAN 77 program and user's guide for the statistical analyses of scatter plots to identify important factors in large-scale simulations.* Albuquerque, NM: Sandia National Laboratories.

Sobell, L. C., S. A. Maisto, A. M. Cooper, and M. B. Sobell. 1982. Comparison of two techniques to obtain retrospective reports of drinking behavior from alcohol abusers. *Addictive Behaviors* 7 (1): 33–38.

Studenmund, A. H. 2006. *Using econometrics: A practical guide.* 5th ed. Upper Saddle River, NJ: Prentice Hall.

U.S. Food and Drug Administration. 2001. Center for Food Safety and Applied Nutrition. The human health impact of fluoroquinolone resistant *Campylobacter* attributed to the consumption of chicken. http://www.fda.gov/downloads/AnimalVeterinary/SafetyHealth/RecallsWithdrawals/UCM042038.pdf.

Vidmar, T. J., and J. W. McKean. 1996. A Monte Carlo study of robust and least squares response surface methods. *Journal of Statistical Computation and Simulation* 54: 1–18.

Warner, R. M. 2008. *Applied statistics: From bivariate through multivariate techniques.* Los Angeles: Sage.

Williams, S., K. Varahramyan, and W. Maszara. 1999. Statistical optimization and manufacturing sensitivity analysis of 0.18 μm SOI MOSFETs. *Microelectronic Engineering* 49 (3–4): 245–261.

World Health Organization. 2006. International Programme on Chemical Safety. *Draft guidance document on characterizing and communicating uncertainty of exposure assessment, draft for public review.* Geneva, Switzerland: World Health Organization. http://www.who.int/ipcs/methods/harmonization/areas/draftundertainty.pdf.

17

Presenting and Using Assessment Results

17.1 Introduction

What do risk managers want from a risk assessment? In best practice, they want answers to their questions in a form they can understand and use for decision making. In general, risk managers want the "answer." What they may not always understand is that we may not know the answer or that the answer can be one of any number of possibilities.

What we all want are good decisions. Patrick Leach (2006) masterfully describes the irresistible tendency to focus on a single number for decision making in his book *Why Can't You Just Give Me the Number?* Numbers, once generated and used, have a way of becoming part of the organization's gospel. An overreliance on single numbers in decision making means that managers are ignoring uncertainty, errors, variability, and ranges of possibilities and that can lead to damaging misjudgments.

Howard Wainer (2009), in his book *Picturing the Uncertain World* said, "The road to advancing knowledge runs through the recognition and measurement of uncertainty rather than through simply ignoring it." If risk analysis is ever to realize its promise, assessors must begin to help managers ask for distributions, i.e., possibilities, rather than "the" number. Information reduces uncertainty. If it does not improve a decision, it is worthless (Savage 2009).

Risk assessment is an information-gathering activity. For risk managers, reading the results of a risk assessment is a very different kind of information-gathering activity. Your risk assessment results need to present information that is observable, treats the relevant uncertainty, can impact a decision, and does not cost the manager more time to find and understand than it is ultimately worth for decision making.

The goal here is to move risk managers, who must make decisions in an uncertain world, away from reliance on a single value. This chapter proceeds first by discussing ways to understand and present the results of risk assessments and then it turns to some thoughts on how to use these results to support decision making. It begins by considering how to examine quantities, probabilities, and relationships in the data and using them to answer risk manager's questions. Visualization of data is briefly considered as a promising field for communicating probabilistic risk assessment results to risk managers and others. The chapter then revisits the topic of decision making under uncertainty, this time considering the need for functional creativity in designing risk metrics to be used to support decision making. The chapter concludes with some rules for decision making under uncertainty that are already in use.

17.2 Understand Your Assessment Output Data Before You Explain It

You need to understand your data before you can explain it. Lest this sound simplistic or condescending, it is absolutely essential that the assessment team explore their data thoroughly and understand its full information content before they begin to present and explain it. The concern is not that assessors do not understand their data so much as they might not appreciate all it has to reveal.

Data can be described narratively or visually. It must be organized in ways that make sense to its audiences. That is likely to mean descriptions at various levels of detail. Different reports are going to be needed for different audiences. Media, the public, peers, decision makers, politicians, stockholders, bosses, all are going to desire different levels of detail. To focus this discussion we'll consider the risk manager as the intended audience for our assessment results.

Summarizing data for others is essential. But there is no substitute for setting out the full details as clearly as can be easily managed by an audience that is charged with decision-making responsibilities. In that delicate balance rests the secret to successfully presenting and using risk assessment results. Too much data can become as much of a hurdle to good decision making as too little.

Learning how to analyze a distribution and extract managerial insights is an art well worth practicing. There are two dimensions to risk data of particular interest: the quantities we are interested in and their probabilities of occurring. It is often easier to get a good sense of your data if you focus first on one, then on the other. It can make understanding the information easier for both assessors and managers.

Those experienced in data analysis will likely have developed their own methodologies for unlocking the secrets to their data. We'll begin our discussion of how to understand and explain risk assessment data with four simple, almost obvious, considerations. All assessors who work with risk data and need to convey their relevant information content to managers should learn to:

1. Examine the quantities
2. Examine the probabilities
3. Examine the relationships
4. Answer the manager's questions

To provide a framework and some sample data for this discussion, two examples are introduced.

17.2.1 Two Examples

We'll use one example to briefly discuss categorical data and a second for all other discussions. The first example involves the vulnerability of dairy production facilities to intentional contamination of their product. It is described at the outset of the next section.

The second example in this chapter is a probabilistic risk assessment for a potential major rehabilitation of a water control structure in need of repair. The design

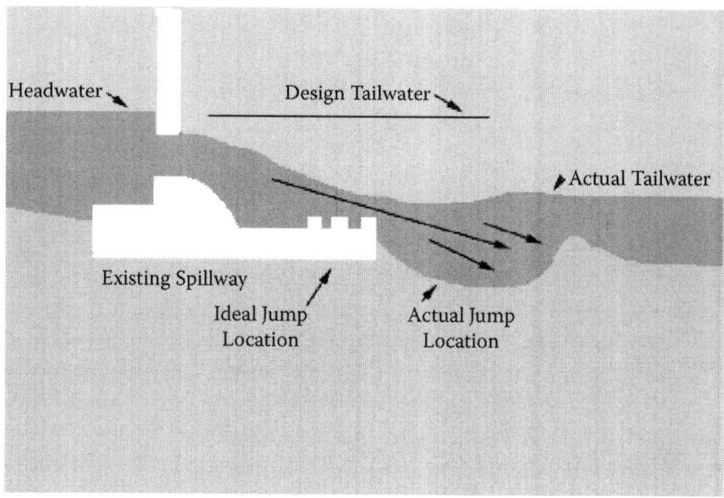

FIGURE 17.1 Existing design flaw in a water control structure.

discharge for this structure jumps beyond the end of the stilling basin and has caused severe erosion, as shown in the actual jump location in Figure 17.1. The erosion has undermined the structure and threatens to cause it to collapse. If the structure fails, unregulated flows will cause extensive erosion and loss of valuable agricultural lands in the project area.

The costs of this project are the costs of correcting this design flaw. This would be done by adding a new weir downstream of the structure, as shown in the proposed modification of Figure 17.2. The benefits are equal to the value of the adjacent land erosion that would be prevented.*

While the details need not concern us, there is substantial uncertainty about every element of this project at the preliminary level of investigation. All costs and quantities are subject to some uncertainty, as are the benefits.

Imagine that risk managers are interested in the answers to the following questions:

1. What is the probability that project costs will exceed $1 million?
2. Identify the cost estimate that has no more than a 20% chance of being exceeded.
3. Identify the cost estimate that has no more than a 10% chance of being exceeded.
4. What is the maximum exposure to cost overruns associated with these costs?
5. What are the most significant contributors to the variations in cost?
6. What is the probability that this project will be economically feasible, i.e., net benefits equal or exceed zero.
7. What is the most likely level of net benefits?

---

* For simplicity this calculation is omitted from the example.

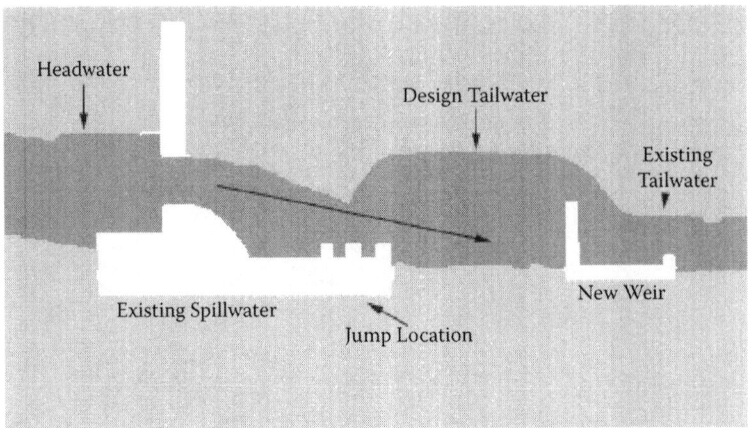

FIGURE 17.2 Proposed modification to water control structure.

Although each question is answered in section 17.6, we'll focus on the answer to question 7 in the chapter's examples to avoid the tedium of repeating the discussion that follows for each answer. All dollar values for this example are present values at a constant price level. Keep in mind that risk assessments can yield a wide range of outputs, and that the simulation results used here focus narrowly on costs and net benefits. This example, chosen for the common familiarity with the costs and benefits of decisions, does not produce all the types of data assessors may encounter in their work. Nonetheless, the approach and methods described are valid for most kinds of risk data.

We'll discuss output quantities in three stages. First, we'll consider categorical quantities. Second, we'll consider nonprobabilistic quantities. Third, and finally, we will look at probabilistic quantities. Our discussion begins with the consideration of categorical data and a return to the dairy production example.

17.3 Examine the Quantities

17.3.1 Categorical Quantities

Let's begin by considering the categorical data produced when considering the vulnerability of dairy production plants to intentional attack on their products. The *CARVER + Shock* vulnerability assessment tool* (Catin and Kautter 2007, USFDA 2009a, 2009b) can be used to determine how well various food processing plants are prepared to resist an intentional contamination threat by an internal (employee) or external (terrorist) attacker. Each processing point (node) in the production facility

* CARVER was a tool developed by the Department of Defense to identify the most vulnerable enemy targets during the cold war period. Post-9/11, the tool was adapted to help antiterrorist strategists to "think like the bad guys" and identify the most vulnerable food sectors in the United States. The tool was subsequently modified to enable food processors to evaluate the steps in their own production process that are most susceptible to attack.

CARVER + SHOCK

Criticality: assesses the public health and economic aspects of an attack
Accessibility: evaluates the attackers' ability to reach the target (and get away unseen)
Recognizability: considers the ease of identifying a target
Vulnerability: analyzes whether an attack will be successful
Effect: estimates the direct loss from an attack, as measured by loss of production
Recuperability: assesses the ability of a system to recover from an attack
Shock: measures the combined health, economic, and psychological effects of an attack within the food industry.

is categorized from 1 (low susceptibility) to 10 (high) for each of seven CARVER + Shock elements (see sidebar) through a series of incisive questions.

Imagine that vulnerability assessments have been conducted for a number of dairy plants and their specific processing steps in the production process, such as receiving materials, storage, mixing, pasteurizing, bottling, distribution, and so on, have been assessed.* Now imagine that we are initially interested in the ratings (categories) for the accessibility and vulnerability of the 144 production nodes assessed across half a dozen or so dairy processors, i.e., we now have 144 categorical estimates of accessibility and 144 estimates of vulnerability for a variety of production nodes in a number of facilities. Accessibility describes the ease with which an attacker can gain access to a node or target. Vulnerability describes the ease with which an attack can be successfully executed once an attacker gains access to the target. A node with high accessibility and high vulnerability is a soft target. Nodes with low accessibility and vulnerability are hard targets. Most nodes are spread between these two extremes.

With this example in mind, let's consider how to examine and present the categorical data produced by a CARVER + Shock vulnerability assessment. When working with categorical or nominal data, a first question to consider is, "Does the order of the data matter?" Explaining why it does or doesn't is the second logical step. When the order of the risk assessment output data matter, such as when there is a sequential logic compelling the results, the simple techniques described here may be of limited value. In the current instance, the order of the data does not matter.

When the order does not matter, one can begin to understand the data better by counting categorical values and calculating percentages. Are there interesting patterns in the data? How many elements fall into each category? What percentage of the total falls in each category? Are the elements evenly distributed across the categories? Do any patterns appear in the distribution of elements across the categories? Are there groupings of elements that are interesting?

Numerical displays and tables can be effectively used to summarize qualitative data. Table 17.1 presents frequency tables for the two variables of interest. These typically provide counts, percentages, and cumulative percentages for all relevant

* This example is based on an actual assessment. The data have been modified to protect the confidentiality of the dairy plants.

TABLE 17.1

Frequency Tables for the Accessibility and Vulnerability Elements of 144 Potential Food Processing Targets in Dairy Processing Facilities

	Accessibility				Vulnerability		
	Frequency	Percent	Cumulative Percent		Frequency	Percent	Cumulative Percent
Valid				Valid			
1	5	3.5	3.5	1	53	36.8	36.8
2	1	0.7	4.2	2	3	2.1	38.9
3	6	4.2	8.3	3	5	3.5	42.4
4	6	4.2	12.5	4	35	24.3	66.7
5	9	6.3	18.8	5	43	29.9	96.5
6	6	4.2	22.9	6	2	1.4	97.9
7	3	2.1	25.0				
8	13	9.0	34.0				
9	7	4.9	38.9	9	1	0.7	98.6
10	88	61.1	100.0	10	2	1.4	100.0
Total	144	100.0		Total	144	100.0	100.0

categories for a variable. A quick glance at these tables informs risk managers that most (61%) of the production nodes are highly accessible with a maximum score of 10. On the other hand, about 2% of all the potential targets have vulnerability scores above 6. The percentages of nodes in each rated category together with the cumulative distribution of categorical ratings provide a quick overview for the individually rated elements of accessibility and vulnerability. These tables can be supplemented with graphical displays like the dot plot of accessibility scores seen in Figure 17.3.

The dot plot is one of the simpler graphs you will find. It shows all the data in a readily understood fashion. It is ideally suited to categorical data where the categories are relatively limited and the data set is not too large. Here the number of nodes with a maximum accessibility score of 10 threatens to dwarf the other data points.

When elements fall into more than a single category, as is the case with the accessibility and vulnerability of production nodes, more questions arise. What are the numbers and percentages of elements in each pair of categories? Are there interesting pairs of categories to consider? Are there relationships between any pairs of categories? Do the numbers of elements in any combinations of categories stand out for any reason?

Cross tabulations or contingency tables are useful displays that enable us to explore relationships among categories for more than one variable. Table 17.2 shows a contingency table for accessibility and vulnerability of production nodes. The lower right-hand corner of the table will show those nodes that are at greatest risk of attack. Likewise, the upper left-hand corner would identify targets that are at very low risk of attack. We see a large number of production nodes that are highly accessible, i.e., many data points fall in the bottom row of the table. Yet 37 of these accessible nodes are virtually invulnerable (the (10,1) cell in the table). About half (78) of the nodes fall in vulnerability categories 4 and 5. Risk managers can use this kind of information to

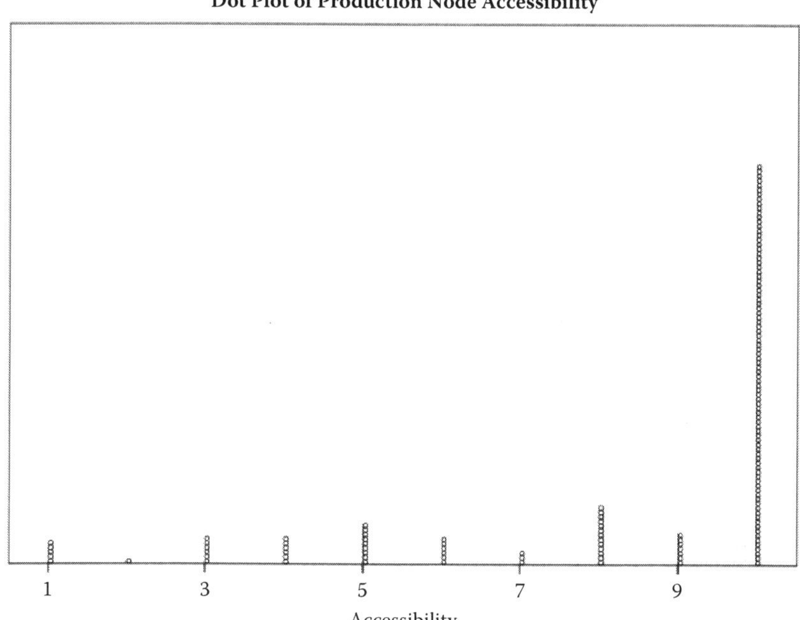

FIGURE 17.3 Dot plot of accessibility categories for dairy production nodes.

aid in their resource allocation decisions by protecting those nodes that are at greatest risk. We see few node targets in obvious need of hardening.

The same questions posed here can be adapted for quantitative data, and more questions can always be added, such as: Are their groupings of data? Are the data symmetrical or do they tail off, in which direction and why? Are there unexpectedly popular or unpopular values? Where do the data center? How widely do they

TABLE 17.2

Cross Tabulation of Production Node Accessibility and Vulnerability

	Vulnerability								
Accessibility	**1**	**2**	**3**	**4**	**5**	**6**	**9**	**10**	**Total**
1	0	2	0	0	3	0	0	0	5
2	0	0	0	0	1	0	0	0	1
3	2	0	0	0	4	0	0	0	6
4	1	0	1	1	3	0	0	0	6
5	2	0	0	4	3	0	0	0	9
6	3	1	0	0	1	0	1	0	6
7	0	0	0	2	1	0	0	0	3
8	6	0	2	0	4	0	0	1	13
9	2	0	0	3	1	0	0	1	7
10	37	0	2	25	22	2	0	0	88
Total	53	3	5	35	43	2	1	2	144

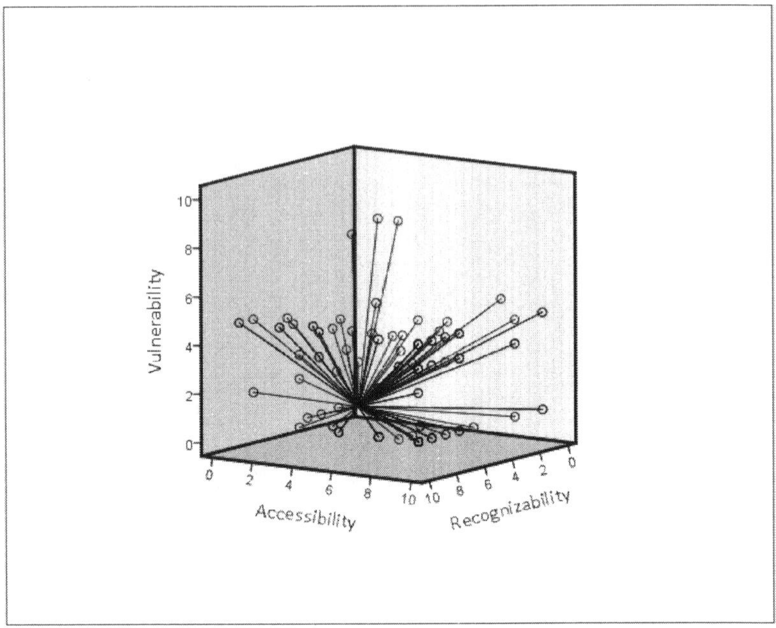

FIGURE 17.4 Three-dimensional plot of recognizability, accessibility, and vulnerability of dairy production nodes.

spread? Which values are most likely? How are the data shaped? Are there any significant thresholds? Can you identify pairs of explanatory (independent) and response (dependent) variables? And what does all of this mean for understanding the risks and managing them? Separating the discussion of quantities based on the qualitative/quantitative distinction is largely artificial. Answers to some of these questions will be found in the discussions that follow.

Figure 17.4 offers a simple three-dimensional graphic that is useful for showing how categorical data tend to be grouped across variables. A third variable, recognizability (see previous sidebar), has been added. A target that is easily recognized, accessible, and vulnerable provides the most attractive target for an attack. The origin point of the figure is recessed into the page, and the data reach out toward the reader. There are three rather clear clusters of vulnerability (the vertical dimension of the figure): one low, one in the middle range, and one small cluster that lies above the others. Within these clusters we find a spread of the other two variables without the same obvious clustering. The points farthest from the origin are those that represent the greatest risk of attack. Only a few nodes stand out as easy targets.

Although these examples do not exhaust the tools available, frequency tables, contingency tables, dot plots, and three-dimensional plots are useful tools for exploring and displaying categorical information. Histograms, scatter plots, and other tools can be equally useful. Examples of these using quantitative data follow later in the chapter.

17.3.2 Nonprobabilistic and Probabilistic Quantities

Let us quickly address nonprobabilistic quantities (point estimates) before turning primarily to probabilistic risk assessment data. If statements about means, medians, and other selected statistical measures or point estimates are presented without a description of the relevant uncertainties attending them, they should be accompanied by narrative descriptions of the risk assessment results so as to expand their utility for decision making. In general, it is wise to avoid presenting quantitative results with no discussion of the uncertainty that attends their estimation. Even in the absence of probabilistic data, a sincere effort should be made to convey the limitations of numerical estimates, which may appear to have more credibility than they in fact have. Do what you can to properly convey the degree of confidence you have in all nonprobabilistic data.

Different point estimates may have different levels of uncertainty and, therefore, confidence associated with them based on model assumptions, available data, calculation methods, and other factors. These differences should be conveyed to the risk manager. The WHO (2006) addresses the importance of wording in communicating assessment results and has suggested sample phrases to communicate the uncertainty attending point estimates. Following the WHO guidance would generate statements like this: "Taking into account the uncertainty that has resulted from the lack of sufficient data, we assume that the exposure of the highest exposed individuals in the population is lower than X with about 66% confidence." When discussing uncertainty even of nonprobabilistic numbers, it is virtually impossible to avoid words that estimate probability in some way. So remember the lessons of Chapter 13 on the use of probability words. Although there is a great deal more that can be said about nonprobabilistic quantities, it will be more convenient to combine the discussion of the quantities.

Many of us like to begin with numerical measures of our data: means, medians, minimums, maximums, and the like. An alternative approach is to begin by trying to get a feel for what the data are like. Because this latter approach may be less familiar, let's begin there. To get a feel for the data, you must first see the data, all the data. Look at it from several perspectives to get a sense of it. Then convey that to decision makers.

17.3.2.1 Graphics

How are the data alike? Where do they tend to cluster? How are the data different? How spread out are they? Look at points when you can. Use a scatter plot or a time trend but do not automatically connect the points, if you connect them at all. Suppress grids and use only a few numbered ticks while you get a feel for your data. Let the evidence speak!

Graphs can be both useful and friendly. When well chosen, they can help us to note the unexpected, and little is more important to understanding one's data. There is no more reason for us to expect one graph to tell all about our data than there is reason to expect one number to reveal all. Be prepared to use multiple graphs. Robert L. Harris (1999) provides an excellent reference for choosing innovative graphics in his book *Information Graphics: A Comprehensive Illustrated Reference.*

A histogram is an obvious place to begin. How do the data appear? Is there one large single peaked group, or do your data separate into groups or clusters? The net benefits in Figure 17.5 show two different distributions of net benefits. The first is

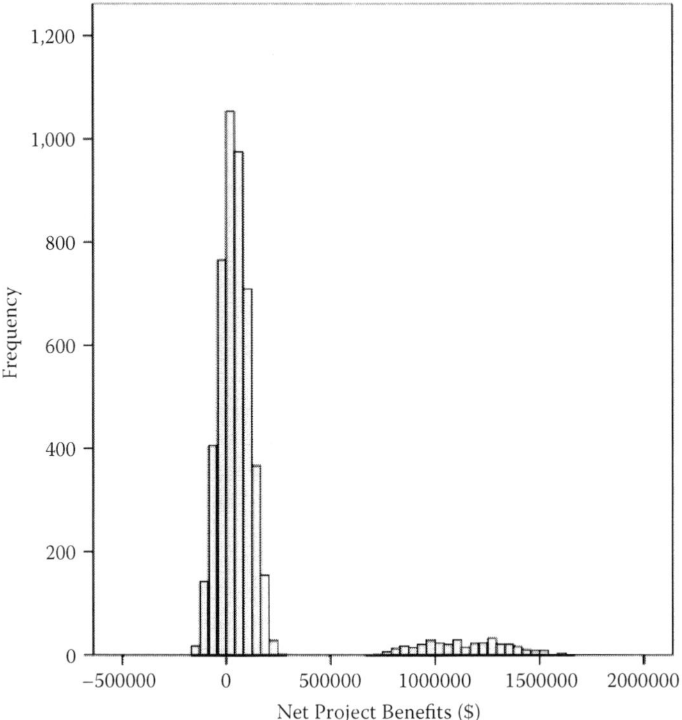

FIGURE 17.5 Histogram of project net benefits showing two distinct clusters of data.

single peaked and roughly symmetrical; it tends to center around $40,000–$50,000 with a range of about $400,000. The second distribution has a much wider spread with a range of about $800,000 or so. It tends to center at about $1.2 million in net benefits.

This distribution strongly suggests two distinctly different kinds of outcomes. We need to learn why that is so. In the present case, the reason for this dual distribution outcome stems from uncertainty about the value of the adjacent land that is being eroded. Values from the left distribution are based on the land staying in agricultural use. Values from the right are associated with a relatively small (10%) chance that the land will receive a zoning change and come into more valuable commercial usage in the near future. Preventing the loss of commercial land produces larger benefits than does preventing the use of agricultural land.

Keep in mind that a histogram can be deceptive, depending on the number of bins or bars used to display the data. Alternative views of the 5,000 simulated values are provided in Figures 17.6 and 17.7.

The box plot clearly shows two distinct clusters of data. The median value (vertical line in box at left) is quite small compared to the potential range of benefits. The dots in the box plot show outliers. These values, which comprise what is effectively the second cluster of data, lie well outside the range of most of their "colleague" points.

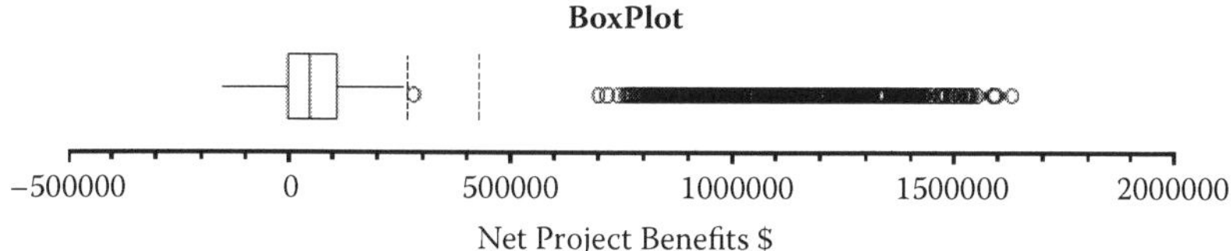

FIGURE 17.6 Box plot of project net benefits.

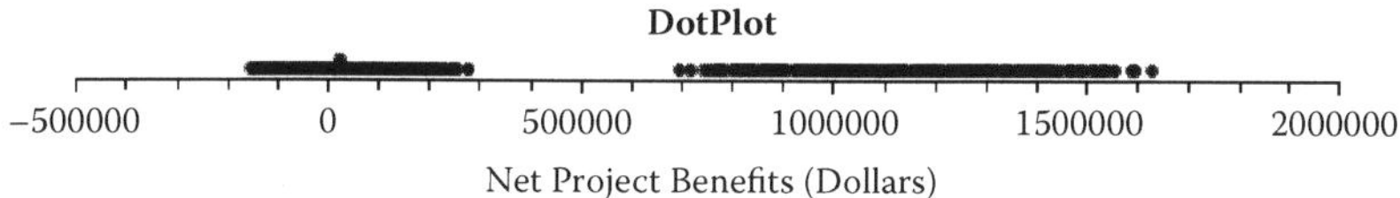

FIGURE 17.7 Dot plot of project net benefits.

The dot plot adds no real insight beyond the two distinct clusters of points, each with their own centers and their own spread of values. The dots here have no vertical dimension with one exception. Dot plots are often more effective with categorical data, as noted previously. Nonetheless, they do effectively show there are some "unpopular" values between the two clusters. They also reveal the lack of any specific "popular" values.

At this point, the assessor may want to decide whether to separate these outputs into two distinct data sets for analysis or not. We know that the cluster on the left corresponds to the prevention of agricultural land loss and the cluster on the right corresponds to the prevention of commercial land loss. Separating them for analysis is a decision that would be made in consultation with the risk manager. It is a simple matter to filter and separate the data using most spreadsheet and data analysis software.

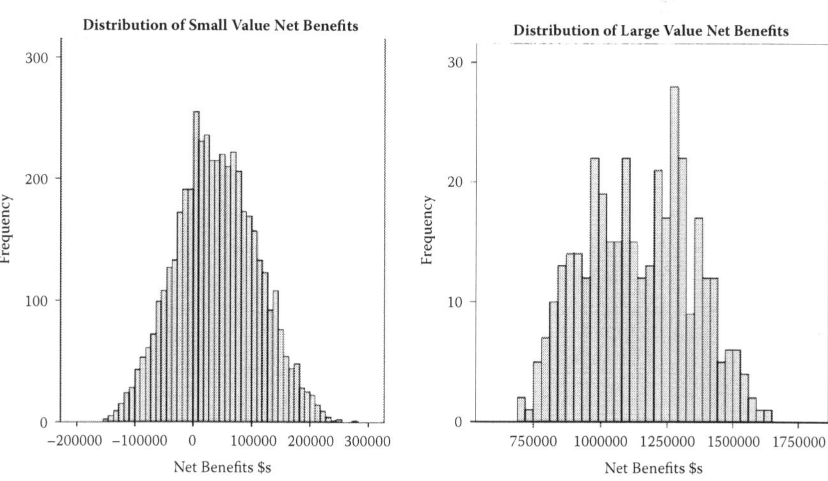

FIGURE 17.8 Two separate histograms showing small-value and large-value clusters of net benefits.

Histograms of the two distributions, obtained by separating the data into its two clusters, are shown in Figure 17.8. Note that the vertical scales differ by an order of magnitude. The distribution of small values is reasonably normal; the other distribution, with a smaller sample, is less obviously so.

Stem-and-leaf plots provide a histogram-like perspective that also shows the data. The plot of the smaller net benefit values in Figure 17.9 shows a histogram in a horizontal orientation. The individual data values are easier to pick out. For example, note how easy it is to find negative values and to observe the actual frequency with which these values occurred.

The far left column shows the actual number of observations in a row (stem and leaf). The first value after the frequency is the stem. Note that its value is in units of $100,000. Each leaf, denominated in $10,000 increments, represents 6 individual cases. Consider the row with a frequency of 460. The stem is 0; this means the dollar value is less than $100,000. There are 6 values in the $60,000s for each leaf of 6 in the plot. Likewise a leaf of 7 means there are 6 values in the $70,000s for each 7 in the plot. An & means there are observations but not enough (i.e., under 6) to complete a leaf.

The cluster of large net benefit values are shown in the stem-and-leaf plot of Figure 17.10. Notice each leaf now represents only one case. The minimum value is about $700,000 and the maximum value is about $1,600,000. The distribution shown in this way suggests less jaggedness than the histogram showed. This is an artifact of the number of bins used in the two different plots, a point to always bear in mind when using histograms.

These figures are generally not well suited to the needs of the general public. However, when communicating complex information to risk managers, it is always useful to show the data if we are going to move decision making away from overreliance on a single value.

What these initial plots do well is to reveal the location, spread or scale, and shape of the data and tendencies to cluster or not. This helps experts notice the unusual when it exists. They also can show outliers that we need to understand and explain. These are important characteristics of assessment outputs that assessors and, perhaps, managers will want to consider during decision making.

17.3.2.2 Numbers

Once you have a feel for the data, numerical summaries can be useful for communicating that feel. Table 17.3 provides a numerical summary of the two clusters and the total data set. When conducting an expert elicitation, the literature cautions against asking for a mean or most likely value first. Anchoring to this mythical "average" value can seriously impede an expert's ability to consider realistic extremes in data. For the same reason, we should not be too quick to present averages when we want managers to move away from single-point decision making. To move away from the "just tell me the number" mentality we must begin to present risk managers with information that discourages that kind of thinking.

The five-number summary provides an easy and concise summary of the distribution of the observations without resorting to using the mean. Consistently reporting these five numbers avoids both anchoring to a mean as well as the need to focus

Large Value Net Benefits Stem-and-Leaf Plot

Frequency		Stem & Leaf
7.00	–1.	4&
20.00	–1.	223
56.00	–1.	000001111
112.00	–0.	88888888888999999999
151.00	–0.	6666666666666667777777777
243.00	–0.	4444444444444444444444555555555555555555
324.00	–0.	222222222222222222222222222222333333333333333333333333
420.00	–0.	0000000000000000000000000000000000011111111111111111111111111111111111
541.00	0.	000000000000000000000000000000000000000000000011111111111111111111111111111111111111111111
474.00	0.	2222222222222222222222222222222222222222223333333333333333333333333333333333333
502.00	0.	444444444444444444444444444444444444444444455555555555555555555555555555555555555555
460.00	0.	66666666666666666666666666666666666666677777777777777777777777777777777777777
375.00	0.	888888888888888888888888888888899999999999999999999999999999999
318.00	1.	00000000000000000000000000000001111111111111111111111
238.00	1.	2222222222222222222223333333333333333333
169.00	1.	4444444444444445555555555555
95.00	1.	6666667777777
68.00	1.	88888889999
36.00	2.	000011
13.00	2.	23
4.00 Extremes (>=243485)		

Stem width: 100000.0
Each leaf: 6 case(s) and denotes fractional leaves.

FIGURE 17.9 Stem-and-leaf plot of small-value net benefits.

Large Value Net Benefits Stem-and-Leaf Plot

Frequency		**Stem & Leaf**
1.00	6..	9
11.00	7..	14566778899
40.00	8..	0001112222333445555555556677777888889999
55.00	9..	0000011112222233334444555566666667777777778888889999999
57.00	10.	000000011111444433333334445555666677777778888999999999999
47.00	11.	00000001111111122233334445555566667777777889999
73.00	12.	0000000000111122222223333344444444555566666667777777778888888888889999999
48.00	13.	000000000011112233344445555555666666677778888899
28.00	14.	0000011112222223344555667888
13.00	15.	1111223335889
1.00	16.	2

Stem width: 100000.0
Each leaf: 1 case(s)

FIGURE 17.10 Stem-and-leaf plot of large-value net benefits.

TABLE 17.3

Five-Number Summary for Net Benefit Values

Item	Small Values	Large Values	All Data
Minimum	$(152,036)	$696,575	$(152,036)
1st quartile	$(8,238)	$979,164	$(3,604)
Median	$38,347	$1,141,616	$45,938
3rd quartile	$88,447	$1,294,851	$103,078
Maximum	$276,460	$1,627,013	$1,627,013

on any one summary statistic. The five-number summary is useful because it provides information about the location of the data with the median. The data's spread is described by the quartiles, and extreme values are estimated by the minimum and maximum observations. The data range and interquartile range (middle 50% of all observations) are easily calculated from the five-number summary.

Each of the five numbers is an order statistic. This makes it easier to imagine data sets, as we see in the table. Look at the medians. It is easy to quickly see how the central location of the three data sets varies. The median and quartile values are resistant statistics, i.e., they are not much influenced by outliers. Nonresistant statistics, like the mean and standard deviation, are heavily influenced by outliers. These five numbers constitute the values needed to define a box plot, as shown in Figure 17.11.

Describing how the data are alike is sometimes a desirable task. We want to summarize the most frequently occurring characteristics of the data using a few numbers that are easily understood and agreed upon. The mean, median, and mode are the most popular measures of your data's central tendency. Try not to rely on the mean and encourage the five-number summary.

Table 17.4 adds a few more commonly calculated descriptive statistics to the previous table. Notice how the extreme values can affect the mean and standard deviation. The overall mean is $123,000 while the separated means are $40,000 and $1,143,000. The overall mean is almost three times as large as the median. Extreme values can have this effect on means. That can make these familiar but nonresistant statistics somewhat misleading to decision makers. Consider relying more on resistant order statistics than on more traditionally reported values.

The standard deviation is useful for helping others understand what constitutes an unusual value for your output of interest. Adding ±2 standard deviations to the mean provides a first cut at identifying unusual values for a single-peaked symmetric distribution; ±3 standard deviations defines a rough cutoff for identifying very

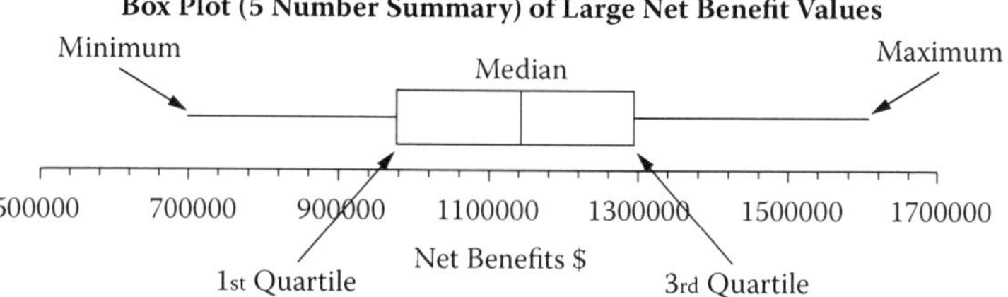

FIGURE 17.11 Box plot and five-number summary of large net benefit values.

TABLE 17.4

Selected Descriptive Statistics for Net Benefits Data

Item	Small Values	Large Values	All Data
Minimum	$(152,036)	$696,575	$(152,036)
1st quartile	$(8,238)	$979,164	$(3,604)
Median	$38,347	$1,141,616	$45,938
3rd quartile	$88,447	$1,294,851	$103,078
Maximum	$276,460	$1,627,013	$1,627,013
Mean	$40,228	$1,143,208	$122,731
Standard deviation	$69,871	$203,235	$302,998
Interquartile range	$96,685	$315,686	$106,682
Range	$428,495	$930,439	$1,779,049
Count	4,626	374	5,000

unusual values. Calculate these values and use them to examine and explain your data.

Another obvious set of values an assessor must understand includes significant thresholds. These will vary depending on the decision context and the questions risk managers have asked, of course, but a few are predictable. Separating good/desirable values from bad/undesirable values will always be important. Minimums and maximums are likely to be important. Values set in policy, zeros, and unusually large or unusually small values may all be important. Using z-scores as a measure of a threshold distance from the mean provides a good first indication of how "usual" such a result is in your distribution. Thresholds within $Z = \pm 2$ would be considered within the usual range of observed values.

CHEBYSHEV'S THEOREM AND THE EMPIRICAL RULE

Chebyshev's theorem: The fraction of any data set lying within k standard deviations of the mean is at least

$$1-\frac{1}{K^2}$$

where k = a number greater than 1.

This theorem applies to all data sets, which includes samples and populations.

The *empirical rule* gives more precise information about a data set than Chebyshev's theorem, but it only applies to a data set that is bell shaped. The empirical rule says:

68% of the observations lie within one standard deviation of the mean.
95% of the observations lie within two standard deviations of the mean.
99.7% of the observations lie within three standard deviations of the mean.

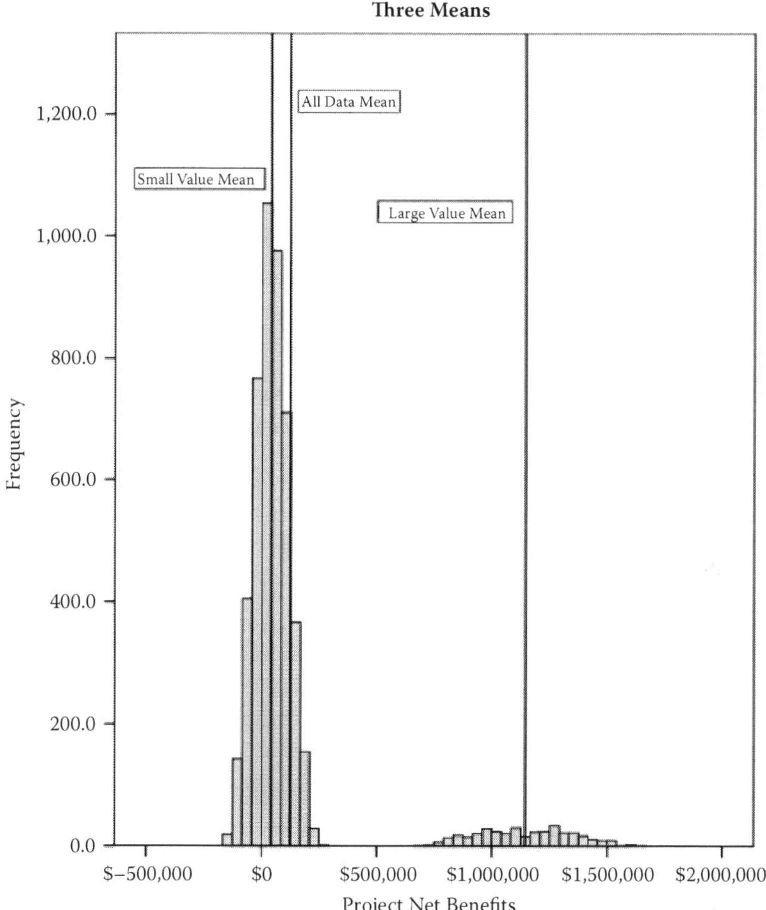

FIGURE 17.12 Locating important risk estimates in a distribution.

If point estimates of any risk measure are used, they should be identified, explained, and their locations in the distribution(s) of results shown. Figure 17.12, for example, shows the three different means of Table 17.3.

In the current case, zero is a significant threshold, separating negative net benefits from positive ones. That this project could produce a net loss in the magnitude of $150,000 is important to know. Likewise it is useful to know that if the land does get rezoned commercial, there is no chance the project will result in a net economic loss. The maximum gain under the agricultural land use scenario is about a quarter of a million dollars compared to $1.6 million for the commercial land-use scenario. For simplicity we have considered only one output from this model. Similar sorts of analyses should be conducted for other outputs as well.

17.4 Examine the Probabilities

Once you understand your quantities you are ready to tackle their probabilities. Let's begin with a brief review of the ways probabilistic data can be presented graphically, as seen in Figure 17.13. The vertical axis of the probability density function (PDF) has no convenient probability-related interpretation. Probabilities are calculated as areas under this curve. The PDF shows the relative likelihoods of the different values in its skew, kurtosis, and overall shape characteristics. It effectively reveals the most likely values and the scale of the output.

The vertical axis of the ascending cumulative distribution function (CDF) shows the probability that a specific value or less will be realized. The CDF shows fractile and median values. It is useful for estimating probability intervals or for making confidence statements. When comparing two or more CDFs, stochastic dominance is easier to see. The survival function, sometimes called the exceedance distribution or descending CDF, shows the probability that a specific value or more will be realized. Different people find one curve easier to understand than another. Experience suggests that the cumulative distribution function may be favored more often than the other two forms, but the best choice will depend on the quantities you are examining and the people with whom you work. The EPA (1997) has suggested that it is useful to display both the PDF and CDF, one above the other, with identical horizontal scales.

17.4.1 Quartiles

The simplest way to begin to address probabilistic information is by using the quartiles developed for the five-number summary. Table 17.3 provides all four quartile values, with the median being the second quartile and the maximum the fourth quartile value. These four values represent four standard points on the cumulative distribution function.

The four quartiles are shown for project net benefits in Figure 17.14. Each quartile comprises one-fourth (1,250) of the observations. There is a 25% chance that net benefits will be –$4,000 or less; a 50% chance they will be below $46,000; and a 75% chance they will be below $103,000.

The more sharply the CDF rises, i.e., the steeper the slope, the more densely concentrated the values are. That is, the scale of the distribution is smaller. When the CDF flattens out, the spread of the distribution is greater. Imagine that the CDF casts a shadow on the horizontal axis. The shorter the shadow, the more concentrated is the distribution.

17.4.2 Probabilities of Thresholds

One category of quantities to consider, mentioned previously, was that of significant thresholds. It is often important to know the likelihood that a threshold will be missed, attained, or exceeded. When there are specific quantity values that we do not want to exceed or fall below, or ranges of values we need to hit, it is useful to estimate the probabilities of these events.

With probabilistic risk assessment, this is relatively simple to do. An obvious threshold for the example project would be the break-even point of $0 net benefits. In a world of limited maintenance resources, risk managers would be very interested in the probability of this particular major rehabilitation project producing negative

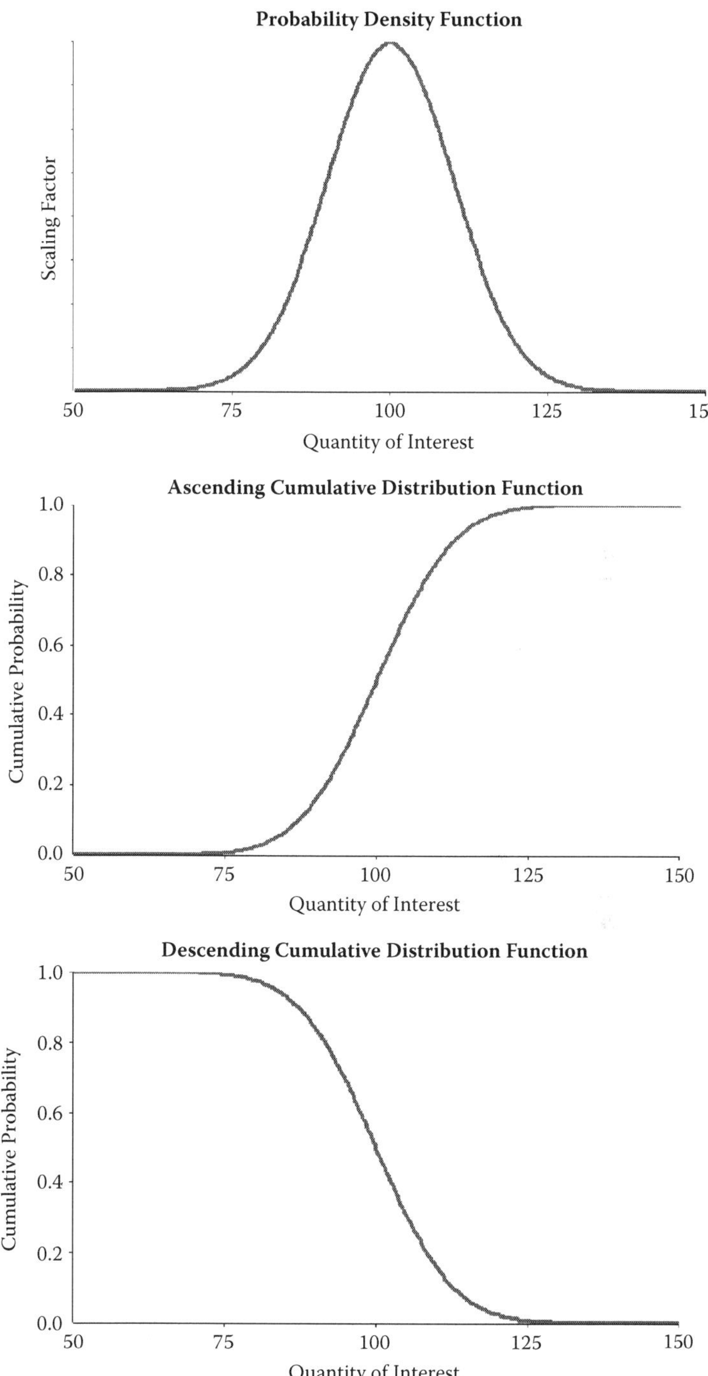

FIGURE 17.13 Review of three common ways to display a distribution of hypothetical values.

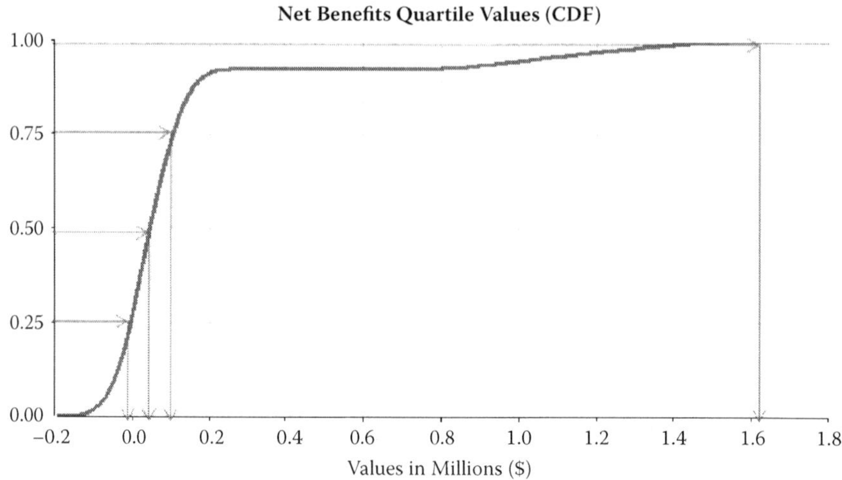

FIGURE 17.14 Quartile values for project net benefits.

net benefits. The distribution of outputs in Figure 17.15 shows that 26.7% of all the simulated values are negative; thus we estimate the probability this project will not produce a positive net return as 27.3%.

Let us, for the sake of an additional illustration, assume that a higher budget authority is considering funding a number of projects, but it cannot fund them all, so they are interested in projects with $1 million or more in net benefits. Figure 17.15 shows that this project has a 5.3% chance of meeting or exceeding that threshold.

Risk managers sometimes find it easier to understand the CDF view of a distribution. The quantity of interest is shown on the horizontal axis with a reasonably intuitive vertical axis that shows percentile/fractile data for the simulation results, which lend themselves well to likelihood estimates. Thus, we can point to the lower arrow in Figure 17.16 and explain that the likelihood of the quantity being 0 or less is given by the cumulative probability, 27.3%, on the vertical axis, which is also shown in the delimiter bar at the top of the graphic. Likewise the probability of net benefits less than $1 million is 94.7%. Using the complementary law of probability, we know the probability of being more than $1 million is 100% – 94.7%, or 5.3%. It is a simple matter to run a simulation with thousands of iterations and then to analyze, sort, and count outputs to estimate the likelihood of virtually any imaginable threshold being met, exceeded, or fallen short of as well as the probability of any interval of values occurring.

17.4.3 Confidence Statements

One of the strengths of risk assessment is its ability to cope with uncertainty. Probabilistic risk assessment techniques enable us to cope with uncertainty and to express our confidence in our results in a quantitative way.

After reporting on quantities of particular interest, it is important to convey to risk managers the degree of confidence that assessors have in their estimates. This can be

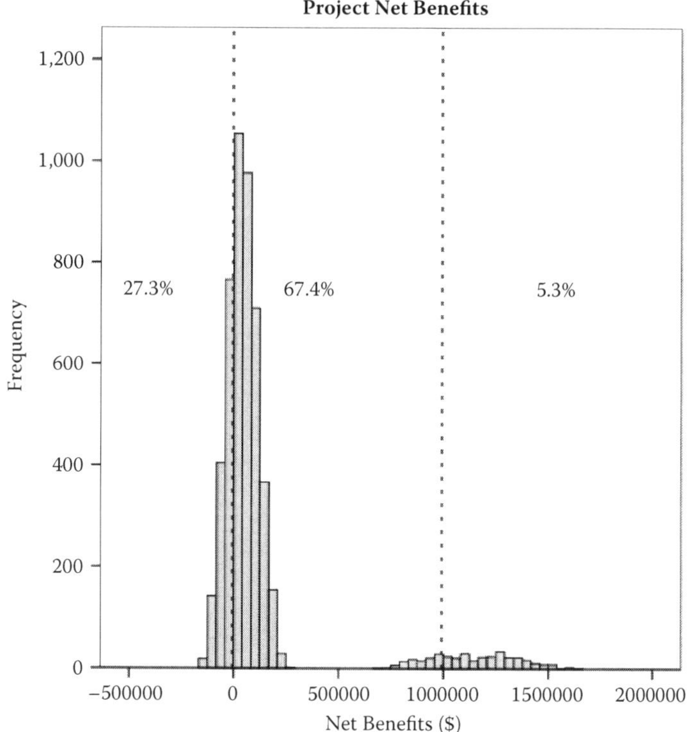

FIGURE 17.15 Project net benefit threshold values.

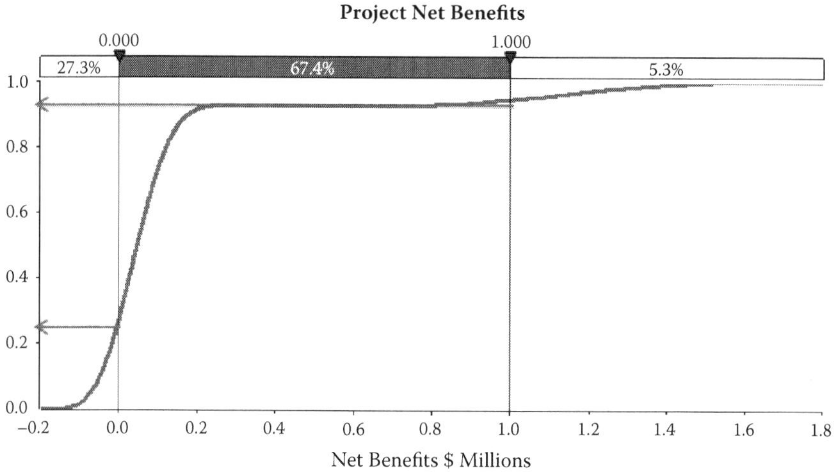

FIGURE 17.16 Project net benefit threshold values in a CDF.

TABLE 17.5

Confidence Statements for Project Net Benefits

Minimum	($152,036)	5th percentile	($71,162)	2.5th percentile	($90,395)
Maximum	$1,627,013	95th percentile	$1,031,349	97.5th percentile	$1,238,284
Range	$1,779,049	90% confidence range	$1,102,511	95% confidence range	$1,328,679

done qualitatively in the narratives of reports and meetings. It is, however, important to convey quantitative messages about quantitative values. Information like that in Table 17.5, for example, enables risk assessors to tell risk managers that we are 90% sure that the eventual value of net benefits lies between –$70,000 and $1,030,000.

Alternatively, we are 95% sure that the true value of net benefits will lie between –$90,000 and $1,240,000. This gives the risk manager a much more vivid understanding of the effects of uncertainty than if you begin by saying our best estimate of net benefits is $123,000. Confidence ranges are probabilistic statements that help diminish overreliance on a single estimate of the output.

The confidence ranges described here are not the same as the confidence intervals calculated for sample statistics. When we speak of being "sure," as we do here, this is used in a rather loose sense. It literally means that 90% (or 95%, etc.) of our results fell between these two numbers. Thus, this confidence statement is based on the assessor's best data and efforts. It is a subjective quantitative measure. The ranges presented should meet the communication needs of the risk assessor and the decision-support needs of the risk manager. If the risk manager is dismayed because the confidence range is too broad for his purposes, this should lead to a discussion of practical options for further reducing the existing uncertainty. In the current example, the obvious choice is to find out what is happening with the rezoning appeal in an effort to pin down the value of the land that could be lost.

17.4.4 Tail Probabilities and Extreme Events

Dealing honestly with uncertainty means moving away from single-value estimates of complex phenomena. Expected values are rarely going to be the only number relevant to risk managers and other decision makers. The value of probabilistic risk assessment is—in the way that its results contribute to our understanding of the problems we face, the complex systems that produce them, and the characterizations of the effects of uncertainty and variability—propagated in the range of results generated by our probabilistic models.

The current example provides an ideal illustration of one of the problems of dealing with expected values. They are not resistant statistics. As noted earlier, the expected value of net benefits is $123,000. Compare that to the median of $46,000 and we can see the influence of extreme values on the mean. Decisions based on the mean may well distort expectations about a project's actual performance.

Let's shift our focus now from net benefits to project costs to facilitate a simpler consideration of distribution tails and extreme events. Total project construction costs are distributed as seen in Figure 17.17.

We see that costs span from $750,000 to $890,000, a range of $140,000. If our only interest is in the mean cost estimate, that is $812,000. That cost will only occur in the

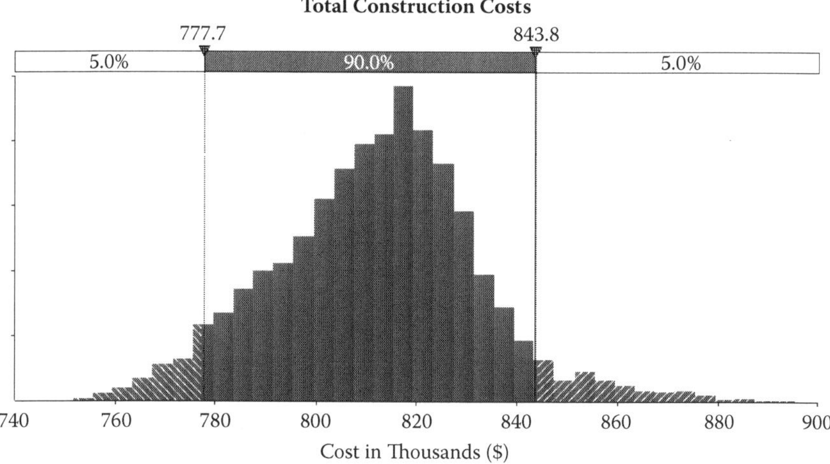

FIGURE 17.17 Distribution of total construction costs.

extremely unlikely event that all of our "best estimates" are realized. The purpose of identifying and addressing uncertainty, however, is to learn what the actual truth might be and what might happen if our best estimates are not realized. How low might costs be, and how high could they rise? This information is embedded in the tails of the distributions.

Costs appear to have an asymmetric distribution, with more values to the left of the single peak than to the right. The tails seem a bit unevenly defined by the 5,000-iteration simulation. If tails values are of interest, it is generally wise to do iterations in the magnitude of 10^4, and if extreme events are of interest at least 10^5.

There are several options for discussing and presenting tail values. Deciding how to define the tails is the starting point. We previously discussed the importance of threshold quantity values, whether high or low. Here we define the tails by percentages and suggest the highest and lowest 5% as starting points.

The threshold values, the 5th and 95th percentiles, for our example are $778,000 and $844,000, respectively. Using the filter feature of your spreadsheet software, it is a simple matter to isolate these clusters of data to analyze and summarize the tail data.

Let's consider the lowest 5% of cost estimates as shown in Figure 17.18. Assessors can now explain to risk managers that if all the uncertainties resolve themselves "favorably" (i.e., producing a lower cost), we are looking at a range of costs roughly from $750,000 to $780,000, with a conditional expected value of about $771,000.

This sort of information is conditional on the fact that actual costs will be somewhere among the lowest 5% of all possible costs, i.e., an optimistic scenario. The distribution shown in the histogram is a new conditional distribution. It isolates and magnifies the tail portion of an assessment output distribution. Conditional information can be informative for risk managers when tail values are important to the decision context. This kind of partitioned analysis can be done for any size left or right tail.

Extreme values can also be important to risk managers, especially when loss of life, human health, and safety are among the outputs of concern. In the case of a cost estimate, the extremes in this example are not quite so compelling as they might be for

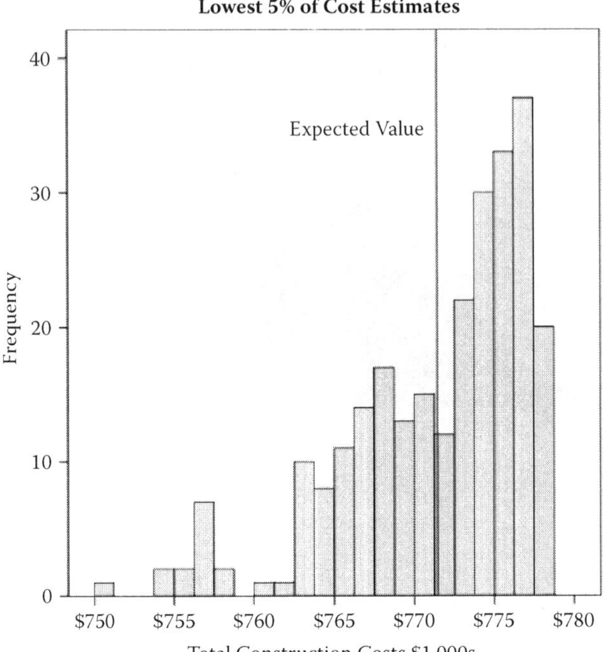

FIGURE 17.18 Distribution of the lowest 5% of cost estimates.

human life and safety. However, it is a relatively simple matter to explore the potential extremes of any given situation with your probabilistic risk assessment model.

Doubling the number of iterations provides a better chance of capturing a more extreme set of high or low input values. The reductions in the low extreme cost estimate going from 10,000 to 100,000 iterations, shown in Table 17.6, are relatively modest. This indicates that we are likely to be zeroing in on a reasonable estimate of this extreme value. Note that high extremes are a little further from the mean than low extremes. The additional iterations add no real precision to the estimate of the mean.

17.4.5 Stochastic Dominance

Stochastic dominance, as used here, refers to a form of ordering for probability distributions. It is a concept developed in decision theory that sometimes enables one to

TABLE 17.6

Extreme Value Estimates of Cost with Varying Numbers of Iterations

Iterations	Minimum	Mean	Maximum
5,000	$750,959	$811,951	$886,142
10,000	$745,355	$811,952	$887,159
100,000	$743,097	$811,926	$903,757

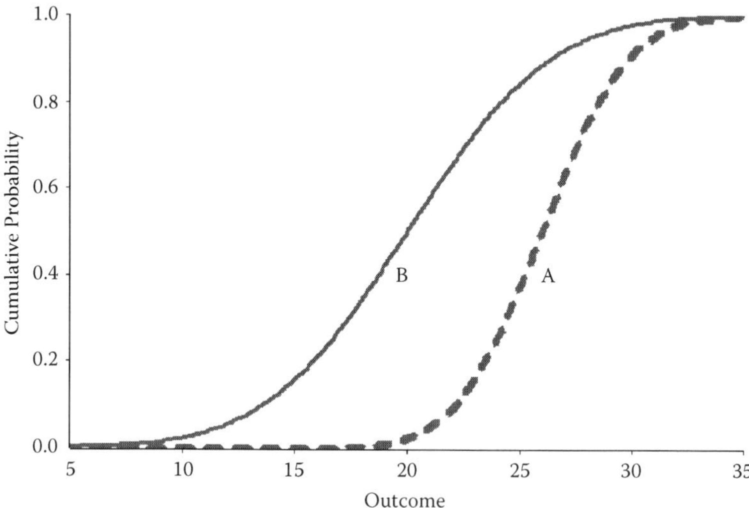

FIGURE 17.19 First-order stochastic dominance.

call one distribution better or more desirable than another. To illustrate this concept, let's revert to some hypothetical outputs associated with two hypothetical risk management options A and B. Let $F_A(x)$ and $F_B(x)$ be the cumulative distribution functions for these two options, and let larger values of x be preferable.

Figure 17.19 demonstrates first-order stochastic dominance, i.e., where $F_A(x) \leq F_B(x)$ for all x. Let the horizontal axis show a hypothetical assessment output where higher values are preferable to lower values. For any given x-value, option A provides a lower probability of that value or a smaller one being realized. Likewise for any given probability (percentile), option A yields a higher outcome. That means for any potential value of the outcome and for any cumulative probability, option A is preferable to option B, and so it dominates them at all points of the curve. Had the example been reversed, with lower values preferable to higher values (such as lives lost, illnesses, or damages sustained), we would say option B dominates A and is preferable.

What happens when the CDFs intersect and there is no first-order dominance? Is it still possible to call one distribution preferable to another? In Figure 17.20, option A (dashed line) has second-order stochastic dominance over B (solid line). To calculate second-order dominance we consider a function, $D(z)$, that is the cumulative difference between the two CDFs. The function is defined as follows:

$$D(z) = \int_{min}^{z} \left(F_B(x) - F_A(x)\right)dx \geq 0, \forall z \tag{17.1}$$

then A is said to have second-order stochastic dominance over B.

The figure shows that A dominates B up to the value, V. From this point on, B dominates A. The areas between the curves is calculated as D_1 and D_2. D_1 is a positive value because $F_B(x)$ is the greater cumulative probability up to V. D_2 is a negative value. The sum of D_1 and D_2 is positive, and so A is said to hold second-order

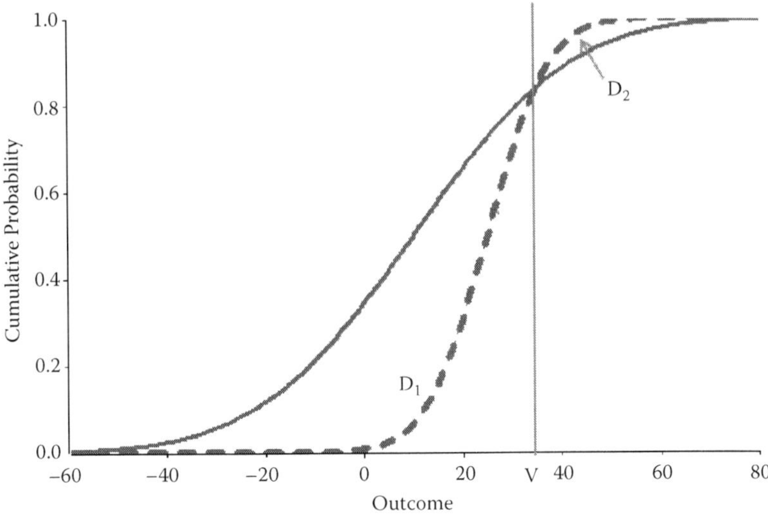

FIGURE 17.20 Second-order stochastic dominance.

dominance over B. While first-order dominance is rather unambiguously superior, second-order dominance is not; it is based on the assumption that the decision maker has a risk-averse utility function. Option A offers the best chance of a preferred outcome. Second-order dominance does not take into account different preferences for large or small outcomes. It is not always convenient to calculate second-order stochastic dominance.

17.5 Examine Relationships

So far we've considered how to explain the quantities and probabilities of individual variables and outputs. Explaining relationships between variables is sometimes more important. Dose-response relationships in hazard characterizations for human health risks, for example, are a common example of a two-variable relationship that is critically important to some risk assessments. Contingency tables, discussed earlier, can be an effective tabular display of relationships between and among variables. In this section we examine tools and methodologies to help you understand and explain relationships between two variables.

17.5.1 Scatter Plots

The single best, simple graphic device for exploring relationships between variables is the scatter plot. Think in terms of exploring relationships between:

- Outputs
- Inputs and outputs
- Significant inputs

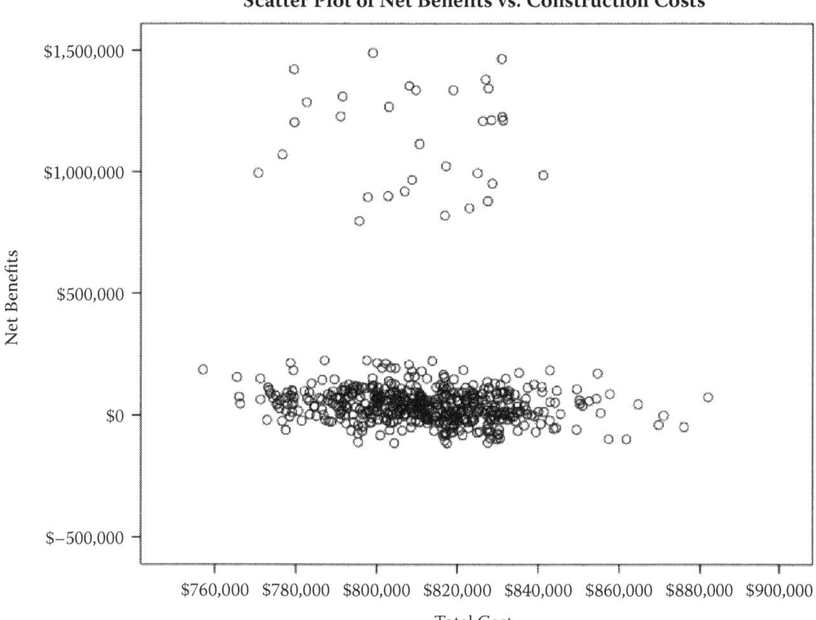

FIGURE 17.21 Scatter plot of project net benefits and construction costs.

Bear in mind that scatter plots do not reveal cause and effect but simple correlation. Consider a few scatter plots* from our example, beginning with the relationship between project costs and net benefits, two outputs of interest shown in Figure 17.21. The plot reveals the two distinct clusters that we identified previously. There appears to be no obvious association between construction costs and net benefits in the top cluster, beyond a vague suggestion of a negative slant to the cloud of points, despite the very obvious fact that benefits − construction costs = net benefits. The bottom cluster, with more data points, does exhibit the expected negative relationship. As costs get larger, net benefits decline.

Now let's have a look at two more outputs: gross benefits and net benefits. Figure 17.22 shows a positive relationship: as gross benefits increase so do net benefits. The relationship is also very tight, indicating a very close association between these two values. Figure 17.22, although far from surprising, was generated to demonstrate the scatter plots' ability to reveal a range of relationships.

To illustrate a relationship between an input and an output, let's examine the relationship between the cost per cubic yard of excavation and the total cost estimate. The upward sloping circular pattern of Figure 17.23 indicates the absence of a strong correlation. A weak positive relationship is confirmed by the correlation coefficient of 0.284. There are enough other cost factors that this one does not exhibit an especially strong association with construction costs. Scatter plots can be one of the more effective tools for identifying sensitive inputs in a risk assessment model.

* Plotting all 5,000 points obliterates a lot of the detail. A random sample of 500 output points was used to create these scatter plots.

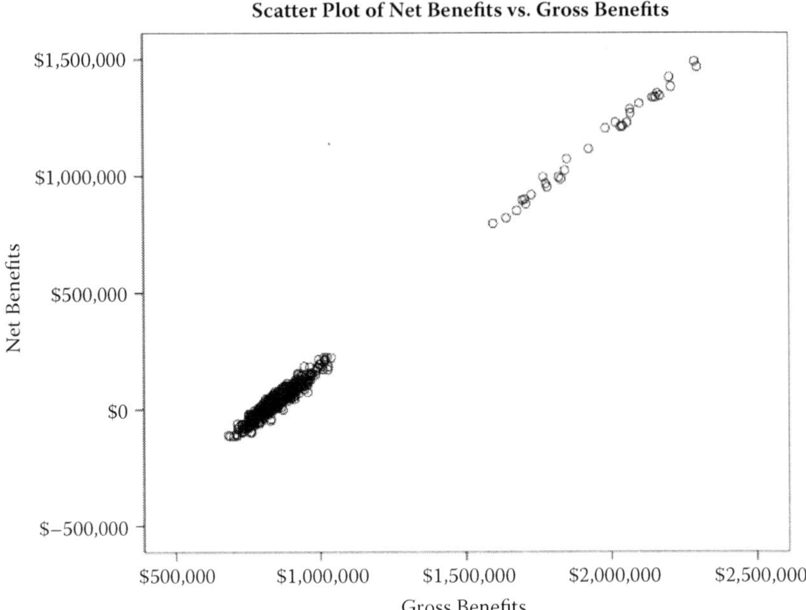

FIGURE 17.22 Scatter plot of project gross benefits and net benefits.

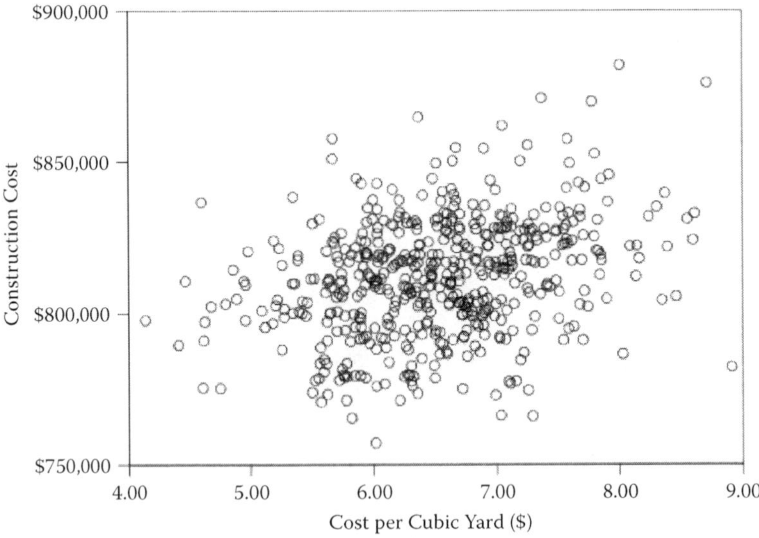

FIGURE 17.23 Scatter plot of excavation costs per cubic yard and excavation quantity in cubic yards.

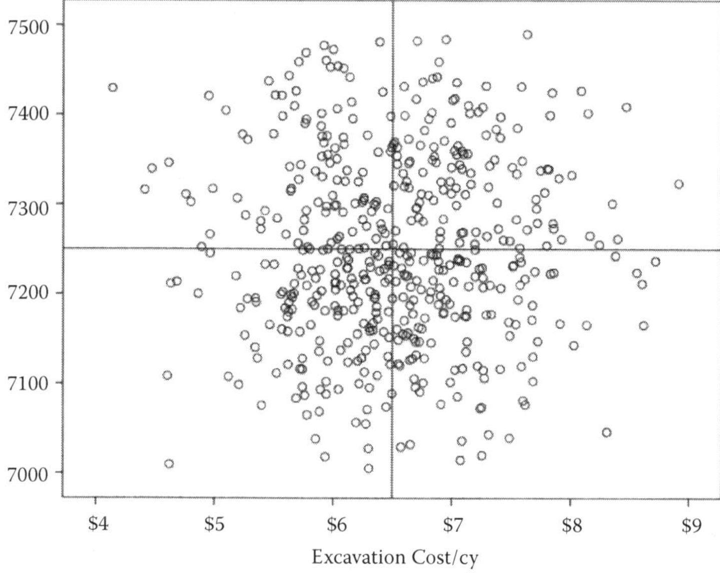

FIGURE 17.24 Scatter plot of excavation quantity and excavation cost.

Mulitple-minis,* a page with a large number of relatively small graphics, can provide a useful summary of the assessor's investigation of the relationships among inputs and outputs.

An example of a scatter plot for two inputs is shown in Figure 17.24. Excavation cost and excavation quantity were treated as independent inputs by cost estimators. This assumption is borne out by the scatter plot. If they were associated, that relationship would have to be reflected in the model. Thus, checking pairs of inputs can be informative not only in surfacing relationships that exist, but in revealing expected relationships that do not exist. If excavation cost and quantity are, in fact, related, this figure would reveal they are not related in the model. Recall from Chapter 10 that sketching the relationships between variables is an important part of model building. Examining the scatter plots of simulation results (inputs and outputs) to verify those relationships can provide a good reality check for your results.

Scatter plots can help reveal overall patterns of relationships between variables. Lines and curves, with their tightness and direction, quickly reveal facts about relationships. It is easy to see how individual points differ from the averages when the averages are identified on the graphs. Unusual points and subclusters are also easy to identify. When clusters of points are found, this invites the assessor to explore what "membership" in the group may be based upon. Use scatter plots to explore, understand, and then explain relationships to risk managers. Remember that not every detail of your exploration and analysis needs to be documented in your main

* The International Shark Attack File provides an excellent example of multiple-minis at http://www.flmnh.ufl.edu/fish/sharks/statistics/pop2.htm.

report. However, all important details should be captured either in support files for the assessment or the assessment's technical appendices if it is formally documented.

17.5.2 Correlation

Correlation coefficients measure the strength of a relationship between a pair of variables. We see changes in variables all of the time in risk assessment. When two variables are changing at the same time, there are three possible relationships among the variables. When higher-than-average values of one variable tend to occur with higher-than-average values of the other variable and lower-than-average values of one variable are associated with lower-than-average values of the other, the variables covary and have a positive correlation.

The second kind of correlation is a negative one. This means the two variables vary inversely or oppositely. Higher-than-average values of one variable occur with lower-than-average values of the other variable and vice versa. The third possibility is that there is no discernible pattern among higher or lower values of one variable with another variable's values.

The correlation coefficient takes a value between −1 and +1. A scatter plot for variables with these minimum and maximum correlations would be perfect straight lines. The sign indicates the direction of the association, and the size of the correlation indicates the statistical strength of the relationship. A coefficient of 0 indicates the absence of a statistical association. Coefficients with an absolute value close to 0 are weak; absolute values closer to 1 are strong.

Two commonly used coefficients are the Pearson and Spearman rank correlation coefficients. The Pearson coefficient is based on the differences from means between paired raw-data values. The Spearman rank coefficient is based on differences between the ranks of paired raw-data values.

The Pearson correlation coefficients (r) for the relationships shown in the preceding scatter plots are shown in Table 17.7. Costs and net benefits have a very small negative relationship, with r = −0.052. The relationship between gross and net benefits is almost a perfectly linear one, as indicated by a coefficient of 0.998. There is a positive relationship between construction costs and the cost of excavation of 0.284. The correlation between excavation quantities and costs is almost 0, with r = −0.031.

TABLE 17.7

Pearson Correlation Coefficient for Scatter-Plot Variables

	Construction Cost	Gross Benefits	Net Benefits	Cubic Yards	Cost/CY
Construction Cost	1.000	.014	−.052[a]	.026	.284[a]
Gross Benefits		1.000	.998[a]	.016	.025
Net Benefits			1.000	.014	.006
Cubic Yards				1.00	−.031[b]
Cost/CY					1.000

[a] Significant at the 0.01 level (two-tailed).
[b] Significant at the 0.05 level (two-tailed).

Correlation tables are easy to produce with most commercial software packages, and they can provide a handy first screening tool to identify potential linear dependencies and independencies among model inputs and outputs.

17.5.3 Reexpression

It can sometimes help to express the data in different ways in order to see relationships. It helps to straighten out the dependence or point scatter as much as possible; it is usually easier to see what is going on in linear relationships. If you have a curvilinear plot, try converting your data using logarithms or roots. This can sometimes make data more linear and easier to understand.

Whatever the reasons, most of us do better understanding addition rather than multiplication. "This plus this" rather than "that times that" is just easier for most people to comprehend. Use logs to convert data and change multiplicative relationships to additive ones. Logs, although foreign to many audiences, sometimes enable you to examine a linear relationship between variables. That is often a more successful way to examine and explain data. Be aware that most people will still need help interpreting data in a log scale.

17.5.4 Comparison

Comparisons are often appealing to less quantitatively oriented people. Sometimes saying Sue is a head taller than Sarah, or Joe is twice as heavy as Rich, is more revealing than the actual numbers. So when you consider comparisons use the analogy: What can you do mathematically to one person (i.e., variable of interest) to make him like another (variable of interest)?

Comparisons are often a matter of difference or ratio. Find out what is different about your two variables. To compare things, explore whether there is anything that can be added or subtracted from the data to make them comparable. Add or subtract a value to compare a value. For example, you might subtract a background illness or mortality rate from your risk estimate to show how much worse (or better) it is from the background level. Alternatively, you might add or subtract some baseline costs or revenues from financial outputs to place them in a more user-friendly perspective. This option will cost $\$x$ more than average cost or earn $\$y$ more than average revenues.

Ratios can sometimes support useful comparisons. Comparing the largest value to the smallest value can sometimes help. Construction costs in our example yield the following ratio $882,142/$750,959 = 1.18. This means the smallest cost estimate plus 18 percent covers all the potential cost possibilities. Other times ratios are not so useful; for example, net benefits range from –$152,035 to $1,627,013 and this ratio does not have the same easy interpretation.

17.6 Answer the Questions

The reason for understanding and explaining one's results is to answer questions and provide information useful for decision making. At the outset of this chapter we posed the following seven hypothetical questions for the example.

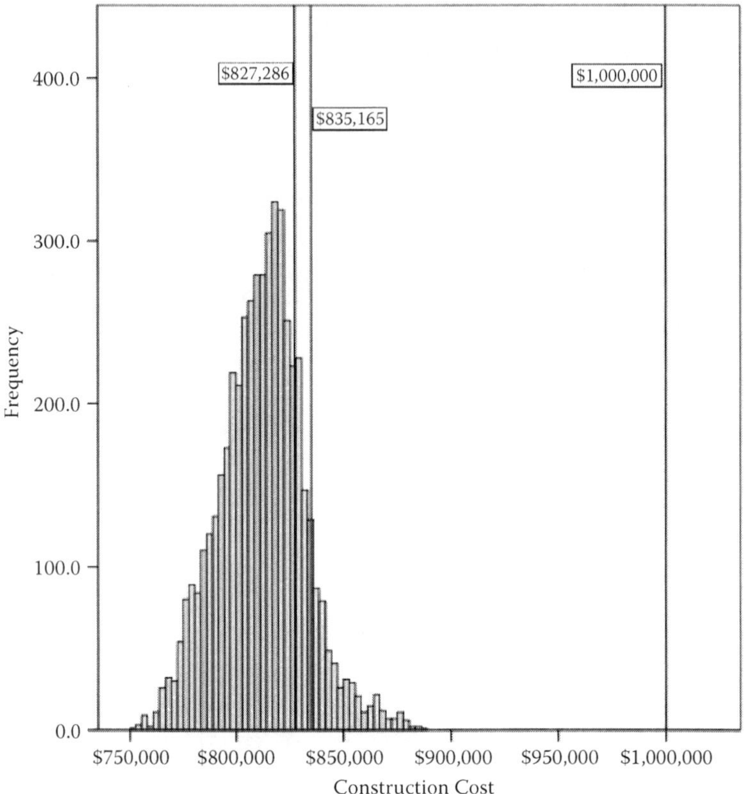

FIGURE 17.25 Distribution of project costs with selected thresholds identified.

1. What is the probability that project costs will exceed $1 million?

   The maximum cost estimate was $886,142. There is no chance the cost of this project will exceed $1 million. Figure 17.25 shows the distribution of costs with $1 million, as well as the 80th and the 90th percentiles identified.

2. Identify the cost estimate that has no more than a 20% chance of being exceeded.

   The 80th-percentile value in the simulation is $827,286. Because 20% of all the simulation values exceeded this amount, we assume there is a 20% chance that costs will exceed this value, as shown in Figure 17.25.

3. Identify the cost estimate that has no more than a 10% chance of being exceeded.

   The 90th-percentile cost estimate is $835,165, as seen in Figure 17.25. There is a 10% chance that costs will exceed this amount. Thus, faced with a decision of which of 5,000 possible cost estimates to use in the budget process, risk managers would decide the probability of a cost overrun they are willing to tolerate. If they can tolerate a 20% chance of a cost overrun, they will use $827,000. If they want to reduce that risk to 10%, they will use $835,000 as the cost estimate.

Costs are rather trivially different in the example. In fact, the comparison here indicated that the maximum cost estimate (based on 5,000 iterations) is only 18% more than the minimum. In projects with larger price tags or more significant uncertainties, the difference between a 20% and 10% chance of an overrun might be substantial.

4. What is our maximum exposure to cost overruns associated with these costs?

   The maximum cost estimate was $886,000, and the 20% and 10% overrun risk costs are $827,000 and $835,000, respectively. Therefore the exposure to cost overruns is $59,000 and $51,000, respectively. If the consequences associated with these chances of an overrun are unacceptable, we will have to refine our analysis.

5. What are the most significant cost uncertainties?

   Borrowing from lessons learned in Chapter 16 on sensitivity analysis, we can list the most significant uncertainties using simple regression analysis, as shown in Table 17.8. To improve the cost estimate and reduce the variation in cost, we're best advised to improve our estimate of the amount of steel sheet pile needed, followed by getting a better idea of real estate costs. If we examined contributions to the variation in net benefits or any other assessment output, we might expect a different list of significant inputs.

6. What is the probability that this project will be economically feasible, i.e., net benefits equal or exceed zero.

   We have seen from the previous analysis that there is a 27.3% chance that this project will produce negative net benefits. Thus, there is a 72.7% chance that the project is economically feasible.

7. What is the most likely level of net benefits?

   We are 90% sure net benefits will be between –$71,000 and $1,031,000, as seen in Figure 17.26. Our best estimate is the median value, $46,000. The range in possible outcomes is substantial. We have learned from our analysis of the data that the single greatest uncertainty in this project evaluation is whether the adjacent eroding land will be in agricultural use, which will result in lower net benefit estimates or if the land is rezoned it will result in larger net benefits with no probability of a negative return. In short, the project may be worth doing to protect commercial land but it is not as clear

TABLE 17.8

Uncertainty Rankings Based on Simple Linear Regression Standardized Coefficients

Cost Input	Standardized Regression Coefficient
Number of pieces of sheet pile	0.695
Lands and damages cost	0.489
Cost of mobilization and demobilization	0.297
Excavation cost/CY	0.293
Tremie concrete cost	0.184
Tremie concrete quantity	0.160

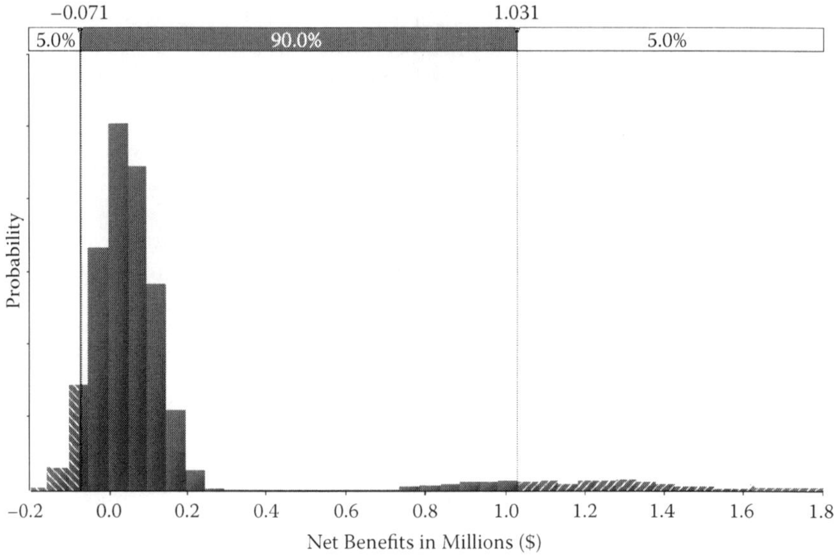

FIGURE 17.26 Distribution of project net benefits with a 90% confidence level.

> that it is worth doing to protect agricultural land. The single most useful thing we could do to reduce significant uncertainty is try to learn whether this land will be rezoned or not.

17.7 Visualization of Data

Technology offers exciting and promising new ways to help us convey the information embedded in data. Data visualization is concerned with the visual representation of information obtained from data. Friedman (2008) describes the main goal of data visualization as:

> To communicate information clearly and effectively through graphical means.... To convey ideas effectively, both aesthetic form and functionality need to go hand in hand, providing insights into a rather sparse and complex data set by communicating its key aspects in a more intuitive way. Yet designers often fail to achieve a balance between design and function, creating gorgeous data visualizations which fail to serve their main purpose—to communicate information.

This balance between design and function opens some exciting possibilities for risk assessors to convey the meaning of their work. In the past we were bound by the creativity of the authors of reports and the tables and graphics they could prepare. Finding new ways, like interactive graphics, to make data available to people so they can examine it themselves in ways that are meaningful to them is the greatest promise of data visualization and its related fields of information graphics, information

DATA GRAPHICS

Data graphics visually display measured quantities by means of the combined use of points, lines, a coordinate system, numbers, symbols, words, shading, and color. The use of abstract, nonrepresentational pictures to show numbers is a surprisingly recent invention....

At their best, graphics are instruments for reasoning about quantitative information. Often the most effective way to describe, explore and summarize a set of numbers—even a very large set—is to look at pictures of those numbers. Furthermore, of all methods for analyzing and communicating statistical information, well-designed data graphics are usually simplest and at the same time the most powerful.

Edward R. Tufte, Introduction to *The Visual Display of Quantitative* *Information*

visualization,* scientific visualization (a three-dimensional variation), and statistical graphics. Inventing interesting and innovative visual displays of complex data, information, and knowledge that can be understood easily and quickly is one of the most promising fields for aiding risk assessors in communicating their work to risk managers and the public.

Graphics developed to convey information and the narratives used to support them are very different from the kinds of graphics and messages we develop for risk communication. The former are a subject of this chapter. Edward Tufte (1983, 1990, 1997) has written a well-received series of books on information graphics. His principles of graphical excellence and integrity are summarized here.

Graphical excellence:

- Well-designed presentation of interesting data is a matter of substance, of statistics, and of design.
- Communicates complex ideas with clarity, precision, and efficiency.
- Gives the viewer the greatest number of ideas in the shortest time with the least ink in the smallest space.
- Almost always is multivariate.
- Tells the truth about the data.

Tufte's six principles of graphical integrity are:

- The physical representation of numbers on graph should be directly proportional to the numerical quantities represented (use consistent size for all numbers).

* Friendly (2009) describes this as the visual representation of large-scale collections of nonnumerical information, such as files and lines of code in software systems, library and bibliographic databases, networks of relations on the Internet, and so forth.

- Avoid graphical distortion and ambiguity with clear, detailed and thorough labeling. Explain the data on the graph itself. Label important events in the data.
- Show data variation, not design variation.
- In time-series displays of money use constant dollar amounts unless you have a specific reason not to.
- The number of information carrying (variable) dimensions depicted should not exceed the number of dimensions in the data.
- Graphics should never depict data out of context.

Tufte is also quite passionate in his views about data ink, i.e., the ink on a graph that represents data. He says good graphics maximize data ink and erase as much nondata ink as possible. He also offers these five points:

- Above all, show the data.
- Maximize the data-ink ratio.
- Erase nondata ink.
- Erase redundant data ink.
- Revise and edit all graphics.

At the time this book was written, the following sites offered useful examples of visual displays of quantitative information and data visualization. It is quite likely that one or more of these may be outdated as you read this. If so, an Internet search on some of the phrases used in this section will undoubtedly produce interesting examples.

Many Eyes presents a set of data visualizations and invites browsing, discussion, and experimentation. http://manyeyes.alphaworks.ibm.com/manyeyes/.

Gapminder shows some innovative ideas for the visual display of data using simple animations of multivariate time series data. http://www.gapminder.org/.

A Taxonomy of Visualizations includes a very handy periodic table of visualization methods complete with roll-over examples. http://www.stat.columbia.edu/~cook/movabletype/archives/2007/01/a_taxonomy_of_v.html.

Information Aesthetics provides an excellent portal of entry into the world of information aesthetics. http://infosthetics.com/.

There are few, if any, current examples of risk assessment utilizing these state of the art methods. They are introduced here to challenge risk assessors to learn and use these exciting new methods for presenting information.

17.8 Decision Making under Uncertainty

The graphic displays and ideas in this chapter are intended to be used to aid and improve risk management decision making under conditions of uncertainty. These decisions may include yes or no decisions on a single action, rating a series of alternatives, ranking the alternatives, or choosing the best option from among a set of

HEURISTICS

Faced with uncertainty, most people revert to the use of rules of thumb that have proven useful to them in similar situations in the past. Cognitive psychology research suggests that subjects make judgments on such inferential rules or heuristics. Although heuristics are frequently used to deal with uncertainty, they are not always valid and can lead to large and persistent biases in decision making. They can be particularly invidious because they are often unknowingly practiced by "experts."

Tversky and Kahneman (1974) offer the following heuristics as sources of bias in judgments. You have already encountered some of these in Chapter 13.

1. *Anchoring and adjustment*: Individuals tend to produce estimates by starting with an initial value and adjusting it to obtain a final answer. The adjustment is typically insufficient. As a result, initial ideas play too large a role in determining final assessments. Experts are prone to use this rule. When faced with uncertainty, they make an initial guess and adjust it up or down, but they rarely venture too far from their first guess or anchor.
2. *Availability*: If it is easy to recall instances of an event's occurrence, that event will tend to be assigned a higher probability than it deserves. People tend to overestimate the probabilities of dramatic events that have recently occurred. This rule may help to explain the often-observed fixation with protecting against a recent low-probability flood of record or a specific terrorist attack.
3. *Coherence and conjunctive distortions*: A good story makes events seem more likely. The probability that a sequence of events will occur often seems higher than it should, especially when the events fit a plausible scenario. The scenario of events required to produce a dam failure, for example, may seem to be far more likely than it is in fact.
4. *Representativeness*: People expect that the essential characteristics of a stochastic process will be represented in any part of the process. Furthermore, people see chance as a self-correcting process in which a deviation from the mean in one direction is offset by a deviation from the mean in another direction. Experts and laymen alike may make too much of a few years of data in trying to understand the totality of complex processes.
5. *Overconfidence*: People, particularly so-called experts, generally ascribe too much confidence to their estimates, thereby underestimating confidence intervals. This rule motivates people to see patterns where none exist, to reinterpret data to be more consistent with their view, and to ignore evidence that contradicts their position.

The message: Heuristics are to be avoided in making decisions under uncertainty.

alternatives. Imagine the risk manager presented with the assessor's best displays now asking, "What do I do with all these data?" The answer is to use it for risk-informed decision making.

Success in risk management is defined by practical and useful solutions for dealing with uncertainty. Risk-informed decision making is the confluence of risk management and risk assessment. It is the risk assessor's job to address the knowledge uncertainty and natural variability in inputs and to convey the significance of the uncertainty in their assessments to risk managers. It is the risk manager's job to address the knowledge uncertainty and natural variability in the assessment results and decision criteria, to take these explicitly into account in decision making, and to manage risks that are not acceptable to at least a tolerable level.

To do this, risk managers need to request and use risk information to aid in their decisions made under uncertainty. This means developing risk-related decision metrics and using rules developed for decision making under uncertainty. In best practice, the risk manager's decision criteria will generally reflect risk metrics implicit in the questions posed to risk assessors as well as the risk management objectives identified in the decision context. To the extent that risk managers become explicit about the risk metrics of interest to them, the decision-making process will be improved.

17.8.1 Risk Metrics

Typical risk metrics vary with the nature of the decision context, but they include such things as mortality, morbidity, and life-safety risk, including such things as the number of lives at risk and social vulnerability. Relative risk, increases or decreases in risk, and odds ratios may also be used. Examples of other values at risk include net economic benefits, financial risks, engineering risk and reliability, and the like. The risks considered and their measurements should include the full range of existing risk, risk reductions, residual risk, risk transformations, risk transfers, and new risks.

Risk-informed decision making is a new enough concept that the most useful risk metrics may have not even been identified as yet. Certainly, we have a good handle on the most obvious risk measures, and many of them have been in use for a long time. The emphasis on risk analysis, however, is young, and clever assessors will hopefully continue to develop new metrics to aid in decision making. One such metric is obtained from the partitioned multiobjective risk method (PMRM) developed by Haimes (1998). The PMRM was developed to respond to the common inadequacy of an expected value as a measure of risk. It develops conditional expected-value functions that represent the risk given that an event of a certain magnitude or frequency has occurred. The method can, for example, be used to isolate one or more damage ranges by specifying a partitioning probability. It then generates a conditional expectation of the damages, given that the damages fall within the identified range. An example follows.

Flooding is a serious risk to life and property in many parts of the world. Floods cause property damage, and reductions in these property damages are a common measure of the benefits to flood risk management measures. These damages are often estimated using the hydroeconomic model shown in Figure 17.27, which is used to estimate the expected annual damages (EAD) associated with flood regimes and risk management options.

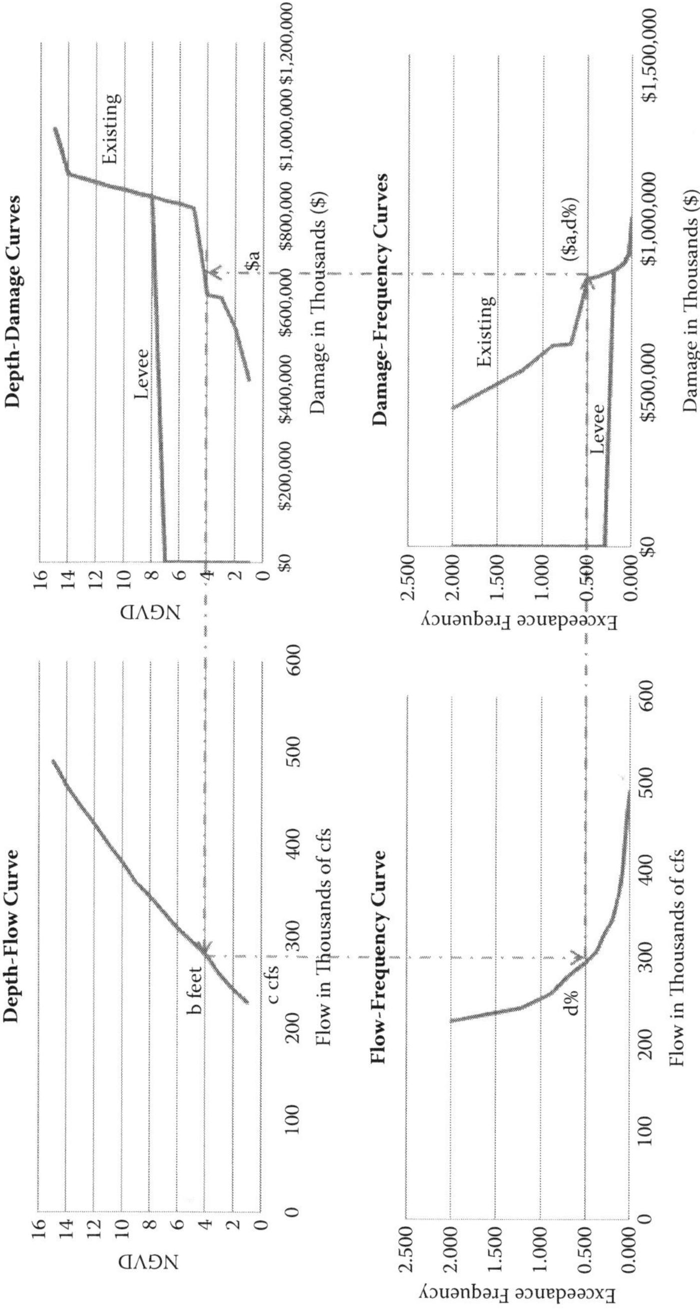

FIGURE 17.27 Hydroeconomic model for estimating expected annual flood damages.

Beginning in the upper right quadrant, property damage is shown to increase as flood depths increase. Moving to the left, we see that increasingly large flows of water (measured in cubic feet per second) are needed to increase flood depths. Moving down a quadrant, the exceedance frequencies of these quantities of water are shown. The lower right quadrant links the three relationships. Choosing any damage amount $*a* from the horizontal axis of the upper right quadrant shows that damage is caused by *b* feet of water (upper left), which occurs with a flow of *c* cubic feet per second of water flow (lower left). That flow occurs with an annual exceedance frequency of *d*%, and so we have $*a* occurring with a frequency of *d*%. Such derived damage-frequency pairs ($*a*, *d*%) trace out the existing-condition damage-frequency curve. When the area under this curve is integrated, it yields an estimate of the expected annual damages (EADs).

The EADs for the example in Figure 17.27 are $12,411,000. The most common interpretation of this value is if the development in the flood plain and its hydraulics and hydrology remained unchanged for a long time (say 10,000 years) and we added the flood damages in constant dollars for each of these 10,000 years (most of these years would be zeros) and then divide the sum by 10,000, we would have an average annual damage of $12,411,000. This is a common flood-risk metric.

If this risk is judged to be unacceptable and it is to be reduced through a risk management option, the model in Figure 17.27 can be used to estimate the effectiveness of the risk management option (RMO). Most RMOs will alter one or more of the first three relationships in the model. A levee is shown in the upper right quadrant of Figure 17.27, and its effect on the estimate of expected annual damages is shown in the lower right quadrant, which is reproduced in Figure 17.28. Damages from floods up to seven feet in depth are reduced to zero by the levee. For the simplicity of the example we ignore the risk of levee failure.

A levee is one option for managing the flood risk. Using the improved-scenario curve (levee) in place of the existing scenario, a new damage-frequency curve is traced out for the levee, as shown in Figure 17.28. The area beneath this curve yields

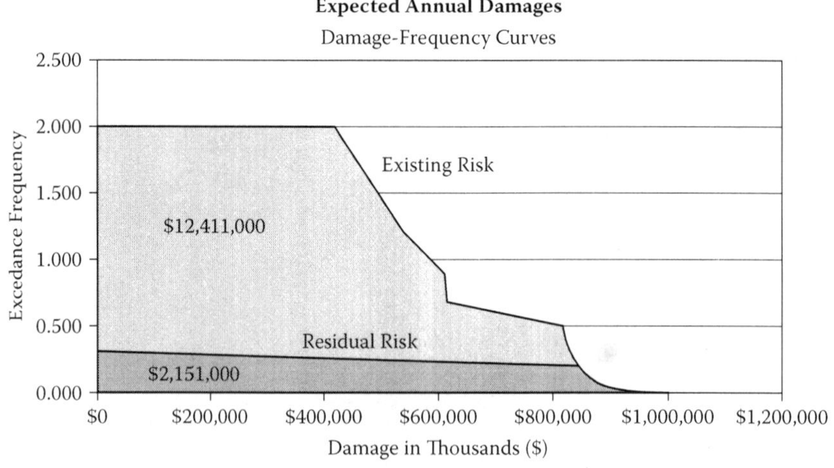

FIGURE 17.28 Damage-frequency curves for existing- and improved-condition scenarios.

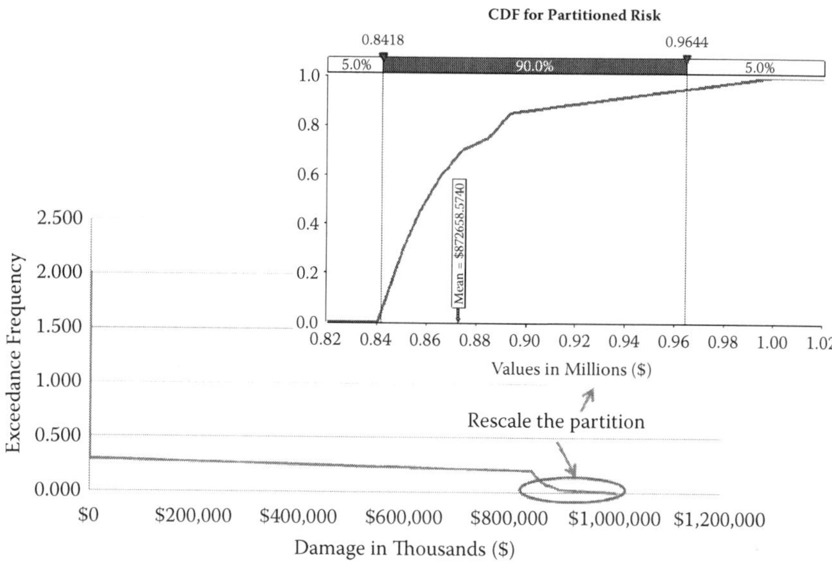

FIGURE 17.29 A risk partition for the hydroeconomic estimation of flood risks.

EAD of $2,151,000. This is a measure of residual damages. A measure of risk reduction is the difference between the EAD estimate for existing risk ($12,411,000) and residual risk ($2,151,000), in this case $10,260,000. This is the standard way of estimating the risk of property damage from flooding and of informing risk managers and the public about this risk.

EAD is now much lower, but what happens to the community when a flood large enough to overtop the levee occurs? Hurricanes Katrina and Rita as well as flooding in the Midwest early in the new century have demonstrated the devastation that can result when levees fail or are overtopped. Risk partitioning is a useful tool for better informing risk managers and the public about extreme risks, and it is an example of a more creative risk metric to better inform decision makers.

Figure 17.29 illustrates the notion. Let us consider that a flood with an annual exceedance frequency of 0.002 or less occurs. Now what is the expected value of damages? This is quite a different metric from the usual EAD. Haimes (1998) provides a rigorous treatment of this concept. In essence, it entails rescaling the probability (vertical axis) partitioned segment of the curve (circled) over the 0 to 1 scale and calculating the expected value of that new distribution. Recall that residual damages are $2,151,000. However, given that an event of equal or greater severity than the 0.002 exceedance frequency flood occurs, then the expected damages are $872,659,000. This provides an entirely different perspective on the residual risk. The likelihood of a flood overtopping the levee is low, but the consequences are devastating. This is not obvious from the more traditional measure of residual damages.

Different RMOs can yield different partitioned risks. For example, a channel or nonstructural flood risk management option might yield a lower traditional estimate of EAD reduced but also a much reduced partitioned risk. Such risk metrics can provide an entirely different risk profile for a flood RMO. The risk assessment community of practice needs to continue to develop and use clever and revealing risk metrics like this in all aspects of risk-informed decision making.

17.8.2 The First Decision

Although there are many different decision contexts and there may in fact be many decisions in any given one of them, we focus narrowly here on a basic recurring question: Is the risk associated with the decision problem acceptable or not? Can we accept this risk? This question is quickly followed by two others. If the existing or perspective risk is not acceptable, what level of risk is tolerable? What is the best way to achieve a tolerable level of risk?

Chapter 3 introduced the notions of acceptable and tolerable risks. The risk manager's first decision is, "Is the existing risk we have assessed one we/the company/our organization/society can accept?" Alternatively, when seeking a gain, the question becomes, "Is the assessed risk of the venture or decision worth assuming, i.e., do the potential gains outweigh the risks of failing to realize them?" We saw in that chapter the following common strategies for answering this and subsequent questions about the tolerable level of risk:

- Policy
- Zero risk
- Weight of evidence
- Precautionary principle
- ALARA (as low as reasonably achievable) principle
- ALOP (appropriate level of protection) principle
- Reasonable relationship
- Safety standards

Any of these may be used to determine the tolerable level of risk.

RISK QUESTIONS

Should we allow the import into the country?
Should we allow the additive into the food?
Should we launch the new product, open the new store, invest in the venture, buy the stock?
Should we administer the flu vaccine?
Should we launch the shuttle today?
Should we reduce the flood damages?
Should we restore the water quality?

17.8.3 Rules for Making Decisions under Uncertainty

No matter which of these strategies the risk manager chooses, there is still the sticky matter of how to handle the uncertainty in the risk assessment outputs. Let's use a new example now, a potential gain. Consider a hypothetical navigation improvement where the policy criterion for decision making is to maximize the net gain to national economic development (NED) of the RMO. Further, suppose the expected value of net NED benefits is shown in Table 17.9.

Using the prescribed criterion, which is the best plan? Clearly, the answer is RMO 1. Under risk-informed decision making, however, decisions will become more multidimensional. We will need to consider more than just the expected value of a decision criterion; we need to consider the risk.

Once the uncertainty in an RMO is considered, decision making gets more complex. Consider the following additional information about the three RMOs in Table 17.9. A five-number summary is added to the means and is offered in Table 17.10 to better characterize the uncertainty about the net benefits for these plans.

The distribution of benefits for these plans are summarized in the frequency distributions and CDFs of Figure 17.30.

The extent of the number line covered varies for each RMO, as does the relative likelihood of any given value being obtained. It is no longer immediately obvious which RMO is best when we begin to consider the relative risk associated with each RMO.

The CDFs present the same data in a different way. There are at least three useful ways to consider the CDF. First, consider the "shadow" of the CDF on the horizontal axis. This identifies the range of values that are possible. RMO 1 has the steepest rise and the shortest shadow. Second, for any given point, e.g., the line demarcating a negative return, the highest lying curve has the greatest probability. Therefore RMO 2 has the highest likelihood of a negative return. If one curve always lies above (or below) the others, this is worth noting. It does not happen here. Third, look at

TABLE 17.9

Net Benefits Accruing to Three Different Risk Management Options

	RMO 1	RMO 2	RMO 3
Expected value of net benefits	$2,519,900	$1,499,306	$374,977

TABLE 17.10

Five Number Summary Plus the Mean for Three RMOs

	RMO 1	RMO 2	RMO 3
Mean	$2,519,900	$1,499,306	$374,977
Minimum	−$3,321,060	−$7,906,015	−$5,023,621
1st quartile	$1,589,976	−$1,949,734	−$1,611,829
Median	$2,573,753	$85,181	$362,086
3rd quartile	$3,501,819	$3,396,973	$2,362,078
Maximum	$7,212,295	$42,288,540	$6,004,597

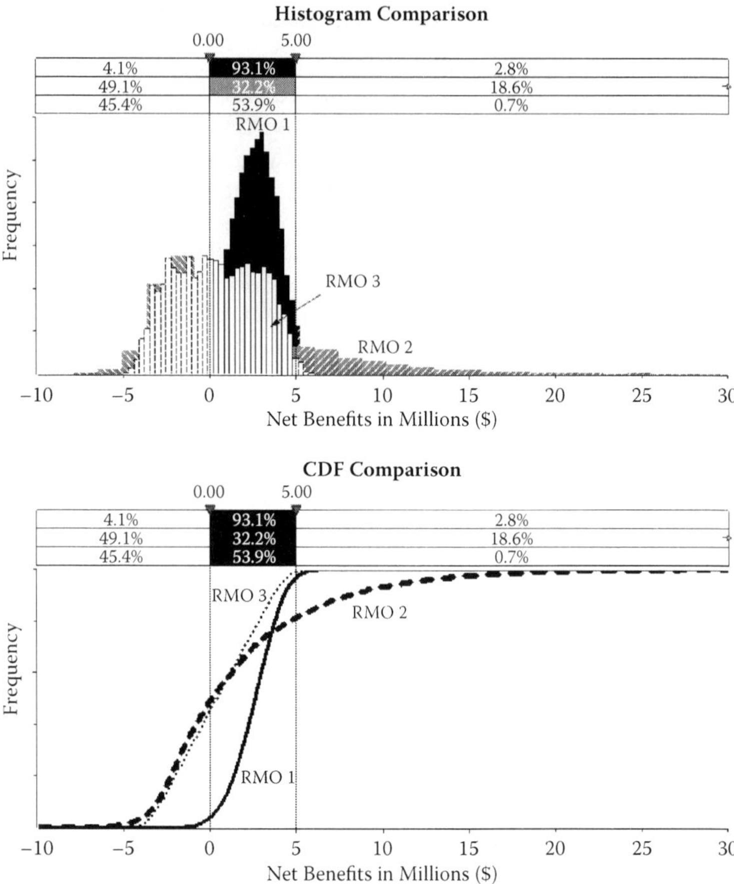

FIGURE 17.30 Frequency distributions and CDFs of net benefits for three navigation RMOs.

which curve lies farthest to the right. If the horizontal axis measures positive/desirable impacts, the better plans lie farther to the right. If the axis measures negative/undesirable results, the better plans lie farther to the left.

Most decision making under uncertainty involves, alternative courses of action, possible events (outcomes, or states of nature), conditional payoffs (results) for the action/event, and unknown probabilities of the events. Figure 17.30 shows that there are no stochastically dominant plans. So how does one make a decision faced with this sort of information? Fortunately, there are a number of rules for making decisions under less-than-perfect certainty that can be used in a situation just like this. They are reviewed in turn in the following subsections.

17.8.3.1 Stochastic Dominance

Figure 17.30 shows that there is no RMO that enjoys first-order stochastic dominance. However, RMO 1 does dominate RMO 3, so there would be no reason to prefer 3

TABLE 17.11

Three States of the World and the Associated Net Benefits for Three RMOs

State of the World	RMO 1	RMO 2	RMO 3
Pessimistic	–$3,321,060	–$7,906,015	–$5,023,621
Most likely	$2,573,753	$85,181	$362,086
Optimistic	$7,212,295	$42,288,540	$6,004,597

when 1 is available. RMO 1 enjoys second-order stochastic dominance over RMO 2. On this basis RMO 1 would be the preferred plan.

Although the following criteria can be applied to the results of a probabilistic risk assessment, they are more traditionally associated with specific states of the world rather than specific results from a simulation model. These states of the world might more usually be estimated from a deterministic scenario analysis. To avoid the proliferation of examples and data, let's imagine that the results of a probabilistic risk assessment have been used to define three different states of the world: pessimistic, most likely, and optimistic. Let Table 17.11 show the relevant values we'll use to demonstrate the remaining criteria.

17.8.3.2 Maximin Criterion

This criterion is often favored by those who are risk averse, i.e., those who have a pessimistic outlook on the future and expect that the worst possible outcome will be realized for each alternative. The alternative that yields the "best" of the worst outcomes, or in this case, maximizes minimum benefits, is chosen.

Under the maximin, or Wald, decision criterion, only the minimum payoffs of each alternative are considered. In our example, RMO 1 would be chosen because its worst possible outcome of –$3.3 million is better than the worst possible outcome of either RMO 2 at –$7.9 million or RMO 3 at –$5.0 million. This approach relies on partial information, i.e., minimums only. If we seek to move decision makers away from overreliance on a single value, this is not the best choice.

17.8.3.3 Maximax Criterion

The maximax criterion is the exact opposite of the maximin criterion. It is based on an optimistic outlook or risk-preferring behavior. This criterion also uses partial information considering only the maximum payoff for each alternative. The alternative that yields the "best" of the best outcomes, or the maximum of all maximum payoffs, is chosen. Under this criterion, RMO 2 is selected because its best possible outcome of $42.3 million is higher than the best possible outcome of either RMO 1 ($7.2 million) or RMO 3 ($6.0 million).

Both of these last two criteria could be easily applied using the results of a probabilistic risk assessment.

17.8.3.4 Laplace Criterion

Whereas the maximin criterion appeals to the cautious and the maximax criterion appeals to gamblers, the Laplace criterion appeals to the risk neutral. In the current example, the Laplace criterion is based on expected values. Assuming the values of Table 17.11 came from a deterministic scenario analysis, the expected value is the weighted sum of the three states of the world. When we lack a more objective set of probability estimates to attach to the different states of the world that could be realized, the Laplace criterion assigns equal probabilities to all states and their outcomes.

Using the revised example data, RMO 1's expected value would become: 1/3 × –\$3.3 million + 1/3 × \$2.6 million + 1/3 × \$7.2 million = \$2.2 million. Likewise, RMOs 2 and 3 have expected values equal to \$11.5 million and \$0.4 million, respectively. RMO 2 is the best plan based on deterministic estimates of states of the world. Had we used a probabilistic risk assessment, the expected value can be read directly from Table 17.10, where RMO 1 has the largest expected value.

17.8.3.5 Hurwicz Criterion

Leonid Hurwicz is said to have suggested a criterion that is a compromise between the maximin and maximax criteria. He used a coefficient of optimism (α) as a measure of the decision maker's optimism. The coefficient ranges from 0 to 1. An $\alpha = 0$ indicates total pessimism (equivalent to the maximin criterion), and $\alpha = 1$ indicates total optimism (equivalent to the maximax criterion). The coefficient of pessimism is thus defined as $1 - \alpha$.

Hurwicz defined the weighted payoff as:

$$\text{Weighted payoff} = \alpha(\text{maximum payoff}) + (1 - \alpha)(\text{minimum payoff})$$

All that remains is to choose a coefficient of optimism and simultaneously a coefficient of pessimism. For this example, let $\alpha = .4$; the weighted payoffs (WP) of the three alternatives in millions are:

$$\text{WP (RMO 1)} = 0.4(\$7.2) + 0.6(-\$3.3) = \$0.9$$

$$\text{WP (RMO 2)} = 0.4(\$42.3) + 0.6(-\$7.9) = \$12.2$$

$$\text{WP (RMO3)} = 0.4(\$6.0) + 0.6(-\$0.4) = -\$0.6$$

RMO 2 is the preferred plan. If the decision maker is unable to determine his or her α, it is possible to determine some critical alpha values (for instance, where weighted payoffs of two alternatives are equal) and ask the decision maker if his or her value is greater or less than these values. Options 1 and 2 have equal outcomes, with a coefficient of optimism of about .1156. If the decision maker thinks the optimistic scenario has more than, say, a 12% chance, then RMO 2 is the preferred option.

17.8.3.6 Regret Criterion

The regret or minimax criterion, is based on the economic concept of opportunity cost. For a given scenario or state of nature, different alternatives may yield different payoffs. The opportunity cost of an alternative for a particular state of nature is the

TABLE 17.12

Identifying Inputs for a Regret Matrix

	States of the World	
Alternatives	**Minimum**	**Maximum**
RMO 1	–$3.3	$7.2
RMO 2	–$7.9	$42.3
RMO 3	–$5.0	$6.0

difference between its payoff and the payoff of the highest-yielding alternative for that state of nature. An example will illustrate this idea. We begin, in Table 17.12, by identifying the minimum and maximum payoff for each alternative, where all values are in millions of dollars.

The next step is to construct the regret matrix of Table 17.13. Look at the minimum payoffs column in Table 17.12. If you choose RMO 1 and a pessimistic state of the world occurs, what is the greatest regret possible? There is none because RMO 1 yields the best possible outcome in a pessimistic outcome. If you chose RMO 2 and pessimistic world results, we would have preferred RMO 1 because with 2 we lose $4.6 million more dollars. This is our regret. So if RMO 3 is chosen, the regret is the difference between the loss with RMO 1 and the loss with RMO 3, or $1.7 million. If RMO 2 is chosen in a minimum world, there is no regret.

If a maximum state of the world is realized, a choice of RMO 1 results in a regret of $35.1 million ($42.3 – $7.2), and so on. In the last column of Table 17.13 we identify the maximum regret possible if we select that RMO. It is the largest number in each row. Once the maximum regret column has been generated, the solution in an uncertain world is simple. Choose the option that minimizes the maximum regret. In this case it is RMO 2. This is the maximum possible opportunity cost of choosing an RMO. Now choose so that you minimize this opportunity cost and you will select RMO 2.

The truth of the matter is that the risk analysis community of practice is still sorting all of this out and figuring out the best ways to use all the information our risk assessments are producing. In the not-too-distant past, risk assessment was the tail wagging the risk analysis dog. An assessment would be done, and someone would have to figure out what to do with it. Risk management's growing ownership of the risk analysis process is relatively recent, and this chapter represents a modest start at trying to figure out how to actually present and then use the information probabilistic risk assessments are capable of producing.

TABLE 17.13

Regret Matrix and the Minimax

	Regret Matrix (millions $)		
	Minimum	**Maximum**	**Maximum Regret**
RMO 1	$0	$35.1	$35.1
RMO 2	$4.6	$0	$4.6
RMO 3	$1.7	$36.3	$36.3
		Minimax	$4.6

17.9 Summary and Look Forward

Quantitative risk assessments, especially probabilistic ones, produce a great deal of data. The starting point for explaining these results to risk managers is understanding the results. Risk assessors must carefully avoid overreliance on any one numerical result. The best way to understand and then explain one's data is to examine the quantities, their probabilities, and the relationships between and among key variables, and then answer the risk manager's question as clearly, concisely, and completely as possible while accounting for the remaining uncertainty.

There are a great many numerical and graphical tools to aid the presentation and understanding of data to risk managers. Be sure to look for ways to convey the nonexistence of "the number." Refuse to give it! The five-number summary and stacked histograms and CDFs, as shown in Figure 17.30, are useful places to begin your search for effective displays. Do not stop there, though. Technology is making a great many new opportunities for displaying the data available to assessors and managers.

We need smart risk assessors to help define useful and new risk metrics. Meanwhile, we look for more practical ways to help decision makers understand how best to use all the data risk assessment makes available in the service of better decisions.

Now the task gets even tougher. How do we explain the information we get from complex assessments to the public? How do we explain all of these numbers? The answer is, maybe you don't need to. Our final chapter focuses on message strategies for risk communication and some dos and don'ts for communicating about risks and in a crisis.

REFERENCES

Catin, M., and D. Kautter. 2007. An overview of the CARVER plus shock method for food sector vulnerability assessments. http://www.fsis.usda.gov/PDF/Carver.pdf.

Friedman, Vitaly. 2008. Data visualization and infographics. *Smashing Magazine*, January 14. http://www.smashingmagazine.com/2008/01/14/monday-inspiration-data-visualization-and-infographics/.

Friendly, Michael. 2009. Milestones in the history of thematic cartography, statistical graphics, and data visualization. http://www.math.yorku.ca/SCS/Gallery/milestone/milestone.pdf.

Haimes, Yacov Y. 1998. *Risk modeling, assessment, and management*. New York: John Wiley & Sons.

Harris, Robert L. 1999. *Information graphics: A comprehensive illustrated reference*. New York: Oxford University Press.

Leach, Patrick. 2006. *Why can't you just give me the number? An executive's guide to using probabilistic thinking to manage risk and to make better decisions*. Gainesville, FL: Probabilistic Publishing.

Savage, Sam. 2009. *The flaw of averages: Why we underestimate risk in the face of uncertainty*. Hoboken, NJ: John Wiley & Sons.

Tufte, Edward R. 1983. *The visual display of quantitative information*. Cheshire, CN: Graphics Press.

———. 1990. *Envisioning information*. Cheshire, CN: Graphics Press.

———. 1997. *Visual explanations: Images and quantities, evidence and narrative*. Cheshire, CN: Graphics Press.

Tversky, A., and D. Kahneman. 1974. Judgment under uncertainty: Heuristics and biases. *Science* 185 (4157): 1124–1131.

U.S. Environmental Protection Agency. 1997. Risk Assessment Forum. *Guiding principles for Monte Carlo analysis*. Washington, DC: U.S. Environmental Protection Agency.

U.S. Food and Drug Administration. 2009a. Strategic partnership program agroterrorism initiative—Q & A. http://www.fda.gov/Food/FoodDefense/FoodDefensePrograms/ucm080897.htm.

U.S. Food and Drug Administration. 2009b. CARVER software. http://www.fda.gov/Food/FoodDefense/CARVER/default.htm.

Wainer, Howard. 2009. *Picturing the uncertain world: How to understand, communicate, and control uncertainty through graphical display*. Princeton, NJ: Princeton University Press.

World Health Organization. 2006. International Programme on Chemical Safety. *Draft guidance document on characterizing and communicating uncertainty of exposure assessment, draft for public review*. Geneva, Switzerland: World Health Organization. http://www.who.int/ipcs/methods/harmonization/areas/draftundertainty.pdf.

18

Message Development

18.1 Introduction

To be successful at risk communication, you need to know how to prepare an effective message. Communication, in a risk analysis context, is not an incidental task. Nor is success at it accidental. Everyone involved in risk analysis needs to know something about risk communication if for no other reason than to keep others from shooting themselves in the foot. Not every organization is going to be able to have risk communication experts on staff. They may not even have much awareness of the importance of the task. That means you may have to bring some of this knowledge to the table.

Risk communication presents unique challenges to communicators for two very specific reasons. One of these is that the content of the messages is often difficult material. Risk, uncertainty, probability, and science are not the easiest things for even eager and interested parties to understand. These are often the subjects of risk communication messages. The other reason risk communication is especially challenging is that the audience is often upset and frightened, experiencing strong emotional noise during the communication process.

If risk communication is to succeed under these circumstances, someone needs to know how to develop an effective message. Developing that message is the focus of this chapter. After a brief review of a basic communication model, we reconsider how the risk communication model differs especially during high-stress situations. From there we turn to message strategies during the four stages of a crisis and to message mapping, a useful technique for developing effective messages.

18.2 Communication Models

After all of humanity's time on the planet, communication remains our greatest challenge. In the arena of risk analysis, information is often mistaken for communication. George Bernard Shaw is quoted as having said, "The single biggest problem in communication is the illusion that it has taken place." Understanding a bit about the basic communication model is a good starting point for understanding how risk communication differs from basic communication.

David Berlo (1960) offered a simple and influential model that suits the current purpose well. Described as the source-message-channel-receiver (SMCR) model, it has four major components as shown in Figure 18.1. The source is a person, organization, or any generator of messages. Berlo offered the message as the central element of his model, which stresses the transmission of ideas. Meaning is encoded into messages.

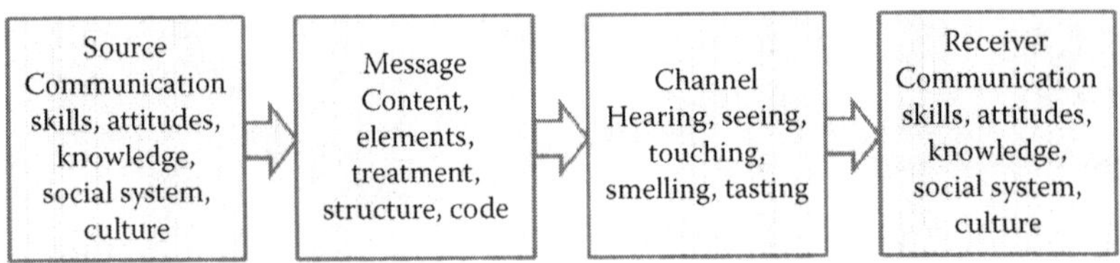

FIGURE 18.1 Berlo's SMCR communication model.

Messages are transmitted through a channel and are then decoded by a receiver. That what is received is not always identical to what was sent has often been the basis for situation comedies and undoubtedly a good portion of the divorce rate. There is a very rich literature describing a great variety of communication models, most far more complex than this one.

Here is what is important for us to take from this basic model. Its underlying premise is that making information available is communicating. In risk communication, this model is tested at several points. If the source has not used a good risk management process, the purpose for communication may not even be clear. The sender's perceived trust and credibility will be critical. Multiple channels must be used to reach the multiplicity of audiences for risk information. Perhaps most significant, the receiver may be highly stressed. People in these situations can lose up to 80 percent of their ability to effectively process information. The message, which may have already been challenging because of its content, needs to be simplified. It is harder for people to hear, understand, and remember in stressful situations.

The developing field of risk communication offers a model that addresses these and other concerns. Two of the most important differences in the risk communication model are that it is a multidirectional model and it actively involves the audience as an information source.

Risk communication is often more than two-way communication because of the complexity of the issues and the number of interested parties. Done well, it is an audience-centered form of communication because it so often is addressing situations of high stress, great concern, or low trust. These situations call for as much attention to the methods of communicating as to the messages. However, it is most likely that risk assessors and risk managers will have substantially more input to the development of the messages than to the more subtle nuances of risk communication strategy and methods. Hence, our focus is on message development.

18.3 The Need for Message Strategies for Risk Communication

My own direct risk communication experience tends to be more internal, and my external communication experiences have been more of the low-stress situations. I am frustrated at times that the risk communication community of practice has not yet devoted more attention to developing strategies in these areas because there is a great need for help there. In fairness, it is possible my frustration stems from my inability to readily discover and uncover the most effective strategies that have been developed. With that caveat, we begin.

Communications between risk managers and assessors would be enhanced by some simple message strategies for clearly communicating problems and opportunities, objectives and constraints, as well as the questions to be answered by risk assessment. There has been great progress along these fronts, but there has been little systematic research on the message strategies that work best for these specific purposes. While many would reasonably suggest that this kind of communication is well documented in the management and communication arts literature, my own experience suggests that the matter of risk is different enough to warrant a closer look.

Another major risk communication stumbling block has been the lack of strategic attention given to how to communicate the findings of risk assessments most effectively. The preceding chapter attests to the limitations in this area. The general practice seems to be to do the work and document it all in a risk assessment; add an interpretative or executive summary for those whom we know will not read it all, which is virtually everyone, and then make it available to anyone who wants it.

There is a pressing need for message development strategies for reporting the results and relevance of risk assessment for several key and recurring audiences that include risk managers, industry, those responsible for implementing the risk management option, consumers, the media, special-interest groups, and the general public, especially those most affected by the risk. How do we explain complex scientific subjects that are highly uncertain or results that are based on probabilistic methods to these audiences in ways that are meaningful to them?

We also need message strategies for risk management decisions. How do we best convey how trade-offs were made, how decisions were arrived at, as well as what those decisions are and who has what specific responsibilities for managing the risks?

Flood risk management provides an excellent opportunity for these kinds of strategies. Although many flood risk management investigations are initiated in response to flood events, they often take a long time to complete, and the public's attention has long since been diverted elsewhere. These extended planning periods are often closer to the low-stress, low-interest circumstances than to the high-stress, high-interest conditions that follow a damaging flood. In addition, the topics are complex and rest on probabilistic concepts we now know are not intuitive.

In the past, the U.S. Army Corps of Engineers, which has primary authority for addressing flood problems in the United States, has done their planning investigations and then sold them based on the "protection" they were expected to provide. A typical situation would be to have had a damaging flood of record and to produce a risk management plan that would provide protection against a recurrence of such a flood.

A simple schematic, shown in Figure 18.2, illustrates the situation. A person living in the so-called 10-year floodplain for a 75-year lifetime is virtually certain to experience one or more floods in that time.* The darker color indicates the probability of one or more floods in a lifetime. A flood with a 1% annual exceedance frequency has, in the past, been called a 100-year event, because in the very long run there would be such a flood on average once in 100 years. (Perhaps you begin to see why message strategies for noncrisis/low-stress situations are needed!) If protective works, say a levee, are constructed to prevent flood damages from the recurrence of such a flood, that reduces our lifetime resident's risk by the amount of the lighter color in the

* To get that value, calculate one minus the binomial probability of zero successes with n = 75 and p = .1.

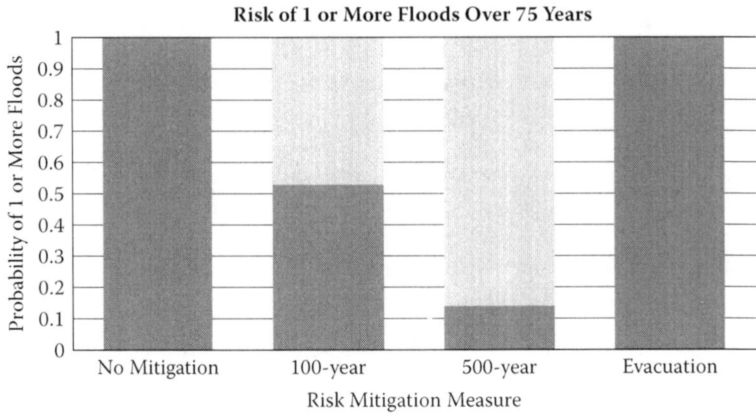

FIGURE 18.2 Residual flood risk for selected risk management options.

second column. That reduction in risk is what the people living in the floodplain were "sold" in the past. The positive effects of the protective works were the focus of the conversation. What was not adequately communicated was the residual risk, the risk that remains even with that levee in place. There is still a 53% chance a person will be flooded one or more times in a lifetime behind this levee.

If the flood of record had an annual exceedance frequency of 0.2%, it has a recurrence interval of 500 years (1/.002) and has been called the 500-year flood. Protection from a flow of this magnitude could require a much higher levee. The risk reduction provided by that levee is shown as the light color of column three. Even though living behind a very high levee, there is still a 14% chance of one or more damaging floods in a lifetime. Permanent evacuation of the floodplain is the only effective way to eliminate the risk of flooding to its residents. This simple discussion does not even consider how a flood risk is transformed when catastrophic failure of protective works, as occurred in New Orleans, is now a possibility. So, perhaps, you see the complexity of the message?

Let's return to the residual risk of columns two and three for a moment. What can be done to manage the residual risk, and whose responsibility is that? These are important risk messages to craft and communicate. One of the principle reasons the so-called 100-year flood has been targeted as a minimum level of flood protection is that if the 100-year floodplain is confined, say to between the levees along a river, then residents no longer live in the 100-year floodplain and are no longer required to buy flood insurance. The figure shows the foolhardy nature of this strategy. The chance of a house fire over 75 years is far less than 53% and everyone with a mortgage is required to have fire insurance, yet flood insurance remains an option.

Residual risks can be addressed in a variety of ways. Homeowners could be required to purchase flood insurance, agree to evacuate when an order is given, and provide a trapdoor escape hatch on the roof of their homes. A great many people who drowned in New Orleans had moved to the highest floor or attic of their homes only to become trapped and drown in rising water.

Not only do we need to formulate management strategies, but we need to formulate message strategies for a great many noncrisis risk communications. Too often there is no effective strategy for communicating this complex information to laypeople in a risk management activity. Strategies for communicating about residual, new, transformed, and transferred risks could be used in a great many situations.

18.4 Crisis Communication

Crisis communication strategies, by contrast, have received a great deal more attention. The National Center for Food Protection & Defense's excellent free teaching resources form the foundation for this section, and the work of the NCFPD is gratefully acknowledged (Food Insight, 2010). Their communication model consists of a sender, receiver, channel, message, feedback, noise, and an environment, i.e., a time and place for the communication. For normal situations, where there is little to no stress, audiences tend to base their trust in the communicator on his or her level of competence and expertise. During a crisis or in other high-stress circumstances, trust factors change, as discussed in Chapter 5. Listening, caring, and empathy become far more important than competence and expertise.

Stress interferes with the function of the basic communication model. The choice of the best sender of a message may change from a low-stress to a high-stress situation. The communicator's effectiveness now depends more on credibility and trust. The audiences (receiver) have a reduced capacity to process complex information, so the messages must be simplified to be effective. Feedback to gauge the public's response becomes more important than ever. Noise, i.e., barriers that interfere with the receiver's ability to process information, can increase during a crisis. Communications systems may be compromised, and even when they are not, stress levels and emotional responses to the situation can present significant barriers to the normal communication model.

This mental noise disrupts the listener's ability to process information to such an extent (see text box) that overcoming the noise must become an explicit counterstrategy for risk communication. This means we need to simplify the message: Offer no more than three message points and use short sentences. Numbers should be used sparingly and carefully. Ideas are often more effectively presented through pictures or graphics during high-stress situations.

Reynolds and Seeger (2005) have identified four stages of crisis communication, shown in Figure 18.3. The pre-event planning for a crisis response takes place in the preparedness stage. Credibility and trust cannot be readily established in the midst of a crisis. They are slow to build and so must be strategically cultivated as part of an organization's crisis preparedness. Educating and informing the public about more complicated information is also better initiated during low-stress times. The preparedness stage is the time to develop message maps, a subject touched on later in this chapter. Messages that may be needed in a crisis can be tested with stakeholders and the public during this stage as well. A crisis is not the time, for example, to begin looking for a translator to reach your multicultural audiences, nor is it the time to start wondering what to say or to whom.

The initial response to a crisis spans the first two days of the event. During this time there may be a lot of unknowns. Uncertainty or "mystery" intensifies the fear. Effective risk communication must begin immediately. You cannot wait until you have all the answers to communicate.

DIFFERENCES BETWEEN COMMUNICATING DURING LOW- AND HIGH-STRESS SITUATIONS

Low: Receiver processes an average of 7 bits of information (e.g., telephone number)
High: Reduced to 3 information bits
Low: Information processed in a linear order (1,2,3)
High: Information processed in order of primacy (1,3,2) or recency (3,2,1)
Low: Information is processed at an average grade level (local newspapers written at eighth-grade reading level)
High: Information is processed at four grade levels lower (e.g., from eighth grade to fourth grade)
Low: Focus is on competence, expertise, knowledge (title, credentials)
High: Focus is on caring, empathy, compassion (metamessages)

***Source*: NCFPD Training Module 3 Message Delivery and Development. (Food Insight, 2010)**

The first 48 hours is a time for responding, not for planning. This need to respond quickly makes the preparedness strategies so very important. It is that preparedness that enables you to respond.

Reynolds and Seeger (2005) offer several principles for the initial response:

- Be first, be right, be credible
- Acknowledge the event with empathy
- Explain the risk and inform the public; tell them what you know, what you don't know, and what you're doing about it
- Commit to continued communication
- Keep communication channels open

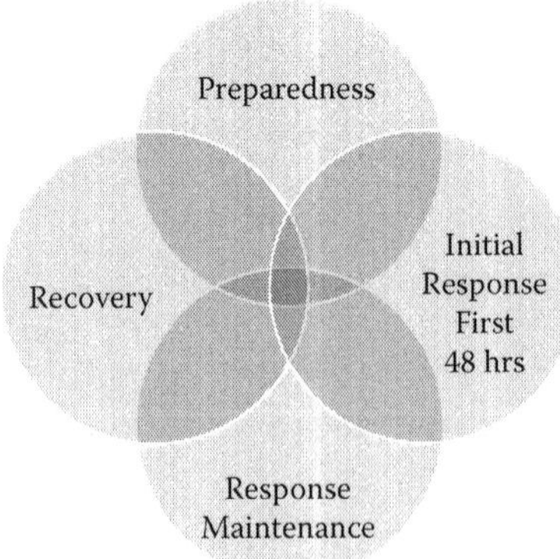

FIGURE 18.3 Reynolds and Seeger's four stages of a crisis.

COMMON MISTAKES MADE IN A CRISIS

Over-reassure
Sound too certain, too confident
Wait too long (Be first...)
Fail to communicate the complexity of decision making or to acknowledge opinion diversity
Try to appear objective by excluding emotions from our messages and metamessages
Treat the public as though they are children
Downplay the mistakes we have made

***Source*: Food Insight, 2010.**

The response maintenance phase of the crisis is a time to continue to build on the response to the crisis. It is important to help the public understand the risks they may face and to provide useful information in a way that empowers people to make their own risk/benefit trade-off decisions. These can be simple messages like: "Stay indoors until the spill has been cleaned up"; "If you are worried, don't eat spinach"; "To be safe, evacuate the area"; and so on.

Subject-matter experts may help at this point by providing background information. The risk communication team needs to listen to their various audiences and stakeholders for feedback and to correct misinformation. This is a time to improve understanding of the problem and to build support for the response and recovery plans.

After the event has passed, the recovery stage begins. It is often accompanied by feelings of relief, celebration, and gratitude for getting through the event. Not everyone will share in these feelings, however, and it is important to continue to express empathy and caring for those who have suffered through the event. It is time to listen to the public and the stakeholders to hear how they perceived your effectiveness during the event. It is important to acknowledge any shortcomings and to explain how they will be rectified in the future. Actions must match the messages, or trust and confidence may be irreparably harmed.

18.5 Message Mapping

Many crisis situations can be anticipated by a risk management agency. Floods and other natural disasters, traffic accidents, food-borne disease outbreaks, terrorist attacks, train wrecks, airline crashes, product recalls, bank failures, toxic spills, and the like are expected events for a variety of organizations. Planning the communications response is often as important as planning the first responders' activities.

Based on how people process information during times of stress, Dr. Vincent Covello (2002) of the Center for Risk Communication has developed the idea of message mapping. Message maps are a good first step for developing risk communication messages during the preparedness stage of a crisis communication. They enable staff to develop effective messages that organize and simplify complex content in a systematic way during periods of low stress.

Covello describes this tool as "a roadmap for displaying detailed, hierarchically organized responses to anticipated questions or concerns. It is a visual aid that provides at a glance the organization's messages for high concern or controversial issues."

A message map provides a framework for organizing information and developing clear and concise messages that anticipate stakeholder questions and concerns. Done in advance of a crisis in a thoughtful way, they offer an organization the opportunity to promote a consistent message and voice In the midst of a crisis, message maps offer a systematic approach for preparing effective risk communication messages. The message map is not the message, however. It is a device that aids in the development of risk communication messages.

Remember, in high-stress situations, people can process three bits of information effectively. So message maps are based on the "rule of threes." In high-stress situations it is best to:

- Present three key messages
- Repeat key messages three times
- Prepare three supporting messages for each key message

This rule is supplemented by Covello's 27/9/3 Rule, which says to use a total of 27 words, delivered in no more than 9 seconds, to deliver three key messages.

These message maps can be used to prepare risk communication messages for emerging or anticipated crises. They are also useful for preparing for media interviews, press conferences, and other public forums. They are a good first step in creating a risk communication message or statement.

The steps for developing message maps are simple. Begin by making a list of all the questions you may be asked to address. Then prepare a message map for each question. In answer to each question you should identify the following:

- The three most important things you would like your audience to know
- The three most important things your audience would like to know
- The three most important things your audience is likely to get wrong unless they are emphasized.

A message map template can be used to organize this information. The key message should be able to stand on its own. The support points provide factual information, visual aids, citations to credible third-party information, and sources of more information.

The NCFPD provides a sample message map, shown in Figure 18.4, that addresses the question: "What should I know about anthrax?" Note that the three key points are expressed in a total of 22 words. There are also three supporting ideas for each key message. The message map content is constructed based on the answers to the bulleted items above.

1. Anthrax is a disease that can affect people and animals.	2. Anthrax occurs naturally in the soil.	3. Anthrax is both preventable and treatable.
1a Anthrax is caused by bacteria that form spores.	2a Anthrax occurs worldwide.	3a Effective vaccines are available for livestock.
1b The spores can be inhaled, swallowed or enter the skin (contact).	2b Spores are resistant to manmade disinfectants.	3b Vaccines for humans are developed and can be used prior to or after exposure.
1c Animals most commonly affected are cattle, sheep, and goats.	2c Anthrax spores can survive for many years in soil without an animal host.	3c Early treatment antibiotics can be effective.

FIGURE 18.4 Sample message map for anthrax scare. (*Source*: Food Insight 2010.)

18.6 Impediments to Risk Communication

There are many issues that affect our ability to communicate about risk. Covello, McCallum, and Pavlova (1989) offer a dozen factors that complicate risk communication. They are well worth understanding for constructing effective messages. First among these is that risk is an intangible concept that the general public does not understand. As a result, their risk-taking or risk-managing decisions can be based on bad information. Second, the public likes simple solutions. They are more likely to take effective action when the action is simple, e.g., "Don't eat spinach." They are less likely to take effective action if the solution is more complex, e.g., "Move out of the floodplain."

Third, the public does not like uncertainty or probability. They much prefer concrete facts upon which to base their decisions. People are prone to drawing inappropriate conclusions from uncertain and probabilistic information. When the media draw the wrong conclusions, the consequences of that faulty information can be magnified. Fourth, the public can react unfavorably to fear.

A fifth confounding factor is people's aversion to a loss of control over their personal well-being. Risks perceived as beyond the control of the individual present a special challenge for message development. When the U.S. Army Corps of Engineers described two dams upstream of Houston in need of repair as "extremely high risk" in February 2010, citizens were concerned despite the fact the Corps said not to be alarmed.* Sixth, the public sometimes doubts scientific predictions. Y2k disasters and flu epidemics that did not materialize have taught people that the scientists and experts can be wrong. A seventh factor is that the risk at issue may simply not be one of the public's priorities. Feelings of invincibility among the young, for example, may render some risk messages ineffective because they cannot imagine their susceptibility to the risk. Related to this is an eighth confounding factor: The public tends to underestimate their personal risk. Bad things happening to other people is not just a defense mechanism of youth.

Ninth, the public hold contradictory beliefs. ("I am not going to get cancer"/ "Everything causes cancer!") Tenth, a majority of Americans lack a strong future orientation. Some version of "live for today and let tomorrow take care of itself" is a common

* "Engineers to Update on Addicks, Barker Dams." http://www.ultimatekaty.com/stories/4406-engineers-to-update-on-addicks-barker-dams.

READABILITY SCORE

For readable messages, experts recommend writing for a seventh- to eighth-grade reading level, i.e. scores around 7 or 8. During high stress situations, scores around 4 are better. Several indices are available for evaluating your written messages.

Flesch-Kincaid Grade Level

$$FKRS = (0.39 \times ASL) + (11.8 \times ASW) - 15.59$$

where FKRS is the Flesch-Kincaid readability score, ASL is the average number of words in a sentence, and ASW is the average number of syllables per word.

Gunning's Fog Index Grade Level

$$GFI = 0.4\ (ASL + PHW)$$

where PHW = percentage of hard words. Hard words are words of three or more syllables that are not (1) proper nouns, (2) combinations of easy words or hyphenated words, or (3) two-syllable verbs made into three with -es and -ed endings.

perspective on the future. The future is less relevant to those in the lower socioeconomic strata of society. It is not easy to get people to change their diet today in exchange for better health in the future.

The last two factors include the public's tendency to personalize new information. When risks are described at societal or aggregated levels above the individual, people have to translate it into terms that are personal. This invites opportunities for misinterpretation. Finally, the public does not understand science. The models, methodologies, and descriptions of risk can be too technical for many people. The facts that science evolves and changes and that risk assessment is iterated just further confuse people. Many people are poorly equipped to understand scientific messages. Armed with an awareness of these factors that complicate risk communication, careful message development becomes an indispensable skill for a risk analysis organization. It is important to write messages that readers can understand (see text box).

18.7 Developing Risk Communication Messages

Developing a risk communication message requires the communicator to integrate all he knows about risk communication. It begins with the goals and desired outcomes of the risk communication. A good message recognizes risk perception as a combination of facts and feelings as well as common human reactions to risk, such as fear, denial,

MESSAGE DEVELOPMENT TEMPLATE

SCENARIO:
COMMUNICATOR ROLE:
COMMUNICATION GOAL:
PREPAREDNESS STRATEGIES:

Key Audience(s)	Medium/ Delivery Mode	Key Questions and/or Messages	Metamessage Strategies
Message Text			

FIGURE 18.5 National Center for Food Protection and Defense message development template. (*Source*: Food Insight 2010.)

and panic. This necessitates careful attention to techniques for communicating in high-stress situations.

The NCFPD team has developed a risk communication message template, shown in Figure 18.5. The scenario for using this template can be found in any of the four phases of a crisis. The strategy for the communication might include such things as relationship building, partnership responsibilities, trust building, changing behavior, coming to agreement, and so on.

The communicator's role and communication goal identify the who and the why of message development. The who may include such roles as subject-matter expert, industry representative, government official, emergency responder, agency spokesperson, university scientist, consultant, and so on. The purposes of the communication are varied and might include increasing awareness of a potential risk, providing background or technical information, executing a response plan, or supporting recovery.

Key audiences need to be identified. Communicators must be aware of the education level, general awareness of the risk or event, as well as relevant demographic and cultural characteristics of the population.

The medium and delivery mode depend in part on the desired impact and influence of your communication. Channels have varying extents of persuasive influence and they reach audiences of varying size (impact) and type. Channel choice often goes back to knowing your audience. Where are they? How do they get their information? How do you reach them? What do they listen to? What do they read? What radio/television stations do they tune into? What other communications do they use? The most commonly used channels include interpersonal communications, television, radio, print, Internet sites, blogs, e-mail, mobile phone, text messaging, tweeting, chat rooms, and the like. Each has its pros and cons. Multiple channel strategies are often most effective.

The key message grows from the message mapping activity. Metamessages are the nonverbal messages you want to send. The substantive information of your message is only part of the message content. All the other things and ways your message communicates comprise the metamessage. Metamessaging should be an intentional strategy that, among other things, reveals your empathy and caring. Whether you decide to come across as confident or tentative, open or not, how human and how official, your degree of candor, openness, and such: These are all strategies you should consider beforehand. Metamessage strategies do not preclude sincerity. Deciding to be sincere does not make one insincere. Deciding to sound sincere while not being sincere does, however.

Best risk communication practice uses as many opportunities as possible to reinforce the metamessage strategy. If you have decided to be more official, then staging may include flags and podiums with official symbols. Dress will be formal and the attendees will include officials at all relevant levels. A more informal and human metamessage strategy will lead you to choose other nonverbal metamessages.

In a risk communication message text you provide information, give people meaningful things to do, and reinforce your metamessage. The information content includes what you know and what you don't know. It should also include what you are trying to do about the situation. The information should also include when you'll provide the next update. The text may come from messages you have mapped.

Your risk message should empower people to make decisions and take actions to help themselves. This is the self-efficacy component of the message. The intent is to help restore a sense of control over an uncertain and threatening situation. You empower audiences by giving people the information they need to make informed decisions and by giving them meaningful things to do.

A good risk message includes three kinds of self-efficacy statements. It tells people the things they must do now. It tells them what they should do to reduce the risk or their exposure to it. And it tells them what they could do to help themselves further. An alternative emergency message development worksheet developed by the Centers for Disease Control is shown in Figure 18.6.

18.8 Summary

Risk analysis is an effective paradigm for decision making under uncertainty. Communicating about risks and communicating uncertainty are especially challenging tasks. This is in part because the subject matter is difficult and in part because the audience for this communication is often in an emotional state. Risk communication entails a great deal more than providing people with information. Best-practice risk communication necessitates that key messages be developed carefully and, as often as possible, well before they are needed.

The external risk communication task has received the lion's share of the scholarly research. There is still a great and growing need for improving the internal communications about risks between assessors and managers and for explaining the results of risk assessments to the general public in low-stress situations where neither their emotions nor their interests are piqued.

Message Development Worksheet for Emergency Communication

First, consider the following:

Audience:	Purpose of Message:	Method of delivery:
❑ Relationship to event ❑ Demographics (age, language, education, culture) ❑ Level of outrage (based on risk principles)	❑ Give facts/update ❑ Rally to action ❑ Clarify event status ❑ Address rumors ❑ Satisfy media requests	❑ Print media release ❑ Web release ❑ Through spokesperson (TV or in-person appearance) ❑ Radio ❑ Other (e.g., recorded phone message)

Six Basic Emergency Message Components:

1. Expression of empathy:

____

____

2. Clarifying facts/Call for Action:

Who____

What____

Where____

When____

Why____

How____

Add information on what residents should do or not do at this time ______________________

____

3. What we don't know:

____

____

4. Process to get answers:

____

____

5. Statement of commitment:

____

____

6. Referrals:
For more information____
Next scheduled update____

Finally, check your message for the following:

Positive action steps Honest/open tone Applied risk communication principles Test for clarity Use simple words, short sentences	Avoid jargon Avoid judgmental phrases Avoid humor Avoid extreme speculation

FIGURE 18.6 CDC emergency message development worksheet. (*Source*: www.health.mo.gov/living/lpha/toolkit/chap7/01.doc accessed 7/2/11.)

Message development strategies are a critical part of effective risk communication. Message mapping is a useful component of these strategies. Know the impediments to successful risk communication and develop messages to avoid them. So, how do you explain the complex distributions, sensitivity analyses, and other details we have discussed throughout this book? The answer may well be that you do not have to; just remember the "rule of threes."

REFERENCES

Berlo, David K. 1960. *The process of communication.* New York: Holt, Rinehart, and Winston.

Covello, Vincent. 2002. Message mapping, risk and crisis communication. Invited paper presented at the World Health Organization Conference on Bio-terrorism and Risk Communication, Geneva, Switzerland. http://www.orau.gov/cdcynergy/erc/Content/activeinformation/resources/Covello_message_mapping.pdf

Covello, Vincent T., David B. McCallum, and Maria Pavlova. 1989. Principles and guidelines for improving risk communication. In *Effective risk communication: The role and responsibility of government and non-government organizations*, ed. Vincent T. Covello, David B. McCallum, and Maria Pavlova, 3–19. New York: Plenum Press.

Food Insight. 2010. Risk communicator training for food defense preparedness, response and recovery: Trainer's overview. http://www.foodinsight.org/Resources/Detail.aspx?topic=Risk_Communicator_Training_for_Food_Defense_Preparedness_Response_Recovery.

Leitch, Matthew. 2010. ISO 31000: 2009 – The new international standard on risk management. *Risk Analysis* 30(6) 887–892.

Purdy, Grant. 2010. ISO 31000: 2009 – The new international standard on risk management. *Risk Analysis* 30(6) 881–886.

Reynolds, B., and M. Seeger. 2005. Crisis and emergency risk communication as an integrative model. *Journal of Health Communication* 10 (1): 43–55.

Appendix A: The Language of Risk and ISO 31000

A.1 Introduction

Let me be a bit less "book formal" to introduce this appendix. The language of risk analysis is messy. The many different fields of risk analysis use their own terms, methods and models. Though the terminology of risk is broadly used in many disciplines, it is often narrowly defined within a field. Chapter 1 only skimmed the surface of the language of risk. The analogy to our human languages is irresistible here. There is something compelling about the way so many disciplines see the emerging importance of making decisions about risky situations, absent the certainty we all prefer for decision making. Notions of risk and dialects to describe it have emerged more or less simultaneously in many disciplines. Each discipline, satisfied with its own risk language dialect, is reluctant to change, if only to preserve the culture that spawned the language.

I am not sure it is fair to say that anyone has truly tried to standardize the language of risk, but from time to time some group or organization or a bold individual will step to the fore with a proposed definition they hope will please all. It never does. As I noted back in Chapter 1, I am under no delusion that the definitions offered in this text will please anyone who was already happy with the terminology they know. It is worth noting, however, that some groups and organizations have succeeded in defining the language of risk for their purposes, and this is good enough. One group that has attempted to standardize the language of risk is the International Organization for Standardization (ISO).

There is, in my opinion, less disagreement over how one does risk analysis than there is over what we call the things we do when we do risk analysis. In these matters there is sometimes significant disagreement. It is especially significant when a major international organization devotes months or years to standardizing the language. Therein lies the origins of this appendix.

While shopping a text around, it is standard practice to have multiple experts in the field review the book proposal. One reviewer of my own proposal argued quite passionately against publishing this book, saying in part: "The point is that risk analysis does NOT comprise risk assessment, risk communication, and risk management. Rather it is only one step, the first, in risk assessment which is, in turn, the next step in a risk management process."

So you can see there is both considerable disagreement and passion about the language of risk. The language presented in the body of this text is consistent with my own experience and that of many colleagues around the world over many years of practice. In support of this point and my own ego, all other reviewers supported the text and its language. Even so, it is not the only way the language is used.

I would like to think that if the keepers of the flame for every risk dialect assembled as one, then perhaps we could get agreement that this "risk thing" we are all grappling with does indeed answer the following questions:

- What's the problem?
- What information do we need to solve it, i.e., what questions do we want risk assessment to answer?
- What can be done to reduce the impact of the risk described?
- What can be done to reduce the likelihood of the risk described?
- What are the trade-offs of the available options?
- What is the best way to address the described risk?
- (Once implemented) Is it working?
- What can go wrong?
- How can it happen?
- What are the consequences?
- How likely is it to happen?
- Why are we communicating?
- Who are our audiences?
- What do our audiences want to know?
- How will we communicate?
- How will we listen?
- How will we respond?
- Who will carry out the plans? When?
- What problems or barriers have we planned for?
- Have we succeeded?

We just never get agreement about what we call the various tasks required to do these things.

A careful review of the language is not the purpose of this book. The U.S. EPA undertook a thoroughgoing review of the language used by federal and international agencies, private-sector organizations, and academics in their microbial risk assessments. This "Thesaurus of Terms Used in Microbial Risk Assessment" is available at http://www.epa.gov/waterscience/criteria/humanhealth/microbial/thesaurus/, and it provides a great example of the Tower of Babel that risk language has become in just this one application. In the background notes to the thesaurus, the authors note: "Currently, various program offices within EPA, as well as other Federal Agencies… utilize terms often unique to the activities or MRA applications for that specific agency. Different Agencies may also have different operating definitions for the same term."

This book has taken the common but far from universal view that risk analysis comprises the three tasks of risk management, risk assessment, and risk communication. The Society for Risk Analysis's home page offers the following definition from its glossary. "Risk analysis is a detailed examination including risk assessment, risk evaluation, and risk management alternatives, performed to understand the nature of unwanted, negative consequences to human life, health, property, or the environment;

an analytical process to provide information regarding undesirable events; the process of quantification of the probabilities and expected consequences for identified risks." Risk communication is not even mentioned! ISO defines risk analysis as a process to comprehend the nature of risk and to determine the level of risk. And so it goes.

I contend that when we strip the language away and look at what we are doing, there is a common core of principles and activities we all recognize and support. Language is our principle means for conveying meaning in the printed form, and so language matters a great deal. Unfortunately, the language of risk is not settled nor is it likely to soon be so. A great deal of time, money, and effort is being invested in training in the many dialects, and no one is anxious to abandon their own in favor of another.

Of all the many dialects in usage, the ISO language is especially important. The ISO describes itself as the global network of the national standards institutes of 162 countries. It is a nongovernmental organization formed to bridge the gap in industrial standards between the world's public and private sectors. Since 1947 it has established over 18,000 international standards. It is a unique amalgam of government and private-sector member institutes whose purpose is to seek consensus on solutions to issues concerning international standards that meet both the requirements of business and society. ISO standards are broadly supported by industry. Much of the current language of risk analysis, as reflected in this book, was developed in the public sector. It differs significantly from the language of ISO 31000. In recognition of the significant differences in the usage of risk terminology and the unique importance of the ISO, this appendix provides an introduction to the risk language of ISO.

Before proceeding to that introduction I offer a personal testimony to the ISO risk management process. One of my recent professional tasks was to help develop a risk management process for a federal agency of the United States that is involved in water resource management. That process is still under way as this is written. After a broad search of as many existing risk management models as we could find, we have used a slightly modified version of the ISO risk management process to embody the three tasks of risk analysis as described in this book. That adapted model, seen in Figure A.1, and its implementation are wholly consistent with the contents of this book. Thus, I am fully convinced of the compatibility of the two dialects, although I am under no delusion that my conviction eases anyone else's discomfort.

A.2 ISO 31000

ISO 31000:2009, *Risk Management—Principles and Guidelines,* was adopted as a standard in 2009 to provide principles and generic guidelines on risk management. Because of its unique standing as a global standard-setting organization, the ISO treatment of risk management is worthy of some attention. A careful reading of the standard and this text would reveal little if any substantive disagreements but many semantic ones.

The standard was devised to be applicable to all forms of risk and would, according to Purdy (2010), contain:

1. One vocabulary
2. A set of performance criteria

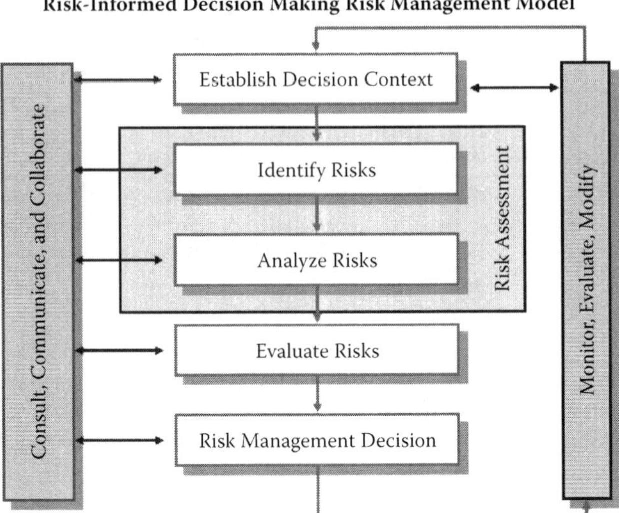

FIGURE A.1 Risk-informed decision making risk management model adapted from ISO 31000 risk management process. (*Source*: Casualty Actuarial Society 2003.)

3. One common, overarching process for identifying, analyzing, evaluating, and treating risks
4. Guidance on how that process should be integrated into the decision-making processes of any organization

Leitch (2010), however, found that some definitions in the ISO documents have meanings different from those of ordinary language and other terms that change their meaning from one place to another. He concluded that "many of the definitions in ISO 31000:2009 are not clear and meaningful, let alone close to the actual usage of the terms." So it is obvious that there are some who believe the language has a long way to travel before it is unified.

ISO 31000 defines risk as the effect of uncertainty on an organization's objectives. In Chapter 1, I have said risk is the chance of an undesirable outcome and went on to explain that uncertainty is the mother and father of that chance. The words are different and some may argue, perhaps successfully, that they have different meanings. I would disagree and find the definitions quite similar in meaning, especially insofar as they both embrace a notion of risk as an outcome of hazards and opportunities. Point by point, it would be possible to take the ISO language and find its counterpart in the content of this text. Unfortunately, the language of risk defies standardization at this point in time. I would like to assure readers that the risk analysis discussed throughout this text is no different in principle and relatively little different in fact from the risk management process described by ISO. It is, however, shockingly different at times in how the language is used. ISO and the world of practitioners this book will appeal to sometimes use the same words to mean different things. Other times they use different words to mean the same thing. On occasion they even manage to use the same words to mean essentially the same things.

RISK DIALECTS ISO AND THIS BOOK

The same words can mean different things: ISO considers risk analysis to be one of the tasks in risk assessment. Others consider risk analysis to be the overarching concept, one task of which is risk assessment.

Different words can mean the same thing: ISO's "level of risk" and this book's "risk estimate."

The same words can mean the same thing: "risk acceptance," for example.

ISO defines risk management as the coordinated activities to direct and control an organization with regard to risk. In this sense it is close to the overarching concept of risk analysis while being simpatico with the notion of a risk management activity introduced in Chapter 3.

Purdy (2010) identifies eleven principles of risk management embodied in the ISO standard. These are that risk management should:

1. Create and protect value
2. Be an integral part of all organizational processes
3. Be part of decision making
4. Explicitly address uncertainty
5. Be systematic, structured, and timely
6. Be based on the best available information
7. Be tailored
8. Take into account human and cultural factors
9. Be transparent and inclusive
10. Be dynamic, iterative, and responsive to change
11. Facilitate continual improvement of the organization

All of these are fully consistent with the principles put forth in this book. While that point-by-point discussion to establish the equivalence of the two dialects may be useful, it is not possible. Doing so would essentially require reproducing the information contained in ISO documents, which are considered proprietary information and are available for a fee from the ISO.

In place of an ISO glossary, I have reproduced the ISO relationships among terms based on their risk management principles and guidelines. Numbering and indentation of the layout of the ISO language is faithful to the structure of the January 4, 2008 draft of ISO/IEC CD 2 Guide 73. The numbering system indicates the terms that are subservient to higher order terms.

- 3.1 Risk
- 3.2 Risk management
    - 3.2.1 Risk management framework
    - 3.2.2 Risk management policy

3.3.8 Monitoring and review
  3.3.8.1 Monitoring
  3.3.8.2 Review
  3.3.8.3 Risk reporting
    3.3.8.3.1 Risk register
    3.3.8.3.2 Risk profile
  3.3.8.4 Risk management audit

This structure provides some useful information about how ISO has organized its thoughts about what this book calls risk analysis. First, notice that risk management is the unifying concept, not risk analysis. Managing risk is a quite suitable focus for many of the organizations that use the ISO principles and guidelines. Many other practitioners rely on risk analysis as the unifying concept for its three components of management, assessment, and communication. Each of these three components is found in the ISO risk management description.

Extracting from the outline presented here, the ISO risk management process consists of five steps and two ongoing processes, as indicated in Figure A.2. Using the abbreviations RC for risk communication, RM for risk management, and RA for risk assessment, the ISO steps are equated to the risk analysis components where the same work is accomplished according to the language used throughout this text. Thus, looking at the ISO risk management process as a whole, it includes risk management, risk assessment, and risk communication and is compatible with the risk analysis view that dominates much of applied practice.

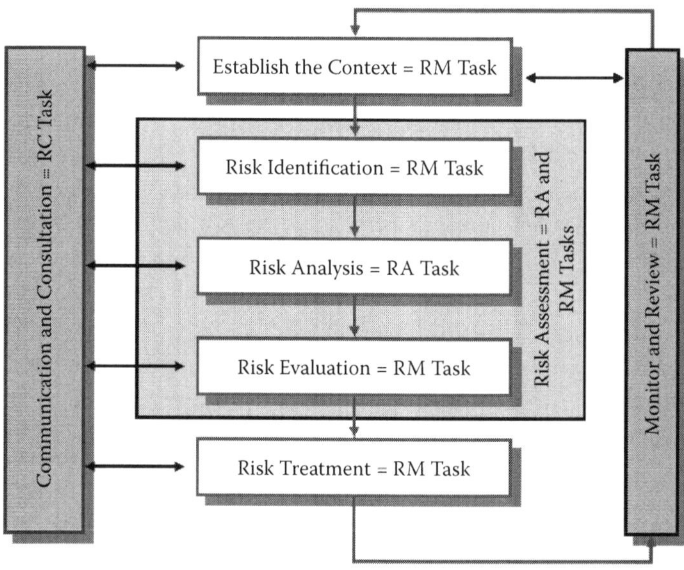

FIGURE A.2 Mapping the ISO risk management process into the three components of risk analysis as described in this book.

Undeniably there are major differences in terminology. As noted earlier, risk analysis is seen as a subset of risk assessment in the ISO terminology. The risk assessment steps contained within the rectangle mix what I and others would call risk management and risk assessment. Despite the differences in terminology, all of the same work gets done in both dialects.

Purdy (2010) has characterized the publication of ISO 31000:2009 and Guide 73:2009 as "a very significant milestone in mankind's journey to understand and harness uncertainty." And so it may be. Twenty-five ISO committee members voted for the standard with only one voting against it. This support was unprecedented in ISO. It is being formally adopted by many states to replace their national standard and is causing other standard-setting bodies to revisit their documents. Before all is said and done, this may be a significant step forward in achieving the four goals set out above by Purdy. If it succeeds in this endeavor, little of the content of this book will be devalued. At the present time and for the foreseeable future, most government agencies and those directly affected by them tend to follow the risk dialect used throughout the body of this text.

I repeat, the language is messy. Not everyone is enamored of the ISO effort, nor should we expect them to be. Leitch (2010) summarized his review by saying ISO 31000:2009:

1. Is unclear
2. Leads to illogical decisions if followed
3. Is impossible to comply with
4. Is not mathematically based, having little to say about probability, data, and models

It is my hope this book contributes some to the development and maturation of the discipline and that readers brought up on the ISO dialect will find it worth the effort to become familiar with the dialect used here.

A.3 Enterprise Risk Management

Enterprise risk management is another term of art that has found common usage in the risk community. The Casualty Actuarial Society has an Enterprise Risk Management (ERM) Committee that in 2003 espoused the ERM process summarized in Figure A.3. It is based on the Australia/New Zealand Risk Standard AS/NZS 4360, which was a precursor to ISO 31000. They define ERM as a "discipline by which an organization in any industry assesses, controls, exploits, finances, and monitors risks from

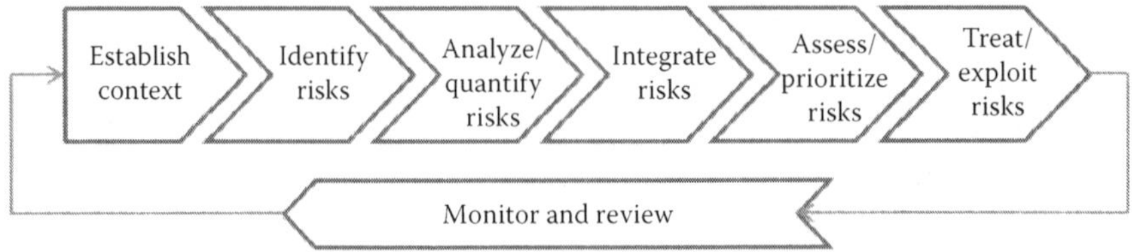

FIGURE A.3 Overview of enterprise risk management process steps.

all sources for the purpose of increasing the organization's short- and long-term value to its stakeholders."

As was the case for ISO 31000, the ERM community practices the same principles and works with the same toolbox. The cubbyholes the concepts are organized into and many of the words by which they are called differ, but substantively there is little to no difference from the practices outlined in this book.

A.4 Observations

Everyone with whom I work uses the risk dialect used throughout this book. It is a fractured language, as I have admitted from the outset. Many others can make the opposite contention that everyone with whom they work uses the ISO dialect. Those who follow the ideas set forth in this book and those who follow the ISO standard are all doing essentially the same things.

I see great value to a standardized language and a process that is firm in its principles yet flexible in the details. Consequently, I would have readily adopted the language of ISO 31000 and ISO Guide 73 except for one detail. That detail is that the ISO language and guidelines are neither broadly recognized, accepted, nor used yet by risk practitioners in government and many industries.

It would seem to me that the language of risk is not only fractured by the many disciplines that have spawned it, but also by the large macroenvironment sectors that use it. The public sector is taken for the moment by the language used in this book. Likewise, their clients and customers are inclined toward that dialect. The so-called Red Book—*Risk Assessment in the Federal Government: Managing the Process* (NRC 1983)—risk assessment model of hazard identification, dose-response, exposure assessment, and risk characterization, for example, has deep roots in many sectors. It will not soon be sacrificed to risk assessment as risk identification, risk analysis, and risk evaluation. Too many people are vested in their own dialects.

On the other hand, the ISO principles and guidelines represent the only risk management model some sectors have ever known. It is my hope that as the language of risk continues to evolve, the principles found in this book will remain useful to all those interested in risk, no matter which dialect they speak.

REFERENCES

Casualty Actuarial Society Enterprise Risk Management Committee. 2003. Overview of enterprise risk management. http://www.casact.org/research/erm/overview.pdf.

International Organization of Standardization. 2009a. *Risk management—principles and guidelines*. Geneva, Switzerland: International Organization of Standardization.

———. 2009b. *Risk management—vocabulary*. Geneva, Switzerland: International Organization of Standardization.

National Research Council. 1983. Committee on the Institutional Means for Assessment of Risks to Public Health. *Risk assessment in the federal government: Managing the process*. Washington, DC: National Academies Press.

Appendix B: Using Palisade's Decision Tools Suite

B.1 Introduction

Palisade's DecisionTools Suite* is an integrated set of programs for risk analysis and decision making under uncertainty that run in Microsoft® Excel. The DecisionTools Suite includes @RISK for Monte Carlo simulation, PrecisionTree for decision trees, and TopRank for "what if" sensitivity analysis. In addition, the DecisionTools Suite comes with StatTools for statistical analysis and forecasting, NeuralTools for predictive neural networks, and Evolver and RISKOptimizer for optimization. All programs work together and are integrated completely with Microsoft® Excel for ease of use and maximum flexibility.

This suite of tools was used to complete the examples found throughout this book. This appendix shifts from an emphasis on risk analysis to how to get started using three of the programs contained in the DecisionTools Suite. These are TopRank, @RISK, and PrecisionTree. Many of the files used in the creation of this book, as well as additional exercises to help you master some of the techniques described, can be found at http://www.palisade.com/bookdownloads/yoe/principles/, a companion site for this book graciously provided by the Palisade Corporation.

The goal of this appendix is to help you learn enough to begin using these analytical tools. It is not the intention of this appendix to teach you all about the software. Once you begin to use these tools, you can explore additional capabilities of the software through the users' manuals, online help, and users' forums.

The appendix begins with a spreadsheet cost-estimate model that will be familiar to you from the text. The model presented initially is a deterministic model. TopRank is the first software introduced. It is software that you might use in the early stages of a risk assessment when using test-case data for your computational model, for example. TopRank performs automated "what-if" sensitivity analysis on your model and its point estimates. It is used to identify the inputs to which the model output (in this instance the total cost estimate) is most sensitive. Used in this way TopRank can narrow your focus before completing a simulation with @RISK. Next, you will be introduced to the basic features of @RISK, Palisade's Monte Carlo process software. Following the

---

* Palisade Corporation is the maker of the market-leading risk and decision analysis software @RISK and the DecisionTools® Suite. Virtually all Palisade software adds in to Microsoft Excel, ensuring flexibility, ease of use, and broad appeal across a wide range of industry sectors. Its flagship product, @RISK, debuted in 1987 and performs risk analysis using Monte Carlo simulation. With an estimated 150,000 users, Palisade software can be found in more than 100 countries and has been translated into five languages. Headquartered in Ithaca, New York, Palisade also maintains offices in London, England; Sydney, Australia; and Rio de Janeiro, Brazil.

introduction to @RISK you will be introduced to the basic tree structuring capabilities of PrecisionTree.

B.2 TopRank

This appendix provides you with enough instruction to begin to make simple use of the TopRank software. To understand the nature of the computations or the meaning of the outputs, you will need to review the content of interest to you in the related chapter of this book. In this section you will learn how to:

1. Identify an output in your model
2. Change the analysis settings
3. Have TopRank identify the inputs to vary
4. Run a what-if analysis
5. Generate results

Figure B.1 presents the spreadsheet cost model prepared in Microsoft Excel. Total project cost, shown in the bottom right cell, is the output of interest in this model. It is computed by multiplying quantities by unit costs (row-wise) and then summing the amounts. Do not be concerned if you are unfamiliar with the nature of the items included in the cost estimate.

Description	Quantity	Unit	Unit Price	Amount
Relocations				
Lower 20 pipeline, 653 + 00	425	LF	$730	$310,250
Remove 8" pipeline, 678 + 00	1,000	LF	$50	$50,000
Total – Relocations				$360,250
Navigation, Ports and Harbors				
Mobe and Demobe	1	LS	$500,000	$500,000
Pipeline Dredging, Reach 1	691,328	CY	$2.78	$1.921,892
Pipeline Dredging, Reach 2	1,022,769	CY	$2.60	$2,659,199
Pipeline Dredging, Reach 3A	1,182,813	CY	$3.16	$3,737,689
Pipeline Dredging, Reach 3B	736,713	CY	$2.76	$2,033,328
Scour Pad, Reach 1	17,550	SY	$25.69	$450,860
Geotubes, 30', Reach 1	1,400	LF	$188.52	$263,928
Geotubes, 45', Reach 1	4,912	LF	$222.18	$1,091,348
Scour Pad, Reach 3	38,750	SY	$25.69	$995,488
Geotubes, 45', Reach 3	13,940	LF	$222.18	$3,097,189
Total – Navigation, Ports and Harbors				$16,750,921
Subtotal				$17,111,171
Engineering and Design	8	%		$1,368,894
Construction Management	6	%		$1,026,670
TOTAL PROJECT COST				$19,506,7324

FIGURE B.1 Deterministic Microsoft Excel spreadsheet cost model.

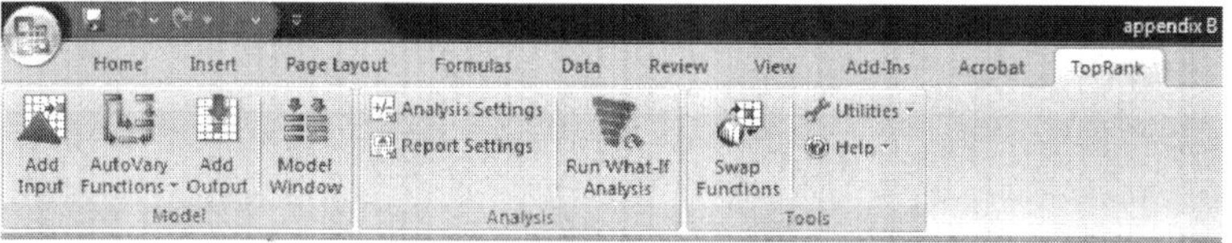

FIGURE B.2 TopRank icons.

TopRank is an Excel add-in that provides a new tab and the features shown in Figure B.2. The sections that follow make reference to these icon functions.

B.2.1 Identify Model Output

The output of interest in this model is the total project cost estimate. You must identify each output of interest to you by placing your cursor in the output cell and selecting the "Add Output" icon (third from left in Figure B.2). When you do, the window shown in Figure B.3 will open. It shows a default name when it opens. The default is chosen by identifying the nearest text to the left of the cell and the nearest text above the cell. You may accept the default or edit it to include the output name you prefer. Repeat this process for every output in your model.

B.2.2 Change the Analysis Settings

The workings of TopRank are explained in Chapter 16. You may want to change the analysis settings before you select the input variables. This is done in the Analysis Group (second from the left on the TopRank tab) using the "Analysis Settings" option seen in Figures B.2 and B.4.

You can vary the inputs by a fixed percentage or a fixed amount. The point estimate in your original model is called the base, and all changes are relative to it. You are able to vary the base value up or down symmetrically or asymmetrically. Unless you know why you want a different distribution, use a uniform distribution. You can control the number of intermediate values between the minimum and maximum by varying the number of steps. For this example, the default values are sufficient, ±10%, a uniform distribution, and five steps. Just know that you can explore wider or narrower ranges and with different shapes over those ranges. In general, you would define the range to reflect your relative uncertainty. If you can estimate a value within ±10%, that is what you use. If you are more uncertain, you might choose ±30% or whatever best bounds your uncertainty.

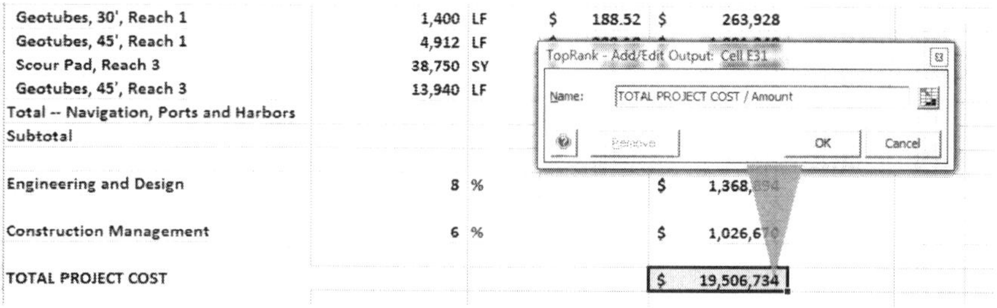

FIGURE B.3 Add output window.

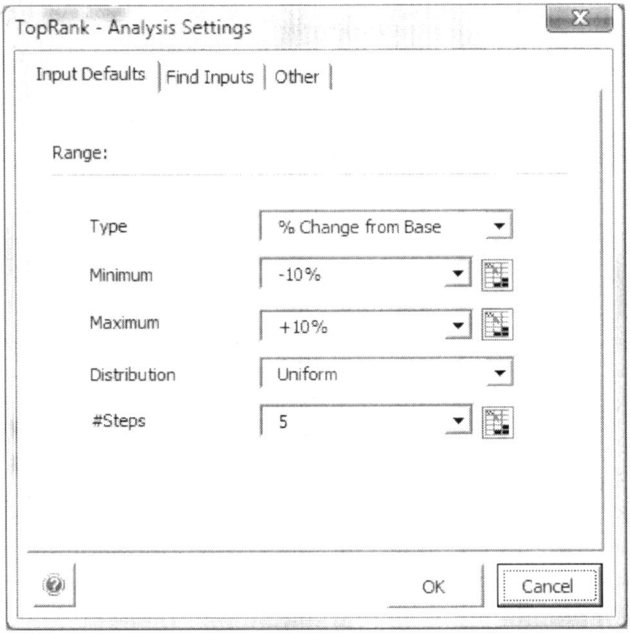

FIGURE B.4 Analysis settings window of TopRank.

B.2.3 Identify the Inputs to Vary

The purpose of your what-if analysis is to identify those inputs that have the greatest potential impact on your output based on the structure of your model. You have two options for identifying inputs. The first is to identify each input as you did the outputs, choosing the "Add Input" icon instead of the add output icon. Alternatively, you can have TopRank automatically identify the inputs to vary based on the structure of your model.

Select the "AutoVary Function" and the "Add AutoVary Functions" as seen in Figure B.5. You will be asked if it is okay to replace the current values in the model (see Figure B.6), so be sure to keep an original copy of any model you might use with TopRank.

Once you select OK, your model will be modified as shown in Figure B.7. The cursor is in cell B15, and the formula window shows that although the base value of 691,328 is shown, it has been replaced by a TopRank formula indicating that this value will be varied by ±10%. Each constant in the spreadsheet that is used to calculate the

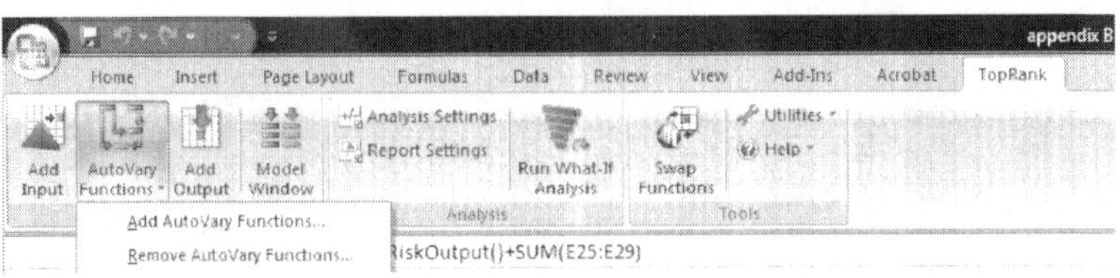

FIGURE B.5 Auto Vary Functions options in TopRank.

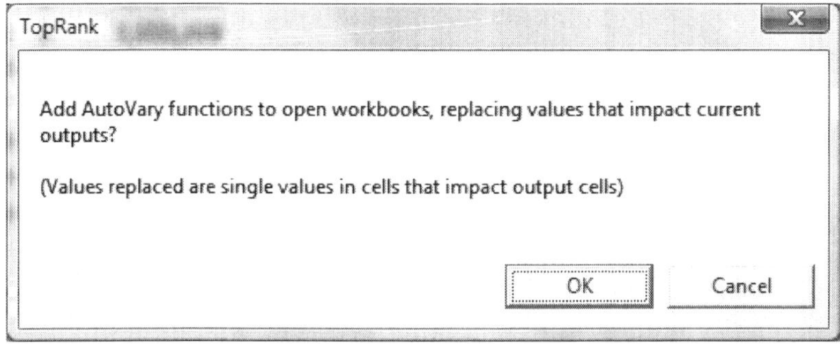

FIGURE B.6 Permission to replace original values with vary functions.

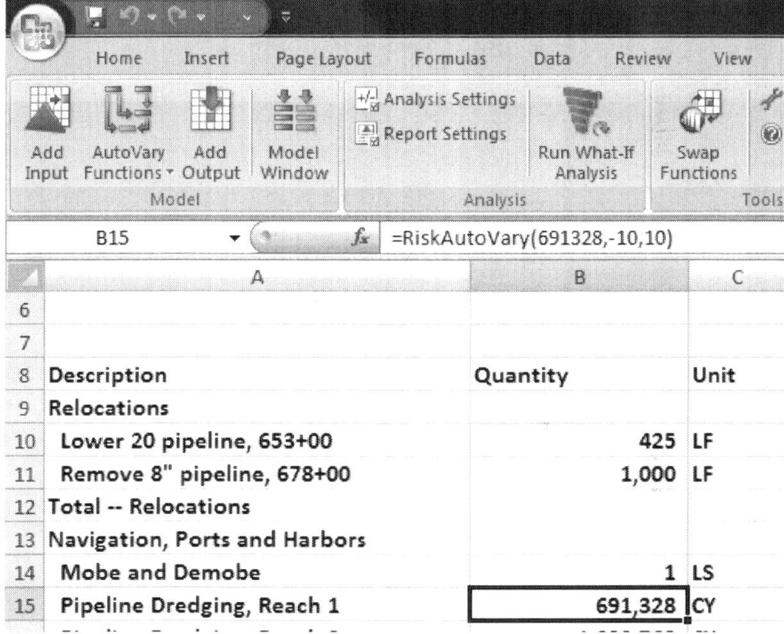

FIGURE B.7 Model cell showing auto vary function in formula window.

output (total project cost) is replaced by such a formula. Notice that values in the Amount column have not been modified, as they consist of the product of two modified values. Your spreadsheet is now ready for what-if analysis.

B.2.4 Run What-If Analysis and Generate Results

Before running your analysis, select the "Reports Setting" icon from the analysis group. This allows you to identify the types of graphs you want, as well as enabling you to identify the outputs and inputs for which you'd like to see results. Select the desired options before you run your what-if analysis. Figure B.8 shows this window.

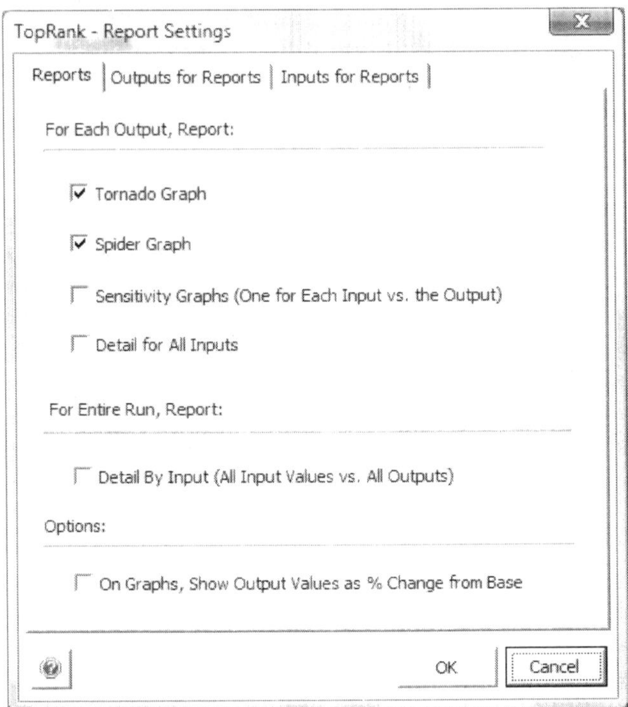

FIGURE B.8 TopRank report settings options.

Notice the "Run What-If Analysis" icon in Figure B.7. Select it to run the analysis. Before the analysis runs, you will see a summary of the analysis parameters as shown in Figure B.9. You will notice the number of outputs and inputs as well as the variation in your inputs and several other details.

Once you select the "Run" option, TopRank will write the results to a new workbook. The number of pages in this workbook will vary according to the report settings you selected. This is where you will receive the kinds of outputs discussed at length in Chapter 16. Figure B.10 shows a partial sample of the output generated using the default settings described in this chapter.

When you are done you can select "AutoVary Functions" and then select "Remove AutoVary Functions" and your model will be restored to its original condition. You now know enough about TopRank to begin to use it. Be sure to explore its other capabilities, especially the multiway what-if analysis that enables you to examine the impacts of combinations of changes in two or more inputs on model outputs. The files used in this section are available for your use at the Palisade link http://www.palisade.com/bookdownloads/yoe/principles/.

B.3 @RISK

@RISK is the Excel add-in that enables you to use the Monte Carlo process in a simulation. With @RISK you can replace any point estimate in your model with a

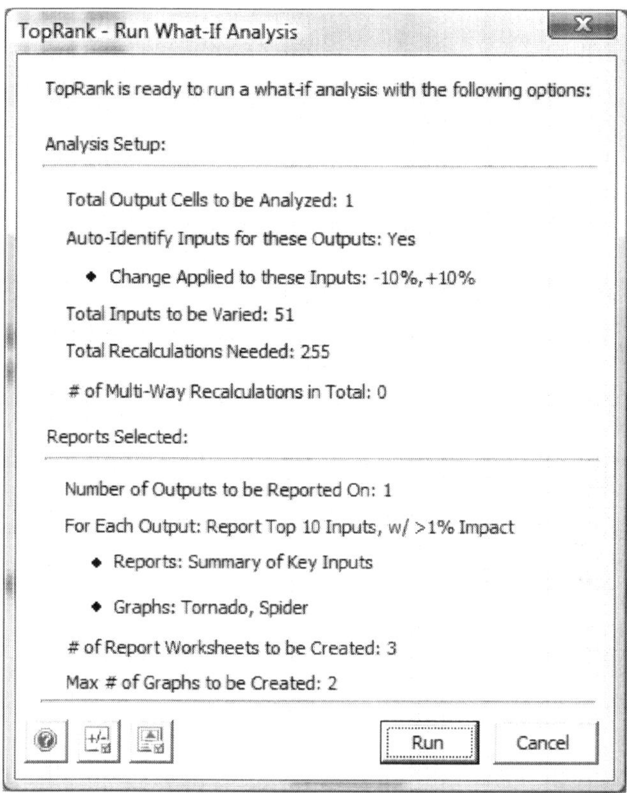

FIGURE B.9 TopRank what-if options.

probability distribution. As with TopRank, this description of @RISK is intended to help you learn how to use the software well enough to get started applying the principles described at length in this book. In this section you will learn how to:

1. Enter a probability distribution into your model
2. Modify that distribution
3. Identify an output in your model
4. Set up and run a simulation
5. Modify a graph
6. Generate reports

B.3.1 Enter a Probability Distribution (and Modify It)

We'll use the same model shown in Figure B.1. For simplicity, we'll work with the triangular distribution. All the files used in this section as well as additional exercises can be found at the Palisade link http://www.palisade.com/bookdownloads/yoe/principles/. Let's begin with a look at the @RISK icons in Figure B.11.

TopRank - Tornado Graph

Performed By: Charles Yoe
Date: Thursday, December 23, 2010 3:53:30 PM
Model: appendix B materials.xlsx
Output: TOTAL PROJECT COST / Amount (E31)
Base Value: 19506734.4384

Tornado Graph of TOTAL PROJECT COST / Amount
Impact by Input

Pipeline Dredging, Reach 3A / Quantity (B17)
CY / Price (D17)
Geotubes, 45', Reach 3 / Quantity (B23)
LF / Price (D23)
Pipeline Dredging, Reach 2 / Quantity (B16)
CY / Price (D16)
Pipeline Dredging, Reach 3B / Quantity (B18)
CY / Price (D18)
Pipeline Dredging, Reach 1 / Quantity (B15)
CY / Price (D15)

19000000 19100000 19200000 19300000 19400000 19500000 19600000 19700000 19800000 19900000 20000000

Value of TOTAL PROJECT COST / Amount

What-If Analysis Summary for Output TOTAL PROJECT COST / Amount
Top 10 Inputs Ranked By Change in Actual Value

				Minimum			Maximum		
				Output		Input	Output		Input
Rank	Input Name	Worksheet	Cell	Value	Change (%)	Value	Value	Change (%)	Value
1	Pipeline Dredging, Reach 3A / Quantity (B17)	OAATA	B17	$ 19,080,638	-2.18%	1064531.7	$ 19,932,831	2.18%	1301094.3
2	CY / Price (D17)	OAATA	D17	$ 19,080,638	-2.18%	2.844	$ 19,932,831	2.18%	3.476
3	Geotubes, 45', Reach 3 / Quantity (B23)	OAATA	B23	$ 19,153,655	-1.81%	12546	$ 19,859,814	1.81%	15334
4	LF / Price (D23)	OAATA	D23	$ 19,153,655	-1.81%	199.962	$ 19,859,814	1.81%	244.398
5	Pipeline Dredging, Reach 2 /	OAATA	B16	$ 19,203,586	-1.55%	920492.1	$ 19,809,883	1.55%	1125045.9

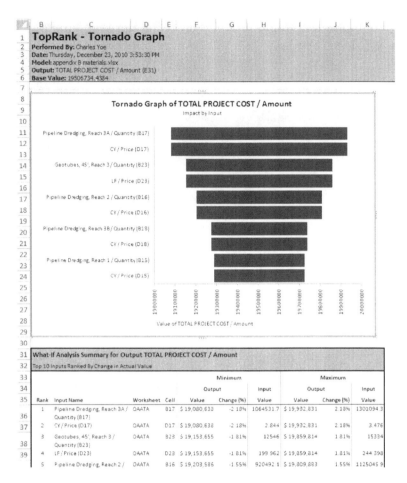

FIGURE B.10 Sample of TopRank tornado graph output.

There are four groupings on the @RISK tab: model, simulations, results, and tools. The model functions are used primarily to set up your model prior to running the simulation, so we will begin there.

Placing your cursor in a cell where you want to insert a distribution, select the "Define Distributions" icon (first on the far left). You will see something like Figure B.12. Note that the define distributions window has several tabs. Figure B.12 has the "All" (show all distributions) tab selected. This is the easiest way to select a distribution when you are starting. Simply scroll up or down to find the distribution you need. Note that there is a grayed-out input name and a cell formula at the top of this window

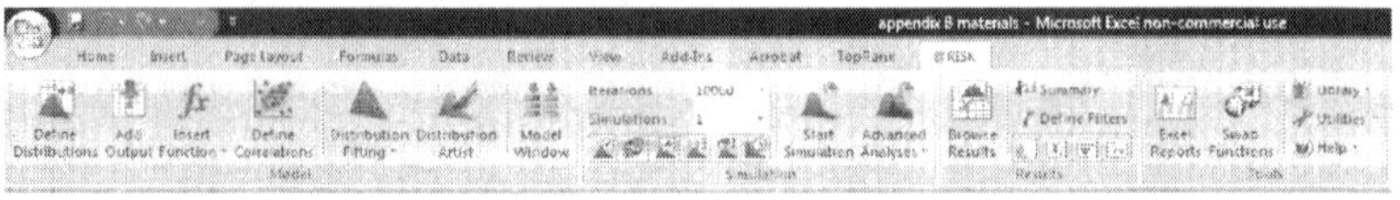

FIGURE B.11 @RISK icons.

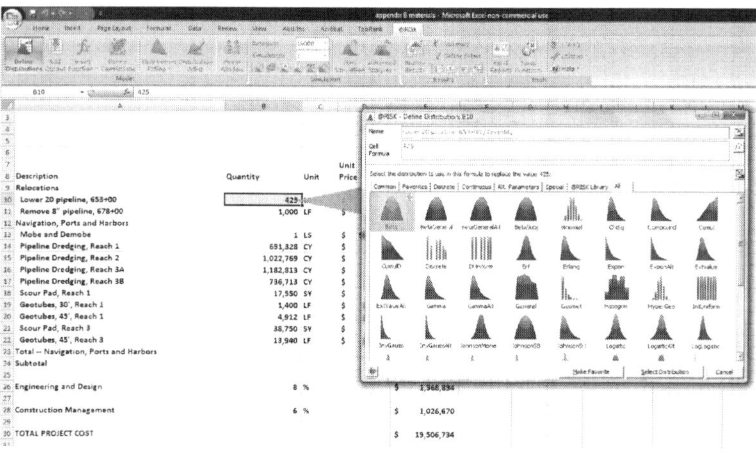

FIGURE B.12 @RISK's available distributions accessed through the define distribution icon.

when you are working with an existing model. If you begin from scratch in a new spreadsheet, these entries will be blank.

Select the triangular distribution (Triang). Each distribution is represented by a typical shape for that family of distributions. Once selected, you will see something like Figure B.13. Whenever you select a distribution, @RISK will display an example of that distribution in the window that pops up. This distribution and the parameters entered (382.5, 425, and 467.5 in this instance) are absolutely meaningless. New risk assessors sometimes think @RISK chooses the distribution for you. It does not. You must do that, and you do it as described in Chapter 12. @RISK uses any preexisting value in the cell as a default base for constructing a distribution to show you what the distribution you chose looks like. Even when you enter a distribution into a blank cell, there are default values provided. They must be replaced by your chosen values.

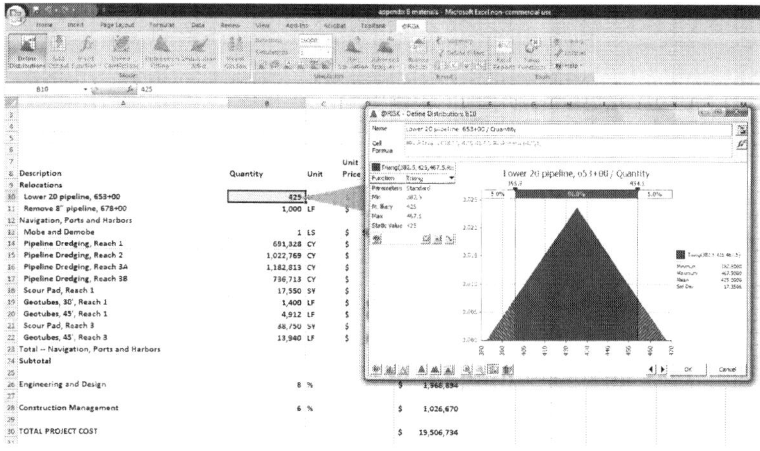

FIGURE B.13 Default view of a triangular distribution with a preexisting cell entry.

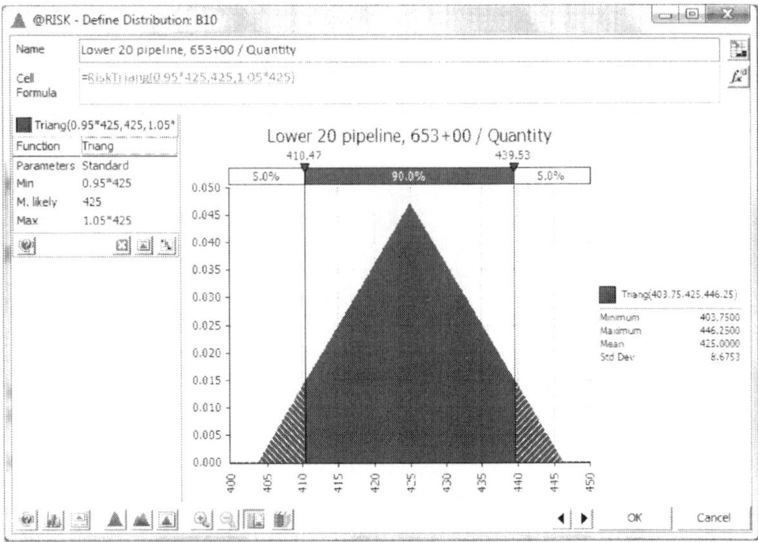

FIGURE B.14 The triangle distribution in @RISK.

Figure B.14 shows a close-up of a typical distribution window. Note the name and cell formula at the top. These can be edited directly in this window. Changing the name does not change your spreadsheet text. Once you learn the @RISK syntax, you can type distribution formulas, like that in the cell formula box, directly into a spreadsheet cell. The bulk of the window is taken up by the distribution graph. To the left of it you will find the distribution parameters. (If you run into unfamiliar terminology, check the body of the textbook for an explanation.) The parameters will vary with the distribution, but they are always found here.

First, note that this is the standard format for entering parameters, as numerical values of the variable. In this example the minimum is considered to be 5% less than the point estimate, and the maximum is 5% more. Simple formulas (without equal signs) are typed into the cell to define the distribution. To the right of the graph are a few statistics.

Look at the distribution parameters area on the left again. If you left mouse click on the cell that says "Standard" just to the right of the word "parameters," you'll see the window in Figure B.15 appear. This is a bit advanced for a simple getting-started

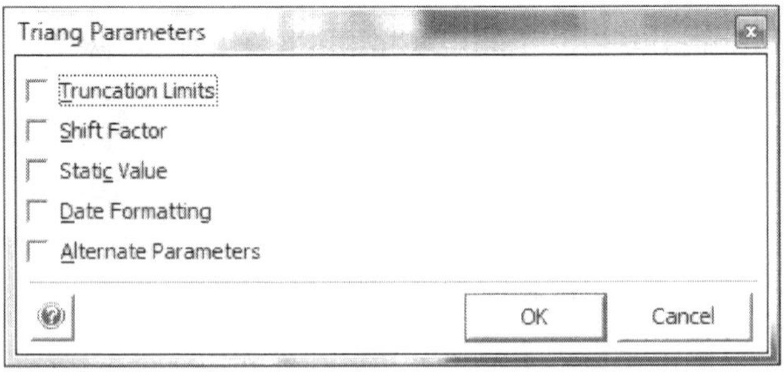

FIGURE B.15 Distribution parameter modification options in @RISK.

introduction, but it is a useful feature that allows you to truncate your distributions or, for example, to define your distribution using percentile values (Alternate Parameters). Remember where this is and you can grow into it later. There are many other icons on the distribution window, but for now, select "Okay" and you have entered a distribution. Repeat this process for every input you will represent with a probability distribution. TopRank can be used to help you decide which distributions may have the greatest impact on your outputs.

B.3.2 Identify an Output

You identify outputs in exactly the same way as you did for TopRank.

B.3.3 Set up and Run a Simulation

After you have entered all the distributions you intend to use in your simulation and identified all your output cells (remember that any intermediate calculations you want to monitor must be identified as outputs), it is time to move to the simulations grouping. Hit F9. Do the numbers in your model change? If not, the Monte Carlo process recalculation is not turned on (it need not be turned on to run a simulation). Find the icon with a pair of dice under the word "simulations" in the simulations group. It is the "Random/Standard Static F9 Recalc" icon, and it enables or disables your ability to iterate your model by striking the F9 key.

For a fast start, put your cursor in an output cell and set the number of iterations to the desired number (say 1,000 for now) and then select the "Start Simulation" icon. The simulation will begin and you can monitor its progress as seen in Figure B.16. If you put the cursor in an output cell you will also see the output distribution forming. If it all happens too fast, increase your iterations to 10,000 while you are learning how the software works.

You have now run a simulation using @RISK. Figure B.17 shows a sample output. It is that simple. Now, we'll double back and explore some more features.

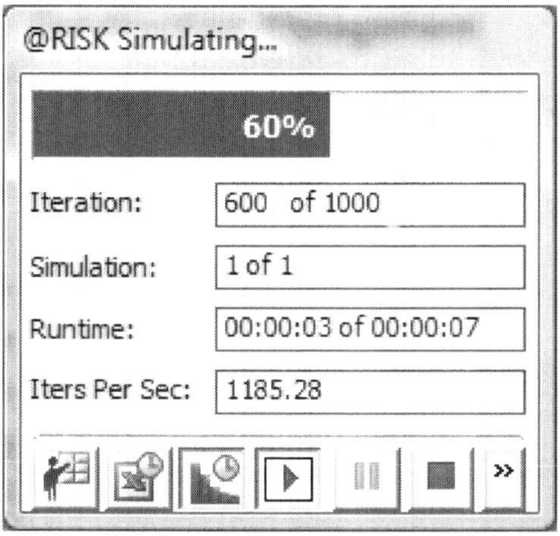

@RISK Simulating...

60%

Iteration:	600 of 1000
Simulation:	1 of 1
Runtime:	00:00:03 of 00:00:07
Iters Per Sec:	1185.28

FIGURE B.16 Simulation progress window in @RISK.

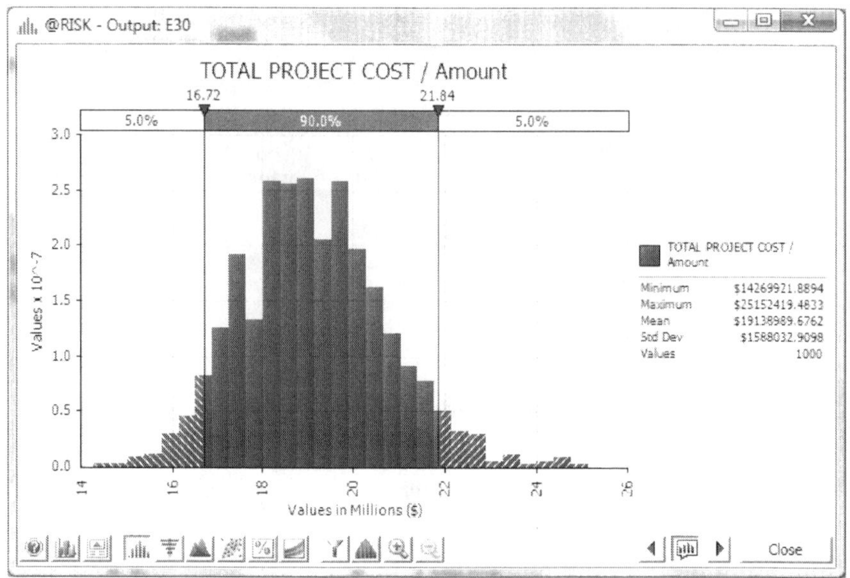

FIGURE B.17 Sample model output distribution in @RISK.

B.3.4 Modifying a Graph

Display your output graph. If it has closed, place your cursor in an output cell and select "Browse Results" from the results group. Right mouse click inside the graph frame and select "Graph Options." You will see the window shown in Figure B.18.

The distributions tab enables you to alter the graph format. Choose a cumulative ascending distribution. Note that when you use the automatic, probability density, or relative frequency options, you can vary the number of bins used to define your graph. As noted in the text, this can have a strong influence on the appearance of your data at times.

The title tab enables you to enter a title and control the font style and color. The x- and y-axis tabs enable you to enter labels on the axes and to control their range as well. You have the option of eliminating grid lines, scale factors, and tick labels if you desire. With the curves tab, you can alter the style, fill, and color of each distribution in your graphic. You can eliminate the legend or modify what appears in it with the legend tab. Delimiters can be removed by the delimiters tab and new markers can be added with the markers tab. Figure B.19 shows a transformation of the information shown in Figure B.17 using the graph options tabs with markers added.

B.3.5 More Results

The small icons on the right side of the results group will make the results of the simulations available to you. The first one, "Simulation Detailed Statistics," provides descriptive statistics and percentiles in 5% increments for every output and input in your model. You may also query your simulation data to learn specific percentile values (e.g., the 99.5 percentile) or to learn the percentile of any given value

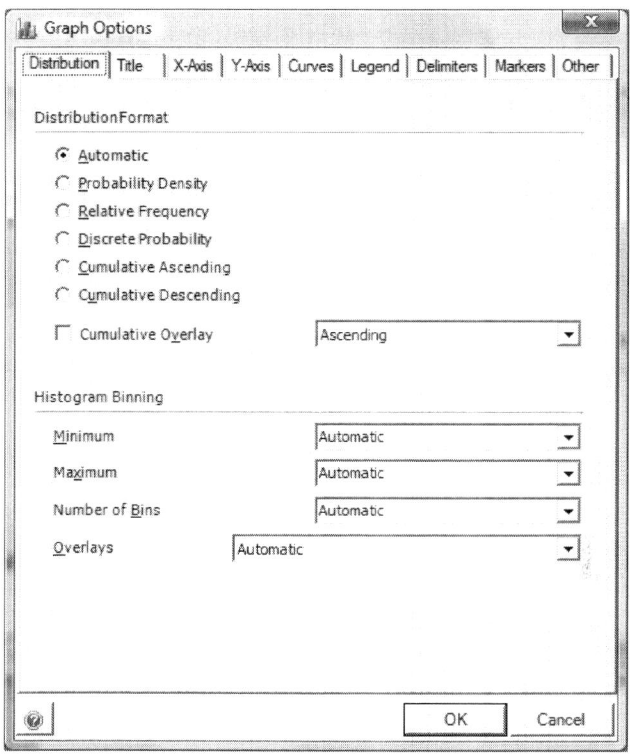

FIGURE B.18 Graph options in @RISK.

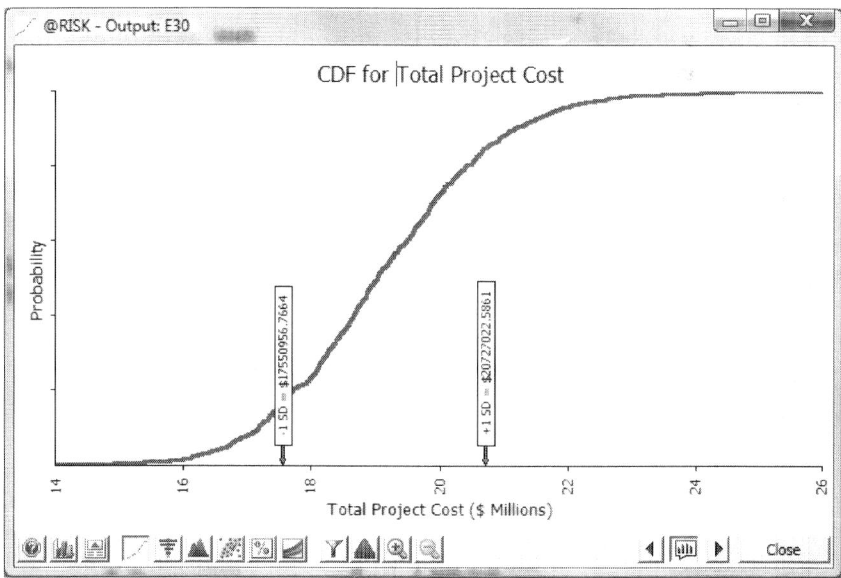

FIGURE B.19 Transformation of the data in Figure B.17.

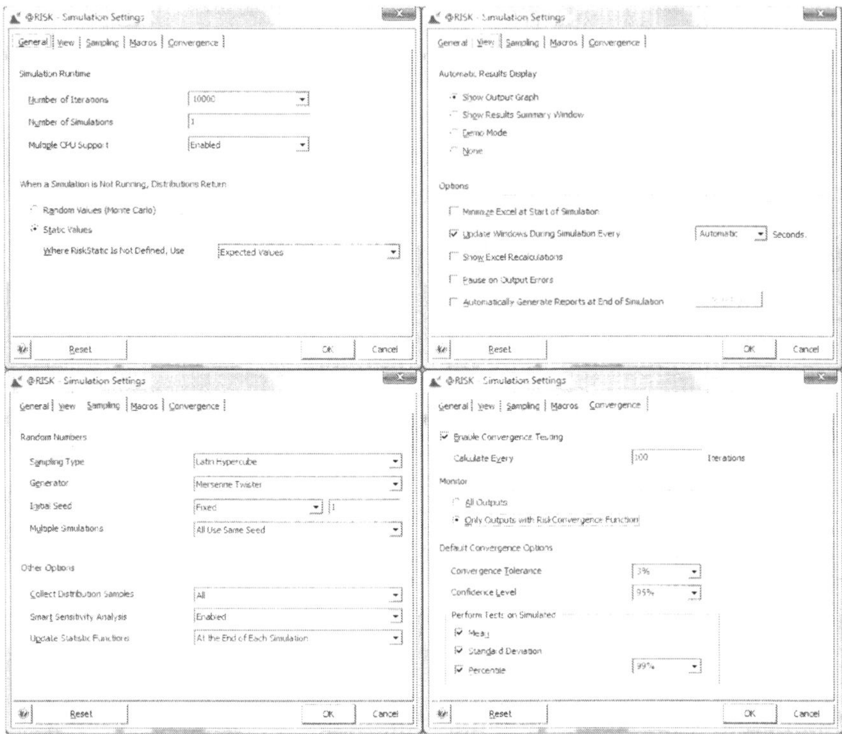

FIGURE B.20 Four of the simulation setting icon tabs in @RISK.

($19,000,000). This feature enables you to describe the CDF for any variable in your simulation. The "Simulation Data" icon makes every iteration value for every input and output available.

B.3.6 Simulation Settings

Now let's look back in the simulations group at the small icons there. The first one, "Simulation Settings," is important to understand. Four of its tabs are illustrated in Figure B.20. The first of these allows you to set the number of iterations and simulations you will run and to turn the Monte Carlo process on or off. These features are also directly available in the simulation group.

The view tab lets you control what you see during a simulation. The "Pause on Output Errors" option can be especially helpful when you are verifying your model. The random numbers options on the sampling tab are useful to know. Generally, you will leave the sampling type on Latin Hypercube; the alternative is Monte Carlo, and these two sampling techniques are discussed in Chapter 14.

You have a choice of many different random number generators. Most users rely on the default generator shown in the figure. Of special interest is the initial seed option. You can choose this seed randomly or use a fixed seed. There may be times when you are running the exact same model with relatively subtle changes in selected inputs that reflect the effect of a risk management option on your issue. In these instances it

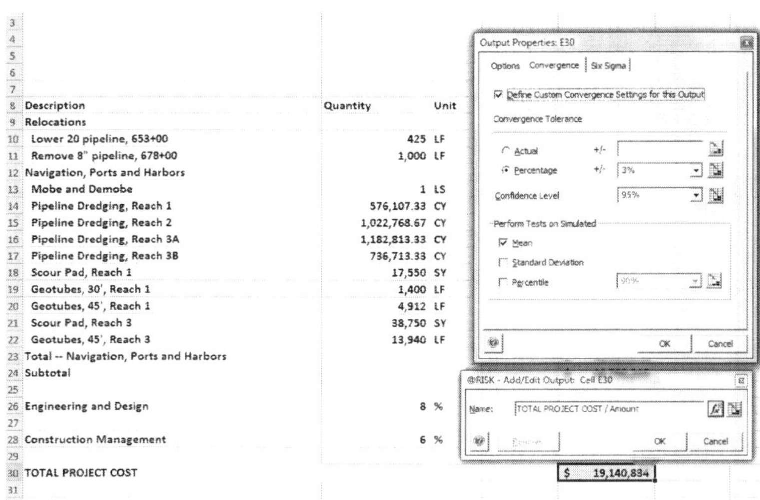

FIGURE B.21 Custom convergence settings for model outputs.

would be useful to compare without and with RMO (risk management option) conditions, all other things being equal. Choosing a fixed seed ensures that your model generates the same random values for unchanged input distributions. Thus, any changes in outputs are attributable to the RMO.

If you use a fixed seed, you have the option of entering the seed. A "1" is shown in Figure B.20. Choose something simple and easy to remember.

Collecting data from your input distributions can slow your simulation down when you have a lot of inputs and a large number of iterations. You have the option (Monitor) of collecting all, some, or none of your input distributions.

The convergence tab gives you an alternative to running your model for a fixed number of iterations. With large and slow models, it may be helpful to just run your model until a few parameters of special interest, i.e., outputs you have identified as RiskConvergence outputs, have stabilized. Figure B.21 shows how to identify these outputs. Simply select the "Add Output" icon when your cursor is in the cell with the output you want to identify. When the familiar window opens with the output name, click on the small function icon (f_x) and the output properties box opens. Choose the convergence tab and check the box to define custom convergence settings. Here you can identify convergence as an absolute amount or a percentage. You can also identify the parameters from the output distribution to use in your convergence monitoring.

Look back at Figure B.20 again. Once the identified statistics stop changing by more than 3% (with a 95% confidence interval) every 100 iterations (you can change this option as well), the simulation will stop. This can save time with some models while providing you with the level of precision you desire. Less commonly, you may be working with a model where you cannot be sure that convergence is reached with, say, 10,000 iterations. In such a case you may want to use convergence rather than a fixed number of iterations.

There are many more useful features of @RISK. Correlations, distribution fitting, and other sensitivity analysis are features you can grow into once you begin to use the software. Exercises explaining these more advanced features are found at the Palisade Web site, http://www.palisade.com/bookdownloads/yoe/principles/, to accompany this text. Before moving on to consider PrecisionTree, note that the "Excel Reports" icon enables you to export results from and descriptions of your model to a Microsoft® Excel spreadsheet.

B.4 PrecisionTree

PrecisionTree supports risk assessment in Microsoft Excel using decision trees and influence diagrams. Decision trees enable you to visually map complex decision problems in an organized and sequential manner. The decision tree software can also be used, along with @RISK, to build powerful event-tree models commonly used in risk assessment as described in Chapter 15.

The description of PrecisionTree, which follows, will introduce you to the use of the software for event-tree models. Overlooking the decision tree capability of this software does not do justice to the full range of features available to you with PrecisionTree. The simple event-tree example that follows does, however, expand the range of applications for this tool.

A new example is used for this discussion. Consider a risk management activity that is concerned about the likelihood of aquatic nuisance species (ANS) reaching and spreading throughout a lake. The necessary sequence of events is that the ANS reach the lake from its originating waterway, that it survive in the lake, and that it find enough life requisites to establish a breeding colony. It would then have to spread from that breeding colony throughout the lake.

In this section you will learn how to:

1. Build a tree model
2. Use @RISK to change the probability inputs in the model
3. Modify the model to use as an event tree
4. Recognize other features of PrecisionTree

B.4.1 Building a Simple Tree Model

Figure B.22 shows the icon set used by PrecisionTree. Notice that PrecisionTree enables you to build and use influence diagrams. This capability is not explored in this discussion, but a discussion of influence diagrams can be found in Chapter 15. The analysis group is likewise not explored. These features are used primarily with PrecisionTree's decision tree applications, which we ignore in order to focus on how one would use the software to build an event tree.

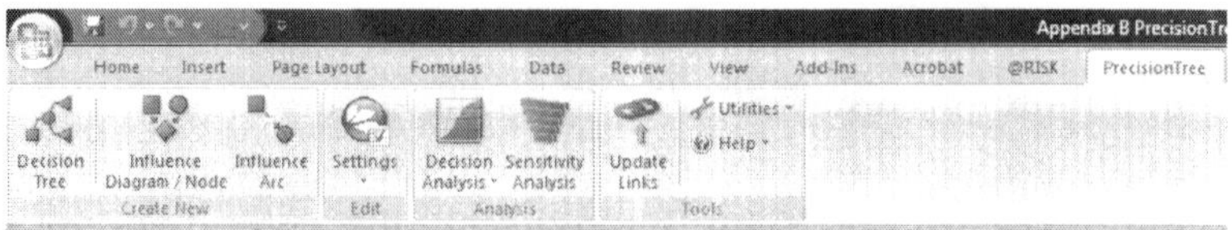

FIGURE B.22 PrecisionTree icon set.

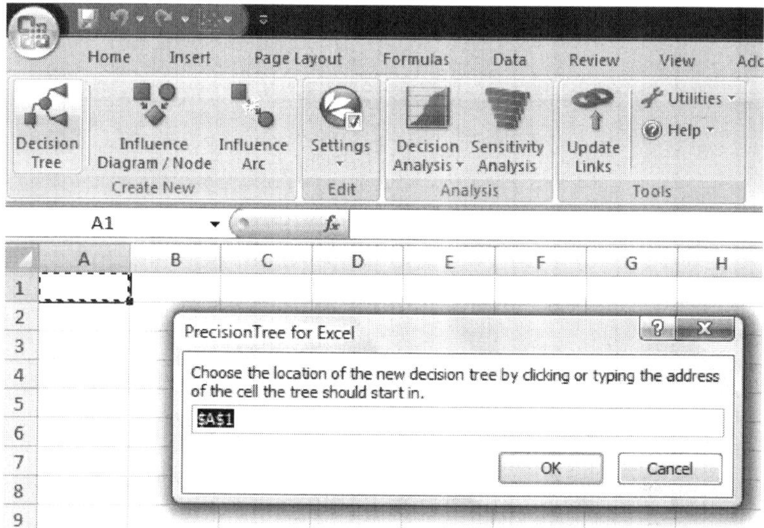

FIGURE B.23 Starting a PrecisionTree model.

To begin a new tree model, place your cursor in a cell and select the "Decision Tree" icon. This will open a box that asks you to identify the cell in which you'd like to begin your model. See Figure B.23. You can use any cell in your spreadsheet. If you have no clear need for another starting point, cell A1 works fine.

After you identify the starting point for your model, a new tree will appear, as seen in Figure B.24. The first thing you will have to do is name your model. Simply click on the box that says "New tree" and the window shown in Figure B.24 will open. Type the name as shown and click "Okay."

Note that there are five tabs in this window. You will want to explore these tabs when you are ready to build and use decision trees. With your model named, you are ready to build your model.

Click on the triangle endpoint to begin to build your model. This is how you expand your model, by clicking on endpoints along the pathways you wish to expand. When you click on an endpoint, the node settings window of Figure B.25 will appear. Notice there are five different kinds of nodes you can add to your model at this point. For an event tree, all nodes will be chance nodes. When the chance node is selected, the default number of branches from the node is two. To add additional branches select the "Branches" tab. Two nodes will be fine for this example. Enter the name for your node. When you click on "okay," you will see the tree expanded as shown at the top of the figure.

There are a few things to notice. First, notice that the boxes on the branches have default names. To change them, click on the box and type in your desired branch names. Note that each branch has 50% on the top of it to designate the probability of following that branch's path. These are default values. Leave them for now; they can be changed later once the structure of your model is determined. Beneath the branch you see a 0; this is another default value. The #VALUE you see to the right of the node is part of the back-solving calculation used by PrecisionTree when solving decision trees. Note that the formula box for Microsoft® Excel that indicates the

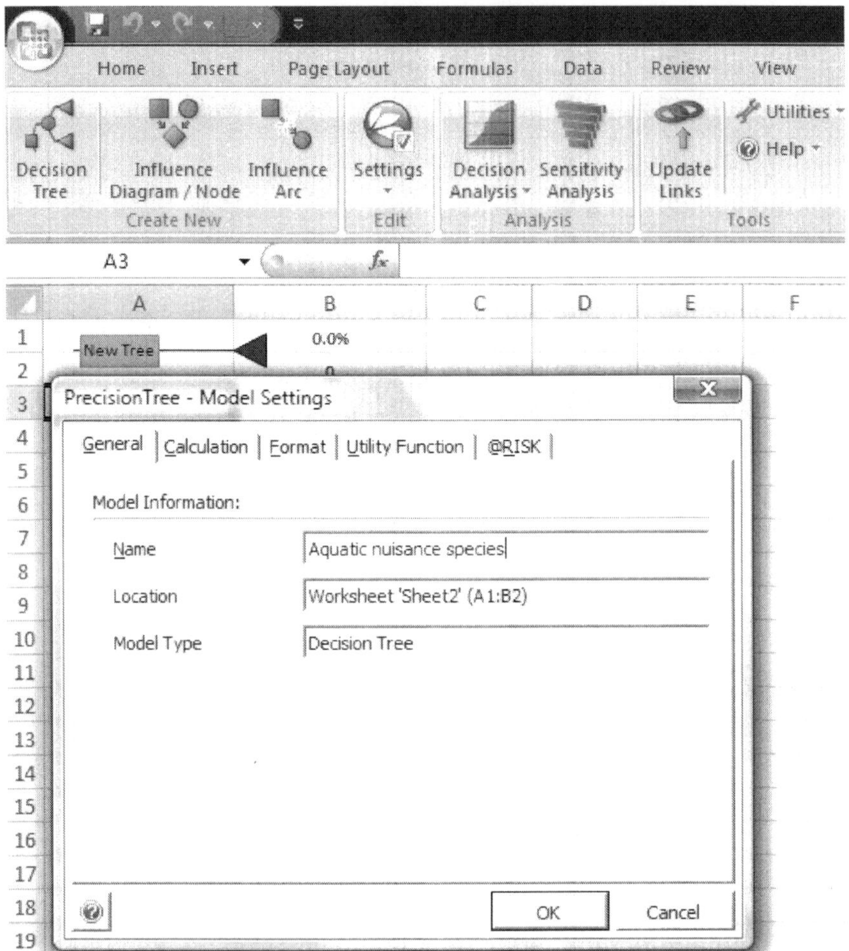

FIGURE B.24 Naming your tree model in PrecisionTree.

calculation for this value is to be found on a protected and hidden worksheet that is part of the software. The other #VALUE entries simply indicate that the model is not yet completed with the open window.

If you repeat this process as often as necessary, you can complete the structure of your model as shown in Figure B.26. Each node's name is given in the form of a question. Note that each of the branches is labeled yes or no, to answer the question. All "no" values result in endpoint nodes. Yes nodes continue, except for the very last one. Notice probabilities are multiplied along each pathway to calculate the probability of arriving at any one given endpoint. These probabilities are based on PrecisionTree's default values, and we now need to return to the task of entering the actual values for our model.

You can change the probability along a branch by entering that cell (B11) and typing in a new value. You must take care to make sure the probability values you enter obey all the relevant laws of probability. PrecisionTree's tree structuring features

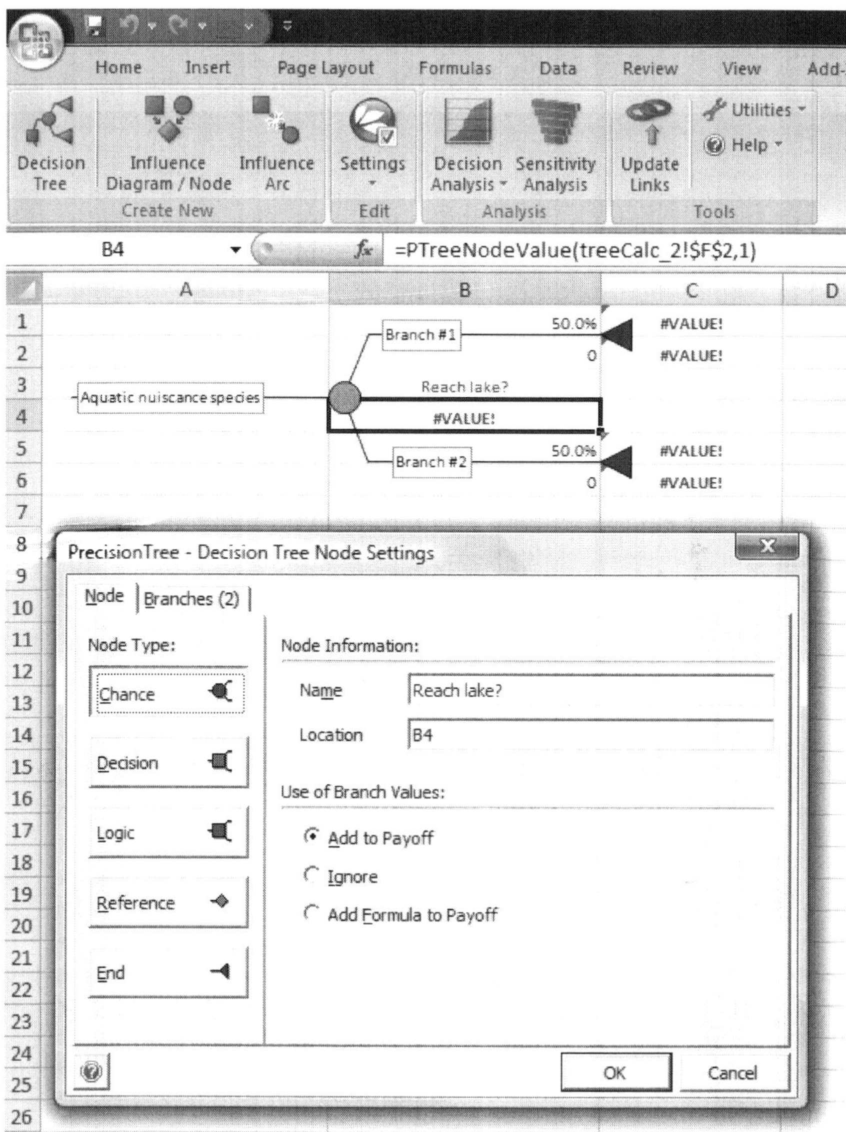

FIGURE B.25 Adding a chance node and two branches to your model.

can be combined with @RISK's Monte Carlo process features to yield a very powerful tool.

Figure B.27 shows a probability value being entered as a distribution. In this instance it is a Pert distribution, where the minimum probability the ANS will actually reach the lake from their originating waterway is .6. The most likely probability of this happening is .7. It is not a certainty that the ANS will arrive, so the maximum probability is estimated to be .85. Because the value for cell B11 will vary according to the distribution entered here, it is important to ensure that the probabilities of the

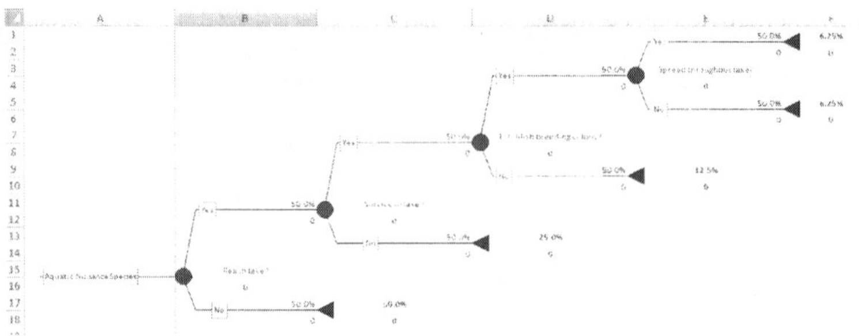

FIGURE B.26 First iteration of a built event tree.

yes and no branches always sum to 1. You must go into cell B17 and type =1−B11. This ensures that B17 always reflects the complement to cell B11. If you use more than two nodes with distributions, be sure you understand how PrecisionTree handles the normalization of probabilities along multiple branches to avoid mistakes.

Figure B.28 shows the model again after probability distributions have been entered for each yes branch in the model. Notice the cursor in the no branch cell and its corresponding formula line. The probability values shown reflect a single random iteration of the probability values. Cell F1 is, nominally, a cell of interest as it displays the probability that the ANS will spread throughout the lake. This cell could be added to the outputs and a simulation could be run to obtain a better answer to the risk manager's question.

If the risk manager had been interested in the numbers of ANS and not just their probabilities, these could have been added to the model on the bottom of the branch. It should be noted that PrecisionTree makes no claim to be event-tree software. Hence, using it as event-tree software requires the user to make careful modifications to the tree model and its internal logic. Examples of these modification can be found in exercises found at the Palisade Web site supporting this text,

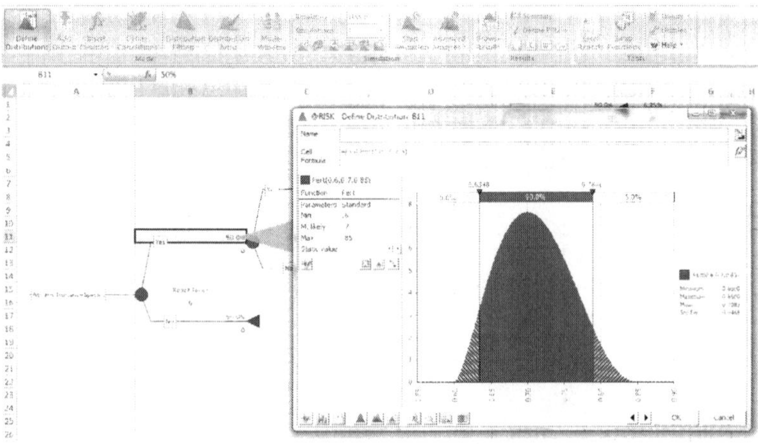

FIGURE B.27 Entering a probability distribution into an event-tree cell.

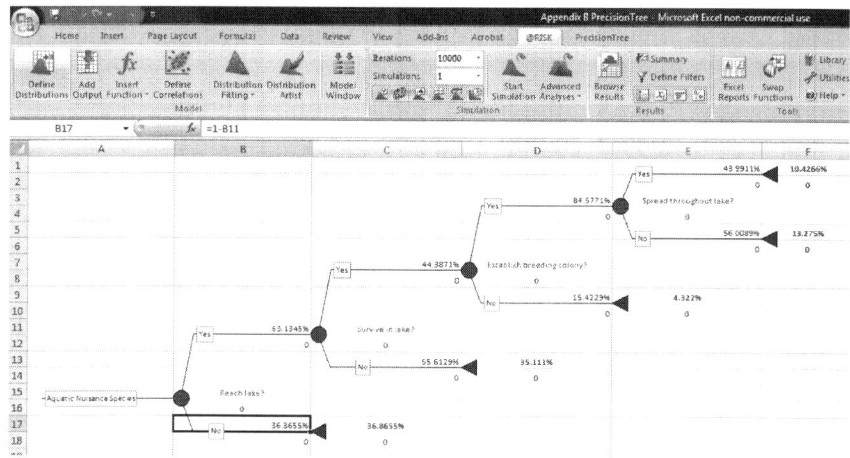

FIGURE B.28 Second iteration of the built model with uncertain probabilities.

http://www.palisade.com/bookdownloads/yoe/principles/. With these brief introductions, you are ready to begin to use and to learn more about the powerful risk assessment tools available in the Palisade DecisionTools Suite.

Index

P

Q

R